Advances in Fundamental Physics

Advances in Fundamental Physics

Editor

Eugene Oks

MDPI • Basel • Beijing • Wuhan • Barcelona • Belgrade • Manchester • Tokyo • Cluj • Tianjin

Editor
Eugene Oks
Auburn University
USA

Editorial Office
MDPI
St. Alban-Anlage 66
4052 Basel, Switzerland

This is a reprint of articles from the Special Issue published online in the open access journal *Foundations* (ISSN 2673-9321) (available at: https://www.mdpi.com/journal/foundations/special_issues/fundamental_physics).

For citation purposes, cite each article independently as indicated on the article page online and as indicated below:

LastName, A.A.; LastName, B.B.; LastName, C.C. Article Title. *Journal Name* **Year**, *Volume Number*, Page Range.

ISBN 978-3-0365-5745-8 (Hbk)
ISBN 978-3-0365-5746-5 (PDF)

Contents

About the Editor

Eugene Oks

He received his Ph.D. degree from the Moscow Institute of Physics and Technology, and later the highest degree of Doctor of Sciences from the Institute of General Physics of the Academy of Sciences of the USSR by the decision of the Scientific Council led by the Nobel Prize winner, academician A.M. Prokhorov. According to the Statute of the Doctor of Sciences degree, this highest degree is awarded only to the most outstanding Ph.D. scientists who founded a new research field of a great interest. Oks worked in Moscow (USSR) as the head of a research unit at the Center for Studying Surfaces and Vacuum, then – at the Ruhr University in Bochum (Germany) as an invited professor, and for the last 30 plus years – at the Physics Department of the Auburn University (USA) in the position of Professor. He conducted research in 5 areas: atomic and molecular physics, astrophysics, plasma physics, laser physics, and nonlinear dynamics. He founded/co-founded and developed new research fields, such as intra-Stark spectroscopy (new class of nonlinear optical phenomena in plasmas), masing without inversion (advanced schemes for generating/amplifying coherent microwave radiation), and quantum chaos (nonlinear dynamics in the microscopic world). He also developed a large number of advanced spectroscopic methods for diagnosing various laboratory and astrophysical plasmas – the methods that were then used and are used by many experimental groups around the world. He recently revealed that there are two flavors of hydrogen atoms, as proven by the analysis of atomic experiments; there is also a possible astrophysical proof – from observations of the 21 cm radio line from the early Universe. He showed that dark matter or at least a part of it can be represented by the second flavor of hydrogen atoms. He published about 550 papers and 11 books, including the books "Plasma Spectroscopy: The Influence of Microwave and Laser Fields", "Stark Broadening of Hydrogen and Hydrogenlike Spectral Lines in Plasmas: The Physical Insight", "Breaking Paradigms in Atomic and Molecular Physics", "Diagnostics of Laboratory and Astrophysical Plasmas Using Spectral Lineshapes of One-, Two, and Three-Electron Systems", "Unexpected Similarities of the Universe with Atomic and Molecular Systems: What a Beautiful World", "Analytical Advances in Quantum and Celestial Mechanics: Separating Rapid and Slow Subsystems", "Advances in X-Ray Spectroscopy of Laser Plasmas", "Simple Atomic and Molecular Systems: New Results and Applications", "Advances in the Physics of Rydberg Atoms and Molecules", and "The Second Flavor of Hydrogen Atoms – the Leading Candidate for Dark Matter: Theoretical Discovery and the Proofs from Experiments and Astrophysical Observations". He is the Chief Editor of the journal "International Review of Atomic and Molecular Physics" and the Editor-in-Chief of the Physical Sciences section of the journal "Foundations". He is a member of the Editorial Boards of seven other journals: 'Symmetry", "American Journal of Astronomy and Astrophysics", "Dynamics", "Open Physics", "Open Journal of Microphysics", "Physics International", and "Current Physics". He is a member of the Reviewers Board of the journal "Atoms". He is also a member of the International Program Committees of the two series of conferences: Spectral Line Shapes, as well as Zvenigorod Conference on Plasma Physics and Controlled Fusion.

Preface to "Advances in Fundamental Physics"

This Special Issue celebrates the opening of a new section of the journal Foundation: Physical Sciences. Theoretical and experimental studies related to various areas of fundamental physics are presented in this Special Issue. The published papers are related to the following topics: dark matter, electron impact excitation, second flavor of hydrogen atoms, quantum antenna, molecular hydrogen, molecular hydrogen ion, wave pulses, Brans-Dicke theory, hydrogen Rydberg atom, high-frequency laser field, relativistic mean field formalism, nonlocal continuum field theories, parallel universe, charge exchange, van der Waals broadening, greenhouse effect, strange and unipolar electromagnetic pulses, quasicrystals, Wilhelm-Weber's electromagnetic force law, axions, photoluminescence, neutron stars, gravitational waves, diatomic molecular spectroscopy, information geometric measures of complexity.

Among 21 papers published in this Special Issue, there are 5 reviews and 16 original research papers.

Eugene Oks
Editor

foundations

MDPI

Article

Telling the Wave Function: An Electrical Analogy

Leonardo Chiatti

Medical Physics Laboratory, ASL VT, Via Enrico Fermi 15, 01100 Viterbo, Italy; leonardo.chiatti@asl.vt.it

Abstract: The double nature of material particles, i.e., their wave and corpuscular characteristics, is usually considered incomprehensible, as it cannot be represented visually. It is proposed to the student, in introductory courses, as a fact justified by quantum interference experiments for which, however, no further analysis is possible. On this note, we propose a description of the wave function in terms of a simple electrical analogy, which reproduces at least some of its essential properties. Our aim is to provide a cognitive representation of an analogical type: starting from a classical context (electrical circuits) and introducing in an appropriate way the notions of "wave" and "particle", we show how typically quantum properties such as delocalization and entanglement emerge in a natural, understandable, and intuitive way.

Keywords: quantum mechanics; particle–wave duality; quantum jump; quantum entanglement

Citation: Chiatti, L. Telling the Wave Function: An Electrical Analogy. *Foundations* **2022**, *2*, 862–871. https://doi.org/10.3390/foundations2040058

Academic Editors: Eugene Oks and Martin Bohner

Received: 4 July 2022
Accepted: 4 October 2022
Published: 8 October 2022

1. Introduction

The wave nature of material particles, conceived by de Broglie in 1923 [1], still appears strongly counterintuitive today although confirmed by all the experiments carried out to verify it. There are at least two aspects of the particle–wave duality that seem contrary to common sense. The first is the "delocalization" of the corpuscle, i.e., the fact that an individual entity is somehow simultaneously present on the entire spatial volume occupied by the wave [2]. Secondly, if we accept, as we do in this study, the standard formulation of quantum mechanics and, therefore, the projection postulate [3], then in a "quantum jump" originating from the interaction with other particles or fields, the corpuscle abruptly changes its state of delocalization. As a particular ideal limit case, the corpuscle can undergo a precise localization in space acquiring, at least in an ephemeral way, an attribute of position. That same attribute is permanently possessed by a "classical" corpuscle.

The difficulties, however, do not end here. Amplitude interference experiments with single particles clearly demonstrate [4–6] the absence of the trajectories attributable to quantum "corpuscles". These experiments have become, at least since the 1960s, the basis of a didactic presentation of the particle–wave duality [7]. A widely used example is that of Young's interferometer with a double slit. In this device, the particle, which manifests itself as a localized corpuscle in its impact on the rear screen, interferes with itself as a consequence of the fact that the absence of trajectories prevents the definition of which slit is actually crossed.

A similar phenomenon occurs with two identical particles emitted in a coherent way, which can hit two distinct detectors (intensity interference). In this case, there may be a situation (H_1) in which the particle emitted by source S_1 hits detector D_1, and the particle emitted by source S_2 hits detector D_2, or there may be another situation (H_2) in which the particle emitted by source S_1 hits detector D_2, and the particle emitted by source S_2 hits detector D_1. The absence of trajectories leads to the undecidability between H_1 and H_2, and therefore to an interference connected with this undecidability [8,9]. This interference is a particular effect of a general phenomenon correlated with the non-factorizability of the wave function of the system consisting of the two particles. This phenomenon is the entanglement [10], and it constitutes a further aspect of the quantum domain that does not seem to admit a classical representation.

We are thus faced with the problem of how to find a viewable representation of the quantum phenomena that we have briefly mentioned. In our opinion, this problem has a well-known historical precedent in the development of non-Euclidean geometries, in particular the elliptical geometry of Riemann [11] and the hyperbolic geometry of Bólyai and Lobačevskij [12,13]. In an attempt to construct a geometry that denies the parallel postulate, we are faced with spaces that are impossible to visualize. Certainly, no one is able to visualize a space where, given a straight line and a point external to it, no parallel to the given straight line passes through that point. Or, an infinite number of parallels pass through it. This impossibility, however, persists only as long as we keep the original Euclidean notions of " straight line" and "point". It is, in fact, well-known that Riemann's elliptical (two-dimensional) geometry admits a representation on the surface of a Euclidean sphere when the appropriate redefinitions of the terms are performed, for example, if the straight lines are interpreted as the maximum circles of the Euclidean narrative [14]. In the same way, it is possible to represent the hyperbolic (plane) geometry on a portion of the Euclidean plane through the construction of Klein [14]. In the simplest version of this construction, the space is the internal region of a circumference (understood in the Euclidean sense); the "straight lines" are (in Euclidean terms) the segments inside the circumference that intersect it in two points, etc. In summary, it is possible to represent the unrepresentable by attributing a new meaning to the terms. By carrying out this operation with care, the "new" can be represented, in a fully viewable way, in the same environment as the "old". It should be emphasized that this operation is analogical, and numerous alternative representations of the same represented structure are possible.

Returning to our specific problem, the application of this type of strategy, first of all, involves the identification of a specific classical setting within which to operate the redefinition of the terms "particle", "wave", and "quantum jump". It is then necessary to show that the entities thus redefined (but easily visualized in the chosen classical setting) behave like the corresponding quantum entities in the context of their relations. The intuitive comprehensibility of the redefined entities is guaranteed by their coexisting "narration" in classical terms.

In attempting to search for a representation of quantum phenomena intended in this way, it seems inevitable to start with the concept of "particle". It is immediately evident that the usual notion of localized entity is unsuitable because it conflicts with delocalization and the absence of trajectories; therefore, something geometrically less constrained is needed. We choose to assimilate the particle to an electromechanical actuator (a sort of relay) that exchanges the energy elements associated with a charge. Recall that a particle can carry different types of charge: certainly, a gravitational charge but also possibly a weak, strong, or electric charge. We will deal with the concept of the charge assimilating it to that of electric charge, in both cases, keeping in mind the more general meaning of the term.

The next problem is how to introduce a spatially delocalized charge. The first idea that arises to mind is that, in each point of the empty space, there is a "charge reservoir" (i.e., a system of capacitors) that exchanges the elements of energy with the actuator, i.e., with the particle. The charge and energy of the particle are thus spatially delocalized because they are derived from the contribution of the capacitors associated with each point in space. The wave function of the particle then measures the local contribution to the total charge of the particle.

Invading the empty space with circuits of electric capacitors may appear as an annoying reminiscence of 19th-century English physics with its "ether of space". It is, therefore, appropriate to repeat that the meaning of our proposal is not to specify what a quantum particle is but how it can be told using the language of classical physics. Our "ether of capacitors" is fictional but functional for that purpose. Its state of charge allows us to define the wave function of the particle.

Having in mind these reasons and this strategy, in Section 2, we move on to examine the basic ideas of our representation and then introduce, in the subsequent Section 3, the relevant definition of the wave function. An electrical analogy of the quantum jump is presented in Section 4, which also discusses the reason why the wave function "lives" in the enlarged configurational space and not in ordinary spacetime. The notion of a multi-particle–wave function is thus introduced, and the meaning of entanglement is illustrated. In Section 5, some issues related to the wave–particle dualism are specified; in particular, the notions of particle and corpuscle in the context of the present representation. The inclusion of the spin is discussed in Section 6. Section 7 summarizes the conclusions.

2. Basic Ideas

Let us now see how to express, in a more precise way, the ideas illustrated in Section 1. For the moment, we limit ourselves to considering the case of the propagation of a single particle with mass m, and only subsequently will we consider the more general case of a system with several particles.

We assume that the propagation of the particle in four-dimensional spacetime is associated with a phenomenon of the polarization of the vacuum, structured as follows: We denote a generic point of the four-dimensional spacetime by $\boldsymbol{x}$, whose coordinates in reference to the rest of the particle are (x, y, z, ct), and we consider the two regions of the light cone (past and future) having vertices in $\boldsymbol{x}$ and extension $\pm L/c$ in t. There are no stringent indications on the value of L; we will assume that $L = \hbar/mc$ is the Compton length of the particle. This assumption seems plausible because it is below this spatial scale that the particle is dissociated into particle–antiparticle pairs, and therefore, the polarization effects are manifested [15]; however, any other physically reasonable choice of L is just as good.

As a consequence of the vacuum polarization associated with the particle, two opposite charges of $+Q_1(\boldsymbol{x})$ and $-Q_1(\boldsymbol{x})$ will be induced in the future light cone of $\boldsymbol{x}$ ($t < t' < t + L/c$), while two opposite charges of $+Q_2(\boldsymbol{x})$ and $-Q_2(\boldsymbol{x})$ will be induced in the past light cone of $\boldsymbol{x}$ ($t - L/c < t' < t$). We pose $Q_1, Q_2 \geq 0$ (the opposite choice is just as good). If we admit the existence of "vacuum capacity" C, dependent only on the type of the particle (electron, muon, etc.), these two charges correspond to two energies $Q_i^2/2Cn, i = 1, 2$. The total energy $(Q_1^2 + Q_2^2)/2Cn = Q^2/2Cn$ is that of a group of $n(\boldsymbol{x})$ capacitors in parallel to the same capacity C brought to the common voltage V. The voltage V is assumed to be independent of $\boldsymbol{x}$. The charge $q = CV$ then depends on the type of the particle. From the usual formalism of capacitors in parallel [16], we have $Q^2/2Cn = nCV^2/2$, from which the relation $Q = nq$ follows.

A group of capacitors in parallel is, therefore, associated with each point-event $\boldsymbol{x}$. The key assumption is that this group contributes, with a part of its charge Q, to the total charge of an actuator (the particle), which we will assume to be q.

If the particle was exactly localized in $\boldsymbol{x}$, its charge would be integrally supplied by the capacitor group present in $\boldsymbol{x}$. In this case, the total energy of the group would vary by an amount of $\pm q^2/2Cn$. The negative sign corresponds to the transfer, by the group, of charge q to the particle; the positive sign corresponds to the transfer, by the particle, of charge q to the group. In the hypothesis of the perfect localization of the particle in $\boldsymbol{x}$, the number of possible energy elements that can be exchanged between the group and the particle is given by the ratio $(Q^2/2Cn)/(q^2/2Cn) = (Q/q)^2 = n^2$.

This suggests how to deal with the more general case of a delocalized particle. The charge exchanged between the group and the particle is, in this case, a fraction (possibly infinitesimal) of q. We can, therefore, assume $\rho(\boldsymbol{x})dxdydz = dn^2(\boldsymbol{x})/A$ as the probability of the presence of the particle in the neighborhood $dxdydz$ of $\boldsymbol{x}$. The dimensionless normalization constant A can be determined according to the relation:

$$\int_{Vol} \rho(x, y, z, t)\; dxdydz \;=\; 1 \tag{1}$$

The contribution to q^2 of the element $dxdydz$ around $\boldsymbol{x}$ is then $q^2\rho(\boldsymbol{x})dxdydz = q^2dn^2/A = dQ^2$. Each point of space contributes, with its own group of capacitors in parallel, to the total charge q of the particle. This result constitutes a description of the delocalization of the particle in classical terms, and we will return to it later. At the Compton scale (the minimum scale at which the wave function is defined [15]), the relation $q^2\rho = dQ^2/(dxdydz)$ becomes $q^2\rho \sim Q^2/L^3$. Since q does not depend on x, the density (ρ) is locally proportional to Q^2.

We must now see how the other actor appears, namely the wave function of the particle.

3. Positional and Impulse Representation

Let us now consider the two complex conjugate functions:

$$\psi = \frac{(Q_1 + Q_2)}{2} + i\frac{(Q_1 - Q_2)}{2} \quad (2)$$

$$\psi* = \frac{(Q_1 + Q_2)}{2} - i\frac{(Q_1 - Q_2)}{2} \quad (3)$$

As we have seen, the probability density ρ is proportional to $Q^2 = Q_1{}^2 + Q_2{}^2$ since q is independent of x. Therefore, ρ is proportional to $\psi\psi^*$. We interpret functions (2) and (3) as the two wave functions, retarded and advanced, of the particle. We note that:

(1) If charges $Q_{1,2}$ are multiplied by a real common factor (k), the probability density (not normalized) is multiplied by k^2, while (2) and (3) are multiplied by k;
(2) If k is complex, the (non-normalized) probability density is multiplied by kk^* when (2) and (3) are multiplied by k and k^*, respectively. This implies that the modulus of (2) and (3) is multiplied by the modulus of k, while the two functions are rotated around the origin of the complex plane by an angle equal to the argument of k, k^*;
(3) From both the proportionality of (2) and (3) to the capacitor charges and the additive nature of the charges, it follows that functions of this type can be summed generating interference effects;
(4) The time inversion $t \rightarrow -t$ implies the exchange $Q_1 \leftrightarrow Q_2$, and then $\psi \leftrightarrow \psi^*$;
(5) Functions (2) and (3) have, of course with reference to the representation discussed here, a clear ontic meaning as charge states of the network of groups of capacitors associated with the particle.

On the other hand, it is possible to derive (2) and (3) with respect to the spacetime coordinates. Dimensionally, the quantities $I_{1,2} = c\partial_\mu Q_{1,2}$, where c is the maximal speed, and $\mu = 0,1,2,3$ is the spacetime coordinate index, are currents. In functions (2) and (3), those that are the eigenfunctions of $c\partial_\mu$ are also the eigenfunctions of the four momentums ($i\hbar\partial_0$, $-i\hbar\partial_i$), $i = 1,2,3$. These eigenfunctions can be superposed, thus generating generic wave packets. It, therefore, becomes possible to replace (2) and (3) with analogous complex functions containing currents instead of charges, thus passing to the momentum representation:

$$\varphi = \frac{(I_1 + I_2)}{2} + i\frac{(I_1 - I_2)}{2} \quad (4)$$

$$\varphi* = \frac{(I_1 + I_2)}{2} - i\frac{(I_1 - I_2)}{2} \quad (5)$$

In general, (4) and (5) can be modeled as a system of n inductors in parallel, with individual inductance M, through which the current $I = (I_1{}^2 + I_2{}^2)^{1/2}$ flows. The energy of the single inductor is $Mi^2/2$, with $i = I/n$ as the current flowing in it. The total inductance of the system is $M_T = M/n$. We, therefore, have [16]:

$$\frac{M_T I^2}{2} = \frac{MI^2}{2n} = \frac{M}{n}\frac{n^2 i^2}{2} = \frac{nMi^2}{2} \quad (6)$$

The representation in terms of inductances is completely specular to that in terms of capacitors, and in this paper, we choose to focus on the latter.

4. Multi-Particle Systems

The independent variable x of functions $Q_1(x)$ and $Q_2(x)$ labels a group of capacitors connected in parallel. Therefore, there is a continuous quadruple infinity of these groups. In a quantum jump, functions (2) and (3) are zeroed, and new ϕ functions of the same type are generated at the output. This means that the capacitors associated with the labels $x = (x, y, z, ct)$, with t = instant of the jump, are discharged and new capacitors associated with new labels of the same type are charged according to the functions ϕ. It is possible to represent the quantum jump $\psi \rightarrow \phi$ with the electrical diagram in Figure 1.

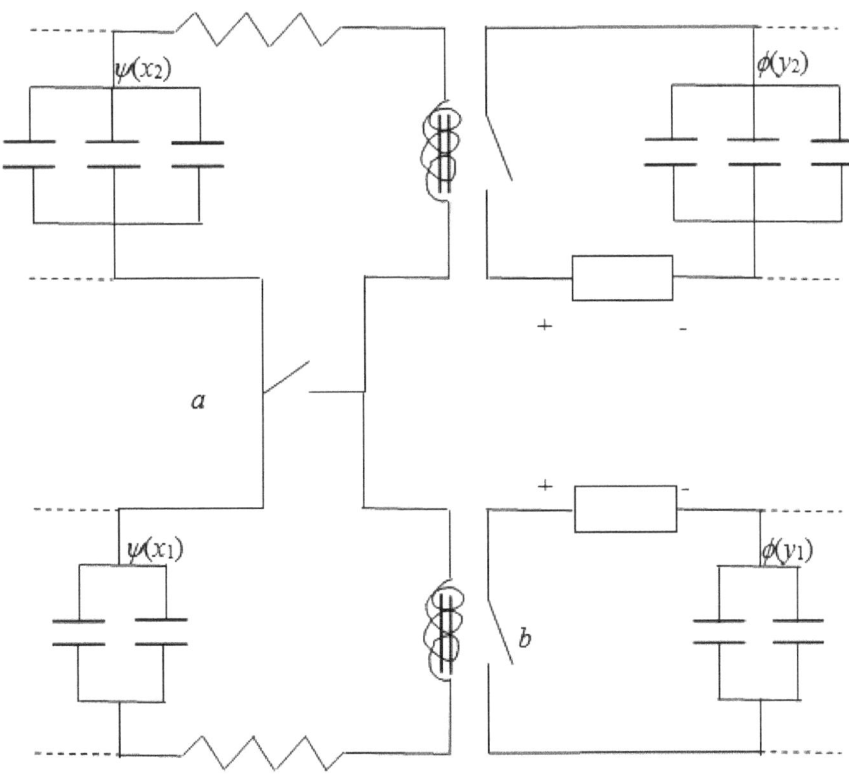

Figure 1. Electrical diagram of the quantum jump $\psi \rightarrow \phi$ (for simplicity, only two pairs of opposing meshes are represented).

Each point-event of spacetime corresponds to a group of capacitors associated with the incoming function ψ and a group of capacitors associated with the outgoing wave function ϕ. There are, therefore, groups $\psi(x_i)$ paired with groups $\phi(y_i)$, as in Figure 1, where the labels x_i, y_i ($i = 1, 2, \ldots$) represent the same point-event, i.e., the same spatial position and same instant in time coinciding with those of the jump.

The homologous ends of all groups $\psi(x_i)$ are connected to the corresponding two ends of switch a. Closing a involves short-circuiting all the groups of capacitors associated with function ψ. The opposite charges are recombined, and ψ is canceled. The closing of a also implies, as an automatic consequence, the closing of other switches such as b. The latter, in turn, implies the charging of all the groups of capacitors associated with function ϕ.

The function of the electromechanical actuator constituting the particle is to keep the circuit associated with function ψ open, that is, to keep switch a open. The energy accumulated in the actuator is, therefore, the potential energy with switch a open. The interaction responsible for the quantum jump provides the energy needed to overcome this potential barrier and close a. At the end of the jump, a is closed, and its potential energy is zero.

It can be seen from the diagram in Figure 1 that each value of the x (y) argument represents a mesh and, therefore, a discharge (charge) line. These labels are unique to the network of groups of capacitors that are discharged or charged. If the network corresponds to a particle, in the sense that it exchanges charge with that particle (actuator) only, then the label is shared by that particle. Two distinct particles, A and B, then have distinct spacetime labels of x_A and x_B. The wave functions associated with them are, respectively, $\phi(x_A)$ and $\varphi(x_B)$, and each describes the state of the charge of the network associated with the corresponding particle. The overall state of the charge of the two quadruple infinity of the groups of capacitors associated with the two particles will be represented by the product $\phi(x_A)\varphi(x_B)$. Normally, in this product, the two functions are considered at the same instant in time, in such a way that the square modulus of the wave function provides the compound probability density of the two particles at that instant.

On the other hand, it is possible to imagine multiple networks and an equal number of actuators that (1) exchange charge with each network and (2) contribute to the potential energy related to the opening of the circuit of each network. In this case, the spatial labels of *all* the networks must be simultaneously assigned to the single actuator (particle). An example can be made by returning to the previous case of the two networks of A and B. Imagine two actuators (particles) to whose charge both networks contribute and each of which contributes to the opening of both circuits A and B. In this case, it is not possible to distinguish the two particles through their network, and one can have an entangled state such as, for example, $\phi(x_A)\varphi(x_B) \pm \phi(x_B)\varphi(x_A)$. This entanglement describes the contributions of the two networks to the opening of the two switches a, inserted, respectively, on network A and network B.

The proposed representation, therefore, allows us to define both single-particle–wave functions and multi-particle–wave functions, the latter both factorizable and entangled. It also supports a viewable model of quantum jump.

5. Corpuscle–Wave Dualism

It is possible to assume that the energy of the actuator constituting the particle associated with the wave function ψ is $q^2/2C$, with the meaning of the symbols already seen in the previous sections. In this hypothesis, this is the minimum energy required to keep switch a open and thus allow the propagation of ψ. It is, therefore, natural to suppose that, for a particle with mass m, the rest energy of the particle is $q^2/2C = mc^2$. This is, in fact, the minimum energy required for an interaction to create an outgoing state containing that particle. The delocalization of the particle described in Sections 2 and 3then corresponds, physically, to the delocalization of its rest energy; the energy stored in each capacitor group represents the local contribution to the rest energy. The relation $q^2/2C = mc^2$ in turn implies $C = 2\pi\varepsilon_0 r_{cl}$, where $\varepsilon_0$and r_{cl} are, respectively, the dielectric constant of the vacuum and the classical radius of the particle; r_{cl}/L is the fine structure constant.

In summary, the quantum wave is the state of the charge of capacitor networks. Each network contributes to the charge and energy of an actuator, which is the particle. In this sense, the particle is "delocalized" on the network. The wave function defines, at the same time, the state of the charge of the network and the delocalization of the particle. Different particles can correspond to different networks, and in this case, the state of the charge of the networks is described by a factorizable wave function. On the other hand, when different networks contribute to powering the same actuator, and the latter acts on all the networks that feed it, entanglement occurs. In both cases, the wave function depends on the spatial coordinates of all the particles involved and not on the generic spacetime coordinates (as it would be for a classical field). The ambient space of the wave function is, therefore, the enlarged space of configurations, not the ordinary four-dimensional space where an observer coordinates the interaction events.

In a quantum jump, there is a sudden change in the wave function: The wave function entering the jump is zeroed by short-circuiting the network of capacitors associated with it; a new capacitor network, associated with a new wave function coming out of the jump, is charged. The net charge q of the particle—if it is a conserved quantity—is transferred from the incoming function to the outgoing one and becomes an additive contribution to the total charge of the particles it describes. This transfer is what we normally refer to with the term "corpuscle". This definition captures a feature of the corpuscular aspect of the matter that is well-known from quantum experiments: its ephemeral nature. In other words, a corpuscle is an event (such as, for example, the impact of an electron on a photographic plate), not a persistent object that can be recognized and traced within the wave function.

The analogical representation proposed here applies to the wave function of systems with one or more particles. It is not applicable to the wave function of idealized systems such as the rotator, the vibrator, etc., because, for these systems, the concept of a charge associated with a (virtual) polarization of the vacuum loses meaning. It must, however, be considered that the real systems corresponding to these idealizations can nevertheless be described as the aggregates of interacting particles, and therefore, their exact treatment leads back to the case examined here.

At this point, we would like to introduce a clarification to avoid misunderstandings. It could be assumed that the vacuum polarization postulated in this paper coincides with the vacuum polarization described by quantum field theory (QFT). This belief can be strengthened by our choice of the value of L as the Compton wavelength of the particle since this is precisely the scale below which the polarization of the QFT vacuum occurs. In reality, there is no obvious identification of the two phenomena.

We have introduced vacuum polarization as an ad hoc postulate, functional to a translation, in classical terms, of the quantum delocalization of the particle, that is, of its wave function. For relativistic reasons (i.e., the polarization of the QFT vacuum and, therefore, the dissociation of the particle into virtual particle–antiparticle pairs, which occurs on the Compton scale), the concept of wave function loses its meaning on the Compton scale [15]. Therefore, the Compton scale is the minimum scale at which the vacuum polarization, *as introduced in this paper*,can be matched to the wave function of a massive particle. For this reason, we choose the value of L as corresponding to the Compton wavelength of the particle.

Our classical description of particle delocalization starts from the vacuum polarization as postulated by us, through the introduction of an entirely fictitious "ether of capacitors", invented only for this purpose. These capacitors exchange, with an actuator, fractions of the *renormalized* charge of the particle.

The polarization of the QFT vacuum around a bare charge leads, in the QFT, to the virtual charges of opposite signs, each of infinite value. In the QFT description, these charges renormalize the bare charge (also infinite), producing a finite effective charge at finite distances. These virtual charges have no obvious relationship with the fractions of the *renormalized* charge considered here, exchanged between the capacitors and the actuator. It is also not clear what this exchange would correspond to in the QFT description. Therefore, although an indirect connection between the polarization of the QFT vacuum and the phenomenon of polarization postulated in this work cannot be absolutely excluded, we limit ourselves to assuming the latter as a simple classical analogy useful for our purposes.

We conclude this section with an observation of resistance R in the series with the capacitor network (Figure 1). We assume the quantum jump, and therefore the discharge of the capacitors, as instantaneous. Strictly speaking, this would require a null time constant $\tau = RC$ (due to the normalization condition, C is substantially referred to as the single capacitor) and, therefore, a zero resistance R and an infinite discharge current $i = V/R = q/RC$. This hypothesis is sufficient for our purposes. If we wanted to contemplate the hypothesis of a finite duration of quantum jumps, the time constant of the discharge must be finite. In such a case, experimental data provide an upper limit of τ, intended as "jump duration"; they then set an upper limit on the value of R and a lower limit on the value of i. The technology of the exploration of the motion of atomic electrons by means of pulsed laser beams had a prodigious development in recent decades, passing from the pulses of the duration of the femtosecond to the pulses of the duration of the attosecond (10^{-18}s) [17]. On this time scale, it is possible to resolve the temporal evolution of atomic orbitals during a transition, but it is not yet possible to resolve the quantum jumps that terminate this transition. The duration of the jumps—if actually finite—must, therefore, be much shorter.

Referring to the electron, the most studied elementary particle, it, therefore, seems reasonable to assume $\tau \sim r_{cl}/c = 0.937 \cdot 10^{-23}$s, and then $C = 2\pi\varepsilon_0 r_{cl} = 1.56 \cdot 10^{-25}$F; $R = \tau/C = 60\ \Omega$; $i = e/\tau = 17{,}100$ A. These results are derived from the currently accepted value for the classical electron radius, $r_{cl} = 2.81 \cdot 10^{-15}$m. However, we do not elaborate here on this aspect, due to its speculative character.

6. Spin

It is possible to construct the row or column vectors whose components are different functions of types (2) and (3), or, respectively, of types (4) and (5). Each component of these vectors is associated with a different network of groups of capacitors in parallel or, respectively, of inductances in parallel, afferent to a single actuator. If the relativistic covariance rules are applied to these vectors, they become spinors. We then have the Pauli spinors in the non-relativistic limit and the Dirac spinors in the general relativistic case. It thus becomes possible, in principle, to extend the present electrical analogy to include the spin of elementary particles and their internal degrees of freedom such as isospin or strangeness. Here, we limit ourselves to an observation on the spin.

A generic wave function of spin $\frac{1}{2}$ takes, in the Pauli algebra, the following form [18,19]:

$$\psi(\vec{x},t) \;=\; \alpha(\vec{x},\, t)\,\psi_z^+ \;+\; \beta(\vec{x},t)\,\psi_z^- \;; \qquad |\alpha\alpha*| \;+\; |\beta\beta*| \;=\; 1\,. \tag{7}$$

where $\vec{x} = (x,y,z)$ is the spatial position, and the spinors ψ_z^+ and ψ_z^- represent fully polarized beams along the z axis, corresponding to the two eigenvalues of the spin projection along that axis. The meaning of (7) in the present representation is that a single actuator (particle) exchanges charge with two distinct networks, each of which is associated with one of the two projections, and keeps them both open. The (complex) coefficients α and β measure the relative intensity of the exchange with the relevant network, in the terms seen in Section 3. The rotation of the frame of reference of an angle χ around a spatial axis $\vec{n}$ (with $\left|\vec{n}\right| = 1$) changes coefficients α and β according to the following law [18,20]:

$$\psi \;\rightarrow\; e^{i\frac{\chi}{2}(\vec{n}\,\cdot\,\vec{\sigma})}\psi \;=\; \left[I\,\cos\frac{\chi}{2} \;+\; i\,(\vec{n}\,\cdot\,\vec{\sigma})\,\sin\frac{\chi}{2}\right]\,\psi \tag{8}$$

For example, if $\alpha(\vec{x},t) = 1$ and $\beta(\vec{x},t) = 0$, a rotation around the z axis gives $\alpha'(\vec{x}',t) = e^{i\frac{\chi}{2}}$, $\beta'(\vec{x}',t) = 0$. This means that the exchange is sensible to rotations. It must be borne in mind that a variation in axes is always associated with a modification of the experimental situation, such as fixing a direction to a magnetic field or reorienting a polarizer. This modification involves a variation in the charge exchange between the particle and the two networks. The directional dependence of the exchange represents, in a certain sense, a directional structure of the particle itself. However, this structure, or the internal "direction" of the particle, has nothing to do with the rotation of a solid object in three-dimensional space. This is consistent with the non-classical nature of the spin [21].

7. Conclusions

In these concluding notes, we would like to try to put this contribution in context. In our opinion, the non-viewability of quantum processes has generated the widespread belief that they are inherently incomprehensible. Quantum theories are often assimilated into formal recipes that are very effective on the predictive level but are ones whose connection with any intelligible ontology remains obscure. Following the historical precedent of non-Euclidean geometries, we tried to circumvent these obscurities by providing a model of quantum behavior in a classical context: that of electrical circuits. As in the case of non-Euclidean geometries, we had to redefine the notions of particle, wave, and corpuscle in this context, moving away from their original classical meaning. By paying this price, we obtained the visualization of quantum entities and processes, guaranteed by the classical nature of the context in which their redefinition was carried out.

We believe that this attempt is located in a sort of middle land between the choice of surrendering to non-visualization (with the consequent problems of conceptual opacity) and the strong choice of determining an ontology of elementary processes, which is the goal of any physical interpretation of the quantum formalism. Although in choosing our representative model, we tried to adhere to the criteria of "sound physics", we find it difficult to seriously believe that space is the equivalent of a cabinet of electric capacitors. Our representation is, therefore, less than a sensu strictu interpretation of quantum formalism such as, for example, Bohm's [22] and relative state [23] interpretations. At the same time, however, it is more than just a surrender to mystery and allows for an analogical narrative of concepts such as the wave function or the particle–wave dualism. Our aim is to facilitate the communication related to quantum processes, through images that can be understood by anyone familiar with the basics of classical physics concerning electrical circuits.

This objective responds to a pedagogical need that has been our strongest motivation and seems particularly urgent to us in a moment like the present one, in which quantum theories have become part of the educational background of the new generations of engineers and technologists, engaged in the development of the amazing technologies that the quantum nature of reality makes possible.

Funding: This research received no external funding.

Institutional Review Board Statement: Not applicable.

Informed Consent Statement: Not applicable.

Data Availability Statement: Not applicable.

Acknowledgments: The author wishes to thank his colleague I. Licata, for the critical revision of the manuscript and many interesting discussions. He also wants to express his gratitude to the referees for their valuable suggestions useful for improving the text.

Conflicts of Interest: The author declares no conflict of interest.

References

1. de Broglie, L. Recherches sur la théorie des quanta. *Ann. Phys.* **1925**, *10*, 22. [CrossRef]
2. Lévy-Leblond, J.M.; Balibar, F. *Quantique (Rudiments)*; Interéditions/CNRS: Paris, France, 1984.
3. Von Neumann, J. *Mathematical Foundations of Quantum Mechanics*; Princeton University Press: Princeton, NJ, USA, 1932.
4. Merli, P.G.; Missiroli, G.F.; Pozzi, G. Electron interferometry with the Elmiskop 101 electron microscope. *J. Phys. E Sci. Instrum.* **1974**, *7*, 729–732. [CrossRef]
5. Merli, P.G.; Missiroli, G.F.; Pozzi, G. On the statistical aspect of electron interference phenomena. *Am. J. Phys.* **1976**, *44*, 306–307. [CrossRef]
6. Taylor, G.I. Interference fringes with feeble light. *Proc. Camb. Philos. Soc.* **1909**, *15*, 114.
7. Feynman, R.; Leighton, R.; Sands, M. *The Feynman Lectures on Physics*; California Institute of Technology: Berkeley, CA, USA, 1965; Volume 3.
8. Hanbury Brown, R.; Twiss, R.Q. A new type of interferometer for use in radio astronomy. *Philos. Mag.* **1954**, *45*, 663–682. [CrossRef]
9. Hanbury Brown, R.; Twiss, R.Q. Correlation between photons in two coherent beams of light. *Nature* **1956**, *177*, 27–29. [CrossRef]
10. Horodecki, R.; Horodecki, P.; Horodecki, M.; Horodecki, K. Quantum entanglement. *Rev. Mod. Phys.* **2009**, *81*, 865. [CrossRef]
11. Gray, J. Bernhard Riemann, Posthumous Thesis 'On the Hypotheses Which Lie at the Foundation of Geometry' (1867). In *Landmark Writings in Western Mathematics 1640–1940*; Grattan-Guinness, I., Corry, L., Guicciardini, N., Cooke, R., Crépel, P., Eds.; Elsevier: Amsterdam, The Netherlands, 2005; pp. 50–520.
12. Lobačevskij, N.I. *New Principles of Geometry with Complete Theory of Parallels. Complete Collected Works*; Kagan, V.F., Ed.; GITTL: Moscow, Russia, 1951; Volume 2.

13. Bolyai, J. *Appendix (The Theory of Space)*; Kárteszi, F., Ed.; Wiley: Amsterdam, The Netherlands, 1987.
14. Non Euclidean Geometry. Britannica. Available online: https://www.britannica.com/science/non-Euclidean-geometry (accessed on 30 May 2022).
15. Davydov, A.S. *Quantum Mechanics*; Elsevier: Amsterdam, The Netherlands, 1965.
16. Halliday, D.; Resnick, R.; Walker, J. *Fundamental of Physics*; Wiley: Hoboken, NJ, USA, 1996.
17. Corkum, P.; Krausz, F. Attosecond science. *Nat. Phys.* **2007**, *3*, 381–387. [CrossRef]
18. Ballentine, L.E. *Quantum Mechanics. A Modern Development*; World Scientific: Singapore, 1998.
19. Fock, V.A. *Fundamentals of Quantum Mechanics*; MIR: Moscow, Russia, 1978.
20. Landau, L.D.; Lifshitz, E.M. *Quantum Mechanics: Non-Relativistic Theory*; Butterworth-Heinemann: Oxford, UK, 1981.
21. Aerts, D.; Sassoli de Bianchi, M. Do spins have directions? *Soft Comput.* **2017**, *21*, 1483–1504. [CrossRef]
22. Bohm, D. A Suggested interpretation of the quantum theory in terms of 'hidden' variables, I and II. *Phys. Rev.* **1952**, *85*, 166–193. [CrossRef]
23. Everett, H., III. 'Relative state' formulation of quantum mechanics. *Rev. Mod. Phys.* **1957**, *29*, 454–462. [CrossRef]

foundations

MDPI

Communication

Experiments on the Electron Impact Excitation of Hydrogen Molecules Indicate the Presence of the Second Flavor of Hydrogen Atoms

Eugene Oks

Physics Department, 380 Duncan Drive, Auburn University, Auburn, AL 36849, USA; goks@physics.auburn.edu

Abstract: In one of our previous papers, we performed a comparative analysis of the experimental and theoretical cross-sections for the excitation of atomic hydrogen by electrons. We found that the theoretical ratio of the cross-section σ_{2s} of the excitation of the state 2s to the cross-section σ_{2p} of the excitation of the state 2p was systematically higher than the corresponding experimental ratio by about 20% (far beyond the experimental error margins). We showed that this discrepancy can be due to the presence of the Second Flavor of Hydrogen Atoms (SFHA) in the experimental gas and that the share of the SFHA in the mixture, required for removing this discrepancy, was about the same as the share of the usual hydrogen atoms. The theory behind the SFHA was based on the standard quantum mechanics—on the second solution of the Dirac equation for hydrogen atoms—and on the experimental fact that the charge distribution inside the proton has the peak at the center of the proton; the term "flavor" was used by the analogy with flavors of quarks. In the present paper, we used the same guiding principles, as employed in that previous study, for the comparative analysis of the experimental and theoretical cross-sections for the excitation of molecular hydrogen by electrons. We found that presumably the most sophisticated calculations, using the convergent close-coupling method involving 491 states, very significantly underestimate the corresponding experimental cross-sections for the two lowest stable triplet states. We showed that if in some hydrogen molecules one or both atoms would be the SFHA, then the above very significant discrepancy could be eliminated. We estimated that it would take such unusual hydrogen molecules to be represented in the experimental gas by the share of about 0.26. This is just by about 40% smaller than the share 0.45 of the SFHA deduced in our previous analysis of the experiment on the electron impact excitation of hydrogen atoms (rather than hydrogen molecules). It should be emphasized that from the theoretical point of view, the share of the unusual hydrogen molecules in any experimental gas and the share of the unusual hydrogen atoms (SFHA) in any experimental gas should not be expected to coincide (it would be the comparison of "apples to oranges", rather than "apples to apples"). In addition, given the roughness of the above estimates, we can state that the results of the present paper reinforce the main conclusion of our previous papers of the very significant share of the SFHA in the experimental hydrogen gases. Thus, the experiments on the electron impact excitation of hydrogen molecules are the fourth type of atomic experiments that proved the existence of the SFHA.

Keywords: electron impact excitation of hydrogen molecules; discrepancy between theories and experiments; second flavor of hydrogen atoms

Citation: Oks, E. Experiments on the Electron Impact Excitation of Hydrogen Molecules Indicate the Presence of the Second Flavor of Hydrogen Atoms. *Foundations* **2022**, 2, 697–703. https://doi.org/10.3390/foundations2030047

Academic Editor: Kazuharu Bamba

Received: 18 July 2022
Accepted: 18 August 2022
Published: 19 August 2022

1. Introduction

The theory behind the Second Flavor of Hydrogen Atoms (SFHA) was based on the standard quantum mechanics—on the second solution of the Dirac equation for hydrogen atoms—and on the experimental fact that the charge distribution inside the proton has a peak at the center of the proton [1]. The term "flavor" was used by analogy with flavors of quarks (see Appendix A). In the same paper [1] the first experimental proof of the existence of the SFHA was presented. Namely, the allowance for the SFHA eliminated a huge discrepancy concerning the linear momentum distribution in the ground state of hydrogen atoms: in its high-energy tail, the distribution derived from the analysis of atomic experiments exceeded the theoretical predictions (made for the usual hydrogen atoms) by several orders of magnitude [1].

The subsequent analysis of the experiments on charge exchange during collisions of low-energy protons with hydrogen atoms yielded the second experimental evidence of the existence of the SFHA [2]. The theoretical cross-sections, calculated with allowance for the SFHA, turned out to agree with the experiments within the experimental error margins—in distinction to the previous calculations made before the theoretical discovery of the SFHA. By the way, this result should be important for atomic codes developed for describing edge plasmas in magnetic fusion devices because charge exchange is a very important atomic process in these plasmas.

The third type of the atomic experiments that proved the existence of the SFHA was the experiments on the electron impact excitation of the n = 2 states of hydrogen atoms [3]. The theoretical ratio of the cross-section σ_{2s} of the excitation of the state 2s to the cross-section σ_{2p} of the excitation of the state 2p turned out to be systematically higher than the corresponding experimental ratio by about 20% (far beyond the experimental error margins). In paper [3] it was shown that this discrepancy can be due to the presence of the SFHA in the experimental gas. The share of the SFHA in the mixture, required for removing this discrepancy, was estimated to be about the same as the share of the usual hydrogen atoms [3].

For atomic physics, the proven existence of the SFHA has fundamental significance in its own right. Nevertheless, it was also found to be important for astrophysics—especially for finding out what is dark matter (i.e., solving the most fundamental cosmological problem). Namely, after Bowman et al. [4] found that the observed absorption signal of the redshifted 21 cm spectral line from the early Universe was about two times more intense than expected from the standard cosmology, meaning that the primordial hydrogen gas was cooler than predicted, Barkana [5] brought up the suggestion that the additional cooling was caused by some unspecified dark matter particles colliding with the hydrogen gas. Then in paper [6] it was shown that the above large discrepancy would be eliminated if it were collisions with the SFHA that caused the additional cooling.

Another astrophysical puzzle was published by Jeffrey et al. [7]: the distribution of dark matter predicted on the basis of Einstein's gravity was not confirmed by the observation—the observed one was smoother. This prompted suggestions on the need for new physical laws that would go beyond Einstein's gravitation. However, in paper [8] it was demonstrated that the perplexing observations by Jeffrey et al. [7] can be explained qualitatively and quantitatively by allowing for the SFHA.

The Occam razor principle favors the SFHA as the possible explanation of dark matter because the SFHA is based on the standard quantum-mechanical Dirac equation and does not resort to new physical laws (in distinction to most other possible explanations of dark matter). Besides, no other hypothesis has the experimental confirmations—in distinction to the SFHA. All of the above reinforced the leading status of the SFHA on explaining dark matter (or at least a part of it).

In the present paper we discuss whether there is yet another experimental proof of the existence of the SFHA—from the fourth type of atomic experiments: the experiments on the electron impact excitation of hydrogen molecules to the lowest triplet states. There are lots of various theoretical approaches on this process—see for example one of the most recent papers [9] and the very extensive list of references therein. We perform our study based on the same principles as in our previous analysis of the experiments on the electron impact excitation of hydrogen atoms [3]. This is explained in the next section.

2. Comparison of the Experimental Cross-Sections with Theories

Let us start by specifying four important points, on which we based our study in paper [3] and which we are going to use in the present study. The first point: in our study of the experiments on the electron impact excitation of hydrogen atoms [3], we chose the first two excited states: 2s and 2p. This was done because for the first two excited states, calculations are simpler (and therefore more reliable) than for higher states.

The second point: the range of energies relatively close to the excitation threshold was not favorable for determining the presence and the share of the SFHA in the experimental gas mixture. This is because in this range, the excitation cross-sections σ_{2s} and σ_{2p} are strongly dominated by so-called "resonances".

The third point: we considered only the theoretical works where both the cross-sections σ_{2s} and σ_{2p} were calculated in the same theoretical approach. The same about the experiments.

The fourth point: after we found about 20% discrepancy between the theoretical and experimental ratio of the cross-sections σ_{2s}/σ_{2p}, the next task was to estimate the percentage of the SFHA in the hydrogen gas mixture required for eliminating this discrepancy. For this purpose we needed the corresponding theoretical cross-section for the SFHA.

The primary feature of the SFHA distinguishing it from the usual hydrogen atoms is that the SFHA only has states of the zero orbital momentum ($l = 0$) both in the discrete and continuous spectra of energies. Therefore, due to the well-known selection rules, the SFHA does not couple to the electromagnetic radiation: the SFHA remains "dark" (except for the 21 cm spectral line resulting from the radiative transition between the two superfine structure sublevels of the ground state). In the discrete spectrum the states of $l = 0$ are called the s-states. [1,6]. The s-states are spherically symmetric.

Theoretical calculations of the electron impact excitation cross-sections most relevant to the SFHA were performed by Poet [10], who considered such excitation for a model hydrogen atom having only spherically symmetric states. By comparing the theoretical results by Poet [10] with the corresponding theoretical results for the usual hydrogen atoms, and combining this with the above 20% discrepancy, we arrived at the conclusion that the SFHA and the usual hydrogen atoms were present in the experimental gas in about the same shares.

Next, we applied the same principles to the analysis of the experimental and theoretical results on the electron impact excitation of hydrogen molecules. First, we chose the first two stable excited electronic *triplet* states of H_2: the state c $^3\Pi_u$ and the state a $^3\Sigma_g{}^+$. The reason for choosing the triplet states is the following. The singlet states can get populated both by the direct excitation and by exchange between the incident electron and one of the molecular electrons. The triplet states can get populated only by the exchange, so that the corresponding theory is simpler for the triplet states. (This is a generalization of principle number one from the study [3] for the case of H_2.)

Second: for avoiding resonances, which complicate the calculations and thus complicate determining the presence and the share of the SFHA in the experimental gas mixture, we chose the range of energies starting from 30 eV.

Third: concerning the choice of the experimental and theoretical works. In the experiments by Wrkich et al. [11] (who improved the previous experimental results by Khakoo and Trajmar [12]), both the excitation cross-section to the state c $^3\Pi_u$ and to the state a $^3\Sigma_g{}^+$ were measured, but only up to the energy of 30 eV, so that only their data at 30 eV is relevant for our purposes (according to the second point above). Therefore, we also chose the experiment by Mason and Newell [13], who covered the energies from 30 eV to 60 eV for the excitation to the state c $^3\Pi_u$, as well as the experiment by Ajello and Shemansky [14], who covered the energies from 30 eV to 60 eV for the excitation to the state a $^3\Sigma_g{}^+$. As for the corresponding theoretical work, dealing with the usual (non-SFHA) hydrogen atoms in the molecule H_2, we choose the (presumably most sophisticated) calculations by Zammit et al. (2017) [9]. In that paper, both the excitation cross-section to the state c $^3\Pi_u$ and to the state a $^3\Sigma_g{}^+$ were calculated by the convergent close-coupling (CCC) method with the total number of states equal to 491—they referred to these calculations as CCC(491).

The comparison of the theoretical CCC(491) results from paper [9] (as well as of some theoretical results from paper [15] included for reasons explained later on), with the experimental results from papers [11,13] in the range from 30 eV to 50 eV (practically the same range as in our study [3]) is presented in Table 1 for the state c $^3\Pi_u$.

Table 1. Comparison of the experimental excitation cross-sections σ (10^{-17} cm^2) to the state c $^3\Pi_u$, deduced from the plots in papers [11,13], with the corresponding theoretical results, deduced from the plots in papers [9,15]. Here N/A stands for "not available".

Energy (eV)	σ Experiment [11]	σ Experiment [13]	σ Theory CCC(491) [9]	σ Theory Lima et al. [15]
30	0.77	0.74	0.38	1.61
40	N/A	0.59	0.19	N/A
50	N/A	0.57	0.10	N/A

From Table 1 it is seen that in the range of incident electron energies from 30 eV to 50 eV, the theoretical CCC(491) results [9] very significantly underestimate the corresponding experimental cross-section: e.g., by a factor of five at 50 eV, by a factor of three at 40 eV, and by a factor of two at 30 eV. At the same time, it is seen that there is a very good agreement between the experimental cross-section by Wkrich et al. [11] at 30 eV (which is the highest energy data point that they measured) with the experimental cross-section by Mason and Newell [13], thus reinforcing the reliability of Mason–Newell results.

Now the question arises: could this huge discrepancy be explained if there were the SFHA in the composition of some hydrogen molecules? (Let's call them "unusual hydrogen molecules"). Physically, what would be the difference in calculating the corresponding theoretical cross-section?

Out of the 491 states, involved in producing the theoretical CCC(491) results, the overwhelming majority of the states would be absent in the unusual hydrogen molecules (because the SFHA has only the s-states). So, how would the dramatic reduction of the states involved in the CCC calculations affect the results? Zammit et al. [9] also provided theoretical results for the CCC involving a lesser number of states. They showed that the decrease of the number of states involved in their calculations yields significantly greater excitation cross-sections than CCC(491).

Here we come to the fourth point in the sequence of steps from paper [3]: the choice of the theoretical calculations of the cross-sections that is the most relevant for the unusual hydrogen molecules. The minimal number of states were used in calculations by Lima et al. [15] (in frames of Schwinger multichannel formulation)—to the best of our knowledge.

From Table 1 it is seen that at the incident electron energy 30 eV (the maximum energy, for which Lima et al. [15] performed their calculations), the CCC(491) result [9] for the usual hydrogen molecules underestimates the corresponding experimental results by a factor of two, while the result from Lima et al. [15] (most relevant for the unusual hydrogen molecules) overestimates the corresponding experimental results by a factor of two. So, if we denote by α the share of the unusual hydrogen molecules in the experimental gas, then from the data at 30 eV it is easy to find that the agreement with the experimental results would be achieved for $\alpha \approx 0.30$ as the solution to the equation $1.61\ \alpha + 0.38\ (1 - \alpha) = 0.755$ (where 0.755 is the experimental value averaged between the corresponding results of the experiments [11,13]).

Now let us proceed to the situation with the excitation to the state a $^3\Sigma_g{}^+$. Table 2 presents the comparison of the experimental results from papers [13,14] with the corresponding theoretical result CCC(491) from paper [9] and with the corresponding theoretical result from paper [15]. It is seen that at the incident electron energy 30 eV (the maximum energy, for which Lima et al. [15] performed their calculations), the CCC(491) result [9] for the usual hydrogen molecules underestimates the corresponding experimental results by a factor of about 1.5, while the result from Lima et al. [15] (most relevant for the unusual hydrogen molecules) overestimates the corresponding experimental results by a factor of about 1.5.

Table 2. Comparison of the experimental excitation cross-sections σ(10^{-17} cm^2) to the state a $^3\Sigma_g{}^+$, deduced from the plots in papers [11,14], with the corresponding theoretical results, deduced from the plots in papers [9,15]. Here N/A stands for "not available".

Energy (eV)	σ Experiment [11]	σ Experiment [14]	σ Theory ccc(491) [9]	σ Theory Lima et al. [15]
30	0.18	0.33	0.18	0.52
40	N/A	0.15	0.081	N/A
50	N/A	0.10	0.054	N/A

The share α of the unusual hydrogen molecules in the experimental gas necessary for achieving the agreement with the corresponding experimental results can be easily estimated from the data at 30 eV as the solution of the equation $0.52\ \alpha + 0.18\ (1 - \alpha) = 0.255$ (where 0.255 is the experimental value averaged between the corresponding results of the experiments [11,14]). It yields $\alpha \approx 0.22$.

For estimating the share of the unusual hydrogen molecules we could not use the data at 40 eV and 50 eV because for these energies there is no data from Lima et al. [15] representing the corresponding theoretical cross-sections for unusual hydrogen molecules.

Both for the state c $^3\Pi_u$ and for the state a $^3\Sigma_g{}^+$, these are rough estimates. The value of α averaged over the corresponding results for the states state c $^3\Pi_u$ and a $^3\Sigma_g{}^+$ is $\alpha = 0.26 \pm 0.04$, so that the scatter is just about 15%. The above results can be interpreted as the possible evidence that in the experimental gas, the shares of the usual and unusual (i.e., the SFHA-based) hydrogen molecules differed just by a factor of three.

In paper [3] where we compared the experimental and theoretical results on the electron impact excitation of hydrogen atoms, the share of the SFHA was found to be approximately 0.45. The corresponding value of 0.26 for the experiments on the electron impact excitation of hydrogen molecules (rather than hydrogen atoms) is less than 0.45 by just about 40%. It should be emphasized that from the theoretical point of view, the share of the unusual hydrogen molecules in any experimental gas and the share of the unusual hydrogen atoms (SFHA) in any experimental gas should not be expected to coincide (it would be the comparison of "apples to oranges", rather than "apples to apples"). In addition, given the roughness of the above estimates, we can state that the results of the present paper reinforce the main conclusion of paper [3] of the very significant share of the SFHA in the experimental hydrogen gases.

3. Conclusions

We performed a comparative analysis of the experimental and theoretical cross-sections for the excitation of molecular hydrogen by electrons. We employed the same set of guiding principles as in our previous analogous study of the electron impact excitation of hydrogen atoms [3]. We found that presumably the most sophisticated calculations by Zammit et al. [9], using the convergent close-coupling method involving 491 states, very significantly underestimate the corresponding experimental cross-sections.

We showed that if in some hydrogen molecules one or both atoms would be the SFHA, then the above very significant discrepancy could be eliminated. We estimated that it would take such unusual hydrogen molecules to be represented in the experimental gas in the share of about 0.26. This is about 40% smaller than the share 0.45 of the SFHA deduced by the corresponding analysis (in paper [3]) of the experiments on the electron impact excitation of hydrogen atoms (rather than hydrogen molecules). It should be emphasized that from a theoretical point of view, the share of the unusual hydrogen molecules in any experimental gas and the share of the unusual hydrogen atoms (SFHA) in any experimental gas should not be expected to coincide. Given the roughness of the above estimates, we can state that the results of the present paper reinforce the main conclusion of paper [3] of the very significant share of the SFHA in the experimental hydrogen gases. Thus, the experiments on the electron impact excitation of hydrogen molecules are the fourth type of the atomic experiments that proved the existence of the SFHA (the three previous types of atomic experiments proving the existence of the SFHA being listed in the Introduction of the present paper).

The rough estimates provided in the present paper are intended to get the message across and to motivate further experimental and theoretical works on this subject.

Funding: This research received no external funding.

Data Availability Statement: All data in included in the paper.

Acknowledgments: The author is grateful to Prof. W. Ubachs for advising him to search for indications of the SFHA in experiments on the molecular hydrogen.

Conflicts of Interest: The author declares no conflict of interest.

Appendix A. On Using the Term "Flavor"

Both the regular and singular solutions to the Dirac equation outside the proton correspond to the same energy. As this means the additional degeneracy, then according to the fundamental theorem of quantum mechanics, there should be an additional conserved quantity. In other words, the situation is that hydrogen atoms have *two flavors*, differing by the eigenvalue of this additional, new conserved quantity: hydrogen atoms have *flavor symmetry* [16].

It is called so by analogy with quarks that have flavors: for example, there are up and down quarks. For representing this particular flavor symmetry, there was assigned an operator of the additional conserved quantity: the isotopic spin I—the operator having two eigenvalues for its z-projection: $I_z = 1/2$ assigned to the up quark and $I_z = -1/2$ assigned to the down quark.

References

1. Oks, E. High-Energy Tail of the Linear Momentum Distribution in the Ground State of Hydrogen Atoms or Hydrogen-like Ions. *J. Phys. B At. Mol. Opt. Phys.* **2001**, *34*, 2235–2243. [CrossRef]
2. Oks, E. Analysis of Experimental Cross-Sections of Charge Exchange between Hydrogen Atoms and Protons Yields another Evidence of the Existence of the Second Flavor of Hydrogen Atoms. *Foundations* **2021**, *1*, 265–270. [CrossRef]
3. Oks, E. Experiments on the Electron Impact Excitation of the 2s and 2p States of Hydrogen Atoms Confirm the Presence of their Second Flavor as the Candidate for Dark Matter. *Foundations* **2022**, *2*, 541–546. [CrossRef]
4. Bowman, J.D.; Rogers, A.E.E.; Monsalve, R.A.; Mozdzen, T.J.; Mahesh, N. An Absorption Profile Centred at 78 Megahertz in the Sky-Averaged Spectrum. *Nature* **2018**, *555*, 67–70. [CrossRef]
5. Barkana, R. Possible Interaction between Baryons and Dark-Matter Particles Revealed by the First Stars. *Nature* **2018**, *555*, 71–74. [CrossRef]
6. Oks, E. Alternative Kind of Hydrogen Atoms as a Possible Explanation of the Latest Puzzling Observation of the 21 cm Radio Line from the Early Universe. *Res. Astron. Astrophys.* **2020**, *20*, 109. [CrossRef]
7. Jeffrey, N.; Gatti, M.; Chang, C.; Whiteway, L.; Demirbozan, U.; Kovacs, A.; Pollina, G.; Bacon, D.; Hamaus, N.; Kacprzak, T.; et al. Dark Energy Survey Year 3 results: Curved-Sky Weak Lensing Mass Map Reconstruction. *Mon. Not. R. Astron. Soc.* **2021**, *505*, 4626–4645. [CrossRef]
8. Oks, E. DES Map Shows a Smoother Distribution of Matter than Expected: A Possible Explanation. *Res. Astron. Astrophys.* **2021**, *21*, 241–245. [CrossRef]
9. Zammit, M.C.; Savage, J.S.; Fursa, D.V.; Bray, I. Electron-Impact Excitation of Molecular Hydrogen. *Phys. Rev. A* **2017**, *95*, 022708. [CrossRef]
10. Poet, R. The Exact Solution for a Simplified Model of Electron Scattering by Hydrogen Atoms. *J. Phys. B Atom. Mol. Phys.* **1978**, *11*, 3081–3094. [CrossRef]
11. Wrkich, J.; Mathews, D.; Kanik, I.; Trajmar, S.; Khakoo, M.A. Differential Cross-Sections for the Electron Impact Excitation of the B $^1\Sigma_u{}^+$, c $^3\Pi_u$, a $^3\Sigma^+$, C $^1\Pi_u$, E, F $^1\Sigma^+$ and e $^3\Sigma_u{}^+$ States of Molecular Hydrogen. *J. Phys. B At. Mol. Opt. Phys.* **2002**, *35*, 4695–4709. [CrossRef]
12. Khakoo, M.A.; Trajmar, S. Electron-Impact Excitation of the a $^3\Sigma_g{}^+$, B $^1\Sigma_g{}^+$, c $^3\Pi_u$, and C $^1\Pi_u$ states of H_2. *Phys. Rev. A* **1986**, *34*, 146–156. [CrossRef] [PubMed]
13. Mason, N.J.; Newell, W.R. The Total Excitation Cross Section of the c $^3\Pi_u$ State of H_2. *J. Phys. B At. Mol. Opt. Phys.* **1986**, *19*, L587–L591. [CrossRef]
14. Ajello, J.M.; Shemansky, D.E. Electron Excitation of the H_2(a $^3\Sigma_g{}^+$–b $^3\Sigma_u{}^+$) Continuum in the Vacuum Ultraviolet. *Astrophys. J.* **1993**, *407*, 820–825. [CrossRef]

15. Lima, M.A.P.; Gibson, T.L.; McKoy, V.; Huo, W.M. Cross Sections for Excitation of the b $^3\Sigma_u{}^+$, a $^3\Sigma_g{}^+$, and c $^3\Pi_u$ States of H_2 by Low-Energy Electrons. *Phys. Rev. A* **1988**, *38*, 4527–4536. [CrossRef] [PubMed]
16. Oks, E. Two Flavors of Hydrogen Atoms: A Possible Explanation of Dark Matter. *Atoms* **2020**, *8*, 33. [CrossRef]

foundations

MDPI

Article

Fundamental Spacetime Representations of Quantum Antenna Systems

Said Mikki [1,2]

[1] Zhejiang University/University of Illinois at Urbana-Champaign Institute (the ZJU-UIUC Institute), Zhejiang University, 718 East Haizhou Road, Haining 314400, China; mikkisaid@intl.zju.edu.cn

[2] Electrical and Computer Engineering Department, University of Illinois at Urbana-Champaign, Engineering Hall MC 266, 1308 West Green Street, Champaign, IL 61820, USA

Abstract: We utilize relativistic quantum mechanics to develop general quantum field-theoretic foundations suitable for understanding, analyzing, and designing generic quantum antennas for potential use in secure quantum communication systems and other applications. Quantum antennas are approached here as abstract source systems capable of producing what we dub "quantum radiation." We work from within a generic relativistic framework, whereby the quantum antenna system is modeled in terms of a fundamental quantum spacetime field. After developing a framework explaining how quantum radiation can be understood using the methods of perturbative relativistic quantum field theory (QFT), we investigate in depth the problem of quantum radiation by a controlled abstract source functions. We illustrate the theory in the case of the neutral Klein-Gordon linear quantum antenna, outlining general methods for the construction of the Green's function of a source—receiver quantum antenna system, the latter being useful for the computation of various candidate angular quantum radiation directivity and gain patterns analogous to the corresponding concepts in classical antenna theory. We anticipate that the proposed formalism may be extended to deal with a large spectrum of other possible controlled emission types for quantum communications applications, including, for example, the production of scalar, fermionic, and bosonic particles, where each could be massless or massive. Therefore, our goal is to extend the idea of antenna beyond electromagnetic waves, where now our proposed QFT-based concept of a quantum antenna system could be used to explore scenarios of controlled radiation of any type of relativistic particles, i.e., effectively transcending the well-known case of photonic systems through the deployment of novel non-standard quantum information transmission carriers such as massive photons, spin-1/2 particles, gravitons, antiparticles, higher spin particles, and so on.

Keywords: quantum antennas; quantum field theory; relativistic quantum mechanics; quantum radiation; propagator; Green's function; quantum engineering; quantum technologies; radiation pattern

Citation: Mikki, S. Fundamental Spacetime Representations of Quantum Antenna Systems. *Foundations* **2022**, *2*, 251–289. https://doi.org/10.3390/foundations2010019

Academic Editor: Michal Hnatič

Received: 15 November 2021
Accepted: 3 February 2022
Published: 2 March 2022

1. Introduction

The main objective of this paper is to develop conceptual, physical, and mathematical foundations for quantum antenna theory based on a very broad approach to generic quantum fields produced and consumed by source and sink systems separated in spacetime. Quantum antenna technology is a recent emerging subfield within the larger and more fluid research area often referred to as quantum engineering, quantum technologies, or just quantum information processing. In particular, and within this subfield, we find that the main intention behind the desire for developing a new "quantum antenna technology" is to serve the needs of current and future quantum *communication* systems [1–4], where information is transmitted using quantum states [5,6], regardless of whether digital data are encoded as classical bits or qbits [1,7–10], with obvious applications to physical-layer security [11–14]. However, the peculiar system known as "the quantum antenna" may also perform functions other than mere information transmission in wireless quantum links, for example, quantum tomography, quantum state estimation, biophotonics, sensing, molecular communications, space exploration, and other applications [13,15–26].

We believe that quantum antenna theory and technology may be viewed as an attempt to synthesize three already established subfields: (1) classical antennas [27,28], (2) optical nanoantennas [29,30], and (3) quantum emitters [29,31]. As such, a quantum antenna is often confused with one of these three topics. For example, occasionally, laser sources or photodiodes are treated as quantum antennas by some authors working in free space optical communications. This is *not* what we understand by the term here. A laser source could constitute a part of the q-antenna system, while a larger part, for example a terahertz or an optical antenna [29,32], may serve as a secondary reflector as in nanoantennas [30]. In addition, an atom emitting a photon after undergoing a transition from excited to ground states is not considered an "antenna" in this paper. Instead, we propose the following general definition:

Definition 1 (Antennas). *By the term antenna, we follow classical antenna theory where the antenna system is defined as an externally controlled spatiotemporal current distribution capable of controlling the spatiotemporal properties of the radiation emitted by the antenna system.*

It will be seen that the most salient point in this definition is the emphasis on the following three features:

1. The source is a controllable current distribution function of *both space and time.*
2. Ultimately, the source current is *externally* controlled.
3. A spatiotemporal current distribution profile serving as a model for a quantum antenna source can control the *radiation* proprieties in both space and time, and usually for near- and far-field scenarios as well.

Hence, the existence of *spatiotemporal*, *externally* controlled source current distributions capable of modifying its produced *radiation* in both space *and* time is the main content of antenna theory.

Clearly enough, mainstream examples in quantum optics such as a laser source by itself, or an isolated atom undergoing a spontaneous emission of light, do not fall under a conceptual umbrella such as the one supplied by Definition 1, at least not naturally. However, such conventional quantum optical sources may still serve as essential *sub*-components of such systems. For example, probably, the most natural method to inject a controlled *time* signal into a radiating nanostructure is through a modulated laser beam. However, controlling the *spatial* distribution of the radiating *current* excited by this time signal would require the use of *additional* methods and objects; for instance, an optical antenna, or an engineered metamaterial, or an optimized array of point sources. Even though the data to be transmitted are encoded into time signals, in general, the quantum antenna system is much more complex than its time excitation method. There is a need, then, to understand the complex and multifaceted nature of the spatiotemporal structure of the radiation field emitted by a generic quantum system at a very broad level. This paper is a contribution toward this goal.

The classical theory of controlled radiation has been extensively studied and developed in applied electromagnetics [27,28,33–35]. On the other hand, the general theory of quantum antennas has not been investigated in depth so far, in spite of the publication in recent years of a number of reports and proposals about the subject, not restricted only to quantum wireless communications but also involving other topics as well, e.g., see [15–21]. Some of the main conceptual and philosophical hurdles that a viable theory of quantum antennas need to overcome include, though by no means are restricted to, the following issues:

1. Understanding what is meant by *quantum radiation*.
2. Understanding the role played by particle emission dynamics in light of the wave-particle duality characteristic of quantum phenomena.

3. Understanding the complex role played by quantum fields, propagators, and Green's functions in quantum radiation.
4. Understanding the role played by many-particle states/interactions in quantum radiation processes.

It should be noted that these conceptual and technical issues are still considered difficult open research fields in both foundational and applied research as they have not been resolved even within standard quantum field theory itself. For instance, it is still not clear what the ultimate meaning of "particle" in quantum fields is [36,37]; questions about the nature of quantum excitations in interacting field theories have been asked in the past [38] and are still being investigated up to date [39]. Therefore, there is a need to re-examine the subject of quantum antennas at a very general and fundamental level, that of developing possible foundations for the topic that may help illuminate current and future open theoretical problems on one hand, and to help devise and evolve new genera of quantum systems and applications on the other hand. A viable candidate for such foundational approach, we believe, is to formulate the entire problem of quantum antennas and quantum radiation using a relativistic spacetime formalism (QFT in this case).

Our intended goals in the present article include a wide spectrum. Most of these objectives are research related and can be classified as part of the emerging area of quantum antenna theory within both quantum physics and quantum engineering. Yet, additionally, other aspects in our work are pedagogical, relating to the wish to reach a wide multidisciplinary audience. We summarize these two flavors of our main goals as follows:

- Primary research objectives of the present article:
  1. Generalizing the concept of antennas beyond acoustic and electromagnetic antennas, the two concepts that have dominated the field so far, by demonstrating how relativistic QFT can be used to formulate a single and unified concept of "quantum radiators" valid for a large number of possible radiation processes in nature.
  2. Providing a concrete illustration of some of the potential algorithmic capabilities of the spacetime formalism of quantum antennas by constructing various possible candidates for radiation pattern functions and gains in the case of the quantum (spin-0) Klein–Gordon q-antennas.
- Secondary (pedagogical) objectives of the present article:
  1. Introducing new applications of fundamental theory (here relativistic quantum mechanics) to different audiences, e.g., quantum engineering and quantum technology research.
  2. Introducing the subject of QFT in a self-sufficient manner by providing detailed appendices explaining how relativistic quantum mechanics is formulated for an audience familiar only with nonrelativistic quantum mechanics.[1]

This paper is organized as follows. In Section 2, the classical theory of antennas is re-examined, and the comparison with the new, more general concept of quantum source radiators is explored at the thematic level in order to prep the reader for the subsequent, more technical quantum-field theoretic treatment. This is followed in Section 3 by a very broad view on the theory of quantum radiators developed based on interacting quantum mechanics, without much emphasis at this stage on the relativistic quantum field theoretic scenario. The purpose is to map out the generic structure of the problem and to highlight the distinction between linear and nonlinear quantum antennas. Starting from Section 4, we narrow down our focus to the special but rich enough special case of the neutral Klein–Gordon field linear quantum antenna, which appears to exhibit many of the salient features of the general quantum antenna system. Then, the abstract formal and physical structure of such a system is explicated in Section 5. To provide more concrete applications of the theory, Section 6 presents a series of examples and constructions aiming at illustrating how one may define radiation pattern measures such as directivity and gain (transmission) coefficients in quantum antenna systems. Finally, we end up with conclusions.

2. Antenna Theory: Classical and Quantum Radiation Scenarios

In analogy to classical antenna theory [27,35,40], we formally define a *quantum antenna* in terms of the mathematical representation of a radiating source term as follows:

Definition 2 (Quantum Antennas). *Within the context of relativistic quantum field theory, a quantum antenna (q-antenna) $\mathcal{J}$ is an operator map of the form*

$$\mathcal{J} : J \to \Psi, \tag{1}$$

where Ψ is a quantum field system produced by the abstract source function

$$J : \mathbb{M}^4 \to \mathbb{K}, \tag{2}$$

which maps the Minkowski spacetime $\mathbb{M}^4$ of special relativity to the set $\mathbb{K} = \mathbb{R}$ (real numbers) or $\mathbb{K} = \mathbb{C}$ (complex numbers). We assume that the function $J(x), x \in D_s \subset \mathbb{M}^4$ is integrable over any spacetime region D_s of interest. We further assume that the source $J(x)$ is compactly supported; i.e., $cl\{D_s\}$ is compact where cl is the closure operator in the standard Euclidean topology on $\mathbb{R}^4$.

Physically, the main property of the q-antenna (source) function $J(x)$ is that it controls the quantum radiation $\Psi(x), x \in \mathbb{M}^4$, emitted by the source system $J(x), x \in D_s \subset \mathbb{M}^4$. The generic configuration itself is shown in Figure 1. In fact, much of this paper will concentrate on understanding the precise nature of how a generic source function $J(x)$ can control the spacetime structure of the quantum field emission $\Psi(x)$ in applications characteristic of the special context of an arbitrary quantum *communication* system (source-channel-detector system). Thus, the article's main scope is the theoretical foundations of the physical layer of the quantum communication link structure, e.g., how a quantum state can be directed to a given spacetime region through the choice of a proper abstract source function. The particular discipline devoted to this problem is the area we dub *quantum antenna theory*, which is a relatively new interdisciplinary research area.

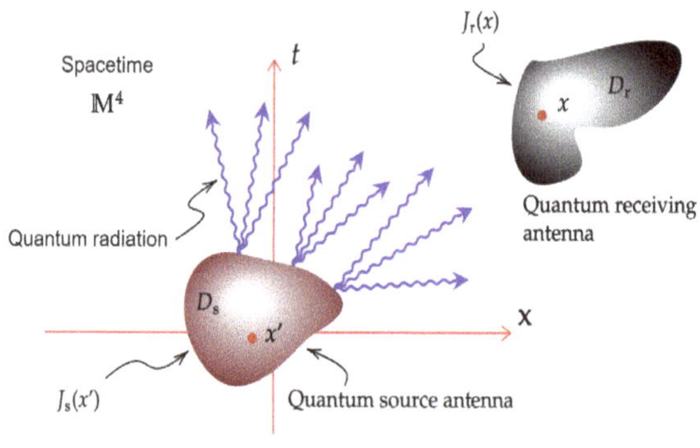

Figure 1. The fundamental configuration of a quantum antenna communication system in spacetime.

Remark 1. *It may be noted here that there is a similarity between the quantum communication problem [7] and the quantum measurement problem in quantum physics [41–43]. The analogy is manifested by the common underlying abstract structure shared by the two distinct processes of communications and generalized quantum measurement [1,8]. Both involve a source of information (often called quantum state preparation in QM), which is localized within a given spacetime region; and a process of destructive consumption of the transmitted information, this time localized at a receiver or detector located at a timelike-distant spacetime region [43,44]. The quantum receiver system's ultimate objective is to extract the maximum amount of information embodied in the received quantum state in QM, or quantum field in QFT, through the judicious utilization of optimized combinations of measurement operations and signal processing algorithms.*

Quantum antenna systems turn out to be considerably more complex to understand and analyze than conventional antennas, where the latter antenna type is often approached within the framework based on *classical* electromagnetism (and sometimes acoustics). A reason behind this noticeable difference in complexity between the two theories might be related to the fact that a proper and completely satisfactory understanding of quantum interactions is ultimately based on relativistic QFT, which is inherently a *many-particle* world picture. That is, in QFT, it is inconsistent to assume the possibility of a single-particle-interaction configuration due to the inherent tension that this assumption will bring with special relativity (SR) and other fundamental principles [38]. On the other hand, a traditional classical antenna is essentially a "one-point physical process" in which a generic source point $x' = (\mathbf{x}', t')$ radiates into another arbitrary observation point $x = (\mathbf{x}, t)$, where the total radiation is just the linear superposition of all contributions emanating from all relevant individual non-interacting spacetime points [45].

The most obvious and rigorous way to understand the classical process of radiation and reception of electromagnetic signals is through the two fundamental Green's functions of the antenna system, the source or current Green's function $\overline{\mathbf{F}}(\mathbf{x}, \mathbf{x}', t - t')$ [28,46–48], and the radiation Green's function $\overline{\mathbf{G}}(\mathbf{x} - \mathbf{x}', t - t')$ [49–51]. These two are essentially defined via their crucial superposition-like integral summation rules (A1) and (A2); see Appendix A, where the Green's function approach to classical antenna theory is briefly reviewed and additional references are given.

However, when we next move toward examining the emerging theory of quantum antenna systems, the natural question is whether this generalized structure of classical antennas persists. It turns out that radiation formulas similar to the Green's function-based expressions (A1)–(A3), Appendix A, can *not* in general be maintained in the case of *quantum* radiation. However, in this paper, we focus on ways to retain as much classical structure as possible in the new quantum antenna theory. This turns out to be possible for the case of *linear* quantum antenna systems, which will be investigated in details below.

To summarize, we find that the following general situation holds in antenna theory:

1. Classical electromagnetic radiation in free space or linear materials, when viewed from the perspective of its ultimate source (external field), is inherently *linear*.[2]
2. On the other hand, quantum antennas involving higher-order processes (many-particle interactions, n-point processes with $n > 1$, etc.) are intrinsically *nonlinear* radiation problems (due to the many-body nature of interactions in quantum field theory).

To be more precise, it can be shown that only *first*-order quantum radiation processes may lead to linear operator radiation relations similar to (A1)–(A3). This is the process that corresponds to what we call *linear quantum antennas*, which will be studied in more details in this paper, while a more complete theory of relativistic nonlinear quantum antennas is relegated to future publications. All other higher-order processes, which are ultimately due to the many-body quantum interactions between source points, can be shown to lead to *nonlinear* contributions to the total probability amplitude of the quantum radiation emitted by the source system.

3. The General Theory of Quantum Antenna Systems

3.1. Preliminary Considerations

According to Definition 2, a q-antenna system consists of a *spacetime source region* D_s together with a current source function defined on this region. The *receiving* q-antenna is located in the compact spacetime region D_r. The overall spacetime configuration is illustrated in Figure 1. To allow information transmission from the source to the receiver regions, the two spacetime domains D_s and D_r must have timelike separation [53].[3] For simplicity, D_s and D_r as spacetime regions are assumed to be factorizable into the forms

$$D_s = S_s \times T_s, \quad D_r = S_r \times T_r, \tag{3}$$

where S_s and S_r are the three-dimensional spatial subdomain components of the original four-dimensional regions $D_{s,r}$ satisfying

$$\mathbb{M}^4 \supset D_{s,r} \supset S_{s,r} \subset \mathbb{R}^3, \tag{4}$$

while $T_{s,r} \subset \mathbb{R}$ is the "temporal component" (timelike slice) of the source/receive regions $D_{s,r}$, respectively. Here, both $D_{s,r}$ and $T_{s,r}$ are assumed to be *compact*, i.e., $D_{s,r}$ and $T_{s,r}$ are closed and bounded sets in the standard Euclidean metric space $\mathbb{R}^3$.[4]

Throughout this section, the quantum field $\phi(x)$ is allowed to refer to a general field with possible spins $0, 1/2, 1, 2$, etc., without worrying much about indices (in the next sections, we work only with scalar fields for simplicity). Moreover, any particular QFT under consideration may itself be of the free (non-interacting) or interacting types. Interacting field theories often require the use of perturbation theory in order to obtain practical results. However, all information about the interacting system is conveniently encoded in the *propagator*, which will be used extensively in our theory below.[5] Therefore, for generality, in this paper, we utilize the propagator concept as the fundamental mathematical carrier of physical information about the generic quantum antenna system in spacetime.

3.2. A Generic Interaction Hamiltonian Description of Quantum Antenna Systems

Here, a high-level view on interacting field theory is provided. Note that a full treatment of interacting field theories is outside the scope of the present paper, whose main goal is to supply the reader with a minimal picture of how eventually quantum antenna theory should be fully formulated when specific physical layouts, often described with their own Hamiltonian, are introduced. On the other hand, starting from Section 4, we work mainly with the scalar field theory (the Klein–Gordon theory) in order to illustrate the general structure of quantum radiation with minimum knowledge of the full details of matter–field interaction mechanisms. Readers interested in more information related to specialized physical layouts may consult numerous other publications, for example those quoted in the Introduction section of this paper.

We assume that all relevant quantum field operators can be described within one and the same large enough combined Hilbert space $\mathcal{H}$, which in the case of QFT is a Fock space $\mathcal{F}$. The total Hamiltonian operator of the system is written as

$$H = H_0 + H_{\text{int}}, \tag{5}$$

where H_0 is the *free* Hamiltonian (non-interacting part that is usually solvable), while H_{int} is the *interacting* Hamiltonian, which is a time-dependent operator that in turn can be expanded into four basic component as follows

$$H_{\text{int}} = H_{\text{in}} + H_s + H_r + H_c. \tag{6}$$

Here, we have the following three categories of interaction Hamiltonian terms:

1. H_{in} captures *intrinsic* interaction in the fundamental quantum field of the q-antenna system, e.g., self interactions such as polynomial Lagrangian terms containing powers of $\phi(x)$ larger than three such as the mainstream interacting ϕ^4-theory [54,56,57].
2. The terms H_s and H_r describe, respectively, the interaction between the source and the receiver antennas on one side and the fundamental quantum field $\phi(x)$ of the q-antenna system on the other side. These interactions should be understood as processes localized within their respective spacetime domains D_s and D_r.
3. Finally, the term H_c corresponds to *channel* interactions and couplings, e.g., coupling of the excited quantum field $\phi(x)$ with scattering objects located within the effective path of an excited quantum particle produced by the source and directed toward the receiver.[6]

A couple of important observations on this decomposition of the interacting Hamiltonian are in order. First, to keep the discussion at the most general level, we assume that the intrinsic interacting Hamiltonian, if nonzero, is present all the time. That is, it becomes an essential ingredient of the fundamental field $\phi(x)$ of the q-antenna system. In this sense, the latter field is defined as precisely that quantum field corresponding to the Hamiltonian

$$H_\phi = H_0 + H_{\text{in}}. \tag{7}$$

On the other hand, the source, receiver, and channel Hamiltonian terms are treated in this theory as *extrinsic* interactions, i.e., external disturbances coupled to the fundamental quantum field $\phi(x)$ associated with the Hamiltonian H_ϕ. As expected in quantum physics, coupling often leads to nonlinear equations of motion [54]. In this sense, the uncoupled fundamental field $\phi(x)$ is perturbatively transformed into a *new* coupled field $\phi'(x)$ with a full Hamiltonian

$$H_{\phi'} = H_\phi + H_{\text{s}} + H_{\text{r}} + H_{\text{c}} \tag{8}$$

solved using perturbation Dyson or path integral expansions [55]. Following standard conventions in QFT, we do not change the notation of the field but always use $\phi(x)$ while clearly stating which Hamiltonian is being used (if needed). In addition, we will not label Hamiltonians by their fields or spacetime arguments unless this is needed.

Remark 2. *Sometimes, it is more convenient to group channel interactions H_{c} and intrinsic field self-interactions H_{in} into one term*

$$H'_{\text{in}} = H_{\text{in}} + H_{\text{c}}. \tag{9}$$

The reason is that there is indeed some similarity between the two types of interactions above. They are both unrelated to the transmitter (source) and receiver terminals and can be considered then as indigenous components of the q-antenna system field itself. However, there are also some differences, since channel couplings are not the same everywhere but are localized at the scattering objects themselves. On the other hand, self-interactions and also intrinsic mutual interactions captured by the term H_{in} are generally "turned on" most of the time. Nevertheless, in some applications, it might be useful to group H_{in} and H_{c} with each other under the rubric of scattering-based interaction processes in order to distinguish them from transmitter and receiver types of interactions (information source and sink).

Let us examine now how the source and the receiver terminals of the system may interact. Working in the Schrodinger picture,[7] Let the state of the source at the end of the interaction interval $T_{\text{s}} = [t'_{\text{s}}, t_{\text{s}}]$ be denoted by $|J, D_{\text{s}}\rangle_{t_{\text{s}}}$. Here, J is a generic symbol for the overall set of (classical) disturbances supported on D_{s} ultimately constituting our proposed formal Definition 3 of the q-antenna current source system. Then, we may write

$$|J, D_{\text{s}}\rangle_{t_{\text{s}}} = P(t'_{\text{s}}, t_{\text{s}}; J, D_{\text{s}})|0_{\text{in}}\rangle, \tag{10}$$

where $P(t'_{\text{s}}, t_{\text{s}}; J, D_{\text{s}})$ is the interaction picture (Dirac) propagator (A18), which is reviewed in Appendix D for more details. That is, the computation of this propagator is based on the substitution $H_{\text{I}} = H_{\text{s}}$ into (A19). The special state labeled $|0_{\text{in}}\rangle$ is the ground state of the *intrinsically interacting* system, i.e., the state satisfying

$$(H_0 + \underbrace{H_{\text{in}} + H_{\text{c}}}_{H'_{\text{in}}})|0_{\text{in}}\rangle = 0, \tag{11}$$

where, as discussed above, we have here opted for generality by including all possible wireless channel scattering effects captured by H_{c} with the field intrinsic self-interaction Hamiltonian H_{in}. Then, the ground state $|0_{\text{in}}\rangle$ represents both the (i) true or actual physical initial state of the q-antenna system *before* interacting with the source J localized in the spacetime domain D_{s}; and (ii) the final state reattained *after* finishing interaction with the receiver localized in the disjoint domain D_r. On the other hand, the *bare* ground state $|0\rangle$ is the ground state of the *free* Hamiltonian H_0 satisfying

$$H_0|0\rangle = 0. \tag{12}$$

An exactly analogous general analysis can be conducted in order to understand the receiving q-antenna interaction problem. The latter is that concerned with what happens after interaction with the detector during the time interval $T_r = [t'_r, t_r]$, change in fields, states, observation outcomes, etc.

3.3. The General Expansion Theorem of Quantum Radiation Fields

We now generalize the concept of quantum source in order to take into account many-point interactions. Let us consider a generic quantum source located in the region $D_s \subset \mathbb{M}^4$.

Definition 3 (Generalized quantum antenna source). *A generalized quantum source is defined as a countable set of real-valued functions J_n of the form*

$$J_n : \underbrace{D_s \times D_s \cdots \times D_s}_{n \text{ times}} \to \mathbb{R}, \quad n \in \mathbb{N}, \tag{13}$$

where D_s is compact. In other words, a generalized quantum antenna source is defined as the set

$$\mathcal{J} = \{J_n, n \in \mathbb{N}\} \tag{14}$$

of all real-valued functions on all product spaces of D_s.

The motivations behind the apparently abstract definition are in fact the actual physical relevance of all the functions J_n mentioned there. It turns out that J_1 represents nothing but a direct linear current source, while all higher-order sources $J_n, n > 1$, can be interpreted as mutual interaction[8] strength factors. To see this, we now present the following fundamental theorem about quantum sources:

Theorem 1 (The q-antenna decomposition theorem). *Let $A(x) \in \mathbb{C}$ be the probability amplitude of the observation (particle annihilation) of the quantum antenna field at a location $x \in \mathbb{M}^4$. Then, it follows that when expressed in terms of the generalized q-antenna source of Definition 3, the amplitude $A(x)$ can be expanded as*

$$A(x) = \sum_{n=1}^{\infty} \int \prod_{l=1}^{n} \mathrm{d}^4 x_l \, G^n(x, x_1, \ldots, x_n) J(x_1, \ldots, x_n), \tag{15}$$

where all integrals are performed within the source region D_s. Here, $J(x_1, \ldots, x_n)$ is identified with J_n in Definition 3. The set of functions G^n are $(n+1)$-point Green's functions where n is a labeling superscript, not a power.

See Appendix E for some background to the decomposition theorem, where a general discussion and additional references are given. A complete and rigorous treatment of Theorem 1 is outside the scope of the present paper, since the proof is fairly lengthy though relatively straightforward, and the details are not needed in order to understand linear quantum antennas (the case mostly discussed in the remaining parts of this paper). A general idea of the scope of the proof is briefly outlined in Appendix E.

We discuss the physical meaning of Theorem 1. For clarity, let us rewrite the general expression (15) as a sum of first-order term (linear radiation) and all higher-order terms:

$$A(x) = \underbrace{\int \mathrm{d}^4x_1 G^1(x,x_1)J(x_1)}_{\text{Linear quantum radiation integral}} + \underbrace{\sum_{n=2}^{\infty}\int\prod_{l=1}^{n}\mathrm{d}^4x_l\, G^n(x,x_1,\ldots,x_n)J(x_1,\ldots,x_n)}_{\text{Nonlinear quantum radiation higher-order integrals}}. \tag{16}$$

The first term on the RHS represents conventional linear quantum radiation processes and will be studied in detail in the remaining parts of this paper. It involves a usual quantum source function $J(x)$ (see Definition 2), with a two-point Green's function $G^1(x,x')$ serving as a "system spacetime transfer function" of the antenna system (or "generalized impulse response" using the terminology of signal processing and the engineering sciences). Note that while $A(x)$ is a complex probability amplitude the like of which is completely absent in the classical world, radiation expressions analogous to the first term on the RHS of (16) do possess some—at least formal—structural similarity to the classical radiation formula (see Section 5.2 for an in-depth comparative analysis between different antenna types).

On the other hand, the remaining terms in (16) involve *mutual* source functions of the form $J(x_1,\ldots,x_N)$. These are *joint interaction* terms describing *coupling phenomena* among the generic points $x_1,\ldots,x_n \in D_s, n > 1$, which may introduce mutual correlation between some or all points of the source systems, which in turn are ultimately explainable as many-body effects[9] as illustrated by the following example.

Example 1 (Classical current source). In order to appreciate why the presence of joint source functions signifies interactions, let us consider the well-known case of a classical current source (i.e., the backreaction of the quantum field on the source is ignored) when such source function $J(x)$ is inserted into the Lagrangian of quantum field theory [14,62]. In that case, one may obtain an exact solution of the interaction problem. It turns out that higher-order processes in this solution can be all expressed as simple direct multiplications of the same source function $J(x)$, i.e., we have:

$$J_n = J(x_1,\cdots,x_n) = \Pi_{l=1}^n J(x_l) = J(x_1)J(x_2)\times\cdots J(x_n). \tag{17}$$

Therefore, in such theory, the second term on the RHS of (16) looks like

$$\int \mathrm{d}^4x_1\mathrm{d}^4x_2\, G^2(x,x_1,x_2)J(x_1)J(x_2), \tag{18}$$

which is essentially nonlinear in the current source. In the more general case, when there is a correlation between x_1 and x_2, the following condition holds:

$$J(x_1,x_2) \neq J(x_1)J(x_2). \tag{19}$$

However, still even in this correlated latter case, one may expand the function $J(x_1,x_2)$ in Taylor series around the uncorrelated case $J(x_1,x_2) = J(x_1)J(x_2)$ in order to understand the general structure of the problem whose general form is now given by the second-order process[10]

$$\int \mathrm{d}^4x_1\mathrm{d}^4x_2\, G^2(x,x_1,x_2)J(x_1,x_2). \tag{20}$$

Nevertheless, the message of this example is clear: A higher-order process, i.e., J_n with $n > 1$, introduces nonlinear contributions to the total probability amplitude $A(x)$ as per (16).

4. Linear Quantum Antenna Systems

4.1. Introduction

The conceptual, physical, and mathematical structure of quantum antennas will now be further investigated in depth by focusing on the first term on the RHS of (16). This restriction is motivated by the desire to keep the presentation as simple as possible. In fact, most of the key concepts and structure to be developed in connection with the main component of the quantum source, the function J_1, can be adapted and also applied to the higher-order components $J_n, n > 1$, although the details are lengthy. For example, a radiation pattern could be constructed for each source mode function J_n similar to the one to be given below (see Section 6), after which the various contributions of all such terms are summed together in order to estimate the total radiation characteristics of the source system.

Our approach is fundamentally based on the idea of propagators and Green's functions in field theory. Similar to the corresponding situation in QFT itself, the concept of Green's function is also fundamental in classical antenna theory [27,28,33,35,49,63,64]. There, one finds that Green's function connects the source (cause) with its produced radiation field (effect), which is a fact that has been frequently exploited as a useful computational tool in both classical electromagnetism [49,63–66] and quantum physics [54,55,67].

However, for quantum radiating systems to behave as antennas, one must track down and separate from out of the total quantum system an ultimate terminal where a purely *classical*—i.e., deterministic—source function is identified. Through such a source function, the quantum source system may then allow for an *external* user control for the purpose of sending information [10,13].[11] For the purpose of this paper, we will show that QFT allows the construction of a very natural and direct source model for q-antenna systems.

In order to simplify the presentation, we break down our method into several steps as follows:

1. Construct a quantum source model resembling the point (infinitesimal source) in classical antenna theory.
2. Using the previous quantum point source model, construct the quantum state radiated by the q-antenna due to arbitrary continuous or discrete source distribution (superposition principle).
3. Construct Green's function of the q-antenna using the previous superposition integral.
4. Introduce and evaluate the q-antenna radiation pattern using Green's function (mostly in the momentum space representation).

References on the Green's function approach to classical antenna theory can be found in [28,46–48]. The infinitesimal source approach is developed in [68]. The momentum space approach to electromagnetic theory and antennas is outlined in [69–73].

For maximum clarity, this program will be carried out throughout the remainder of the present paper for the special case of a neutral Klein–Gordon field $\phi(x)$ (the standard spin-0 scalar field theory, Section 4.2). In addition, and as already mentioned above, for emphasis on simplicity, we focus on the linear quantum antenna case. In fact, the key concepts introduced below through the Feynman propagator for the construction of a quantum radiation pattern are essentially the same whether the radiation regime under consideration is linear or nonlinear.

4.2. The Klein–Gordon Field Theory

In relativistic quantum field theory [38,54,55], everything takes place in the four-dimensional Minkowski spacetime [51,53], denoted by $\mathbb{M}^4$, which is a linear vector space endowed with a special metric, the Lorentz metric $g_{\mu\nu}$, see Table A1 in Appendix B, where the relativistic notation and key quantum formulas in the natural unit systems are reviewed. (In what follows, we work in the natural unit systems where c, the speed of light, and $\hbar$, the Planck constant, are both reduced to unity ($c = \hbar = 1$); see Appendix C). Without loss of generality, and as mentioned earlier, the q-antenna quantum field $\phi(x)$ is assumed to be that of scalar massive theories.[12] To further simplify the mathematical manipulations, we focus on the neutral massive Klein–Gordon field theory, which is outlined in Appendix F. For massive particles (spin-0 particles in our scalar theory example), the dispersion relation (A27) governs the behavior of particles emitted when the q-antenna system is used in applications such as quantum molecular communications. On the other hand, massless ($m = 0$) particles models "scalar photons" or photons with polarization ignored, since in this case, the Klein–Gordon equation reduces to the (scalar) wave equations.[13] In general, adding spin to the theory does not involve any major changes in the main conceptual ideas related to q-antennas and hence will be left for future work.

We also further note that the field modal expansion (A26) is intrinsically Lorentz invariant, even though the spectral integration performed there is carried with respect to the non-Lorentz invariant volume measure d^3x because the standard method of *normalizing factor* has been already employed in our formulation.[14] In other words, the spacetime q-antenna theory developed here is fully relativistic. A breakdown of relativistic invariance, as exemplified by choices of spatiotemporal decomposition (slicing) of the source or receiver regions $D_{\mathrm{s,r}}$, e.g., as in (3), can be introduced later as an external restriction enforced by hand to simplify the calculations and the presentation.

4.3. An Elementary Model for Point Quantum Particle Excitation

In our model, a fundamental spacetime quantum field $\phi(x)$ is associated with the q-antenna.[15] We start by provisionally identifying $\phi(x)$ as the "quantum source field" of the q-antenna system, i.e., the fundamental quantum field of the system directly produced by the source $J(x), x \in D_{\mathrm{s}}$. Roughly speaking, this terminology indicates that this quantum field plays a double role:

1. It emits quantum particles (massive particles when $m \neq 0$ and scalar Klein–Gordon particles when $m = 0$) at particular spacetime positions.
2. Once generated, the quantum field $\phi(x)$ would somehow "propagate" the quantum particle in space and time.

Equivalently, we say that the quantum field $\phi(x)$ enjoys the double role of being both the producer and propagator of the quantum radiation particle.

Next, let us hit the vacuum state $|0\rangle$ with the q-antenna source field $\phi(x')$ operator; i.e., we wish to excite a quantum particle at the position $x' = (t', \mathbf{x}')$. Using (A26) and the standard facts $a_{\mathbf{p}}|0\rangle = 0$ and $a_{\mathbf{p}}^{\dagger}|0\rangle = |\mathbf{p}\rangle$, we compute

$$\phi^{\dagger}(x')|0\rangle = \int_{\mathbf{p}\in\mathbb{R}^3} \frac{\mathrm{d}^3p}{(2\pi)^{3/2}} \frac{e^{\mathrm{i}p_{\mu}x'^{\mu}}}{(2\omega_{\mathbf{p}})^{1/2}} |\mathbf{p}\rangle. \tag{21}$$

Therefore, we managed to set up a superposition of *outgoing particles*, i.e., radiation quanta *leaving* the position $\mathbf{x}'$ at time t'. This may explain why $\phi(x)$ was duped the quantum source field of the q-antenna: The field literally "creates" many quantum particles radiating away from the initial source location. (The latter source plays the role of the "initial state" in quantum mechanics, though it should be noted that QFT does not use the same concept of the state).

Furthermore, with the help of the expansion (21), we are now in possession of a satisfactory understanding of the composition of the quantum states radiated by this point-like quantum source: they are essentially superpositions of pure momentum states $|\mathbf{p}\rangle$, with scaling factors (momentum state excitation strength) given by the total factor of the integrand of (21) multiplying each such momentum ket.[16]

Let us now compute how much probability amplitude there is in the new excited q-antenna state (21) when an observer tries to measure the q-antenna's radiation field's momentum. If the observation momentum eigenvalue is $\mathbf{q}$, then we form the relativistic bra

$$\langle q| = (2\pi)^{3/2}(2\omega_{\mathbf{p}})^{1/2}\langle \mathbf{q}|, \tag{22}$$

through which one may estimate the amplitude $A(q)$ by calculating the matrix element

$$A(q) := \langle q|\phi^{\dagger}(x')|0\rangle = \int_{\mathbf{p}\in\mathbb{R}^3} \frac{\mathrm{d}^3 p}{(2\pi)^{3/2}} \frac{e^{\mathrm{i}p_\mu x'^\mu}}{(2\omega_{\mathbf{p}})^{1/2}} \langle q|\mathbf{p}\rangle. \tag{23}$$

Using the basic orthogonality relation between momentum eigenstate [57]

$$\langle \mathbf{q}|\mathbf{p}\rangle = \delta(\mathbf{q} - \mathbf{p}), \tag{24}$$

the integral (23) can be readily evaluated, yielding

$$A(q) = e^{\mathrm{i}q_\mu x'^\mu} = e^{\mathrm{i}(\omega_{\mathbf{q}} - \mathbf{q}\cdot\mathbf{x}')}. \tag{25}$$

This is a deceptively simple-looking relation, but it underlies a powerful structure that enjoys considerable importance in the theory of q-antennas. From the engineering point of view, the expression (25) may be shown to lead to the emergence of the classical antenna array factor when moving into the quantum context, hence the ability to shape the quantum radiation emitted by q-antennas using techniques borrowed from what is essentially classical antenna theory.

From the physics point of view, (25) says that the complex probability amplitude of finding the radiated particle emitted at the spacetime point x' at the four-momentum state $|q\rangle$ is simply $\exp(\mathrm{i}q_\mu x'^\mu)$. Furthermore, the relation (25) also confirms the provisional interpretation proposed above that the state $\phi^{\dagger}(x')|0\rangle$ may be viewed as a one-particle quantum field excitation state "localized" at the spacetime point x'. The reason is that such interpretations is reminiscent of the standard relation

$$\langle \mathbf{p}|x\rangle = \exp(\mathrm{i}\mathbf{p}\cdot\mathbf{x}) \tag{26}$$

in nonrelativistic quantum mechanics, suggesting a "quasi-localization" of the particle's momentum state at x. From (25), we obtain $\Pr\{q = q'\} = 1$ for any four-momentum range $q \in \mathbb{M}^4$. This is completely natural since, as in nonrelativistic quantum mechanics, momentum eigenkets are maximally nonlocalized. Since we have just established that the state $\phi^{\dagger}(x)|0\rangle$ represents what, within QFT, corresponds to a pure one-particle momentum state, total nonlocalizablity of the conjugate position parameter is expected. However, note that in standard perturbative QFT, it is very difficult to mathematically describe the complete localization of particles. For an in-depth discussion of this problem, see the footnotes.[17]

4.4. *The Feynman Propagator of Quantum Antennas*

So far, we have succeeded in modeling the process in which a q-antenna emits a particle at a specific spacetime point $x' = (t', \mathbf{x}')$. We also explicated the momentum-space composition of the emitted one-particle radiation state and found that it is comprised of a superposition of multiple outgoing one-particle momentum states. Moreover, we estimated the probability amplitude of measuring a certain (generic) momentum in this radiated state. However, for practical applications, there is a need to actually compute the effective coupling between the source on one side and generic observation spacetime points located either in the near- or far-zone on the other side. To do so, we will make use of one of the most powerful methods in QFT, the *Feynman propagator* [38,54,55,67]. Effectively, this will also directly provide us with the two-point Green's function $G^1(x, x')$ of the q-antenna system corresponding to the first term on the RHS of (15).

The importance of moving to a mathematical description based on propagators stems from the fact that in QFT, one often finds that exciting the ground state at a specific spacetime point does *not* automatically imply that the radiated particle will reach *every* point in the far-zone with *significant* probability. To ensure that the quantum wireless or molecular communication link's receiver has access to the radiated particles with significant probability of detection, we need to compute the probability amplitude of measuring a particle at a generic spacetime point x away from the source point x'. Since in QFT, there are no measurement operators as in nonrelativistic quantum mechanics, we may model the observer's interactions with the receiver by the process of particle *annihilation* [38]. This is one of the key ideas of quantum measurement in QFT to be adopted in this paper.

In order to mathematically implement this idea, we compute the two-point Green's function of the q-antenna system, which is defined as follows:

Definition 4 (The two-point q-antenna Green's function). *The q-antenna two-point (or one-particle) Green's function, denoted by $G_q(x, x')$, is defined as the probability amplitude of the process that a particle created at the point $x' \in D_s$ in the source region, while the system was initially at the ground state $|0\rangle$, will be annihilated at the later point $x \in D_r$ in the receiver (detection, observation) region, after which the q-antenna system will return again to the ground state.*

Theorem 2. *The correct relativistic expression for $G_q(x, x')$ is given by the formula*

$$G_q(x, x') = \langle 0|\mathcal{T}\phi(x)\phi^{\dagger}(x')|0\rangle, \tag{27}$$

where $\mathcal{T}$ is the time-ordering operator defined by (A17).

Proof. This follows directly from Definition 4, the quantum field expansion (A26) in terms of creation and annihilation operators, and the definition of the time-ordering symbol (A17). The time ordering operator is included in order to automatically deal with both particle and antiparticle emission.[18] □

Remark 3. *Definition 4 is inspired by the propagator-based quantum field-theoretic formalism of condensed matter physics, see in particular Mattuck's elegant formulation in [60], which influenced our approach here. The same quantity $G_q(x, x')$ is constructed in perturbative QFT as the fundamental tool for computing scattering cross-sections in experiments involving fundamental particle interactions [38,54,67] and the ground state energy in many-body condensed matter physics [59–61]. In the expression (27), $\mathcal{T}$, the time-ordering operator, is inserted in order to automatically ensure that fields with later time components x^0 are always placed to the left of earlier ones.*

Theorem 2 provides a characterization of the q-antenna Green's function expressed directly in spacetime. However, in most applications of QFT, it is the *momentum* space representation that often proves to be *the* most useful space to do calculations in.[19] As will become more evident below, this is also the case in q-antenna theory. We would like then to derive an expression for the propagator in momentum space. In fact, this is a quite straightforward process: taking the conjugate of (21), multiplying the result with (21), then making use of the basic position-momentum eigenket orthogonality relation

$$\langle \mathbf{q}|\mathbf{p}\rangle = \delta(\mathbf{q} - \mathbf{p}), \tag{28}$$

the q-antenna Green's function readily evaluates to

$$G_q(x, x') = \int_{\mathbf{p}\in\mathbb{R}^3} \frac{d^3p}{(2\pi)^3} \frac{e^{-ip_\mu(x^\mu - x'^\mu)}}{2\omega_{\mathbf{p}}}. \tag{29}$$

This spectral expansion of the q-antenna Green's function is very important and will be illustrated in several examples, while its structure is investigated in depth below.

Remark 4. *It should be noted that in all momentum space integrals of the form (29), a rigorous treatment would require that we place a step function of the form $\Theta(x^0 - x'^0)$ before the integral (or integrand) in order to transition from the time-ordered form (27) to the final expression (29). Here, we are effectively focusing on the causal or retarded radiation problem where it is understood that* *particles observed at x possess a clock time x^0 that is later with respect to the spacetime creation point x', whose internal clock starts ticking at x'^0; i.e., $x^0 > x'^0$ and $\Theta(x^0 - x'^0) = 1$. If antiparticles are to be included, this restriction is not needed. The advantage of using the Feynman propagator (among many other things) is that it naturally and economically leads to a direct and efficient computational formulation of the problem of quantum radiation, where no distinction between particles and antiparticles is required, since the propagator (27) can handle both at the same time.*

Remark 5. *The integral representation of Green's Function (29) can also be put in the following $(\mathbf{x}, t)$-form more suitable for future use in antenna theory:*

$$G_q(\mathbf{x}-\mathbf{x}', t-t') = \int_{\mathbf{k}\in\mathbb{R}^3} \frac{d^3k}{(2\pi)^3} \frac{e^{-i|\mathbf{k}|(t-t')}}{2\sqrt{|\mathbf{k}|^2+m^2}} e^{i\mathbf{k}\cdot(\mathbf{x}-\mathbf{x}')}, \tag{30}$$

where the dispersion relation (A27) was utilized. Moreover, the natural-unit replacement $\mathbf{p} \to \mathbf{k}$ from Table A1 was used. For the computational evaluations of Green's functions in terms of special functions, see Appendix H.

4.5. Generalization to Multiple Discrete and Continuous and Sources

Suppose that now, we apply the quantum source field again but at a *different* spacetime point, say x_2', while we rename x' in (21) as x_1'. Furthermore, we note that each excitation of the vacuum by a field $\phi(x')$ localized at x' can be controlled by some position-dependent "scaling factor", say $J(x')$. This is again the abstract source representation of the q-antenna in the spirit of Definition 2. Then, the total quantum source field is

$$J(x_1')\phi(x_1') + J(x_1')\phi(x_2'). \tag{31}$$

Again, if we hit the vacuum state by this new operator, the radiated quantum state can be written as

$$[J(x_1')\phi(x_1') + J(x_2')\phi(x_2')]|0\rangle = \int_{\mathbf{p}\in\mathbb{R}^3} \frac{d^3p}{(2\pi)^{3/2}} \frac{J(x_1')e^{ip_\mu x_1'^\mu} + J(x_2')e^{ip_\mu x_2'^\mu}}{(2\omega_{\mathbf{p}})^{1/2}} |\mathbf{p}\rangle, \tag{32}$$

where we used (21). Physically, the expression (32) means that two clusters of outgoing waves (quantum particles) are emitted, one emanating from the spacetime point $x_1'^\mu = (t_1', \mathbf{x}_1')$, while the other cluster is directly radiated from $x_2'^\mu = (t_2', \mathbf{x}_2')$. In all cases, note that the linearity of the quantum source field operator $\phi(x')$, as manifested by the expression $\phi(x')|0\rangle$ in (21), is the ultimate basis behind the applicability of the principle of superposition in q-antenna systems.

It is straightforward to generalize the q-antenna's radiation Formula (32) to the generic scenario of an arbitrary number of N discrete point sources located at the spacetime points x_n', each with its own excitation strength specified by $J(x_n'), n = 1, 2, \ldots, N$. This leads to the expression

$$J_q(x')|0\rangle = \int_{\mathbf{p}\in\mathbb{R}^3} \frac{d^3p}{(2\pi)^{3/2}} \frac{\sum_{n=1}^N J(x_n')e^{ip_\mu x_n'^\mu}}{(2\omega_{\mathbf{p}})^{1/2}} |\mathbf{p}\rangle, \tag{33}$$

where

$$J_q(x') := \sum_{n=1}^N J(x_n')\phi(x_n') \tag{34}$$

is the effective (discrete) quantum source distribution operator.

It should become clear now how to generalize from discrete to continuous sources. Let us expand the continuous classical current source function $J(x')$ in terms of a finite number of spacetime (four-dimensional) Dirac delta functions $\delta(x')$ as follows:

$$J(x') = \sum_{n=1}^N J(x_n')\delta(x' - x_n'), \tag{35}$$

where the source domain D_s, the spacetime region upon which the source is supported, is in this case the discrete set

$$D_s := \{x'_1, x'_2, \ldots, x'_N\}. \tag{36}$$

Using the sifting property of the Dirac delta function, it is evident that (33) can be rewritten as

$$J_q(D_s)|0\rangle = \int_{\mathbf{p}\in\mathbb{R}^3} \frac{\mathrm{d}^3 p}{(2\pi)^{\frac{3}{2}}(2\omega_{\mathbf{p}})^{\frac{1}{2}}} \int_{x'\in D_s} \mathrm{d}^4 x' J(x') e^{\mathrm{i}p_\mu x'^\mu_n} |\mathbf{p}\rangle, \tag{37}$$

where the quantum source operator is given by the integral

$$J_q(D_s) := \int_{x'\in D_s} \mathrm{d}^4 x' J(x') \phi^\dagger(x'). \tag{38}$$

While (33) and (37) are formally equivalent in the special case of discrete sources, it is the *second* form (37) that is needed for writing down the correct expression corresponding to the continuous source case. Indeed, when the set D_s becomes dense (which implies $N \to \infty$), the relation (37) continues to hold.

Remark 6. *In deriving (37), we implicitly assume that the order of the dx'- and d^3p- integrals can be exchanged. This can be justified relatively easily when we impose the condition that D_s is compact, which is an assumption we make throughout this paper that still covers most practical antenna systems. The details of the rigorous proof are lengthy and will not be given. For a comprehensive discussion of the rigorous mathematical theory behind representing integrals (continuum sums) over quantum kets in QFT, see [54,55,67,79].*

To summarize, we have managed so far to construct two fundamental types of quantities behaving as "sources" in q-antenna theory:

1. The classical source function

$$J(x') : D_s \subset \mathbb{M}^4 \to \mathbb{R}. \tag{39}$$

2. The quantum source operator

$$J_q(D_s) : \mathcal{D}_s \to \mathcal{O}, \tag{40}$$

where

$$\mathcal{D}_s := \{D \subset \mathbb{M}^4 |\, D \text{ is open, } \mathrm{cl}(D) \text{ is compact in } \mathbb{R}^4\} \tag{41}$$

is the set of all open subsets in the Minkowski spacetime $\mathbb{M}^4$ whose topological closure is compact in $\mathbb{R}^4$. On the other hand, $\mathcal{O}$ is defined as the space of all operators acting on elements of the Fock (occupation state Hilbert space representation [54,55]) space of the q-antenna system.[20]

Remark 7 (The source design problem in quantum antenna theory). *The classical source $J(x')$ is needed in order to delimit the actual q-antenna configuration in spacetime. As in classical antenna theory, an antenna source is considered known if its geometric support regions D_s is known in addition to the value of the source $J(x')$ at each point $x' \in D_s$. The situation is quite similar here. The q-antenna designer is interested in obtaining bounds or information on both D_s and $J(x')$ in order to attain certain radiation states. This is the design problem that is usually solved by means of optimization methods.*

Once the classical source $J(x')$ is fixed, one can employ (38) in order to immediately construct the quantum source operator J_q, which depends (among other things) on the geometrical and topological structure of the source support domain D_s.

Remark 8. *It should be noted that the operator J_q depends on the classical function $J(x')$ defined on the entire spacetime regions D_s. Thus, formally speaking, one should write that operator as $J_q[D_s, J(x')]$. However, to simplify the presentation, this is avoided, since it should always be clear from the context which specific classical source function $J(x')$ is involved with the quantum source operator J_q.*

The probability amplitude $A(x)$ of receiving a particle/wave at the spacetime position x can now be readily computed in exactly the same way we did with the point quantum source at x', i.e, the amplitude $G_q(x-x')$ in (27), simply by adding the various contributions coming from all source points $x' \in D_s$. The calculations are

$$\begin{aligned} A(x) &= \langle 0|\mathcal{T}\phi(x)J_q(D_s)|0\rangle = \langle 0|\mathcal{T}\phi(x)\int_{x'\in D_s} \mathrm{d}^4x'\; J(x')\phi^\dagger(x')|0\rangle \\ &= \int_{x'\in D_s} \mathrm{d}^4x' \langle 0|\mathcal{T}\phi(x)\phi^\dagger(x')|0\rangle J(x'), \end{aligned} \tag{42}$$

where (37) was employed in order to write down the second equality in (42). Note that we assume that x and x' are well separated from each other such that there is no overlap between the observation spacetime region D_r and points $x' \in D_s$ in the source domain. This allows as to freely move the time-ordering operator $\mathcal{T}$ so to obtain (with negligible error) the third equality in (42) above, making our radiation formulas ultimately valid in the *exterior* region of the q-antenna system.

Finally, with the help of (27), we arrive at our main q-antenna radiation amplitude formula summarized by the following theorem:

Theorem 3. *The probability amplitude of a continuous source q-antenna system $J(x), x \in D_s \subset \mathbb{M}^4$, where the closure of D_s is compact in $\mathbb{R}^4$, is given by the following superposition integral*

$$A(x) = \int_{x'\in D_s} \mathrm{d}^4x' G_q(x-x')J(x'), \tag{43}$$

where $G_q(x-x')$ is the q-antenna Green's function in Definition 4.

Remark 9. *We may rewrite (43) in a form more familiar to antenna engineers by recruiting the notation $x = (t, \boldsymbol{x})$ and $x' = (t', \boldsymbol{x}')$, viz., in the "$(\boldsymbol{x}, t)$-format", allowing us to restate (43) as*

$$A(\mathbf{r}, t) = \int_{\mathbf{x}'\in S_s}\int_{t'\in T_s} \mathrm{d}t'\; \mathrm{d}^3\mathbf{x}'\; G_q(\mathbf{x}-\mathbf{x}', t-t')J(\mathbf{x}', t'), \tag{44}$$

where S_s and T_s are the spatial and temporal components of the four-dimensional spacetime source region D_s; see (3). Thereby, we find that the spatial support domain of the q-antenna's source system, i.e., its spatial extension properties, is captured by S_s, while T_s is the time interval of the application of the source.

Remark 10. *The relation $D_s = S_s \times T_s$ holds locally in D_s. The reason is that for attaining wide generality, we are already implicitly permitting D_s to posses a manifold structure in which, locally speaking, each point $x' \in D_s$ is homeomorphic (topologically isomorphic) to $\mathbb{R}^d, d = 2, 3, 4$. That is, from the topological viewpoint, the source region $D_s \subset \mathbb{M}^4$ can be said to be of type (0 + 1) for q-antenna point sources; type (1 + 1) for one-dimensional sources such as wire or loop antenna; type (2 + 1) for surface radiators; or full (3 + 1) type for volumetric sources; and so on.*

Remark 11. *Either one of the two expressions (43) or (44) may be used to express the probability amplitude of receiving a particle at a specific location $x = (t, \boldsymbol{x})$ when radiated by a source distribution written as either $J(x')$ or $J(\boldsymbol{x}', t')$. Both are Lorentz invariant, but the form (43) expresses that more clearly. On the other hand, the form (44) may be utilized when a concrete frame of reference (lab frame) is used such that an unambiguous decomposition of the source region D_s into spatial* *and temporal part can be made (this is in fact the case in most practical scenarios). Physically, the radiation formulas (43) or (44) express quantum radiation by a source $J(x)$ in terms of the q-antenna Green's function, which happens in this particular case to coincide with the Feynman propagator of QFT.*

5. On the General Structure of Radiation Processes in Linear Quantum Antenna Systems

Before moving next to the construction of practical definitions for the q-antenna directivity and gain patterns, we pause for a moment in order to provide a deeper insight into the nature of quantum radiation in q-antenna systems based on the new q-antenna Green's function $G_q(x-x')$ whose main concept was introduced earlier via Definition 4.

5.1. The General Structure of the Quantum Antenna Propagator Process

If we insert (29) into (43), interchanging the order of integrals, the following spectral (momentum-space) result is obtained.

Theorem 4. *The momentum space probability amplitude $A(q)$ in Theorem 3 can be expanded in the spectral domain using the following integral formula:*

$$A(x) = \int_{\mathbf{p}\in\mathbb{R}^3} \frac{d^3 p}{(2\pi)^3\, 2\omega_{\mathbf{p}}} \int_{x'\in D_s} \mathrm{d}^4 x'\, J(x') e^{ip_\mu x'^\mu} e^{-ip_\mu x^\mu}. \tag{45}$$

Here, the spectral integration is performed with respect to all real momenta $\boldsymbol{p}$ in $\mathbb{R}^3$.

Proof. In order to obtain (45), we assume that $\mathrm{cl}(D_s)$ is compact in $\mathbb{R}^4$ in order to make sure that a needed interchange of limiting operations can be justified. The rest of the proof is immediate. □

The anatomy of the quantum radiation's Green's function's spectral expansion (45) is illustrated in Figure 2. The various fundamental sub-processes composing the overall process of quantum radiation can be formally identified as follows:

1. We first must form the correct relativistic sum over all allowable momentum states. This is accomplished by the Lorentz invariant integral operator

$$\int_{\mathbf{p}\in\mathbb{R}^3} \frac{\mathrm{d}^3 p}{(2\pi)^3} \frac{1}{2\omega_{\mathbf{p}}}. \tag{46}$$

2. Each momentum state $|\mathbf{p}\rangle$ will be summed over all possible source locations $x' \in D_s$ in the source region via the integral operator

$$\int_{x'\in D_s} \mathrm{d}^4 x'. \tag{47}$$

This step is also relativistic, since $D_s \subset \mathbb{M}^4$ and $\mathrm{d}^4 x'$ are Lorentz invariant.[21]

3. The next crucial step is to multiply by the factor $\exp(ip_\mu x'^\mu)$. This will trigger the production of a quantum wave (particle) emanating from x' and spreading radially away from the point source.
4. Finally, in order to observe the radiation field at a distance, we multiply the wave produced at x' by a propagation factor $\exp(-ip_\mu x^\mu)$. This will guarantee that the field has been effectively propagated and absorbed at the observation location x.

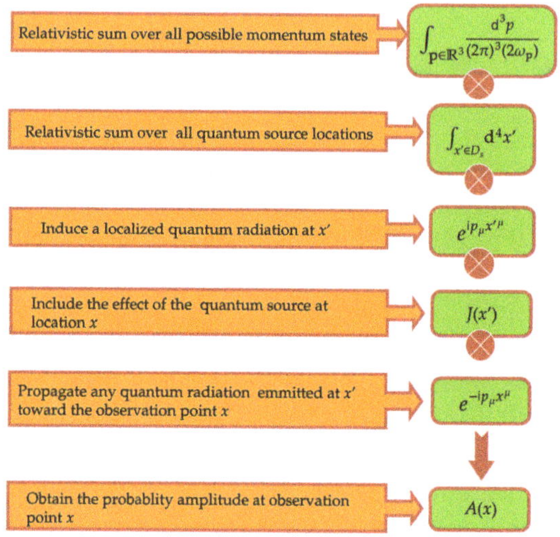

Figure 2. The anatomy of the q-antenna's propagator-based quantum radiation process.

Remark 12. *It may be seen from the above algorithmic construction of the q-antenna propagator in the spectral domain that the problem of quantum radiation acquires a very intuitive and concrete structure when viewed from the spectral (momentum space) domain's perspective. This observation will be exploited in Section 6 when we explore candidate expressions for the q-antenna radiation pattern.*

Remark 13. *The probability amplitude $A(x)$ can be seen as a superposition of "quantum plane waves" each of the form $\exp(ip_\mu x^\mu)$. This is in fact somehow similar to the plane-wave (Weyl) expansion of electromagnetic fields radiated by continuous sources [63,64,80–82]. As in the electromagnetic case, each point source will emit a continuum of plane waves (some of them are evanescent, and the others are pure propagating), with wavevector/momentum specified by $\mathbf{p}$ and frequency $\omega_{\mathbf{p}}$. The total sum of all these waves will produce an effective field moving gradually away from the source and approaching the far zone of the antenna by first going through the near zone. (The evanescent mode character of the momentum space representation becomes directly visible when we transform the above three-dimensional spectral integrals evaluated with respect to $\mathbf{p} \in \mathbb{R}^3$ to equivalent four-dimensional integrals now performed with respect to $p \in \mathbb{M}^4$, see for example [57]).*

5.2. Comparative Analysis of the Three Fundamental Types of Antennas

At this point, it is instructive to give a bird's eye view on the various genera of antenna theories available to us so far. *Classical antennas* (c-antennas) involve excitation with an external electric field $\mathbf{E}^{\mathrm{ex}}$, which in turn induces a current on the antenna via the current Green's function [28,46–48].[22] This is illustrated in Figure 3a. Here, it is essential to note that the "system input" is a classical field, while the "system output" is *also* the classical radiated fields $\mathbf{E}$ and $\mathbf{H}$. On the other hand, the *quantum-optics* approach to defining *quantum antennas*, which is the one currently most often mentioned in connection with applications to quantum communications [1,2,13], treats the input as a classical source while the output is a quantum state $|\alpha(t)\rangle$ [10,13,62,83]. This is illustrated in Figure 3b, where the q-antenna current Green's function $\overline{\mathbf{F}}_q$ is *still* a *classical* function, just like the induced radiating current $\mathbf{J}_q(\mathbf{x}')$.

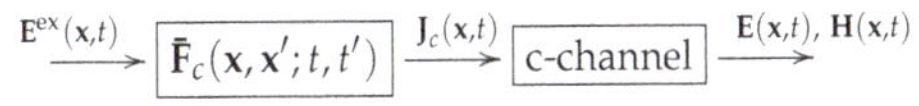

(**a**) Electromagnetics-based c-antennas.

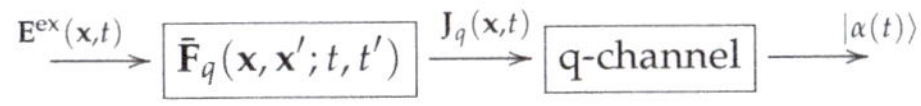

(**b**) Quantum-optics approach to q-antennas.

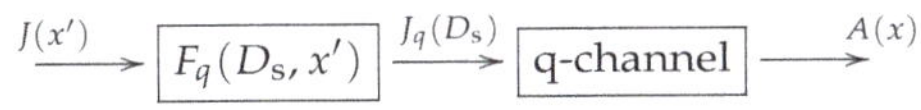

(**c**) Relativistic-QFT approach to q-antennas.

Figure 3. The fundamental operational modes of c- and q- antennas. Both types requires a current Green's function (ACGF) $\bar{\mathbf{F}}$ [28] to connect classical excitation field $\mathbf{E}^{\text{ex}}$ with a classical induced current $\mathbf{J}$. Note that in both c- and q-antennas, the induced currents $\mathbf{J}_c$ and $\mathbf{J}_q$, respectively, are classical. However, the difference resides mainly in the "output" or the radiation state, which is classical (quantum) in c- (q-) antennas.

Consider now the *third* type of antennas shown in Figure 3c. The input to the system $J(x')$ (exactly like the previous two antenna types) is still a classical source, but now the radiating source $J_q(D_s)$ is an *operator* that can describe quantum particles emission (quantum radiation) from within the q-antenna's source spacetime region D_s. In fact, we may rewrite (38) in the revealing integral form

$$J_q(D_s) := \int_{x' \in \mathbb{M}^4} d^4x' F_q(D_s, x') J(x'), \tag{48}$$

where the q-antenna current Green's function in this case is simply given by

$$F_q(D_s, x') := \begin{cases} \phi^\dagger(x'), & x' \in D_s, \\ 0, & x' \notin D_s. \end{cases} \tag{49}$$

Clearly, this is quite different from the two cases depicted in Figure 3a,b. In the case Figure 3c, captured by the expression (49), the current Green's function itself is a *quantum field*, i.e., an operator-valued function on spacetime. The source J_q reproduced by the Green's function superposition integral is also an operator, and the ultimate "output" coming out from the relativistic QFT-based antenna system is the probability amplitude $A(q)$ of annihilating a particle at some generic observation point $x \in D_r \subset \mathbb{M}^4$.

5.3. On the Causal Spacetime Structure of Radiation Emitted by Quantum Antenna Systems

In both special and general relativity, two events x and x' with timelike distance, i.e., $|x - x'|^2 > 0$, can be causally connected [51,53,84]. This has a direct and obvious implication for the general spacetime theory of q-antennas developed here.

1. Consider a point source case. Figure 4a indicates the future lightcone $C_{x'}$ of the event located at x', i.e., the apex of the cone in the spacetime diagram given therein. Since we assume for simplicity that the operational principle behind our q-antenna-based communication link's receiver is based on the process of annihilating the radiated particle at the observation point x, it follows that *only receivers located inside the antenna causal lightcone $C_{x'}$ can receive information from the point source.*
2. For potential receivers located *outside* the antenna causal cone, it is not possible to transmit information at all, unless one admits some superluminal mechanism to be used for sending information, which is currently not accepted by majority of scientists.

These two observations above are enough to determine the causal structure of q-antennas for the case of point sources. However, since arbitrary sources can be constructed from assembling clusters of point sources, the argument can be expanded, as will be shown next.

For the case of continuous sources, the situation is qualitatively similar to the discrete scenario discussed above, but the detailed content of the antenna's causal domain becomes somehow more complicated, since the situation strongly depends on the geometry of the source region in the former case. In Figure 4b, we schematically illustrate the problem when a generic continuous source $J(x')$ is applied in the spacetime region $D_s \subset \mathbb{M}^4$. Since any current source function $J(x')$ can be expanded into a continuous sum of point sources, as per the sifting property of delta functions

$$J(x') = \int_{x \in \mathbb{M}^4} d^4x \, J(x)\delta(x' - x), \tag{50}$$

one may then attempt to construct a causal lightcone for each point $x' \in D_s$, which is again denoted by $C_{x'}$. A *continuum* of causal cones is shown in Figure 4b where all cones appear to be aligned in parallel in spacetime because we ignore gravitational effects.

We now introduce the following construction of the antenna's horizon causal structure, which is valid for both classical and quantum antennas:

Definition 5 (The antenna causal domain). *The antenna causal domain associated with a given source region D_s, which we denote by $C(D_s)$, is defined as simply the fusion (set-theoretic union) of all individual causal cones $C_{x'}$ based on events located inside the source region $x' \in D_s$. That is, according to the recipe*

$$C(D_s) := \bigcup_{x' \in D_s} C_{x'}. \tag{51}$$

The part of $C(D_s)$ that is solely due to future lightcones $C^+_{x'}$ is called the antenna future causal domain. Similarly, the components of $C(D_s)$ due to contributions emanating from past lightcones $C^-_{x'}$ lead to the antenna past causal domain. Then, we have

$$C^+(D_s) := \bigcup_{x' \in D_s} C^+_{x'}, \quad C^-(D_s) := \bigcup_{x' \in D_s} C^-_{x'}, \tag{52}$$

and

$$C(D_s) = C^+(D_s) \cup C^-(D_s). \tag{53}$$

Note that the same constructions can be laid out for the receiving q-antenna case when the detector is in the receive domain D_r.

Remark 14. *Note that it is still possible to incorporate gravitation into our model, since the effect of gravitational fields is mainly to tilt the lightcones locally, where the tilting is directly determined by the gravitational potential/metric tensor $g_{\mu\nu}$ [41]. For applications of q-antennas in deep space communications, these gravitational effects may have to be taken into account. Consequently, the constructions provided by Definition 5 may become useful in astronomical and cosmological applications of either classical or quantum antennas. That is especially true when the large-scale structural impact of the gravitational field on a planned solar or even future interstellar communication links is important.*

Remark 15. *The causal domain $C(D_s)$ is a function of the source region D_s. Moreover, it should be emphasized that in general, the antenna causal domain, as defined above, need not constitute a simple cone in itself. In addition, in principle, the receivers located inside $C(D_s)$ can receive information transmitted by radiation emitted from inside D_s. On the other hand, noncausal receivers, e.g., see some cases schematically depicted in Figure 4b, can never receive information from an antenna system whose source is supported by the region D_s.*

Overall, the analysis above has highlighted an organic interlinking of spacetime with causality within the context of generic quantum communication systems utilizing quantum antennas. This suggests the need to pay closer attention to the *global* (i.e., topological) structure of spacetime domains when such future quantum technologies are incorporated in the analysis, design, and construction of long-distance space communication systems at the solar, extra-solar, astronomical, or even cosmological scales.

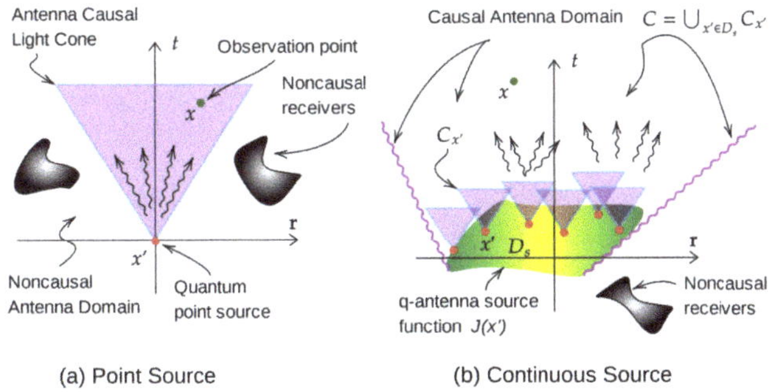

Figure 4. The fundamental causal structure of the q-antenna radiation process. For simplicity, we only show future lightcones.

6. Quantum Antenna Radiation Patterns: Basic Constructions

6.1. Introduction

In contrast to classical antennas, q-antenna systems are intrinsically stochastic. The ultimate goal of a candidate q-antenna theory is to supply rules and guidelines for estimating *probabilities* of potential detection (reception) processes ideally enacted somewhere within the near or far spacetime zones of the quantum sources. In conventional antenna theory, certain quantities are fundamental for analyzing and designing functioning actual devices able to radiate power in real-life settings. These often include directivity, gain (preferably with a measure of radiation efficiency), and the array factor [27,33,35]. As will be illustrated in this section, it turns out that in q-antennas, there also exist close counterparts for many of the conventional characterization measures/concepts already in use in -c-antennas. However, there are also fundamental differences between c- and q-antennas in terms of the physical interpretation and meaning of the results thus obtained since, clearly, electromagnetic and quantum radiation are very distinct (though related) physical processes. For the purpose of constructing working definitions for directivity and gain functions suitable for q-antennas, we explore next some possible antenna radiation pattern constructions made available to us by the relativistic QFT approach proposed above.

6.2. The Probability Law of Producing Radiated Quantum States

From the general rules of QFT [54,56,57], the probability of measuring a particle with four-momentum q within the range $Q' \subset \mathbb{M}^4$ is given by the integral of the square of the probability amplitude $A(q)$:

$$\Pr\{q \in Q\} = \frac{1}{\alpha} \int_{q \in Q'} \mathrm{d}^4 q \, |A(q)|^2 = \frac{1}{\alpha} \int_{\mathbf{q} \in Q} \frac{\mathrm{d}^3 q}{(2\pi)^3} \frac{|A(\mathbf{q})|^2}{2\omega_{\mathbf{q}}}, \tag{54}$$

where $Q \subset \mathbb{R}^3$ is the projection of $Q' \subset \mathbb{M}^4$ onto $\mathbb{R}^3$, and we made use of (A34) to derive the second equality in (54). The number $\alpha \in \mathbb{R}^+$ is a normalization constant defined by (57) inserted in order to ensure that the probability of any event is between 0 and 1.

Remark 16 (Relativistic normalization). *In writing the second equality in (54), a delta function*

$$\delta(p^2 - m^2), \quad p^0 > 0, \tag{55}$$

is included in the d^4p-integral (54), i.e., the mass shell condition expressed by the dispersion relation (A27); see Appendix G for more details on this background.

Remark 17 (Probabilistic normalization). *The correct relativistic normalization of the four-momentum kets $|p\rangle$ and $|q\rangle$, e.g., see relations such as (A38), which we have been already utilizing from the outset, does not automatically ensure the required probability normalization condition*

$$\Pr\{q \in \mathbb{M}^4\} = \frac{1}{\alpha} \int_{q \in \mathbb{M}^4} \mathrm{d}^4 q |A(q)|^2 = 1. \tag{56}$$

The latter is in fact a consequence of the normalization in (54) by

$$\alpha := \int_{q \in \mathbb{M}^4} \mathrm{d}^4 q |A(q)|^2, \tag{57}$$

whenever the integral converges.

The normalization condition (56) is not a consequence of the completeness relation of the Lorentz momentum states (A38), but it should be enforced by hand in order to extract useful probability statements from the theory [85]. However, it should be stressed that the normalization integral (56) for the q-states is not convergent in the usual sense since *exact* continuous momentum states such as $|q\rangle$ or $|\mathbf{q}\rangle$ are unormalizable [52,54,86]. For more detailed information about the proper rigorous mathematical theories that can deal with problems involving continuous states, see [37,55,67,79]. In practice, we seldom need to directly compute continuous state representation, since one always works with "discretized states", where momentum or wavevectors are measured in a finite range [2,87]. The set of all such finite-range states is normalizable [54]. Therefore, whenever expressions such as (56) are encountered, they should be understood in the above of sense of being approximated by discrete sum after which the normalization to unity becomes correct. Nevertheless, for some choices of $A(q)$, for example the point source with $A(q) = 1$, the integral (56) diverges regardless of whether one works with continuous or discretized state representations. This is an example of the persistent and well-known problem of ultraviolet (UV) divergences in QFT [38]. However, this problem is not a major issue in q-antenna theory, since in practical settings, one often requires a smoothly switchable source. In such types of source systems, the Fourier amplitudes of the spacetime source function decay fast enough in momentum space in order to secure the convergence of infinite spectral integrals such as (56). For in-depth discussion of this scenario backed by several examples taken from the area of high-energy physics, see Coleman's lectures [38].

We next give a few examples to illustrate this formulation.

Example 2 (General quantum source's radiation probability law at a particular momentum). First, we construct a suitable probabilistic representation of a point source radiation function. Introduce the Euclidean ball

$$Q(\mathbf{q}_0, \varepsilon) := \{\mathbf{q} \in \mathbb{R}^3, |\mathbf{q} - \mathbf{q}_0| < \varepsilon\}, \tag{58}$$

where $\mathbf{q}_0 \in \mathbb{R}^3$, while $\varepsilon \in \mathbb{R}^+$ is very small. (The norm $|\cdot|$ is that inherited from the standard Euclidean metric on $\mathbb{R}^3$). The probability of measuring a momentum $\mathbf{q}$ in the ball $Q(\mathbf{q}_0, \varepsilon)$ is given by

$$\Pr\{q \in Q(\mathbf{q}_0, \varepsilon)\} = \frac{1}{\alpha} \int_{\mathbf{p} \in Q(\mathbf{q}_0, \varepsilon)} \frac{\mathrm{d}^3 q}{(2\pi)^3} \frac{|A(\mathbf{q})|^2}{2\omega_{\mathbf{q}}} \simeq \frac{(4/3)\pi\varepsilon^3}{\alpha(2\pi)^3} \frac{|A(\mathbf{q}_0)|^2}{2\omega_{\mathbf{q}_0}} = \frac{\varepsilon^3}{12\pi^2\alpha} \frac{|A(\mathbf{q}_0)|^2}{2\omega_{\mathbf{q}_0}}, \tag{59}$$

where $(4/3)\pi\varepsilon^3$ is the volume of the ball $Q(\mathbf{q}_0, \varepsilon)$ centered at $\mathbf{q}_0$ with radius ε. Therefore, the expression (59) gives the probability of measuring a momentum $\mathbf{p} = \mathbf{q}_0$ after exciting the ground state $|0\rangle$ of the q-antenna to a higher-energy state $|q_0\rangle$. It is completely general, regardless of how the vacuum state was excited.

Example 3 (Point source quantum radiation probability model). Let us apply the rule (54) to the single quantum source probability amplitude (25). By treating (25) as a special case of (54), we find that for a point quantum source firing at the spacetime point x', the probability amplitude is $A(q) = 1$; hence, the total radiation probability is given by

$$\Pr\{\mathbf{q} = \mathbf{q}_0\} \simeq \frac{1}{\alpha} \frac{\varepsilon^3}{12\pi^2 \omega_{\mathbf{q}_0}} = \frac{\varepsilon^3}{12\pi^2 \alpha \sqrt{|\mathbf{q}_0|^2 + m^2}}, \tag{60}$$

after using the dispersion relation (A27). Since the expression (60) depends only on the amplitude of the momentum $\mathbf{q}_0$, not its direction, we conclude that the q-antenna with a single scalar quantum source is an *isotropic* radiator since it radiates its various four-momentum states $|q\rangle$ in the same manner in all directions. This is not very surprising, since the quantum field of the q-antenna is scalar, and in classical antenna theory, a scalar point source is also isotropic [27].

Remark 18. *From the antenna viewpoint, the examples above suggest that the quantum state of the q-antenna is inherently global and hence extendable everywhere. Therefore, we can not guarantee that the q-antenna has effectively radiated into every spacetime point until a concrete measurement process is performed at some position x, after which one may determine the probability of actual detection there using the receive antenna model with the help of the q-antenna Green's function in Theorem 2. Such a process requires introducing a more sophisticated approach to measure probability amplitudes when the observation point is included as a parameter in the system. This subject will be taken up again in Section 6.4.*

6.3. Constructing the Quantum Antenna Directivity Pattern

Our goal here is to estimate the directive properties of a continuous source $J(x')$ located in the spacetime region D_s. The probability amplitude of measuring the four-momentum q is denoted by $A[q; J(x'), D_s]$ but will be abbreviated to $A(q)$ whenever there is no confusion about the source. Our main tool here is the following theorem.

Theorem 5. *Consider a q-antenna system $J(x), x \in D_s \subset \mathbb{M}^4$. Let the corresponding probability amplitude be $A(q)$. Then, the following formula holds:*

$$|A(q)|^2 = |J(\mathbf{q}, \omega_{\mathbf{q}})|^2, \tag{61}$$

where $J(q)$ is the spacetime Fourier transform of the radiating current source (A45).

Proof. See Appendix I. □

Remark 19. *Since there is a one-to-one (injective) mapping between the Lorentz four-momentum states $q \in \mathbb{M}^4$ and the conventional three-momentum vectors $\boldsymbol{q} \in \mathbb{R}^3$ via the dispersion relation (A27), we may just write $A(\boldsymbol{q})$ instead of $A(\boldsymbol{q}, \omega_{\boldsymbol{q}})$.*

Theorem 6. *For the situation described in Theorem 5, the quantum radiation probability amplitude can be expressed as:*

$$\Pr\{\boldsymbol{k} \in Q\} = \frac{1}{\alpha} \int_{\boldsymbol{k} \in Q} \frac{d^3k}{(2\pi)^3} \frac{|J(\mathbf{k})|^2}{2\omega_{\mathbf{k}}}, \tag{62}$$

for a generic three-dimensional region $Q \subset \mathbb{R}^3$. Here, $J(\boldsymbol{k}) := J(\boldsymbol{k}, \omega)$ is the momentum space Fourier transform of the source defined by (A45); see also Remark 19 on the reduced notation in momentum space used here.

Proof. Use the probability law (54) in (61) and make the replacement $\mathbf{q} \to \mathbf{k}$. □

An immediate application of Theorem 6 is in expressing the total energy radiated by a quantum source directly in terms of the classical source function $J(x)$, as will be illustrated through the next example.

Example 4 (Total radiated energy and momentum of a q-antenna in momentum space). The total energy radiated by the antenna in a momentum space region $Q \subset \mathbb{R}^3$ can be obtained by multiplying the probability of particle production in the infinitesimal momentum space volume d^3k by the particle's energy, which is $\hbar\omega_{\mathbf{k}}$ (or just $\omega_{\mathbf{k}}$ in natural units), and then integrating. With the help of (62), this procedure yields:

$$\mathcal{E}[Q] = \frac{1}{2}\int_{\mathbf{k}\in Q} \frac{\mathrm{d}^3k}{(2\pi)^3} |J(\mathbf{k})|^2, \tag{63}$$

where $\mathcal{E}$ is the total energy. Similarly, the expected value of the total radiated momentum can be estimated via the formula

$$\mathcal{P}[Q] = \frac{1}{2}\int_{\mathbf{k}\in Q} \frac{\mathrm{d}^3k}{(2\pi)^3} \frac{|J(\mathbf{k})|^2\mathbf{k}}{\sqrt{|\mathbf{k}|^2 + m^2}}, \tag{64}$$

where use was made of the fact that the momentum of the particle excitation associated with the **k**th state is $\hbar\mathbf{k}$ (or $\mathbf{k}$ in natural units).

Remark 20. *It is interesting to note that the momentum space energy pattern of the q-antenna (63) is similar to the energy directivity of a classical source in generic medium. For example, see how general formulas of directivity for classical radiators were constructed recently in momentum space for generic homogeneous nonlocal domains [71–73].*

In order to put relation (62) into further practical use, we transform momentum representations into spherical coordinates so we may obtain some information about the directive properties of quantum radiating sources as illustrated by the following example.

Example 5 (Radiation angular directivity pattern of q-antennas in momentum space). Let us express the wavevector $\mathbf{k}$ in terms of the spherical angular coordinates θ and φ by using the standard transformation

$$\hat{k} = \hat{k}(\Omega) = \hat{x}_1 \cos\varphi \sin\theta + \hat{x}_2 \cos\varphi \sin\theta + \hat{x}_3 \cos\theta, \tag{65}$$

where

$$\Omega := (\theta, \varphi), \quad \hat{k}(\Omega) := \frac{\mathbf{k}}{|\mathbf{k}|}. \tag{66}$$

Here, $\hat{x}_i, i = 1,2,3$, constitute a set of three mutually orthogonal Cartesian unit vectors. The angles $0 < \varphi < 2\pi, 0 < \theta < \pi$ determine the direction of the unit vector $\hat{k}$, or where the emitted quantum particle's momentum $\mathbf{p} = \hbar\mathbf{k}$ is pointing in three-dimensional space. Since physically and intuitively, one would expect that the directions in three-dimensional position space along which momenta tend to maximally flow also correspond to the directions toward which most of the particle ensemble's energy and momentum are directed, we may use Theorem 6 in conjunction with (66) to estimate the quantum source's directive properties in generic three-dimensional scenarios. A simple way to achieve this is by re-expressing (62) using (65) and (66), resulting in

$$\Pr\{k \in [k_{\min}, k_{\max}],\ \Omega \in \Omega_0\} = \frac{1}{\alpha} \int_{k\in[k_{\min},k_{\max}]} \int_{\Omega\in\Omega_0} \mathrm{d}k\,\mathrm{d}\Omega\, \frac{k^2 \sin\theta}{(2\pi)^3} \frac{\left|J\left[k, \hat{k}(\Omega)\right]\right|^2}{2\sqrt{k^2 + m^2}}, \tag{67}$$

where

$$k := |\mathbf{k}|, \quad d\Omega := d\theta d\varphi. \tag{68}$$

The angular set Ω_0 in (67) is the three-dimensional (solid) angular sector inside where one is interested in characterizing the radiating quantum antenna system.

Remark 21. *It is important not to confuse $k \in \mathbb{R}^+$, defined by (68), with the four-vector $k = k^\mu \in \mathbb{M}^4$ (Appendix B). In this section, we use the shorthand notation (68) in order to simplify the presentation.*

Inspired by the expression (67), we may then introduce the following definition of momentum space directivity for quantum antennas:

Definition 6 (Quantum antenna directivity in momentum space). *The momentum space directivity of a q-antenna with source function $J(x), x \in D_s$ is defined as the following angular function*

$$\mathcal{D}(\varphi,\theta;k) := \frac{1}{\alpha_k}\frac{k^2 \sin\theta}{(2\pi)^3}\frac{\left|J\left[k,\hat{k}(\varphi,\theta)\right]\right|^2}{2\sqrt{k^2+m^2}}, \tag{69}$$

where

$$\alpha_k := \frac{1}{4\pi}\int_{4\pi} d\Omega \frac{k^2 \sin\theta}{(2\pi)^3}\frac{\left|J\left[k,\hat{k}(\varphi,\theta)\right]\right|^2}{2\sqrt{k^2+m^2}} \tag{70}$$

is a positive momentum-dependent probability noramalization constant.

Expressed in terms of the momentum-space-type directivity (69), the total probability in (62) may be be put into the more compact form

$$\Pr\{k \in [k_{\min}, k_{\max}],\ \Omega \in \Omega_0\} = \int_{k\in[k_{\min},k_{\max}]} dk \frac{\alpha_k}{\alpha}\int_{\Omega\in\Omega_0} d\Omega\, \mathcal{D}(\varphi,\theta;k). \tag{71}$$

The last expression explains the physical motivation behind Definition 6. Moreover, the following two remarks explain more about the engineering background to directivity concepts in antenna theory motivating the above definition itself.

Remark 22. *It is evident then from (71) that for a given $\hat{k}$, the function $\mathcal{D}(k,\hat{k})$ is the angular probability density in the momentum space variable k. On the other hand, the mathematical dependence of $D(k,\hat{k})$ on the angles φ and θ provides information on how the emitted quantum particles tend to flow along different directions in space. This is why D does indeed behave as a momentum-space radiation pattern (probability per unit momentum per solid angle).*

Remark 23. *The positive number α_k is inserted into (69) in order to ensure that the total probability of radiation at all angles, evaluated at a single radial momentum value k, is equal to that of an isotropic source.[23] Intuitively, α_k represents the total radiation angular density emitted by the q-antenna, so the ratio $\mathcal{D}$ given by (69) is the relative radiation intensity along one direction with respect to a standard isotropic source*

$$\mathcal{D}_0 := \frac{P_0}{4\pi}, \tag{72}$$

where P_0 is the constant radiation angular probability of such an isotropic reference antenna whose radiation angualr density is $\mathcal{D}_0$. That is, we have

$$\int_{\Omega\in 4\pi} d\Omega\, \mathcal{D}(\varphi,\theta;k) = 4\pi\mathcal{D}_0, \tag{73}$$

as expected from a typical directivity expression [35]. Physically, the ratio $4\pi\alpha_k/\alpha$ in (71) represents the fraction of the total quantum radiation contained in a sphere with radius k in momentum space *relative to the total radiation obtained by including contributions coming from all values of the momentum magnitude $k := |\mathbf{k}|$. The ability of a source to direct power along certain directions is measured by angle-dependent generalizations of such total ratios. Then, Definition 6 of the q-antenna directivity is a natural generalization of the corresponding definition in classical antenna theory as traditionally presented in texts such as [27,35].*

6.4. The Probability Law of Receiving Radiated Quantum States: Source-Receiver Coupling Gain Estimation

Consider the configuration shown in Figure 5. The quantum antenna communication system is characterized by the *gain functional*

$$\mathcal{G} : \mathcal{X} \times \mathcal{X} \to \mathbb{C}, \quad \mathcal{G} = \mathcal{G}[J_s, J_r], \tag{74}$$

which is a bilinear functional in the source and receiver currents $J_s(x') \in \mathcal{X}, x' \in D_s$, and $J_r(x) \in \mathcal{X}, x \in D_r$, respectively. Here, $\mathcal{X}$ is the space of real-valued functions on the four-dimensional domains $D_{s,r} \subset \mathbb{M}^4$. The complex number $\mathcal{G}$ gives the probability amplitude of information transmission (coupling amplitude) between source and receiver. This gain plays the role of "transmission coefficient" or the "coupling coefficient" often used in conventional electromagnetic communication systems, e.g., see [33,88]. From the fundamental theory of quantum antennas developed above, we may easily deduce the general expression of this functional by simply treating the receiver as a source with the "quantum reverse" process of that of the transmitter, i.e., the dual of the transmitter problem, e.g., by converting kets to bras, taking adjoints, complex conjugates, etc. Since for *real* sources and receiver currents $J_{r,s}(x')$, the problem is fully reciprocal,[24] we can immediately write down the q-antenna system gain expression as follows:

$$\mathcal{G} = \langle 0| \int_{x \in D_r} d^4x \int_{x' \in D_s} d^4x' \, J_r(x) \mathcal{T} \phi(x) \phi^\dagger(x') J_s(x') |0\rangle, \tag{75}$$

where (43) was used to realize the transmitter, and a similar form was adapted for the receiver.

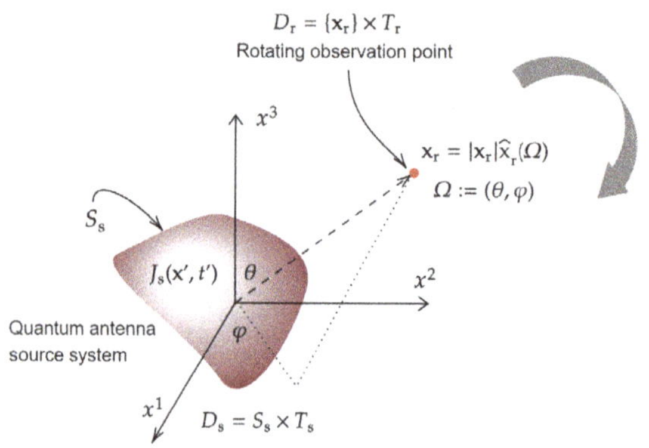

Figure 5. Angular gain measurementconfiguration scenario for a generic quantum antenna source system $J_s(\mathbf{x}, t), \mathbf{x} \in S_s, t \in T_s$.

Example 6 (Construction of practical angular gain pattern in q-antenna systems). We will use the general relativistic expression (75) in order to construct an angular expression characterizing quantum radiation for possible use in practical settings. To do that, we first need to break relativistic covariance in order to simplify the problem and facilitate calculations. The most natural choice for a preferred frame is to use a coordinate system at rest with either the lab frame of the transmitter or the receiver. We choose the source frame below. Consider a standard test receiver to be used for performing quantum measurement on the radiation emitted by the source $J_s(\mathbf{x}', t')$, where $\mathbf{x}' \in S_s$, and S_s is the spatial support of the source region. The receiver spatial region is S_r, and its strength is given by $J_r(\mathbf{x}, t)$. Both t' and t' will be measured with a single clock at rest in the source lab frame. Let the test (receiver) current be reduced to a concentrated delta source

$$J_r(x) = J_r(\mathbf{x}, t) = \delta(\mathbf{x} - \mathbf{x}_r) J_r(t), \quad \forall t \in T_r. \tag{76}$$

That is, we measure the response at a sharp spatial location $\mathbf{x}_r$, while we impose the receiver function restriction $J_r(t)$ on the time measurement interval T_r. We can use (66) to express the position vectors as

$$\mathbf{x}_r = |\mathbf{x}_r| \hat{x}_r(\Omega), \tag{77}$$

where the unit angle-only dependent unit vector $\hat{\mathbf{x}}_r$ is defined by

$$\hat{x}_r(\Omega) := \frac{\mathbf{x}_r}{|\mathbf{x}_r|}. \tag{78}$$

By substituting the relations (76), (77), and (78) into the gain Formula (75), the following angular expression is obtained:

$$\mathcal{G}(\Omega; |\mathbf{x}_r|) = \int_{\mathbf{x}' \in S_s,\, t' \in T_s} \mathrm{d}^3\mathbf{x}' \mathrm{d}t'\, G_q^a(\Omega; |\mathbf{x}_r|, x') J_s(x'), \tag{79}$$

where

$$G_q^a(\Omega; |\mathbf{x}_r|, x') := \int_{t \in T_r} \mathrm{d}t\, G_q(|\mathbf{x}_r| \hat{x}_r(\Omega) - \mathbf{x}'; t - t') J_r(t), \tag{80}$$

and the notation of (27) was used. The integral (80) can be computed numerically for known current source distributions $J_s(x)$ by using the special functions representation of the Feynman propagator $G_q(x - x')$ given in Appendix H.

Remark 24. *We note that the q-antenna system gain Function (79) depends on the distance between the source and the receiver in addition to the angles. In general, one studies the asymptotic behavior of $G_q^a(\Omega; |\mathbf{x}_r|, x')$ in the long distance limit $|\mathbf{x}_r| \to \infty$, with the hope that this behaves as $|\mathbf{x}_r|^{-n}$, with n often a small integer (usually 1 or 2). From the asymptotic limits of the special functions used in Appendix H, some possible relations could be derived using the bessel function large argument approximation, which in turn then might be further utilized in order to eliminate the dependence on the radial distance in the gain pattern (79). However, such detailed computational considerations are outside the scope of this paper, which is mainly focused on the general fundamental theory.*

7. Conclusions

We developed foundations for quantum antenna theory using the relativistic framework of quantum field theory (QFT) with emphasis on the source–sink (transmitter–receiver) model for potential applications to quantum communication systems. The QFT-based q-antenna is the most general type of antenna systems because it includes both the classical and quantum-optics-based q-antennas as special cases. In addition, relativistic q-antennas can accommodate a wider range of physical processes, since the same theory may deal with massive or massless particles and for different spins. These types of q-antennas, the most general to date, appear to enjoy the distinction of exhibiting a current Green's function that is itself a quantum field, which we dubbed before the quantum source of the antenna system, i.e., the field $\phi(x)$ as per (49). We have illustrated the theory mostly with the specific example of the neutral Klein–Gordon field because of its simplicity and importance for building the general solution to other, more complex problems. In particular, we have shown how in the linear quantum antenna model, one may directly construct quantum radiation directivity and gain measures that somehow resemble their counterparts in classical antenna theory. A future research based on the present theory would proceed to construct more concrete radiation models for quantum fields such as the Proca field (massive electromagnetism), the Maxwell–Dirac field (quantum electrodynamics), the graviton field (gravitational antennas), and possibly other types. Then, the present paper attempted to provide a first sketch of a possible general theory that may blend all these distinct phenomena, where our key idea was that relativistic QFT supplies precisely a promising such unifying theoretical framework.

Funding: This research received no external funding.

Institutional Review Board Statement: Not applicable.

Informed Consent Statement: Not applicable.

Data Availability Statement: Not applicable.

Conflicts of Interest: The author declares no conflict of interest.

Abbreviations

The following abbreviations are used in this manuscript:

QFT	Quantum field theory
QM	Quantum mechanics
SR	Special relativity
c-antenna	Classical antenna
q-antenna	Quantum antenna
q-radiation	Quantum radiation
q-state	Quantum state
q-source	Quantum source
c-source	Classical source
EM	Electromagnetic/Electromagnetics
ACGF	Antenna current Green's function

Appendix A. Classical Antenna Theory

Let the source or current Green's function be the dyadic tensor $\overline{\mathbf{F}}(\mathbf{x},\mathbf{x}',t-t')$ [28,46–48,89]. The forward Green's function is the standard retarded radiation Green's function $\overline{\mathbf{G}}(\mathbf{x}-\mathbf{x}',t-t')$ [49]. Then, we have the following two fundamental relations [28,90]:

$$\mathbf{J}(\mathbf{x},t)=\int_S \mathrm{d}^2x'\,\mathrm{d}t'\,\overline{\mathbf{F}}(\mathbf{x},\mathbf{x}',t-t')\cdot\mathbf{E}^{\mathrm{ex}}(\mathbf{x}',t'), \tag{A1}$$

$$\mathbf{E}(\mathbf{x},t)=\int_S \mathrm{d}^3x'\,\mathrm{d}t'\,\overline{\mathbf{G}}(\mathbf{x},\mathbf{x}',t-t')\cdot\mathbf{J}(\mathbf{x}',t'), \tag{A2}$$

where $\mathbf{E}^{\mathrm{ex}}(\mathbf{x},t)$ is the excitation electric field, $\mathbf{J}(\mathbf{x},t)$ is the radiating antenna current on its compact surface S supporting the electromagnetic boundary conditions, and $\mathbf{E}(\mathbf{x},t)$ is the radiated field. For simplicity, we assume perfect-electric conducting (PEC) antennas where the magnetic field does not contribute to the current Green's function. Another radiating Green's functions similar to $\overline{\mathbf{G}}(\mathbf{x},\mathbf{x}',t-t')$ is needed in expressions such as (A2) in order to obtain the magnetic field $\mathbf{B}(\mathbf{x},t)$, which is always present in any radiation problem beside the electric field; see [89–91] for more details and applications.

The relation (A1) captures the antenna's fundamental Mode A, where a source field $\mathbf{E}^{\mathrm{ex}}(\mathbf{x},t)$ produces a radiating current via the current Green's function $\overline{\mathbf{F}}(\mathbf{x},\mathbf{x}',t-t')$, which in turn generates the radiated fields $\mathbf{E}(\mathbf{x},t)$ and $\mathbf{B}(\mathbf{x},t)$ throughout the region exterior to the surface S. To complete the antenna-based wireless communication system, a third Mode C, where a receiving (Rx) antenna, placed at some distance from S, interacts with the radiated field in order to produce an observable receive port current $\mathbf{J}_{\mathrm{rx}}(\mathbf{x})$ by means of the formula

$$\mathbf{J}_{\mathrm{rx}}(\mathbf{x},t)=\int_{S_{\mathrm{rx}}} \mathrm{d}^2x'\mathrm{d}t'\,\overline{\mathbf{F}}_{\mathrm{rx}}(\mathbf{x},\mathbf{x}',t-t')\cdot\mathbf{E}(\mathbf{x}',t'), \tag{A3}$$

where $\overline{\mathbf{F}}_{\text{rx}}(\mathbf{x}, \mathbf{x}', t-t')$ is the receive current Green's function, which is generally different from the transmit (Tx) current Green's function $\overline{\mathbf{F}}(\mathbf{x}, \mathbf{x}', t-t')$. The three relations and Green's functions in (A1)–(A3) fully describe the antenna system in classical electromagnetic theory [92], with obvious generalization to magnetic field interactions included in essentially the same logic [90].

Appendix B. The Relativistic Four-Vector Formalism

In special relativity, everything takes place in the four-dimensional Minkowski spacetime, denoted by $\mathbb{M}^4$, a linear vector space with a special metric, the Lorentz metric, given by

$$g_{\mu\nu} = \begin{pmatrix} 1 & 0 & 0 & 0 \\ 0 & -1 & 0 & 0 \\ 0 & 0 & -1 & 0 \\ 0 & 0 & 0 & -1 \end{pmatrix}. \tag{A4}$$

We follow the standard convention of denoting spacetime indices using Greek letters $\mu = 0, 1, 2, 3$, with the time-like component always given the 0th index and the space-like part denoted by Roman letter. Four vectors with upper and lower indices are dual vectors. The converse of the metric tensor $g_{\mu\nu}$ is $g^{\mu\nu}$, i.e., $g_{\mu\nu}g^{\mu\nu} = \delta^{\nu}_{\mu}$, where δ^{ν}_{μ} is the Kronecker delta function. Note that summation over repeated indices is implied (the Einstein repeated index convention). In general, we raise or lower indices by contracting with the fundamental tensor in relations such as $a_\mu = g_{\mu\nu}a^\nu$ or $a^\mu = g^{\mu\nu}a_\nu$. From this we can see, for example, that $a^0 = a_0$, and $a^i = -a_i$. The dot product between two four vectors a^μ and b^μ is defined as $a^\mu b_\mu := a^\mu g_{\mu\nu} b^\nu = g_{\mu\nu} a^\mu b^\nu$. We also write this as $a \cdot b := a^\mu b_\mu$ where a and b stands for the four-vectors a^μ and b^μ. In particular, the all-important inner product $p \cdot x$ between four-momentum p^μ and spacetime position x^μ, given by

$$p \cdot x = p^\mu x_\mu = \omega t - \mathbf{k} \cdot \mathbf{x}, \tag{A5}$$

will be used frequently. Note that in $\mathbb{M}^4$, the proper "differential operators four-vectors" are given by $\partial^\mu = (\partial/\partial t, -\nabla)$ and $\partial_\mu = (\partial/\partial t, \nabla)$. Table A1 provides a compact summary of the main relativistic formulas.

Table A1. List of main basic relativistic relations in the Natural Units.

Quantity	Expression
$\mathbf{p}$ (particle momentum)	$\mathbf{p} = \mathbf{k}$
$E_{\mathbf{p}}$ (particle energy)	$E_{\mathbf{p}} = \omega_{\mathbf{p}} = \omega_{\mathbf{k}}$
Relativistic dispersion relation	$E^2_{\mathbf{p}} = \mathbf{p}^2 + m^2$
∂^μ (four-vector differential operator)	$\partial^\mu = (\partial/\partial t, -\nabla)$
x^μ (position four-vector)	$x^\mu = (t, \mathbf{x})$
k^μ (relativistic wavevector)	$k^\mu = (\omega, \mathbf{k})$
p^μ (photon four-momentum vector)	$p^\mu = (E_{\mathbf{p}}, \mathbf{p}) = k^\mu$
$g_{\mu\nu}$ (Lorentz metric tensor)	$g_{\mu\nu} = \text{diag}(1, -1, -1, -1)$
$p^\mu x_\mu$ (four-vector inner product)	$p^\mu x_\mu = g_{\mu\nu} p^\mu x^\nu = \omega t - \mathbf{k} \cdot \mathbf{x}$

Appendix C. Natural Units

In relativistic quantum field theory, it is customary to use a system of units in which $\hbar = 1, c = 1$. Here, the action S is dimensionless. Mass, energy, wavevector, and momentum have the same dimension, which is the inverse of the dimension of length and time:

$$[\text{mass}] = [\text{energy}] = [\text{momentum}] = [\text{length}]^{-1} = [\text{time}]^{-1}. \tag{A6}$$

Therefore, the physics of the world is reduced to measurements in two units, length and time. Moreover, in four dimensions, the creation and annihilation operators have the dimension of an inverse energy, i.e., $[a_{\mathbf{p}}] = [a^\dagger_{\mathbf{p}}] = [\text{energy}]^{-1}$.

Appendix D. Dirac Interaction Picture

Consider a Hamiltonian H that can be written as

$$H = H_0 + H'(t), \tag{A7}$$

where H_0 is the *free* Hamiltonian of the system (time-independent function), while $H'(t)$ is the *interaction* (time-dependent) Hamiltonian, which is usually assumed to be turned on at a specific time point. In the Dirac picture, our goal is to *decouple* the free dynamics of the system, i.e., that governed by a Hamiltonian of the form $H = H_0$ (no interaction or $H' = 0$), from the pure interaction component, i.e., that which is solely due to the term H' in the actual full Hamiltonian (A7). Now, let us define the following time-dependent operator

$$O_{\mathrm{I}}(t) := e^{\mathrm{i}H_0 t} O_{\mathrm{S}} e^{-\mathrm{i}H_0 t}, \tag{A8}$$

where O_{S} is the Schrodinger picture (time-independent) operator. Differentiating (A8) with respect to time, we obtain the Heisenberg equation of motion

$$\mathrm{i}\frac{\mathrm{d}}{\mathrm{d}t} O_{\mathrm{I}}(t) = [O, H_0]. \tag{A9}$$

Therefore, the interaction picture operator $O_{\mathrm{I}}(t)$ evolves in time according to the exact same law corresponding to the case when the full Hamiltonian is just the free part, i.e., $H = H_0$.

What about the evolution of the interaction picture state, which will be denoted by $\psi_{\mathrm{I}}(t)$? To find out, note that all different pictures, the Schrodinger, Heisenberg, and Dirac, must agree on probability amplitudes. Therefore, for any two pairs of Schrodinger and Dirac states $\psi_{\mathrm{S}}(t), \varphi_{\mathrm{S}}(t)$, and $\psi_{\mathrm{I}}(t), \varphi_{\mathrm{I}}(t)$, respectively, the matrix elements in both pictures must be identical. Formally, we express this by the condition

$$\langle \psi_{\mathrm{S}}(t) | O_{\mathrm{S}} \varphi_{\mathrm{S}}(t) \rangle = \langle \psi_{\mathrm{I}}(t) | O_{\mathrm{I}}(t) \varphi_{\mathrm{I}}(t) \rangle = \left\langle \psi_{\mathrm{I}}(t) e^{-\mathrm{i}H_0 t} \middle| O_{\mathrm{S}} e^{-\mathrm{i}H_0 t} \varphi_{\mathrm{I}}(t) \right\rangle. \tag{A10}$$

Clearly, this implies that we should define the Dirac (interaction) state by

$$|\psi_{\mathrm{I}}(t)\rangle := e^{\mathrm{i}H_0 t} |\psi_{\mathrm{S}}(t)\rangle. \tag{A11}$$

To find out how this state evolves, we differentiate with respect to time, obtaining

$$\mathrm{i}\frac{\mathrm{d}}{\mathrm{d}t} |\psi_{\mathrm{I}}(t)\rangle = \left(-H_0 + \mathrm{i}\frac{\mathrm{d}}{\mathrm{d}t} \right) |\psi_{\mathrm{S}}(t)\rangle = H' |\psi_{\mathrm{S}}(t)\rangle \tag{A12}$$

where we have made use of the Schrodinger equation

$$\mathrm{i}\frac{\mathrm{d}}{\mathrm{d}t} |\psi_{\mathrm{S}}(t)\rangle = H |\psi_{\mathrm{S}}(t)\rangle \tag{A13}$$

and the form (A7). Finally, using the definition (A11) of the interacting picture state, the relation (A12) becomes

$$\mathrm{i}\frac{\mathrm{d}}{\mathrm{d}t} |\psi_{\mathrm{I}}(t)\rangle = H_{\mathrm{I}}(t) |\psi_{\mathrm{I}}(t)\rangle, \tag{A14}$$

where the Dirac picture interacting Hamiltonian H_{I} is given by

$$H_{\mathrm{I}}(t) := e^{\mathrm{i}H_0 t} H'(t) e^{-\mathrm{i}H_0 t}. \tag{A15}$$

Let us summarize the main features of this Dirac picture:

1. In the Schrodinger picture, the state evolves in time according to the full Hamiltonian while the operators are constant.
2. In the Heisenberg picture, the state is constant (time-independent), but the operator evolves according to the full Hamiltonian.

3. In the Dirac picture, both state and operators evolve with time. However, the time evolution is decoupled into two distinct and independent components. First, all interaction (Dirac) picture operators evolve according to the free Hamiltonian as per the corresponding Heisenberg Equation (A9). Second, the Dirac state $|\psi_I(t)\rangle$ evolves independently according to the dynamic Equation (A14).

Therefore, the main advantage of the Dirac picture is that we can concentrate on the essential aspects of interaction as encoded in the Schrodinger-like dynamic Equation (A14). In this equation, Schrodinger or Heisenberg states and operators can be obtained using the usual transformation formula. On the other hand, note that operators in the Dirac picture, including in particular the interaction Hamiltonian (A15), all evolve according to the free evolution law (A9). In this case, this becomes

$$i\frac{d}{dt}H_I(t) = [H_I(t), H_0], \tag{A16}$$

which says that all quantum fields inside H' are to be evolved under the free Hamiltonian H_0 in order to obtain the dynamics of $H_I(t)$.

Let us finally solve the dynamical evolution Equation (A14). To do so, introduce the *time-ordering symbol* $\mathcal{T}$ defined by

$$\mathcal{T}[A(t_1)B(t_2)] := \begin{cases} A(t_1)B(t_2), & t_1 > t_2, \\ B(t_2)A(t_1), & t_2 > t_1. \end{cases} \tag{A17}$$

Then, the general solution of (A14) can be expressed in terms of the *evolution operator* $P(t_2, t_1)$, which evolves that initial state $|\psi(t_1)\rangle$ to the final state $|\psi(t_2)\rangle$ via the operator relation

$$|\psi_I(t_2)\rangle = P(t_2, t_1)|\psi_I(t_1)\rangle, \tag{A18}$$

where

$$P(t_2, t_1) := \mathcal{T}e^{-i\int_{t_1}^{t_2} dt H_I(t)}. \tag{A19}$$

The reader may verify that (A18) and (A19) indeed solves (A14) by direct substitution. Note that all operators inside the ordered exponential symbol $\mathcal{T}$ commute with each other.

Appendix E. On the Background to Theorem

The complete and most general proof of Theorem 1 can be obtained by utilizing a generalized framework, that of *algebraic quantum field theory* [37,78]. The expansion (15) can be shown to be derivable from a suitable perturbative algebraic quantum field theory, e.g., the recent approach [93]. The advantage of choosing such a method is that one does not need to assume a concrete Lagrangian from the beginning but rather proceed to work directly with algebras of quantum fields and then use the structure of these algebras in order to construct the entire theory, including the quantum states themselves, which are generated internally. However, since the mathematical details are extensive, the full treatment will be presented in a separate paper. Nevertheless, note that special cases of Theorem 1 have already appeared repeatedly in the literature, though in quite different applications and within distinct contexts. For example, a special case of (15) seems to have been discussed by Coleman in his analysis of perturbation calculations in scattering theory [38]. Moreover, a less general form of the expansion (15), known as the Volterra series, is often presented in several textbooks on QFT when discussing the evaluation of the vacuum expectation persistence function in terms of the higher-point Green's (correlation) functions of the quantum field, e.g., see [38,54–58,67].

Appendix F. The Neutral Klein–Gordon Field Theory

A massive neutral Klein–Gordon field with mass m can be fully captured by a real scalar field $\phi(x)$ whose Lagrangian density is given by [38,54,56,75]

$$\mathcal{L}(x) = \frac{1}{2}\partial^{\mu}\partial_{\mu}\phi(x) + \frac{1}{2}m^{2}\phi(x)^{2}, \tag{A20}$$

with the corresponding action integral being

$$S = \int \mathrm{d}^{4}x\, \mathcal{L}(x). \tag{A21}$$

The Euler–Lagrange equation of motion

$$\frac{\delta S}{\delta\phi(x)} = 0 \tag{A22}$$

yields the field equation

$$\partial^{\mu}\partial_{\mu}\phi(x) + m^{2} = 0, \tag{A23}$$

which is the relativistic wave equation of a particle the Klein–Gordon equation [38,75].

Note that within the framework of standard quantum field theory (QFT), fields are promoted to *operators* expanded in terms of plane-wave modes of the form

$$\text{operator} \times \exp(-\mathrm{i}p_{\mu}x^{\mu}), \tag{A24}$$

where the "operator" is either creation $a^{\dagger}_{\mathbf{p}}$ or annihilation $a_{\mathbf{p}}$ operator (for particles or antiparticles). Note further that by convention, the plane wave

$$\exp(-\mathrm{i}p_{\mu}x^{\mu}) = \exp[-\mathrm{i}(E_{\mathbf{p}}t - \mathbf{p}\cdot\mathbf{x})] \tag{A25}$$

is taken to encode an *incoming* wave/particle with momentum $\mathbf{p}$/wavevector $\mathbf{k}$ and energy $E_{\mathbf{p}}$/frequency ω [38]. In the unified language of QFT, we say that the plane wave encodes a particle/antiparticle in a pure momentum state $\mathbf{p}$. Within this convention, the energy/frequency of a particle/wave is *always positive*.[25]

By applying the canonical quantization algorithm [38] (see review in Appendix G), the quantum field $\phi(x)$ may be expanded into a continuous sum of spacetime modes (plane waves) as follows

$$\phi(x) = \int_{\mathbf{p}\in\mathbb{R}^{3}} \frac{\mathrm{d}^{3}p}{(2\pi)^{3/2}} \frac{1}{(2\omega_{\mathbf{p}})^{1/2}} \left[a_{\mathbf{p}}e^{-\mathrm{i}p_{\mu}x^{\mu}} + a^{\dagger}_{\mathbf{p}}e^{\mathrm{i}p_{\mu}x^{\mu}}\right], \tag{A26}$$

with the dispersion relation (in natural units)

$$E_{\mathbf{p}} = \omega_{\mathbf{p}} = \omega_{\mathbf{k}} = +\sqrt{|\mathbf{p}|^{2} + m^{2}} = +\sqrt{|\mathbf{k}|^{2} + m^{2}}. \tag{A27}$$

The creation and annihilation operators $a^{\dagger}_{\mathbf{p}}$ and $a_{\mathbf{p}}$, respectively, obey the standard canonical commutation relations

$$[a_{\mathbf{p}}, a_{\mathbf{p}'}] = 0,\ [a^{\dagger}_{\mathbf{p}}, a^{\dagger}_{\mathbf{p}'}] = 0,\ [a_{\mathbf{p}}, a^{\dagger}_{\mathbf{p}'}] = \delta(\mathbf{p} - \mathbf{p}'), \tag{A28}$$

where δ is the Dirac delta function.

Appendix G. The Relativistic Field-Theoretic Canonical Quantization Algorithm

The general canonical quantization algorithm is shown in Algorithm A1, where we leave the nature of the field (scalar, vector, spin type, tensor, etc.) unspecified. In what follows, the detailed quantization algorithm applies to our main type of fields in this paper, i.e., the massive spin-0 scalar field $\phi(x)$. Recall that in nonrelativistic quantum theory, a generic operator can be Fourier expanded in terms of creation and annihilation operators $a^\dagger_{\mathbf{p}}$ and $a_{\mathbf{p}}$, respectively. For instance, a generic scalar quantum field $\phi(x)$ is expected to be written as

$$\phi(x) = \int_{\mathbf{p}\in\mathbb{R}^3} \frac{\mathrm{d}^3 p}{(2\pi)^3} \left[a_{\mathbf{p}} e^{-\mathrm{i}p_\mu x^\mu} + a^\dagger_{\mathbf{p}} e^{\mathrm{i}p_\mu x^\mu}\right]. \tag{A29}$$

Unfortunately, clearly, this expression is not Lorentz invariant because of the use of the non-Lorentz invariant integration measure $\mathrm{d}^3 p$. We review below how Lorentz-invariant quantum states $|p\rangle$ may be redefined so that Fourier expansions like (A29) can be made manifestly covariant.

In order to expand the quantum field into a proper continuum of plane-wave modes of the form $\exp(\pm \mathrm{i}p_\mu x^\mu)$, we will need to integrate over all four-vector momenta p^μ, i.e., perform four-dimensional integrals over $\mathbf{p}$ and $E_{\mathbf{p}}$ (or equivalently $\mathbf{k}$ and ω). However, for massive particles in general, and Proca waves in particular [94], there is a definite relation between momentum/wavevector and energy/frequency, so the integral over $\mathrm{d}^3 p$ might appear at first sight to be essentially reducible to $\mathrm{d}^3 p$, while $E_{\mathbf{p}}$ is computed from $E^2_{\mathbf{p}} = \mathbf{p}^2 + m^2$. However, the problem is that the differential element $\mathrm{d}^3 p$ is not Lorentz invariant, so there is a need to automatically enforce Lorentz invariance in our quantization rules. In this paper, we adopt the computationally efficient method of applying *normalizing factors* to quantum states right from the beginning. The main idea is to produce Lorentz-invariant momentum quantum states and use them to expand the quantum fields.

The essence of the method of normalizing factors is to perform a transformation of the form

$$\int_{\mathbf{p}\in\mathbb{R}^3} \mathrm{d}^3 p \longrightarrow \int_{p\in\mathbb{M}^4} \mathrm{d}^4 p \tag{A30}$$

taking us from the non-Lorentz invariant space $\mathbb{R}^3$ to Minkowski space $\mathbb{M}^4$, where $\mathrm{d}^4 p$ is clearly invariant since $p := p^\mu = (p^0, \mathbf{p})$. However, note that the mass shell condition

$$p^2 := p^\mu p_\mu = m^2 \tag{A31}$$

forces this integration to remain on the cone defined by equations of this type, i.e., a four-submanifold embedded into $\mathbb{M}^4$ whose equation is (A31). Then, we may write

$$\mathrm{d}^4 p = \mathrm{d}^3 p \delta\left(p^2 - m^2\right)\Theta\left(p^0\right), \tag{A32}$$

where $\Theta(\cdot)$ is the Heaviside unit step function, which is inserted by hand in order to prevent the appearance of negative energies. It can be shown that [38,76]

$$\delta\left(p^2 - m^2\right)\Theta\left(p^0\right) = (1/2E_{\mathbf{p}})\delta(E - E_{\mathbf{p}})\Theta(E_{\mathbf{p}}), \tag{A33}$$

and hence, we conclude

$$\int_{p\in\mathbb{M}^4} \frac{\mathrm{d}^4 p}{(2\pi)^3} = \int_{\mathbf{p}\in\mathbb{R}^3} \frac{\mathrm{d}^3 p}{(2\pi)^3} \frac{1}{2E_{\mathbf{p}}}, \tag{A34}$$

where the factor $1/(2\pi)^3$ was intentionally inserted in order to make the final integral looks like an inverse Fourier transform (this inserted factor will be compensated for shortly when we define the normalizing factors of the Lorentz-invariant state $|p\rangle$). Therefore, we have managed then to reduce a four-dimensional integration in $\mathbb{M}^4$ into a regular volume integral in $\mathbb{R}^3$. Note that by writing the RHS of in (A34) without $\delta(E - E_{\mathbf{p}})\Theta(p^0)$, it is to be implicitly understood that – algorithmically speaking – every appearance of E or ω in the various possible expressions placed to the right of the integral operator sign $\int \mathrm{d}^3 p$ should be automatically replaced by $+E_{\mathbf{p}}$ or $+\omega_{\mathbf{k}}$.

A Lorentz-invariant momentum state $|p\rangle$ should be constructed from the standard three-vector momentum states $|\mathbf{p}\rangle$, which obey the normalization rule

$$\langle \mathbf{p}|\mathbf{p}'\rangle = \delta(\mathbf{p}-\mathbf{p}'). \tag{A35}$$

The most obvious construction of the Lorentz-invariant momentum state then is

$$|p\rangle := (2\pi)^{3/2}(2E_{\mathbf{p}})^{1/2}|\mathbf{p}\rangle. \tag{A36}$$

For example, the inner product of such states is

$$\langle p|p'\rangle = (2\pi)^3(2E_{\mathbf{p}})\delta(\mathbf{p}-\mathbf{p}'). \tag{A37}$$

The definition (A36) can be readily used to yield the desired Lorentz-invariant completeness relation

$$\mathbb{1} = \int_{\mathbf{p}\in\mathbb{R}^3} \underbrace{\frac{\mathrm{d}^3 p}{(2\pi)^3}\frac{1}{2E_{\mathbf{p}}}}_{\text{Lorentz measure}} |p\rangle\langle p|, \tag{A38}$$

where $\mathbb{1}$ is the unity operator. The computational utility of the method of normalizing factors stems from the fact that a completeness relation originally holding for all states $\{|p\rangle, p \in \mathbb{M}^4\}$ in the full four-dimensional Minkowski space is now reduced to carrying out integration in the regular Euclidean space $\mathbb{R}^3$. However, the price to be paid for such simplification is that a momentum-dependent factor $1/2E_{\mathbf{p}}$ must be inserted into the integrand of the reduced integral, which slightly complicates the evaluation of various related expressions such as probability amplitudes.

Finally, we can create the Lorentz state by applying a new normalized creation operator $b^\dagger_{\mathbf{p}}$ to the ground or vacuum state $|0\rangle$, i.e., $|p\rangle = b^\dagger_{\mathbf{p}}|0\rangle$ and $b_{\mathbf{p}}|0\rangle = 0$. Based on (A36), we expect

$$b^\dagger_{\mathbf{p}} = (2\pi)^{3/2}(2E_{\mathbf{p}})^{1/2}a^\dagger_{\mathbf{p}}, \qquad b_{\mathbf{p}} = (2\pi)^{3/2}(2E_{\mathbf{p}})^{1/2}a_{\mathbf{p}}. \tag{A39}$$

The field expansion now, based on (A38), becomes

$$\phi(x) = \int_{\mathbf{p}\in\mathbb{R}^3} \frac{\mathrm{d}^3 p}{(2\pi)^3}\frac{1}{2E_{\mathbf{p}}}\left[b_{\mathbf{p}}e^{-\mathrm{i}p_\mu x^\mu} + b^\dagger_{\mathbf{p}}e^{\mathrm{i}p_\mu x^\mu}\right], \tag{A40}$$

which after using (A39) leads to

$$\phi(x) = \int_{\mathbf{p}\in\mathbb{R}^3} \frac{\mathrm{d}^3 p}{(2\pi)^{3/2}}\frac{1}{(2E_{\mathbf{p}})^{1/2}}\left[a_{\mathbf{p}}e^{-\mathrm{i}p_\mu x^\mu} + a^\dagger_{\mathbf{p}}e^{\mathrm{i}p_\mu x^\mu}\right]. \tag{A41}$$

Algorithm A1 The Canonical Quantization Algorithm (General Formulation).

1: Start with a Lagrangian density $\mathcal{L}$.
2: Find the momentum density. From this knowledge, construct the Hamiltonian.
3: Promote the field and the corresponding momentum variables to operator-valued fields. Apply the standard commutation relations.
4: Expand the quantum fields into sums of plane waves with creation and annihilation operator amplitudes.
5: Evaluate the Hamiltonian operator in terms of the creation and annihilation operators and put in in the normal ordering form.

Appendix H. On the Numerical Evaluation of the Propagator

For the scalar massive or massless field theory considered above, the Green's Function (29) can be evaluated in closed analytical forms after regulating the integral by inserting small imaginary number $\mathrm{i}\epsilon$ at proper locations in the integrand in order to ensure convergence (causality consideration, e.g., see [54,55]). The final expressions are given by [56]

$$G_q(x,x') = \begin{cases} \frac{im}{8\pi}\frac{H_1^{(2)}(m|x-x'|)}{m|x-x'|}, & m \neq 0, \\ \frac{-1}{4\pi^2}\frac{1}{|x-x'|^2 - i\epsilon}, & m = 0. \end{cases} \tag{A42}$$

Here, the distance $|x-x'|$ is computed in Minkowski spacetime $\mathbb{M}^4$ with the metric tensor $g_{\mu\nu}$, i.e, the relation

$$|x-x'|^2 = (t-t')^2 - |\mathbf{r}-\mathbf{r}'|^2. \tag{A43}$$

The Hankel function of the second kind $H_\nu^{(2)}$ is defined as

$$H_\nu^{(2)}(x) := J_\nu(x) \pm \mathrm{i}N_\nu(x), \tag{A44}$$

where J_ν and N_ν are the Bessel and Neumann functions, respectively [95,96].

Appendix I. Proof of Relation

Proof of Relation (61), when there is a source $J(x'), x' \in D_{\mathrm{s}}$, then the actual radiated (quantum) state of the q-antenna system is obtained by applying the operator

$$\int_{x' \in D_{\mathrm{s}}} \mathrm{d}^4x' J(x')\phi^\dagger(x')$$

on the ground state $|0\rangle$ in order to create the following one-particle excited state

$$\int_{x' \in D_{\mathrm{s}}} \mathrm{d}^4x' J(x')\phi^\dagger(x')|0\rangle.$$

To measure the probability of having a four-momentum state $|q\rangle$ in the previous excited q-antenna state, we compute the source probability amplitude $A[q; J(x'), D_{\mathrm{s}}]$ as follows:

$$\begin{aligned} &A[q; J(x'), D_{\mathrm{s}}] \\ &:= \langle q| \int_{x' \in D_{\mathrm{s}}} \mathrm{d}^4x' J(x')\phi^\dagger(x')|0\rangle = \int_{\mathbf{p}\in\mathbb{R}^3} \frac{\mathrm{d}^3p}{(2\pi)^{3/2}(2\omega_{\mathbf{p}})^{1/2}} \int_{x' \in D_{\mathrm{s}}} \mathrm{d}^4x' J(x') e^{\mathrm{i}p_\mu x'^\mu} \langle q|\mathbf{p}\rangle \\ &= \int_{\mathbf{p}\in\mathbb{R}^3} \frac{\mathrm{d}^3p}{(2\pi)^{3/2}(2\omega_{\mathbf{p}})^{1/2}} J(p)\delta(\mathbf{q}-\mathbf{p})(2\pi)^{3/2}(2\omega_{\mathbf{p}})^{1/2} = J(q)|_{q^0=\omega_{\mathbf{q}}} = J(\mathbf{q},\omega_{\mathbf{q}}), \end{aligned}$$

where

$$J(k) := \int_{x \in D_{\mathrm{s}}} \mathrm{d}^4x\, J(x) e^{\mathrm{i}k_\mu x'^\mu} = \int_{\mathbf{x}\in\mathbb{R}^3} \mathrm{d}^3x \int_{t\in\mathbb{R}} \mathrm{d}t\, J(\mathbf{x},t) e^{\mathrm{i}(\omega t - \mathbf{k}\cdot\mathbf{x})} \tag{A45}$$

is the four-dimensional (Minkowski) Fourier transform in spacetime. The second line in (A45) follows from using the momentum-space expansion of the source field $\phi(x')$ given by (A26), while in the third line, we used the normalized four-momentum state inner product formula

$$\langle q|\mathbf{p}\rangle = (2\pi)^{3/2}(2\omega_{\mathbf{p}})^{1/2}\delta(\mathbf{q}-\mathbf{p}), \tag{A46}$$

then proceeded to evaluate the trivial resulting integral over a delta function.

Notes

1 Other possible long-term aims behind the spacetime theory of q-antennas proposed below include the stimulation of fruitful collaboration between theoreticians, especially those working on problems related to foundations, and applied quantum physicists and engineers, whose attention is often more focused on algorithmic and physical-layer applications, e.g., quantum communications, cryptography, computing, and so on.

2 Maxwell's equations in vacuo are exactly linear [49,51]; a photon does not self-interact with itself [52]. As in classical antenna theory, in the proposed quantum antenna theory given here, all nonlinearities are relegated to production of the source $J(x)$ itself. Well-known examples include gun diodes (microwaves) and laser devices (optics), where the diode itself is nonlinear but the relation between the current or field as external source and the fields radiated into vacuum is linear. The physical nonlinear processes behind the source function $J(x)$ itself are outside the scope of the proposed theory.

3 Otherwise, relativistic causality would preclude information transfer [51]. In general, there is an agreement in the literature that entanglement-based quantum communication links cannot transmit information at superluminal speeds even though the quantum correlations between entangled states persists at spacelike separated terminals [2,4].

4 This is a realistic assumption in our model, in conformity with the common practical situation where typical classical or quantum sources are supported by bounded spatial domains and radiate within a finite time interval while practical measurement times are also bounded.

5 The propagators coincide with well-known Green's functions in the case of free fields. For interacting field theories, the propagators are not in general known, but viable approximations can be estimated using perturbation theory, in which the free-field Green's function is used as a fundamental building block in order to compute more complex higher-order interaction processes [54–60].

6 By *effective path*, we just mean the spacetime trajectory in the Feynman's path integral expansion of the propagator that contributes most to the total probability amplitude, e.g., see [57].

7 Cf. Appendix D.

8 Interaction is a more general concept than quantum correlation, since two uncorrelated objects could interact, where in this case, the interaction terms are just the multiplication of the strengths of each process while the two remain, at least stochastically speaking, independent. An example illustrating this will be given shortly.

9 Nowadays, such many-body effects are approached in the modern literature on condensed-matter physics, e.g., through the elaborate and elegant language of many-point correlation or Green's functions in QFT [59–61].

10 In QFT, integrals such as (20) are handled using four-point Green's functions of the form $G^{(4)}(x_3, x_4, x_1, x_2)$, where the latter is called the four-point correlation (Green's) function [54,55,57]. In our case, we just choose $x_3 = x_4 = x$, since x is the common observation point of the receiving q-antenna system.

11 The realization of the need to eventually differentiate a purely classical source function from within any stochastic system (including quantum systems) was originally proposed within the context of quantum optics in the 1960s [62].

12 For the purposes of illustrating the main ideas of q-antenna systems, this assumption simplifies considerably the presentation, but the main ideas related to q-antennas are unchanged when more complicated field theories are considered such as spin-1 and spin-1/2 theories.

13 For additional information on the physical processes modeled by a Klein–Gordon quantum field, see [74,75].

14 For a more detailed discussion of relativistic QFT, see [38,76].

15 The generation of $\phi(x)$ itself is not treated here for simplicity. However, note that computing the quantized fields of coupled matter–field systems is a fairly well-developed area in the physics and engineering literature, mostly using the methods of perturbation theory [38,54,55,67,77]. On the other hand, in this paper, our main focus is on how to deploy an *already* given or generated quantum field $\phi(x)$ in order to construct the radiation pattern and the Green's function of a q-antenna system for use in applications in controlled radiation of quantum states.

16 When reworked in the full momentum space $p \in \mathbb{M}^4$, the integral (21) becomes even more interesting, both computationally and conceptually, since one can show then that point source excitations lead to the production of virtual (off-mass-shell) particles [38,57]. However, we will not make use of these expansions in the present paper though they are expected to play an important role in developing the near-field theory of radiating quantum source systems.

17 The concept of particle localization in QFT is difficult both philosophically and mathematically, and several approaches have been proposed in the literature so far, apparently with no universal agreement on the ontological status of particles in field quantization. Such more advanced issues do not affect the practically oriented theory of quantum antennas developed in this paper. For some in-depth discussion of localization in field theory, see [37,78].

18 Recall that in QFT, antiparticles are interpreted as particles moving *backward* in time [57,58].

19 Momentum space means either $\mathbf{p} \in \mathbb{R}^3$ or $p \in \mathbb{M}^4$. In this paper, whenever the term momentum space is invoked, it is to be understood that we will mostly work with the former version, i.e., in three dimensions.

20 In this paper, we do not consider the possible case of unbounded source domains such as infinite current sheets and lines.

21 Cf. Appendix G.

22 Cf. Appendix A.

23 Indeed, this is how directivity is defined in classical antenna theory, e.g., see [35]. Note further that $\mathcal{D}$ in the LHS of (69) was directly expressed in terms of the spherical angles φ, θ, in order to emphasize the spatial angular character of this momentum space function. See [71,72] for more details about the momentum space approach to directivity.

24 See Coleman's discussion of the generic detection process in high-energy physics as given in [38].

25 Indeed, negative energy/frequencies obtained as solutions to the massive particle dispersion equation $E^2 = \mathbf{p}^2 + m^2$ are often reinterpreted as antiparticles and plane waves of the form $\exp(+ip_\mu x^\mu)$ are taken to represent *outgoing* antiparticles with momentum $\mathbf{p}$ and energy $E_\mathbf{p}$ [38].

References

1. Helstrom, C.W. *Quantum Detection and Estimation Theory;* Academic Press: Cambridge, MA, USA; Cambridge University Press: Cambridge, MA, USA, 1976.
2. Cariolaro, G. *Quantum Communications;* Springer: Berlin/Heidelberg, Germany, 2016.
3. Imre, S.; Azs, F. *Quantum Computing and Communications—An Engineering Approach;* Wiley: Hoboken, NJ, USA, 2005.
4. Nielsen, M.; Chuang, I.L. *Quantum Computation and Quantum Information;* Cambridge University Press: New York, NY, USA, 2010.
5. Barnett, S. Optical demonstrations of statistical decision theory for quantum systems. *Quantum Inf. Comput.* **2004**, *4*, 450–459. [CrossRef]
6. Barnett, S.M. *Quantum Information;* Oxford University Press: Oxford, UK, 2009.
7. Helstrom, C.W.; Liu, J.W.S.; Gordon, J.P. Quantum-mechanical communication theory. *Proc. IEEE* **1970**, *58*, 1578–1598. [CrossRef]
8. Holevo, A. Statistical decision theory for quantum systems. *J. Multivar. Anal.* **1973**, *3*, 37–394. [CrossRef]
9. Holevo, A.; Giovannetti, V. Quantum channels and their entropic characteristics. *Rep. Prog. Phys. Phys. Soc.* **2012**, *75*, 046001. [CrossRef]
10. Mikki, S. A quantum MIMO architecture for antenna wireless digital communications. *Prog. Electromagn. Res. C* **2019**, *93*, 143–156. [CrossRef]
11. Assche, G. *Quantum Cryptography and Secret-Key Distillation;* Cambridge University Press: Cambridge, MA, USA, 2006.
12. Bernstein, D.; Buchmann, J.; Dahmen, E. (Eds.) *Post-Quantum Cryptography;* Springer: Berlin/Heidelberg, Germany, 2009.
13. Mikki, S. Quantum antenna theory for secure wireless communications. In Proceedings of the 14th European Conference on Antennas and Propagation (EuCAP), Copenhagen, Denmark, 15–20 March 2020.
14. Mikki, S.; Herde, M. Analysis and design of secure quantum communication systems utilizing electromagnetic Schrodinger coherent states. *Quantum Eng.* **2021**, *3*, e72. [CrossRef]
15. Farahani, J.N.; Pohl, D.W.; Eisler, H.-J.; Hecht, B. Single quantum dot coupled to a scanning optical antenna: A tunable superemitter. *Phys. Rev. Lett.* **2005**, *95*, 017402. [CrossRef]
16. Filter, R.; Mühlig, S.; Eichelkraut, T.; Rockstuhl, C.; Lederer, F. Controlling the dynamics of quantum mechanical systems sustaining dipole-forbidden transitions via optical nanoantennas. *Phys. Rev. B* **2012**, *86*, 035404. [CrossRef]
17. Slepyan, G.Y.; Boag, A. Quantum nonreciprocity of nanoscale antenna arrays in timed Dicke states. *Phys. Rev. Lett.* **2013**, *111*, 023602. [CrossRef]
18. Kremer, P.E.; Dada, A.C.; Kumar, P.; Ma, Y.; Kumar, S.; Clarke, E.; Gerardot, B.D. Strain-tunable quantum dot embedded in a nanowire antenna. *Phys. Rev. B* **2014**, *90*, 201408. [CrossRef]
19. Liu, J.; Zhou, M.; Ying, L.; Chen, X.; Yu, Z. Enhancing the optical cross section of quantum antenna. *Phys. Rev. A* **2017**, *95*, 013814. [CrossRef]
20. Fitzgerald, J.M.; Azadi, S.; Giannini, V. Quantum plasmonic nanoantennas. *Phys. Rev. B* **2017**, *95*, 235414. [CrossRef]
21. Mikki, S. Quantum antenna theory. In Proceedings of the 2017 IEEE AP-S Symposium on Antennas and Propagation and USNC-URSI Radio Science Meeting, San Diego, CA, USA, 9–14 July 2017.
22. Müller, C.; Combes, J.; Hamann, A.R.; Fedorov, A.; Stace, T.M. Nonreciprocal atomic scattering: A saturable, quantum Yagi-Uda antenna. *Phys. Rev. A* **2017**, *96*, 053817. [CrossRef]
23. Komarov, A.; Slepyan, G. Quantum antenna as an open system: Strong antenna coupling with photonic reservoir. *Appl. Sci.* **2018**, *8*, 951. [CrossRef]
24. Mikhalychev, A.; Mogilevtsev, D.; Slepyan, G.Y.; Karuseichyk, I.; Buchs, G.; Boiko, D.L.; Boag, A. Synthesis of quantum antennas for shaping field correlations. *Phys. Rev. Appl.* **2018**, *9*, 024021. [CrossRef]
25. Liberal, I.N.; Ederra, I.N.; Ziolkowski, R.W. Quantum antenna arrays: The role of quantum interference on direction-dependent photon statistics. *Phys. Rev. A* **2018**, *97*, 053847. [CrossRef]
26. Control of a quantum emitter's bandwidth by managing its reactive power. *Phys. Rev. A* **2019**, *100*, 023830. [CrossRef]
27. Schelkunoff, S.A.; Friss, H.T. *Antennas: Theory and Practice;* Chapman & Hall: New York, NY, USA; London, UK, 1952.
28. Mikki, S.; Antar, Y. *New Foundations for Applied Electromagnetics: The Spatial Structure of Fields;* Artech House: London, UK, 2016.
29. Novotny, L. *Principles of Nano-Optics;* Cambridge University Press: Cambridge, MA, USA, 2012.
30. Arisheh, A.A.; Mikki, S.; Dib, N. A subwavelength-laser-driven transmitting optical nanoantenna for wireless communications. *IEEE J. Multiscale Multiphys. Comput. Tech.* **2020**, *5*, 144–154. [CrossRef]

31. Ghamsari, M.S. Chip-scale quantum emitters. *Quantum Rep.* **2021**, *3*, 615–642. [CrossRef]
32. Yaduvanshi, R. *Terahertz Dielectric Resonator Antennas for High Speed Communication and Sensing from Theory to Design and Implementation*; Institution of Engineering and Technology: Stevenage, UK, 2021.
33. Geyi, W. *Foundations of Applied Electrodynamics*; Wiley: Hoboken, NJ, USA, 2010.
34. Chew, W.; Jin, J.-M.; Michielssen, E.; Song, J. (Eds.) *Fast and Efficient Algorithms in Computational Electromagnetics*; Artech House: Boston, MA, USA, 2001.
35. Balanis, C.A. *Antenna Theory: Analysis and Design*, 4th ed.; Inter-Science: Wiley: Hoboken, NJ, USA, 2015.
36. Auyang, S. *How Is Quantum Field Theory Possible*; Oxford University Press: New York, NY, USA, 1995.
37. Haag, R. *Local Quantum Physics: Fields, Particles, Algebras*; Springer: Berlin/Heidelberg, Germany, 1992.
38. Coleman, S. *Quantum Field Theory: Lectures of Sidney Coleman*; World Scientific: Singapore, 2019.
39. Jaeger, G. The elementary particles of quantum fields. *Entropy* **2021**, *23*, 1416. [CrossRef]
40. Schelkunoff, S.A. A Mathematical Theory of Linear Arrays. *Bell Syst. Tech. J.* **1943**, *22*, 80–107 [CrossRef]
41. Penrose, R. *The Road to Reality: A Complete Guide to the Laws of the Universe*; Vintage Books: New York, NY, USA, 2007.
42. Omnes, R. *Understanding Quantum Mechanics*; Princeton University Press: New York, NY, USA, 1999.
43. *The Interpretation of Quantum Mechanics*; Princeton University Press: Princeton, NJ, USA, 1994.
44. *Quantum Philosophy: Understanding and Interpreting Contemporary Science*; Princeton University Press: Princeton, NJ, USA, 1999.
45. Harmuth, H. *Antennas and Waveguides for Nonsinusoidal Waves*; Academic Press: Orlando, FL, USA, 1984.
46. Mikki, S.; Antar, Y. The antenna current Green's function formalism—Part I. *IEEE Trans. Antennas Propag.* **2013**, *9*, 4493–4504. [CrossRef]
47. The antenna current Green's function formalism—Part II. *IEEE Trans. Antennas Propag.* **2013**, *9*, 4505–4519.
48. On the Fundamental Relationship Between the Transmitting and Receiving Modes of General Antenna Systems: A New Approach. *IEEE Antennas Wirel. Propag. Lett.* **2012**, *11*, 232–235. [CrossRef]
49. Schwinger, J. *Classical Electrodynamics*; Perseus Books: Reading, MA, USA, 1998.
50. Jackson, J. *Classical Electrodynamics*; Wiley: New York, NY, USA, 1999.
51. Landau, L.D. *The Classical Theory of Fields*; Butterworth Heinemann: Oxford UK, 2000.
52. Dirac, P.A.M. *The Principles of Quantum Mechanics*; Clarendon Press: Oxford, UK, 1981.
53. Bohm, D. *The Special Theory of Relativity*; Routledge: London, UK; New York, NY, USA, 2006.
54. Zeidler, E. *Quantum Field Theory I: Basics in Mathematics and Physics*; Springer: Berlin/Heidelberg, Germany, 2009.
55. *Quantum Field Theory II: Quantum Electrodynamics*; Springer: Berlin/Heidelberg, Germany, 2006.
56. Kleinert, H. *Particles and Quantum Fields*; World Scientific: Singapore, 2016.
57. Greiner, W.; Reinhardt, J. *Field Quantization*; Springer: Berlin/Heidelberg, Germany, 1996.
58. Zee, A. *Quantum Field Theory in a Nutshell*; Princeton University Press: Princeton, NJ, USA, 2010.
59. Coleman, P. *Introduction to Many-Body Physics*; Cambridge University Press: Cambridge, UK, 2015.
60. Mattuck, R. *A Guide to Feynman Diagrams in the Many-Body Problem*; Dover Publications: New York, NY, USA, 1992.
61. Altland, A.; Simmons, B. *Condensed Matter Field Theory*; Cambridge University Press: Cambridge, UK, 2010.
62. Klauder, J.; Sudarshan, E.C.G. *Fundamentals of Quantum Optics*; Dover Publications: Mineola, NY, USA, 2006.
63. Felsen, L. *Radiation and Scattering of Waves*; IEEE Press: Piscataway, NJ, USA, 1994.
64. Chew, W.C. *Waves and Fields in Inhomogenous Media*; Wiley: Hoboken, NJ, USA, 1999.
65. Tai, C.-T. *Dyadic Green Functions in Electromagnetic Theory*; IEEE Press: Piscataway, NJ, USA, 1994.
66. Jentschur, U. *Advanced Classical Electrodynamics: Green Functions, Regularizations, Multipole Decompositions*; World Scientific: Singapore, 2017.
67. Zeidler, E. *Quantum Field Theory III: Gauge Theory*; Springer: Berlin/Heidelberg, Germany, 2011.
68. Mikki, S.; Kishk, A. Theory and Applications of Infinitesimal Dipole Models for Computational Electromagnetics. *IEEE Trans. Antennas Propag.* **2007**, *55*, 1325–1337. [CrossRef]
69. Nonlocal eLectromagnetic Media: A Paradigm for Material Engineering. Passive Microwave Components and Antennas; InTech. 2010. Available online: https://www.intechopen.com/chapters/10714 (accessed on 26 January 2022).
70. A symmetry-based formalism for the electrodynamics of nanotubes. *Prog. Electromagn. Res.* **2008**, *86*, 111–134. [CrossRef]
71. Mikki, S. Theory of electromagnetic radiation in nonlocal metamaterials—Part I: Foundations. *Prog. Electromagn. Res. B* **2020**, *89*, 63–86. [CrossRef]
72. Theory of electromagnetic radiation in nonlocal metamaterials—Part II: Applications. *Prog. Electromagn. Res. B* **2020**, *89*, 87–109. [CrossRef]
73. Mikki, S. Exact derivation of the radiation law of antennas embedded into generic nonlocal metamaterials: A momentum-space approach. In Proceedings of the 14th European Conference on Antennas and Propagation (EuCAP), Copenhagen, Denmark, 15–20 March 2020; pp. 1–5.
74. Bjorken, J.; Drell, S.D. *Relativistic Quantum Fields*; McGraw-Hill: New York, NY, USA, 1965.
75. Nastase, H. *Classical Field Theory*; Cambridge University Press: New York, NY, USA, 2019.
76. Greiner, W. *Relativistic Quantum Mechanics: Wave Equations*; Springer: Berlin/Heidelberg, Germany, 2000.
77. Garrison, J.C.; Chiao, R. *Quantum Optics*; Oxford University Press: Oxford, UK, 2014.
78. Streater, R.F.; Wightman, A.S. *PCT, Spin and Statistics, and All That*; Princeton University Press: Princeton, NJ, USA, 2000.

79. Neumann, J. *Mathematical Foundations of Quantum Mechanics*; Princeton University Press: Princeton, NJ, USA, 2018.
80. Keller, O. *Quantum Theory of Near-Field Electrodynamics*; Springer: Berlin/Heidelberg, Germany, 2011.
81. Mikki, S.; Antar, Y. A theory of antenna electromagnetic near field—Part I. *IEEE Trans. Antennas Propag.* **2011**, *59*, 4691–4705. [CrossRef]
82. Mikki, S.; Antar, Y.M.M. A theory of antenna electromagnetic near field—Part II. *IEEE Trans. Antennas Propag.* **2011**, *59*, 4706–4724. [CrossRef]
83. Glauber, R. *Quantum Theory of Optical Coherence: Selected Papers and Lectures*; Wiley-VCH: Weinheim, Germany, 2007.
84. Penrose, R. *Techniques of Differential Topology in Relativity*; Society for Industrial and Applied Mathematics: Philadelphia, PA, USA, 1972.
85. Feynman, R. *The Feynman Lectures on Physics, Volume III: Quantum Mechanics*; Basic Books: New York, NY, USA, 2011.
86. Bohm, D. *Quantum Theory*; Dover Publications: New York, NY, USA, 1989.
87. Louisell, W. *Radiation and Noise in Quantum Electronics*; Krieger Pub. Co.: Huntington, NY, USA, 1977.
88. Collin, R.E. *Field Theory of Guided Waves*; Wiley: Hoboken, NJ, USA, 1991.
89. Mikki, S.; Antar, Y. A rigorous approach to mutual coupling in general antenna systems through perturbation theory. *IEEE Antennas Wirel. Commun. Lett.* **2015**, *14*, 115–118. [CrossRef]
90. Mikki, S. Generalized current Green's function formalism for electromagnetic radiation by composite systems. *Prog. Electromagnet. Res. B* **2020**, *87*, 171–191. [CrossRef]
91. Mikki, S.; Clauzier, S.; Karimi, M.; Shamim, A.; Antar, Y. Slot antenna array synthesis using the infinitesimal dipole model technique: Theory and experiment. *Int. J. Microw. Comput. Aided Eng.* **2020**, *30*, e22402 [CrossRef]
92. Mikki, S. The antenna spacetime system theory of wireless communications. *Proc. R. Soc. A* **2019**, *475*, 20180699. [CrossRef]
93. Rejzner, K. *Perturbative Algebraic Quantum Field Theory: An Introduction for Mathematicians*; Springer: Cham, Switzerland, 2016.
94. Mikki, S. Proca metamaterials, massive electromagnetism, and spatial dispersion. *Ann. Phys.* **2021**, *533*, 2000625. [CrossRef]
95. Hassani, S. *Mathematical Physics: A Modern Introduction to Its Foundations*; Springer: Cham, Switzerland, 2013.
96. Appel, W. *Mathematics for Physics and Physicists*; Princeton University Press: Princeton, NJ, USA, 2007.

foundations

MDPI

Communication

Experiments on the Electron Impact Excitation of the 2s and 2p States of Hydrogen Atoms Confirm the Presence of Their Second Flavor as the Candidate for Dark Matter

Eugene Oks

Physics Department, 380 Duncan Drive, Auburn University, Auburn, AL 36849, USA; goks@physics.auburn.edu

Abstract: For the excitation of the $n = 2$ states of hydrogen atoms due to electron impact, we compared the experimental and theoretical ratios of the cross-sections σ_{2s}/σ_{2p}. We found this theoretical ratio to be systematically higher than the corresponding experimental ratio by about 20%—far beyond the experimental error margins. We suggest that this discrepancy can be explained by the presence of the Second Flavor of Hydrogen Atoms (SFHA) in the experimental hydrogen gas. The explanation is based on the fact that, in the experiments, the cross-section σ_{2s} was determined by using the quenching technique—by applying an electric field that mixed the 2s and 2p states, followed by the emission of the Lyman-alpha line from the 2p state. However, the SFHA only had the s-states, so the quenching technique would not count the excitation of the SFHA in the 2s state and, thus, lead to the underestimation of the cross-section σ_{2s}. We estimates the share of the SFHA in the experimental hydrogen gas required for eliminating the above discrepancy and found this share to be about the same as the share of the usual hydrogen atoms. Thus, our results constitute the third proof from atomic experiments that the SFHA does exist, the first proof being related to the experimental distribution of the linear momentum in the ground state of hydrogen atoms, and the second proof being related to the experimental cross-section of charge exchange between hydrogen atoms and low-energy protons.

Keywords: electron impact excitation of hydrogen atoms; discrepancy between theories and experiments; second flavor of hydrogen atoms

Citation: Oks, E. Experiments on the Electron Impact Excitation of the 2s and 2p States of Hydrogen Atoms Confirm the Presence of Their Second Flavor as the Candidate for Dark Matter. *Foundations* **2022**, 2, 541–546. https://doi.org/10.3390/foundations2030036

Academic Editors: Kazuharu Bamba and Charalampos Moustakidis

Received: 30 May 2022
Accepted: 24 June 2022
Published: 26 June 2022

1. Introduction

The theoretical discovery of the Second Flavor of Hydrogen Atoms (SFHA) in [1] was followed by the first experimental proof of their existence. Namely, the high-energy tail of the linear momentum distribution in the ground state of hydrogen atoms (the distribution being derived from the analysis of atomic experiments) was greater than the theoretical prediction for the usual hydrogen atoms by several orders of magnitude. The allowance for the SFHA eliminated this huge discrepancy [1].

The second piece of experimental evidence of the existence of the SFHA was obtained by analyzing experiments on charge exchange during collisions of low-energy protons with hydrogen atoms [2]. Namely, the allowance for the SFHA brought the corresponding theoretical cross-sections into agreement with the experiments within the experimental error margins.

The proven existence of the SFHA has importance for atomic physics in its own right. However, it has also turned out to have significant astrophysical consequences, including for the most fundamental problem of cosmology: finding out what dark matter is, as explained below.

Bowman et al. [3] reported a perplexing observation of the redshifted 21 cm spectral line from the early Universe. The absorption signal turned out to be about two times more intense than expected from the standard cosmology, thus indicating an additional cooling of the primordial hydrogen gas. Barkana [4] suggested that the additional cooling was due to collisions with some unspecified dark matter particles. In [5], it was demonstrated that, if the additional cooling was caused by collisions with the SFHA, this huge discrepancy would be removed. This outcome suggested the SFHA as a candidate for dark matter.

Jeffrey et al. [6] reported that the observed distribution of dark matter in the Universe was found to be smoother than the predictions based on Einstein's gravitation. This puzzle induced suggestions that new physical laws are needed: a non-Einsteinian gravity. However, in [7], it was shown that the allowance for the SFHA explains the puzzling observational results from Jeffrey et al. [6] not only qualitatively, but also quantitatively.

It should be emphasized that the theory behind the SFHA is the standard quantum-mechanical Dirac equation. So, the qualitative and quantitative explanations of the perplexing observations by Bowman et al. [3] and by Jeffrey et al. [6] obtained by using the SFHA did not introduce any new physical laws (in distinction to the overwhelming majority of other hypotheses on the nature of dark matter) and, therefore, are favored by the principle of Occam's razor. All of this solidified the status of the SFHA as a leading candidate for dark matter (or at least for a part of it).

In the present paper, we provide yet another experimental proof of the existence of the SFHA from the third type of atomic experiment: experiments on the electron impact excitation of hydrogen atoms to the states of the principal quantum number, $n = 2$. There are many different theoretical approaches to this process—see, e.g., [8–17] (listed in alphabetical order) and the references therein. In our analysis, we limit ourselves to the corresponding experimental and theoretical work where both the cross-section σ_{2s} of the excitation of the 2s state and the cross-section σ_{2p} of the excitation of the 2p state were determined *within the same experiment or within the same theoretical approach*. Then, we compare the experimental and theoretical ratios of the cross-sections σ_{2s}/σ_{2p}. We show that this theoretical ratio is systematically higher than the experimental ratio by about 20% (far beyond the experimental error margins). We explain that the presence of the SFHA in the experimental hydrogen gas could be responsible for this discrepancy and estimate the share of the SFHA in the mixture that would be sufficient to eliminate this discrepancy.

2. Comparison of the Experimental Ratios of the Cross-Sections with Theories

Let us first outline the idea. We consider a gas of hydrogen atoms representing a mixture of the SFHA and the usual hydrogen atoms. Further, we consider the excitation of these hydrogen atoms from the ground state to the 2s and 2p states due to electron impact. The experimental measurements of the cross-section σ_{2p} for the excitation to the 2p state are determined by observing the emission of the Lyman-alpha line from the 2p state to the ground state. As for the experimental measurements of the cross-section σ_{2s} for the excitation to the 2s state, they are determined by using the quenching technique: by applying an electric field that mixes the 2s state with the 2p state and then observing the emission of the Lyman-alpha line from the 2p state to the ground state—see, e.g., [18–20] (listed in alphabetical order).

The central point is the following. In the mixture of the SFHA with the usual hydrogen atoms, both the SFHA and the usual hydrogen atoms can be excited to the 2s state. However, after applying the electric field, the mixing of the 2s and 2p states (followed by the emission of the Lyman-alpha line) occurs only for the usual hydrogen atoms. This is because the SFHA has only s-states, so they do not contribute to the observed Lyman-alpha signal. Therefore, measurements of the cross-section σ_{2s} that are conducted in this way should underestimate this cross-section compared to its actual value, while the cross-section σ_{2p} should not be affected by the presence of the SFHA. Consequently, by comparing the experimental ratio σ_{2s}/σ_{2p} with the corresponding theoretical ratio, it should be possible to find out whether the SFHA was present in the hydrogen gas used in the experiments and to estimate the percentage of the SFHA in that hydrogen gas.

The 2s and 2p states are chosen for the following reasons. From the experimental viewpoint, for $n > 2$, the quenching electric field would mix not only s- and p-states, but also states with a higher angular momentum. From the theoretical viewpoint, calculations for $n = 2$ states are simpler than for $n > 2$ states. Therefore, the 2s and 2p states represent the simplest (and, thus, most reliable) test bed from both the experimental and theoretical viewpoints.

Various types of calculations of the absolute cross-section σ_{2s} yield significantly different results—up to a factor of two [8]. Various types of calculations of the absolute cross-section σ_{2p} also yield significantly different results. Therefore, for the stated purpose of our study, we limit ourselves to theoretical papers where both σ_{2s} and σ_{2p} were calculated within the same approach, and we focus on the corresponding σ_{2s}/σ_{2p} ratio within each theoretical approach. In this way, the scatter of the σ_{2s}/σ_{2p} ratios calculated with different theoretical approaches should be noticeably smaller than the scatter of the absolute cross-sections.

Guided by this principle, we determined the theoretical σ_{2s}/σ_{2p} ratio from the values of σ_{2s} and σ_{2p}, which were calculated at three different energies of the incoming electrons by Whelan et al. [16] by using close coupling with the pseudostate basis within the 13-state approximation. We also determined the theoretical σ_{2s}/σ_{2p} ratio from the values of σ_{2s} and σ_{2p}, which were calculated at four different energies of the incoming electrons by Whelan et al. [16] by using the second Born approximation. Then, we determined the corresponding experimental σ_{2s}/σ_{2p} ratio from the values of σ_{2s} and σ_{2p} presented in a paper by Callaway and McDowell [18], which is the latest (to the best of our knowledge) and most accurate discussion of experiments where both σ_{2s} and σ_{2p} were measured in the range of energies given below. (These values of σ_{2s} and σ_{2p} were also cited by Whelan et al. [16].) The results are presented in Table 1.

Table 1. Comparison of the experimental ratios of the cross-sections σ_{2s}/σ_{2p} that we determined on the basis of the paper by Callaway and McDowell [18] with the corresponding theoretical ratios that we determined on the basis of the paper by Whelan et al. [16].

Energy (eV)	σ_{2s}/σ_{2p} from Close Coupling with Pseudostates in 13-State Approximation	σ_{2s}/σ_{2p} from 2nd Born Approximation	Average of These Two Theories	Experimental Ratio σ_{2s}/σ_{2p}	Ratio of the Average Theoretical Value to the Experimental One
35	N/A	0.097	0.097	0.079	1.23
41.65	0.0933	0.0912	0.092	0.076	1.21
50	0.0802	0.0851	0.083	0.070	1.19
54.4	0.0774	0.0828	0.080	0.067	1.19

It can be seen that the average theoretical σ_{2s}/σ_{2p} ratio consistently exceeded the corresponding experimental ratio by about 20% over the entire experimental range of energies. This difference was far beyond the experimental error margin, which was 9% or less. At first glance, this might seem to indicate that about 20% of the hydrogen gas used in the experiments was of the SFHA. However, the actual percentage of the SFHA was much higher, as explained below.

The SFHA differs from the usual hydrogen atoms not only by the fact that the quenching of the 2s state of the SFHA does not work because of the absence of the 2p state (as already noted above), but also in terms of the value of the cross-section of the excitation to the 2s state.

Indeed, for the usual hydrogen atoms, the contribution to the excitation cross-section $\sigma_{2s,usual}$ originates not only from the direct transition of 1s–2s, but also from numerous cascade transitions via the intermediate states with a higher angular momentum. In distinction, for the SFHA, the contribution to the excitation cross-section $\sigma_{2s,SFHA}$ originates only from the direct 1–2s transition because there are no states with a higher angular momentum, so $\sigma_{2s,SFHA}$ should be significantly smaller than $\sigma_{2s,usual}$.

If α is the share of the SFHA in a hydrogen gas mixture, then the effective theoretical cross-section is

$$\sigma_{2s,eff} = \alpha\, \sigma_{2s,SFHA} + (1 - \alpha)\, \sigma_{2s,usual}. \quad (1)$$

The experimental cross-section observed by using the quenching technique is

$$\sigma_{2s,exp} = (1 - \alpha)\, \sigma_{2s,usual}. \quad (2)$$

Consequently, the ratio of the effective theoretical cross-section to the experimental cross-section is

$$\sigma_{2s,eff}/\sigma_{2s,exp} = 1 + [\alpha/(1 - \alpha)]\, \sigma_{2s,SFHA}/\sigma_{2s,usual}. \quad (3)$$

From Equation (3), the ratio of the share of the SFHA to the share of the usual hydrogen gas in the experimental mixture can be represented in the form

$$\alpha/(1 - \alpha) = [\sigma_{2s,eff}/\sigma_{2s,exp} - 1]\, [\sigma_{2s,usual}/\sigma_{2s,SFHA}]. \quad (4)$$

From the analysis in the preceding part of this paper, we found the first factor in the right side of Equation (4) to be

$$\sigma_{2s,eff}/\sigma_{2s,exp} - 1 \approx 0.2. \quad (5)$$

Now, let us estimate the second factor on the right side of Equation (3).

In a paper by Poet [14], the author provided analytical results for the excitation cross-section σ_{2s} for a model where the wave functions of the hydrogen states were spherically symmetric. In other words, the target was a hydrogen atom with only s-states. This means that the results obtained by Poet [14] are applicable to the SFHA.

In a paper by Bhatia [9], the author compared his calculations of the excitation cross-section $\sigma_{2s,usual}$, which was obtained with the variational polarized orbital method, with the corresponding results from Poet [14], that is, with $\sigma_{2s,SFHA}$. It can be seen that, for the values of the energy (of the incoming electrons) closest to the experimental range of the energies from [18], the $\sigma_{2s,usual}/\sigma_{2s,SFHA}$ ratio was about 4. Consequently, from Equation (4), the ratio of the share of the SFHA to the share of the usual hydrogen gas in the experimental mixture can be estimated as

$$\alpha/(1 - \alpha) \approx 0.8. \quad (6)$$

In other words, in the hydrogen gas used in the experiments discussed by Callaway and McDowell [18], the SFHA and the usual hydrogen atoms were represented by about equal shares. Thus, our results constitute the third proof from atomic experiments (this time, from experiments on the excitation of the $n = 2$ states of atomic hydrogen due to electron impact) that the SFHA does exist.

We note that the $\sigma_{2s,usual}/\sigma_{2s,SFHA}$ ratio grows as the incident electron energy increases, as can be seen in Table 1 of Bhatia's paper [9]. Therefore, one can expect that, as the incident electron energy increases, the experimental σ_{2s}/σ_{2p} ratio should become closer to unity when keeping the product on the right side of Equation (4) approximately constant, since it is reasonable to expect that the ratio of the share of the SFHA to the share of the usual hydrogen gas in the experimental mixture is independent of the incoming electron energy. As the experimental σ_{2s}/σ_{2p} ratio would become closer to unity, it would fall within the experimental error margins. Thus, to determine the presence and the share of the SFHA in an experimental gas mixture, one should not use an electron beam with too high of an energy.

The range of energies below the ionization threshold is also not favorable for determining the presence and the share of the SFHA in experimental gas mixtures. This is because, in this range, the excitation cross-sections σ_{2s} and σ_{2p} are strongly dominated by so-called "resonances", which are temporary negative hydrogen ions consisting of a proton and two highly correlated electrons—see, e.g., the paper by Williams [21] and the references therein. However, there is no "second flavor" of a negative hydrogen ion. This is because there is no singular solution (outside the proton) for two highly correlated electrons, in distinction to the singular solution (outside the proton) of the Dirac equation for hydrogen atoms, the solution representing the SFHA. Therefore, in this range of energies, the SFHA practically does not contribute to both σ_{2s} and σ_{2p}.

Thus, the range of incoming electron energies that is most favorable for determining the presence and share of the SFHA in an experimental gas mixture seems to be the range presented in Table 1.

3. Conclusions

For the excitation of the $n = 2$ states of hydrogen atoms due to electron impact, we compared the experimental and theoretical ratios of the cross-sections σ_{2s}/σ_{2p}. We found that this theoretical ratio is systematically higher than the experimental ratio by about 20% (far beyond the experimental error margins) over the entire range of the energies of the incoming electrons used in the experiment by Callaway and McDowell [18].

We suggested that this discrepancy can be explained by the presence of the SFHA in the experimental hydrogen gas. This explanation is based on the fact that, in the experiments, the cross-section σ_{2s} was determined by using the quenching technique—by applying an electric field that mixed the 2s and 2p states, followed by the emission of the Lyman-alpha line from the 2p state. However, the SFHA only has s-states, so the quenching technique would not count the excitation of the SFHA in the 2s state and, thus, lead to the underestimation of the experimental cross-section σ_{2s}.

We estimated the share of the SFHA in an experimental hydrogen gas required to eliminate the above discrepancy and found this share to be about the same as the share of the usual hydrogen atoms. Thus, our results constitute the third proof from atomic experiments that the SFHA does exist—this time, from experiments on the excitation of the $n = 2$ states of atomic hydrogen due to electron impact. This is also important because the SFHA is the leading candidate for dark matter (or at least for a part of it).

Funding: This research received no external funding.

Data Availability Statement: All data is included in the paper.

Conflicts of Interest: The author declares no conflict of interest.

References

1. Oks, E. High-Energy Tail of the Linear Momentum Distribution in the Ground State of Hydrogen Atoms or Hydrogen—Like Ions. *J. Phys. B At. Mol. Opt. Phys.* **2001**, *34*, 2235–2243. [CrossRef]
2. Oks, E. Analysis of Experimental Cross-Sections of Charge Exchange between Hydrogen Atoms and Protons Yields another Evidence of the Existence of the Second Flavor of Hydrogen Atoms. *Foundations* **2021**, *1*, 265–270. [CrossRef]
3. Bowman, J.D.; Rogers, A.E.E.; Monsalve, R.A.; Mozdzen, T.J.; Mahesh, N. An Absorption Profile Centred at 78 Megahertz in the Sky-Averaged Spectrum. *Nature* **2018**, *555*, 67–70. [CrossRef] [PubMed]
4. Barkana, R. Possible Interaction between Baryons and Dark-Matter Particles Revealed by the First Stars. *Nature* **2018**, *555*, 71–74. [CrossRef]
5. Oks, E. Alternative Kind of Hydrogen Atoms as a Possible Explanation of the Latest Puzzling Observation of the 21 cm Radio Line from the Early Universe. *Res. Astrophys.* **2020**, *20*, 109. [CrossRef]
6. Jeffrey, N.; Gatti, M.; Chang, C.; Whiteway, L.; Demirbozan, U.; Kovacs, A.; Pollina, G.; Bacon, D.; Hamaus, N.; Kacprzak, T.; et al. Dark Energy Survey Year 3 results: Curved-Sky Weak Lensing Mass Map Reconstruction. *Mon. Not. R. Astron. Soc.* **2021**, *505*, 4626–4645. [CrossRef]
7. Oks, E. DES Map Shows a Smoother Distribution of Matter than Expected: A Possible Explanation. *Res. Astron. Astrophys.* **2021**, *21*, 241–245. [CrossRef]
8. Bhatia, A.K. Excitation of the 2S State of Atomic Hydrogen by Electron Impact. *Atoms* **2018**, *6*, 7. [CrossRef]
9. Burke, P.G.; Schey, H.M.; Smith, K. Collisions of Slow Electrons and Positrons with Atomic Hydrogen. *Phys. Rev.* **1963**, *129*, 1258–1274. [CrossRef]
10. Callaway, J. Scattering of Electrons by Atomic Hydrogen at Intermediate Energies: Elastic Scattering and n = 2 Excitation from 12 to 54 eV. *Phys. Rev. A* **1985**, *32*, 775–783. [CrossRef]
11. Callaway, J.; Unnikrishnan, K.; Oza, D.H. Optical-Potential Study of Electron-Hydrogen Scattering at Intermediate Energies. *Phys. Rev. A* **1987**, *36*, 2576–2584. [CrossRef] [PubMed]
12. Kamali, M.Z.M.; Ratnavelu, K.; Zhou, Y. Electron Impact Excitation of 2p and 3p States of Hydrogen at Intermediate Energies. *Eur. Phys. J. D* **2008**, *46*, 267–279. [CrossRef]

13. Odgers, B.R.; Scott, M.P.; Burke, P.G. Analysis of the Resonant States of H^- Associated with the $N = 2$ and $N = 3$ Thresholds of Atomic Hydrogen. *J. Phys. B At. Mol. Opt. Phys.* **1995**, *28*, 2973–2984. [CrossRef]
14. Poet, R. The Exact Solution for a Simplified Model of Electron Scattering by Hydrogen Atoms. *J. Phys. B Atom. Mol. Phys.* **1978**, *11*, 3081–3094. [CrossRef]
15. Scott, M.P.; Scholz, T.T.; Walters, H.R.J.; Burke, P.G. Electron Scattering by Atomic Hydrogen at Intermediate Energies: I. Integrated Elastic, 1s-2s and 1s-2p Cross Sections. *J. Phys. B At. Mol. Opt. Phys.* **1989**, *22*, 3055–3077. [CrossRef]
16. Whelan, C.T.; McDowell, M.R.C.; Edmunds, P.W. Electron Impact Excitation of Atomic Hydrogen. *J. Phys. B At. Mol. Phys.* **1987**, *20*, 1587–1598. [CrossRef]
17. Williams, J.F.; Mikosza, A.G. Atomic Hydrogen Revisited: Electron Impact Excitation to H(2p) at 54 eV. *J. Phys. B At. Mol. Opt. Phys.* **2006**, *39*, 4113–4122. [CrossRef]
18. Callaway, J.; McDowell, M.R.C. *Comments At. Mol. Phys.* **1983**, *13*, 19.
19. Kauypila, W.E.; Ott, W.R.; Fite, W.L. Excitation of Atomic Hydrogen to the Metastable $2^2S_{1/2}$ State by Electron Impact. *Phys. Rev. A* **1970**, *1*, 1099–1108. [CrossRef]
20. Kosehmieder, H.; Raible, V.; Kleinpoppen, H. Resonance Structure in the Excitation Cross Section by Electron Impact of the 2s State in Atomic Hydrogen. *Phys. Rev. A* **1973**, *8*, 1365–1368. [CrossRef]
21. Williams, J.F. Resonance Structure in Inelastic Scattering of Electrons from Atomic Hydrogen. *J. Phys. B At. Mol. Opt. Phys.* **1998**, *21*, 2107–2116. [CrossRef]

foundations MDPI

Article

Conditions for Scalar and Electromagnetic Wave Pulses to Be "Strange" or Not

Peter Saari [1,2,*] and Ioannis M. Besieris [3]

1 Institute of Physics, University of Tartu, W. Ostwaldi 1, 50411 Tartu, Estonia
2 Estonian Academy of Sciences, Kohtu 6, 10130 Tallinn, Estonia
3 The Bradley Department of Electrical and Computer Engineering, Virginia Polytechnic Institute and State University, Blacksburg, VA 24060, USA; besieris@vt.edu
* Correspondence: peeter.saari@ut.ee

Abstract: Vector-valued electromagnetic waves for which the integral of the electric field over time is zero at every location in space were characterized as "usual" by Bessonov several decades ago. Otherwise, they were called "strange". Recently, Popov and Vinogradov studied conditions leading to usual waves using a spectral representation. Their main result is that pulses of finite energy in free space are usual and, consequently, bipolar. However, they do not exclude the possibility of the existence of finite-energy strange pulses, although quite exotic, in a vacuum. Our emphasis in this article is to examine what the relevant necessary and sufficient conditions are for usual and strange waves, particularly for scalar pulses. Illustrative examples are provided, including spherical symmetric collapsing pulses, propagation-invariant, and the so-called almost undistorted spatiotemporally localized waves. Finally, source-generated strange electromagnetic fields are reported.

Keywords: space-time couplings; spatiotemporal; ultrafast optics; bipolar pulses; few-cycle pulses; free-space wave equation; space-time wave packets; nondiffracting localized waves

Citation: Saari, P.; Besieris, I.M. Conditions for Scalar and Electromagnetic Wave Pulses to Be "Strange" or Not. *Foundations* **2022**, *2*, 199–208. https://doi.org/10.3390/foundations2010012

Academic Editor: Eugene Oks

Received: 26 December 2021
Accepted: 4 February 2022
Published: 7 February 2022

1. Introduction

Progress in the generation and application of femtosecond and attosecond electromagnetic (EM) pulses have stimulated theoretical research of few-cycle and near-cycle localized exact solutions of the free-space wave equation (see, e.g., [1,2] and references therein). A strong space-time coupling, inherent to such pulses in focusing, makes the classical well-known approximations inapplicable. Moreover, with the decrease in the duration of optical pulses, there is a growing need to go beyond the quasi-monochromatic limit, and spectral representations of fields lose their advantages over direct time-domain ones. There is an ultimate milestone on the road away from the narrow-band limit—where the temporal spectrum of the pulse extends to the frequency scale origin. This causes a merging of positive-frequency and negative-frequency Fourier components, and as a result, the concept of analytic signal with its Hilbert-transform-related real and imaginary parts remains the only sound alternative for determining the envelope of a pulse.

If the spectrum vanishes at the frequency $\omega = 0$, the pulse is bipolar. Several decades ago Bessonov [3] introduced the term "strange" for waves whose electric field does not obey the equality

$$\vec{S}_{\mathbf{E}}\left(\vec{r}\right) \equiv \int_{-\infty}^{\infty} \vec{E}(\vec{r},t)dt = 0 \tag{1}$$

and called "usual" all waves satisfying Equation (1) at every location $\vec{r}$. It is obvious that usual waves are necessarily bipolar.

In a recent paper [4], Popov and Vinogradov studied conditions leading to Equation (1) using a spectral representation. Their main result is that pulses of finite energy in free space are usual and, consequently, bipolar. However, they do not exclude the possibility of the existence of finite-energy strange pulses, although quite exotic, in a vacuum. In other words, they have not proved that the finiteness of pulse energy is a sufficient condition for a pulse to be usual (not strange). Of course, it is not a necessary condition, as a bipolar plane wave pulse can already readily be a usual one.

We will neither question the results of the paper [4] nor prove the sufficiency of the energy condition. Instead, our aims here are to analyze what the relevant necessary and sufficient conditions are and to study the problem with several sample waves.

Specifically, we first consider spherically symmetric pulses converging to and thereafter, at positive times, diverging from a focus at the origin, i.e., pulses described by a wavefunction

$$\psi\left(\vec{r},t\right) = \frac{1}{r}[f(t+r/c) - f(t-r/c)], \tag{2}$$

where $f(s)$ is an arbitrary nonsingular function depending on the spherical radial coordinate $r = \sqrt{x^2+y^2+z^2}$ and time t, with c being the speed of light in a vacuum.

Second, we consider axially symmetric propagation-invariant pulses [5–10] whose wavefunction and/or energy density depend on the axial coordinate and time solely through the propagation variable $\zeta = z - vt$, where v is the group velocity of the pulse. These fields exhibit pronounced space-time coupling. From a theoretical point of view it is interesting that the second type of pulses can be obtained from spherically symmetric ones via relativistic boosts and/or complexifying the axis z [5,6,8]. Finally, we study some exotic pulses described by sophisticated solutions of the wave equation.

2. Methods

We primarily analyze scalar waves for the following reasons:

1. Every component of the electric (and magnetic) field is a scalar-valued field that obeys the wave equation. Hence, in order to judge for a chosen wavefunction $\psi\left(\vec{r},t\right)$ whether the corresponding EM pulse is strange or not, it is sufficient to evaluate the integral

$$S_\psi\left(\vec{r}\right) = \int_{-\infty}^{\infty} \psi(\vec{r},t)dt. \tag{3}$$

2. If EM field vectors are derived by the standard procedure of constructing the magnetic vector potential or the Hertz vector from $\psi\left(\vec{r},t\right)$, even simple expressions of $\psi\left(\vec{r},t\right)$ may result in too cumbersome ones for the EM field vectors, and the integral of Equation (1) may be difficult to evaluate. Moreover, as the procedure involves taking derivatives with respect to time and/or spatial coordinates, a strange $\psi\left(\vec{r},t\right)$, i.e., one with property $S_\psi\left(\vec{r}\right) \neq 0$, generally results in a usual EM field, i.e., $\vec{S}_E\left(\vec{r}\right) = 0$.
3. The notion of strangeness also applies to wave fields that are scalar valued by their physical nature, e.g., sound waves.

It should be pointed out that our approach does not mean resorting to a scalar approximation for EM fields.

In order to determine whether an electric field is strange or not, one can avoid integration according to Equation (1), which in most cases is a computationally difficult task. Instead, one can make use of the expression $\vec{E}\left(\vec{r},t\right) = -\partial\vec{A}\left(\vec{r},t\right)/\partial t$ of the electric field derived from a magnetic vector potential in the Coulomb gauge and consider the vector potential at $t = \pm\infty$ instead of evaluation the integral of the electric field. It is convenient to derive the vector potential from a scalar wavefunction as $\vec{A}\left(\vec{r},t\right) = \nabla \times \left[\psi\left(\vec{r},t\right)\vec{a}_z\right]$, where $\vec{a}_z$ is the unit vector along the z-axis. In the case of cylindrical symmetry, only the azimuthal component remains, and it takes the simple form $A_\varphi(\rho,z,t) = -\partial\psi(\rho,z,t)/\partial\rho$.

In the case of spherical symmetry, the unit vector $\vec{a}_z$ needs to first be expressed in a spherical coordinate system, and then for the curl, only the azimuthal component remains, taking a similar simple form $A_\varphi(r,\theta,t) = -\sin\theta\partial\psi(r,\theta,t)/\partial r$. Hence, if one of the following equalities is fulfilled at every location, in the case of cylindrical or spherical symmetry, respectively,

$$\lim_{t\to\infty}\frac{\partial}{\partial\rho}[\psi(\rho,z,t)-\psi(\rho,z,-t)]=0,\lim_{t\to\infty}\frac{\partial}{\partial r}[\psi(r,\theta,t)-\psi(r,\theta,-t)]\sin\theta=0, \tag{4}$$

the corresponding electric field is not strange. Equation (4) has been successfully applied for the examples in Section 3.

2.1. Evaluation of the Wave Pulse Energy

The (total) energy of a physically realizable pulse is a time-independent spatial integral (over the whole space) of the energy density, which in the case of an EM wave is given by the well-known expression with squares of strengths of electric and magnetic fields. Energy density of a scalar field is frequently defined as the square of the wavefunction (or modulus squared for complex-valued fields). However, the spatial integral of the latter will not be used below for establishing whether the EM pulse corresponding to a given wavefunction has finite or infinite energy.

For scalar-valued wave fields, another definition of the energy density exists [11], which is consistent with the energy conservation law and the Poynting theorem. It is given by Equation (5) below. We will call the spatial integral of W the *Mandel-Wolf total energy* for brevity. Conditions in the spectral domain—analogous to those in [4]—for a scalar field to have finite energy and be usual are discussed in Appendix A.

$$W=\frac{1}{2}\left[\frac{1}{c^2}\left(\frac{\partial}{\partial t}\psi\right)^2+\nabla\psi\cdot\nabla\psi\right]. \tag{5}$$

In order to establish whether a chosen wavefunction gives a pulse of finite energy or not, we used two different packages of scientific calculation for symbolic integration—or, if it turned out to be impossible—numerical integration.

2.2. Time-Domain Representation

As an alternative to using the Fourier expansion of the field as done in [4], for the constituents of the field one may take the singular propagator $D(r,t)$ (sometimes called the Riemann or Schwinger function)

$$D(r,t)=\frac{1}{4\pi rc^2}[\delta(r/c-t)-\delta(r/c+t)]\equiv G_+(r,t)-G_-(r,t), \tag{6}$$

where $\delta(.)$ denotes the Dirac delta and $G_\pm$ are the causal (retarded) and anticausal (advanced) Green functions, respectively. The function $D(r,t)$ represents a spherically symmetric delta-shaped pulse wave, first (at negative times t) collapsing to the origin (the right term) and then (at positive times t) expanding from it.

With this propagator as an elementary constituent, any solution to the three-dimensional homogeneous wave equation can be expressed as the following convolution integral over the whole 3D space:

$$\psi\left(\vec{r},t\right)=\int\left[D(R,t)h\left(\vec{r}'\right)+\frac{\partial}{\partial t}D(R,t)g\left(\vec{r}'\right)\right]d\vec{r}'. \tag{7}$$

Here, $R=\left|\vec{r}-\vec{r}'\right|$ and the distributions $g(.)$ and $h(.)$ are determined by the initial conditions—the field "snapshot" at the time origin moment $g\left(\vec{r}\right)=\psi\left(\vec{r},0\right)$ and

$h(\vec{r}) = \partial/\partial t\, \psi(\vec{r},t)\Big|_{t=0}$. However, unlike the solution of a radiation problem, since D contains not only the retarded Green function but the advanced one as well, g and h describe a distribution of fictitious Huygens-type sources, i.e., sources coupled with sinks of the same strength.

3. Results

3.1. General Conditions for a Pulse to Be "Usual"

3.1.1. Sufficient Conditions

With the help of Equation (7), we can easily find conditions under which the field is not "strange", i.e., conditions of vanishing of the time integral in Equation (3), or—keeping in view the text before it—also conditions for fulfillment of Equation (1). Since $D(R,t)$ is an odd function with respect to time, the first term of the integrand in Equation (7) in any case does not contribute to $S_\psi(\vec{r})$. As the δ-function equals zero at infinity, the time integral of the second term vanishes if $R = \left|\vec{r} - \vec{r}'\right|$ remains finite in the spatial integration. Hence, any wave field that is spatially bounded at a certain instant of time (when the pulse is focused) cannot be "strange". Any such pulse, if it has no unphysical singularities, has finite energy irrespective of which of the abovementioned definitions of the pulse energy has been taken. Hence, the sufficient condition formulated here is consistent with the one found by Popov and Vinogradov [4]. The second term also vanishes if $g(\vec{r}) = \psi(\vec{r},0) = 0$ in the whole space. This condition means that any nontrivial $\psi(\vec{r},t)$ must be an odd function of time. Therefore, such oddness is another sufficient condition for a pulse to be "usual". The results of the next subsection demonstrate accordance with this condition.

3.1.2. Necessary and Sufficient Conditions

For a field to be usual (not strange), the integral in Equation (1) or (3) must vanish everywhere. Therefore, we obtain the necessary condition that it must vanish at the origin $\vec{r} = 0$. In this case, we write $R = \left|\vec{r}'\right| \equiv r$, omitting the prime for simplicity. After integration from $t = -T$ to $t = T$ and subsequently taking the limit $T \to \infty$ in accordance with Equation (3), the second term in Equation (7) can be transformed in spherical coordinates as follows:

$$\begin{aligned} &\left[\int D(R,t) g(r,\theta,\varphi) R^2 \sin\theta dR d\theta d\varphi\right]_{-T}^{T} = \\ &= \frac{1}{4\pi c^2}\left\{\int r^{-1}\begin{bmatrix} \delta(r/c-T) - \delta(r/c+T) - \\ -\delta(r/c+T) + \delta(r/c-T) \end{bmatrix} g(r,\theta,\varphi) r^2 \sin\theta dr d\theta d\varphi\right\} = \\ &= 2\frac{1}{4\pi}\int g(cT,\theta,\varphi) T \sin\theta d\theta d\varphi = 2T\langle g\rangle_{cT}. \end{aligned}$$

Here, g_{cT} denotes the average value of the function g on the surface of a sphere with radius cT and center at the origin. Therefore, a necessary condition for a wave field to be usual is $\lim_{T\to\infty} T\langle g\rangle_{cT} = 0$.

Generally, this means that $g(r,\theta,\varphi)$ must asymptotically vanish as $r \to \infty$ faster than $1/r$. Of course, the surface average might be zero irrespective to such asymptotic behavior of $g(r,\theta,\varphi)$ if the latter is bipolar on the surface, and due to some symmetry the regions of opposite sign cancel each other. However, if applying the condition for a sphere whose center is shifted from the origin, the symmetry would disappear and the condition would be fulfilled due to only the aforementioned asymptotic behavior of $g(r,\theta,\varphi)$.

The transformation of the integral carried out above remains valid for an arbitrary point $\vec{r} \neq 0$ in Equation (7). The reason for this is that whatever the point in the field is, its radial vector is fixed while $|\vec{r}|$ runs together with cT to infinity. Therefore, $\vec{r}$ can be neglected in the expression $R = \left|\vec{r} - \vec{r}'\right|$. Consequently, the necessary and sufficient condition for a wavefield to be "usual" can be stated as follows. The wavefunction $g(r,\theta,\varphi)$

at $t = 0$ must vanish asymptotically as $r \to \infty$ faster than $1/r$; in other words, the following equality must be fulfilled:

$$\lim_{r\to\infty} r\psi(r,\theta,\varphi,0) = 0. \quad (8)$$

Application of the criterion in Equation (8) is especially appropriate when $\psi\left(\vec{r},t\right)$ contains multivalued complex functions and direct temporal integration according to Equation (3) may be hampered due to crossing the branch-cut lines.

3.2. Spherically Symmetric Pulses: Some Examples

Such pulses, the general form of the wavefunction of which is given by Equation (2), are the simplest to analyze. If $\psi\left(\vec{r},t\right)$ in the form of Equation (2) is used to construct a vector magnetic potential, simple relations between $f(s)$ and EM field vectors and EM pulse energy have been derived in [12] and references therein.

3.2.1. Even and Odd Lorentzians

We studied the functions $f_e(s) = 1/\left(1+s^{2n}\right)$ and $f_o(s) = s/\left(1+s^{2n}\right)$ (see Equation (2)) where $n = 1, 2, \ldots$. Results concerning the strangeness of such fields are as follows:

(a) If $f(s)$ is an even function, $\psi\left(\vec{r},t\right)$ is odd with respect to time (hence, automatically not strange) and $\vec{E}(\vec{r},t)$ is even, but nevertheless not strange. The magnetic field is odd and, hence, not strange.
(b) If $f(s)$ is odd, $\psi\left(\vec{r},t\right)$ is even with respect to time (but nevertheless not strange) and $\vec{E}(\vec{r},t)$ is odd, i.e., automatically is not strange. The magnetic field is even, but still not strange.

For these pulses the Mandel–Wolf energy and EM energy are both finite.

3.2.2. Error Function

The function $f(s) = \text{erf}(s)$ is odd, and consequently, $\psi\left(\vec{r},t\right)$ is even. However, it is strange and it is not square-integrable. Nevertheless, its Mandel–Wolf energy and EM energy are both finite. In accordance with point (b) above, $\vec{E}(\vec{r},t)$ is odd, i.e., automatically not strange. The magnetic field is an even function of time but still not strange.

3.3. Propagation-Invariant Pulses: Some Examples

3.3.1. Superluminal X-Waves

Inspired by the quotation "Therefore, the possibility of the existence of strange and unipolar pulses in a vacuum, although quite exotic, remains" from [4], we turned to the so-called X-waves, which were first intruduced in [5,13] and then studied in numerous papers; see [6–9] and references therein.

The so-called fundamental axisymmetric X-wave is given by

$$\psi(\rho,z,t) = \frac{a}{\sqrt{[a+i(z-vt)]^2+\rho^2\gamma^{-2}}}, \quad (9)$$

where $\gamma = \left(v^2/c^2-1\right)^{-1/2}$ is the superluminal version of the Lorentz factor, including the velocity $v > c$ of the pulse; a is a positive parameter that determines the width of the pulse; and $\rho = \sqrt{x^2+y^2}$. This wavefunction is commonly considered as a white-spectrum superposition of Bessel beams.

Referring to Equation (7), it can be also derived as Liénard–Wiechert potentials for Lorentzian distributions of fictitous "charges" (sources and sinks) flying with the constant velocity v along the axis z, i.e., the distributions being $a\left[(z-vt)^2+a^2\right]^{-1}\delta(x)\delta(y)$ (for the imaginary part of $\psi(.)$ and $(z-vt)\left[(z-vt)^2+a^2\right]^{-1}\delta(x)\delta(y)$ (for the real part) [14]. The real part of Equation (9) respresents a strange scalar field and the imaginary part an ordinary (usual) scalar field. At the origin, $\psi(0,0,t)=a/(a-ivt)$, the temporal spectrum of the real part is proportional to $\exp(-a|\omega|/v)$ while that of the imaginary part is proportional to $\mathrm{signum}(\omega)\exp(-a|\omega|/v)$. Hence, although both spectra have their maxima at infinitesimally small frequencies, the spectrum of the imaginary part lacks the constituent at $\omega=0$ exactly, which is in accordance with the established difference in strangeness of the real and imaginary parts.

If the EM vector fields are derived from Equation (9) by standard procedures involving the vector magnetic potential or the Hertz vector, then, due to taking derivatives in the course of these procedures, the EM field turns out to be usual.

A specific procedure to obtain an EM field avoiding the derivatives is to derive the following complex-valued Riemann–Silberstein vector

$$\begin{aligned}\vec{F}\left(\vec{r},t\right)&=\frac{ic^2\sqrt{(v/c)^2-1}}{v^2Q\left(\vec{r},t\right)}\frac{P_-\left(\vec{r},t\right)}{P_+\left(\vec{r},t\right)}\vec{a}_x-\frac{c\sqrt{(v/c)^2-1}}{vQ\left(\vec{r},t\right)}\frac{P_-\left(\vec{r},t\right)}{P_+\left(\vec{r},t\right)}\vec{a}_x+c^2\frac{(v/c)^2-1}{v^2Q\left(\vec{r},t\right)}\vec{a}_z;\\ Q\left(\vec{r},t\right)&=\sqrt{\left((v/c)^2-1\right)\rho^2+(a+i(z-vt))^2};\\ P_\pm\left(\vec{r},t\right)&=\sqrt{1\pm\frac{a+i(z-vt)}{Q\left(\vec{r},t\right)}}\end{aligned}\tag{10}$$

which arises from a superposition of vector-valued Bessel beams. It can be expressed as $\vec{F}\left(\vec{r},t\right)=\sqrt{\varepsilon_0/2}\left(\vec{E}+ic\vec{B}\right)$, where ε_0 is the permittivity of free space and $\vec{E}$ and $\vec{B}$ are real fields obeying the homogeneous Maxwell equations. The z-component of $\vec{F}\left(\vec{r},t\right)$ is essentially the infinite-energy superluminal fundamental X wave, as one can see by inspection of Equations (9) and (10). Hence, the real part of the electric field is a strange field, whereas the magnetic field is a usual field. The EM wave pulse energy is infinite quite analogously to the case of plane waves. This is understandable because the X wave can be thought of as a superposition of plane wave pulses directed along a conical surface.

3.3.2. Subluminal Arctan-Wave

The expression in Reference [6]

$$\begin{gathered}\psi\left(\vec{r},t\right)=\frac{1}{\sqrt{\rho^2+\gamma^2(z-vt)^2}}\tan^{-1}\left(\frac{\sqrt{\rho^2+\gamma^2(z-vt)^2}}{a+i\gamma(v/c)\left(z-\frac{c^2}{v}t\right)}\right);\\ \gamma=1/\sqrt{1-(v/c)^2},\ v<c,\end{gathered}\tag{11}$$

where again ρ is the polar radial coordinate, is a subluminal localized wave that is relatively undistorted upon propagation depending on the value of the positive free parameter a. The real part of $\psi\left(\vec{r},t\right)$ has a finite Mandel–Wolf total energy and is a strange field; however, its imaginary part is normal (usual). Additionally, the finite-energy corresponding electromagnetic fields, constructed within the framework of either a Coulomb gauge or a vector Hertz potential, are usual fields.

3.3.3. Luminal Localized Wave

In cylindrical coordinates, the azimuthally symmetric expression

$$\psi\left(\vec{r},t\right)=\frac{1}{\sqrt{4b^2\rho^2+\left[-b^2+(a_1+i\varsigma)(a_2-i\eta)+\rho^2\right]^2}},\tag{12}$$

where $\zeta=z-ct$, $\eta=z+ct$ are the characteristic variables of the one-dimensional scalar wave equation in vacuum and $a_{1,2}$ and b are positive free parameters, is a spatiotemporally localized extended splash-mode nonsingular solution to the (3 + 1)-dimensional scalar wave equation under the condition $a_1 a_2 - b^2 > 0$. It can be derived from a superposition of Bessel–Gauss focus wave modes (FWM). For $b = 0$, it reduces to the ordinary first-order splash-mode first derived by Ziolkowski (see, e.g., [5,6]); the latter is not strange. The scalar wave field in Equation (12) is not strange, and the Wolf–Mandel total energy of its real part is finite. The electric and magnetic fields arising from a vector potential $\vec{A}\left(\vec{r},t\right) = \nabla \times \left[\psi\left(\vec{r},t\right)\vec{a}_z\right]$ within the framework of a Coulomb gauge have been examined. The electric and magnetic fields are usual. The total electromagnetic energy is finite.

The reason why the "strangeness" integral of Equation (3) turns out to be zero can be understood from Figure 1.

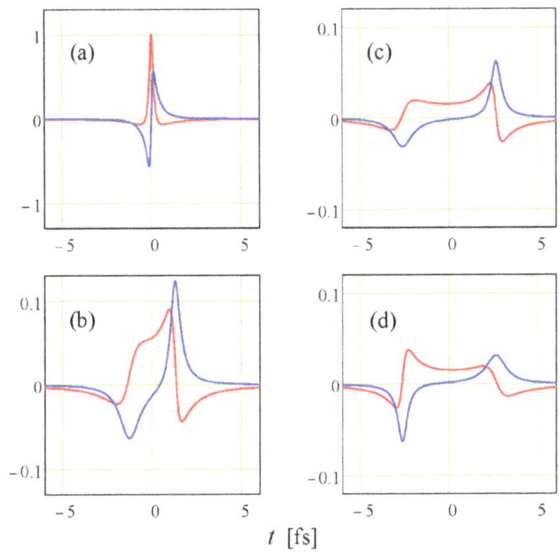

Figure 1. Dependencies of the real part (red) and imaginary part (blue) of the wavefunction in Equation (12) on time (for specificity—in the optical femtosecond domain) along the propagation axis (where $\rho = 0$). Ordinate scales are normalized to $Re\psi(0,0)$ (see (**a**)), but notice the change of the scale in (**b**–**d**). Spatial locations: (**a**) $z = 0$ µm; (**b**) $z = 0.4$ µm; (**c**) $z = 0.8$ µm; (**d**) $z = -0.8$ µm. Values of the parameters: $a_1 = 0.1$ µm, $a_2 = 0.2$ µm, $b = 0.1$ µm.

At first glance, the real part is unipolar (à la that of the X wave); i.e., the real field seems to be strange. However, closer inspection of Figure 1a indicates that the peak appears on the negative-polarity background, which makes the area under the curve equal to zero, as is the case with the imaginary part. Plots (a) and (b) show that outside the origin, the pulse splits into two counterpropagating ones. Comparison of plots (c) and (d) shows that the real part of the wavefunction is even with respect to simultaneous inversion of the sign of the variables z, t and the imaginary part is odd with respect to the same inversion. It should be mentioned that neither numerical computation of the "strangeness" integral Equation (3) nor plotting of the wavefunction can be properly accomplished by straightforward application of Equation (12) due to presence of a brach cut in the square root function of a complex-valued argument.

3.4. Strange Fields Generated by Sources

3.4.1. Bonnor Fields

In cylindrical coordinates, we identify two regions of space: $\rho \geq a$ (outside) and $\rho \leq a$ (inside). Motivated by Bonnor's work [15], we specify scalar and vector potentials in the two regions:

$$\begin{aligned} &\Phi_0(\rho,\phi,z,t) = e^{-(t-z)^2}\cos\phi/\rho,\ A_0(\rho,\phi,z,t) = \Phi_0(\rho,\phi,z,t)\vec{a}_z; \\ &\Phi_i(\rho,\phi,z,t) = e^{-(t-z)^2}\left(\tfrac{4\rho^2}{a^3} - \tfrac{3\rho^3}{a^4}\right)\cos\phi,\ A_i(\rho,\phi,z,t) = \Phi_i(\rho,\phi,z,t)\vec{a}_z. \end{aligned} \tag{13}$$

Here, normalization with the speed of light in a vacuum equal to unity has been used. These potential fields satisfy the Lorentz condition. Additionally, the continuity $\Phi_0(a,\phi,z,t) = \Phi_i(a,\phi,z,t)$ should be noted.

In the region $\rho \leq a$, the electric volume charge and current densities are determined as follows:

$$\begin{aligned} \rho_{vi} &= -\left(\nabla^2 - \tfrac{\partial^2}{\partial t^2}\right)\Phi_i = -\tfrac{12e^{-(t-z)^2}(a-2\rho)\cos\phi}{a^4}, \\ \vec{J}_{vi} &= -\left(\nabla^2 - \tfrac{\partial^2}{\partial t^2}\right)\vec{A}_i = \rho_{vi}\vec{a}_z. \end{aligned} \tag{14}$$

The total charge in this region equals zero. No charges exist for $\rho \geq a$.

The electric and magnetic fields in the two regions are given by the expressions

$$\begin{aligned} \vec{E}_0 &= -\nabla\Phi_0 - \partial\vec{A}_0/\partial t = \tfrac{e^{-(t-z)^2}\cos\phi}{\rho^2}\vec{a}_\rho + \tfrac{e^{-(t-z)^2}\sin\phi}{\rho^2}\vec{a}_\phi; \\ \vec{B}_0 &= \nabla\times\vec{A}_0 = -\tfrac{e^{-(t-z)^2}\sin\phi}{\rho^2}\vec{a}_\rho + \tfrac{e^{-(t-z)^2}\cos\phi}{\rho^2}\vec{a}_\phi; \\ \vec{E}_i &= \tfrac{e^{-(t-z)^2}\rho(-8a+9\rho)\cos\phi}{a^4}\vec{a}_\rho + \tfrac{e^{-(t-z)^2}\rho(4a-3\rho)\sin\phi}{a^4}\vec{a}_\phi; \\ \vec{B}_i &= -\tfrac{e^{-(t-z)^2}\rho(4a-3\rho)\sin\phi}{a^4}\vec{a}_\rho - \tfrac{e^{-(t-z)^2}\rho(8a-9\rho)\cos\phi}{a^4}\vec{a}_\phi; \end{aligned} \tag{15}$$

These are transverse electromagnetic (TEM) structures propagating along the z-direction with the normalized speed of light in vacuum. Both the electric and magnetic fields are strange. The total energy associated with these fields is finite.

3.4.2. Single-Cycle Dipole Electromagnetic Fields

Wang et al. [16] have derived single-cylce electromagnetic fields generated by an oscillating elecric dipole oriented along the x-direction. We examined these fields for "strangeness". The electric field is strange but not the magnetic field.

4. Conclusions

Motivated by the recent work of Popov and Vinogradov [4], we have examined the relevant conditions for usual and strange waves, particularly for scalar pulses. General necessary and sufficient conditions for scalar and electromagnetic field pulses possessing spatiotemporal coupling to be "usual" or "strange" have been derived in the time domain, and their application has been illustrated. For various examples of luminal and rather peculiar subluminal and superluminal scalar waves, it has been shown that even if they are strange, the corresponding finite-energy electromagnetic fields constructed from the scalar fields from either a magnetic potential within the framework of a Coulomb gauge or a vector Hertz potential are usual. Finally, source-generated strange electromagnetic fields have been reported.

Author Contributions: Investigation, P.S. and I.M.B.; writing—original draft preparation, P.S. and I.M.B.; writing—review and editing, P.S. and I.M.B. All authors have read and agreed to the published version of the manuscript.

Funding: This research received no external funding.

Institutional Review Board Statement: Not applicable.

Informed Consent Statement: Not applicable.

Data Availability Statement: Not applicable.

Conflicts of Interest: The authors declare no conflict of interest.

Appendix A

Let $\psi\left(\vec{r},t\right)$ be a real function satisfying the homogeneous scalar wave equation in free space. The energy transport equation for such a wavefunction is given by

$$\nabla\cdot\vec{S}+\frac{\partial}{\partial t}W=0, \tag{A1}$$

where

$$W=\frac{1}{2}\left[\frac{1}{c^2}\left(\frac{\partial}{\partial t}\psi\right)^2+\nabla\psi\cdot\nabla\psi\right],\ \vec{S}=-\frac{\partial}{\partial t}\psi\,\nabla\psi \tag{A2}$$

are, respectively, the real energy density in units J/m^3 and the real energy flow vector (W/m^2).

Let the real field $\psi\left(\vec{r},t\right)$ be expressed in terms of its spatial Fourier spectrum as follows:

$$\psi\left(\vec{r},t\right)=\frac{1}{(2\pi)^3}\int d\vec{k}\,\Psi\left(\vec{k},t\right)e^{i\vec{k}\cdot\vec{r}}. \tag{A3}$$

Then, the total energy density can be written as follows:

$$W_{total}=\int d\vec{r}\;W\left(\vec{r},t\right)=\frac{1}{2}\frac{1}{(2\pi)^3}\int d\vec{k}\left[\frac{1}{c^2}\dot{\Psi}\left(\vec{k},t\right)\dot{\Psi}^*\left(\vec{k},t\right)+k^2\Psi\left(\vec{k},t\right)\Psi^*\left(\vec{k},t\right)\right], \tag{A4}$$

where the dot indicates differentiation with respect of time. Bearing in mind the dispersion relationship $-k^2+\omega^2/c^2=0$, we assume, next, the form

$$\Psi\left(\vec{k},t\right)=\Psi_-\left(\vec{k}\right)e^{ikct}+\Psi_+\left(\vec{k}\right)e^{-ikct}. \tag{A5}$$

Then, we have

$$W_{total}=\frac{1}{2}\frac{1}{(2\pi)^3}\int d\vec{k}\,k^2\left[\left|\Psi_-\left(\vec{k}\right)\right|^2+\left|\Psi_+\left(\vec{k}\right)\right|^2\right]. \tag{A6}$$

Next, let the wavefunction $\psi\left(\vec{r},t\right)$ be defined as $\psi\left(\vec{r},t\right)=\partial\phi\left(\vec{r},t\right)/\partial t$ and consider the integral

$$\begin{aligned}S_\psi\left(\vec{r}\right)&=\int_{-\infty}^{\infty}\psi\left(\vec{r},t\right)\,dt=Lim|_{T\to\infty}\int_{-T}^{T}\psi\left(\vec{r},t\right)\,dt\\&=Lim|_{T\to\infty}\left[\phi\left(\vec{r},T\right)-\phi\left(\vec{r},-T\right)\right]\end{aligned} \tag{A7}$$

The wavefunction $\phi\left(\vec{r},T\right)$ is expressed as

$$\phi\left(\vec{r},T\right)=\int d\vec{k}\,e^{i\vec{k}\cdot\vec{r}}\left[\Phi_-\left(\vec{k}\right)e^{ikcT}+\Phi_+\left(\vec{k}\right)e^{-ikcT}\right]. \tag{A8}$$

According to the Riemann–Lebesgue theorem, $S_\psi\left(\vec{r}\right)$ vanishes provided that

$$\int_0^\infty dk\,k^2\left|\int d\widehat{\Omega}\,\Phi_\pm\left(\vec{k}\right)e^{i\vec{k}\cdot\vec{r}}\right| \tag{A9}$$

converges. However, $\Phi_{\pm}\left(\vec{k}\right) = \mp\Phi_{\pm}\left(\vec{k}\right)/ck$. Therefore, $S_{\psi}\left(\vec{r}\right)$ vanishes and the field is usual provided that the integral

$$\int_0^{\infty} dk\, k \left| \int d\widehat{\Omega}\, \Psi_{\pm}\left(\vec{k}\right) e^{i\vec{k}\cdot\vec{r}} \right| \tag{A10}$$

converges.

References

1. Lekner, J. *Theory of Electromagnetic Pulses*; Morgan & Claypool Publishers: San Rafael, CA, USA, 2018.
2. Kiselev, A.P. Localized Light Waves: Paraxial and Exact Solutions of the Wave Equation (a Review). *Opt. Spectrosc.* **2007**, *102*, 603–622. [CrossRef]
3. Bessonov, E.G. On a class of electromagnetic waves. *Zh. Eksp. Teor. Fiz.* **1981**, *80*, 852–858.
4. Popov, N.L.; Vinogradov, A.V. Free Space Strange and Unipolar EM Pulses: Yes or No? *Foundations* **2021**, *1*, 169–174. [CrossRef]
5. Ziolkowski, R.W.; Besieris, I.M.; Shaarawi, A.M. Aperture realizations of exact solutions to homogeneous-wave equations. *Opt. Soc. Am. A* **1993**, *10*, 75–87. [CrossRef]
6. Besieris, I.; Abdel-Rahman, M.; Shaarawi, A.; Chatzipetros, A. Two fundamental representations of localized pulse solutions to the scalar wave equation. *Prog. Electromagn. Res.* **1998**, *19*, 1–47. [CrossRef]
7. Salo, J.; Fagerholm, J.; Friberg, A.T.; Salomaa, M.M. Unified description of nondiffracting X and Y waves. *Phys. Rev. E* **2000**, *62*, 4261–4275. [CrossRef] [PubMed]
8. Saari, P.; Reivelt, K. Generation and classification of localized waves by Lorentz transformations in Fourier space. *Phys. Rev. E* **2004**, *69*, 036612. [CrossRef]
9. Hernandez-Figueroa, H.E.; Recami, E.; Zamboni-Rached, M. (Eds.) *Non-Diffracting Waves*; Wiley: New York, NY, USA, 2013.
10. Kondakci, H.E.; Abouraddy, A.F. Diffraction-free space-time light sheets. *Nat. Phot.* **2017**, *11*, 733–740. [CrossRef]
11. Mandel, L.; Wolf, E. *Optical Coherence and Quantum Optics*; Cambridge University Press: New York, NY, USA, 1995; p. 288.
12. Artyukov, I.A.; Dyachkov, N.V.; Feshchenko, R.M.; Vinogradov, A.V. Collapsing EM wave—A simple model for nonparaxial, quasi-monochromatic, single and half-cycle beams. *Phys. Scr.* **2020**, *95*, 064006. [CrossRef]
13. Lu, J.; Greenleaf, J.F. Nondiffracting X waves—Exact solutions to free-space scalar wave equation and their finite aperture realizations. *IEEE Trans. Ultrason. Ferroelectr. Freq. Control* **1992**, *39*, 19–31. [CrossRef] [PubMed]
14. Saari, P. Localized waves in femtosecond optics. In *Ultrafast Photonics*; Miller, A., Reid, D.T., Finlayson, D.M., Eds.; Institute of Physics Publishing: Bristol, UK, 2004; pp. 317–340.
15. Bonnor, W.B. Solutions of Maxwell's equations for charge moving at the speed of light. *Int. J. Theor. Phys.* **1969**, *2*, 373–379. [CrossRef]
16. Wang, Z.; Lin, Q.; Wang, Z. Single-cycle electromagnetic pulses produced by oscillating electric dipoles. *Phys. Rev. E* **2003**, *67*, 016503. [CrossRef] [PubMed]

foundations

MDPI

Article

Big Rip Scenario in Brans-Dicke Theory

Sasmita Kumari Pradhan [1,2], Sunil Kumar Tripathy [3,*], Zashmir Naik [1], Dipanjali Behera [3] and Mrutunjaya Bhuyan [4,*]

1 School of Physics, Sambalpur University, Jyoti Vihar, Sambalpur 768019, India; sasmita.gita91@gmail.com (S.K.P.); zashmir@gmail.com (Z.N.)
2 Department of Physics, Centurion University of Technology and Management, Bhubaneswar 751009, India
3 Department of Physics, Indira Gandhi Institute of Technology, Sarang, Dhenkanal 769146, India; dipadolly@rediffmail.com
4 Center for Theoretical and Computational Physics, Department of Physics, Faculty of Science, University of Malaya, Kuala Lumpur 50603, Malaysia
* Correspondence: tripathy_sunil@rediffmail.com (S.K.T.); bunuphy@um.edu.my (M.B.); Tel.: +91-8280295995 (S.K.T.); +60-1137057605 (M.B.)

Abstract: In this work, we present a Big Rip scenario within the framework of the generalized Brans-Dicke (GBD) theory. In the GBD theory, we consider an evolving BD parameter along with a self-interacting potential. An anisotropic background is considered to have a more general view of the cosmic expansion. The GBD theory with a cosmological constant is presented as an effective cosmic fluid within general relativity which favours a phantom field dominated phase. The model parameters are constrained so that the model provides reasonable estimates of the Hubble parameter and other recent observational aspects at the present epoch. The dynamical aspects of the BD parameter and the BD scalar field have been analysed. It is found that the present model witnesses a finite time doomsday at a time of $t_{BR} \simeq 16.14$ Gyr, and for this scenario, the model requires a large negative value of the Brans-Dicke parameter.

Keywords: cosmological constant; generalised Brans-Dicke theory; Big Rip

Citation: Pradhan S.K.; Tripathy, S.K.; Naik, Z.; Behera, D.; Bhuyan, M. Big Rip Scenario in Brans-Dicke Theory. *Foundations* **2022**, 2, 128–139. https://doi.org/10.3390/foundations2010007

Academic Editor: Eugene Oks

Received: 16 December 2021
Accepted: 11 January 2022
Published: 17 January 2022

Publisher's Note: MDPI stays neutral with regard to jurisdictional claims in published maps and institutional affiliations.

1. Introduction

Late-time cosmic acceleration is one of the most bizarre and unsolved problems in modern cosmology. In scalar field cosmological models, the late-time cosmic acceleration issue is predominantly attributed to an exotic dark energy (DE) form that corresponds to a cosmic fluid having low energy density, as well as negative pressure. This is usually understood through a quantity dubbed as the equation of state (EoS) parameter $\omega_D = \frac{p}{\rho}$, where p represents the DE pressure and ρ symbolises the dark energy density. The dark energy with a negative pressure corresponds to a negative EoS parameter. Despite several attempts made by astronomers and cosmologists, the experimental determination of ω_D remains challenging. Its precise estimation at the present epoch along with the knowledge of its development over a long period may unravel the mystery of the dark energy whose nature and origin remains speculative so far. In the ΛCDM model, the cosmological constant Λ with $\omega_D = 1$ plays the role of dark energy. However, in canonical scalar field models, quintessence fields or phantom fields shoulder the burden for the late-time cosmic speed-up, while the EoS parameter for the quintessence field lies in the range $-\frac{2}{3} \leq \omega_D \leq -\frac{1}{3}$ [1–3], which for the phantom fields, becomes $\omega_D < -1$ [4]. However, the EoS parameter as constrained from recent observational data favours a phantom phase in the Universe with $\omega_D < -1$ [5], while constraints from the CMB data in the nine-year WMAP survey suggest that $\omega_D = -1.073^{+0.090}_{-0.089}$ [6], a combination of the CMB data with Supernova data, predicts $\omega_D = -1.084 \pm 0.063$ [7]. Other constraints on the EoS parameter include $\omega_D = -1.035^{+0.055}_{-0.059}$ from Supernova cosmology project [8], $\omega_D = -1.03 \pm 0.03$ from recent Planck 2018 results [9] and from Pantheon data $\omega_D = -1.006 \pm 0.04$ [10]. In phantom dark energy models, the energy conditions are usually violated, and the

Universe may witness a blowing-up of the curvature of space-time at a finite time, leading to the dissolution of the whole material Universe into pieces. This picture of the finite future of the Universe, dubbed as the Big Rip singularity, concerns the recent cosmological research [11,12]. The finite-time future singularity leads to inconsistencies which led to different proposals in recent times, including the quantum effects to delay singularity, possible gravity modification, or the coupling of dark energy and dark matter [13,14].

In recent times, because of the concern regarding the ultimate fate of the Universe for phantom accreted dark energy with $\omega_D < -1$, a number of cosmological models have been used to investigate the Rip cosmologies based on the general relativity (GR) and modified gravity. Darbowski et al. studied the Big Rip singularity and the ultimate cosmic fate [15,16], and Granda and Loaiza showed the occurrence of Big Rip for kinetic and Gauss-Bonnet coupling [17]. The classical and quantum fate of the Big Rip cosmology has been studied by Vasilev et al. [18]. Within the framework of $f(T)$ gravity, Hanafy and Saridakis presented a cosmological model where the Universe may last forever in a Pseudo Rip scenario [19]. Recently, Ray et al. studied the Big Rip and some Pseudo Rip cosmological models in the context of an extended gravity theory [20], and Pati et al. investigated the possible occurrence of Rip scenarios in an extended symmetric teleparallel theory [21]. Big Rip singularity is a unique singularity that possibly occurs in a phantom scenario violating the Null Energy Condition. The occurrence of Big Rip singularity, for which the energy density and the scale factor of the Universe diverge, dissolutes the bound system and ultimately leads to the tearing up of the Universe in finite time. Such a scenario has become a major concern for cosmologists. As such, the present study is aimed at investigating the possibility of the occurrence of a Big Rip scenario within the framework of the generalised Brans-Dicke (GBD) theory. The GBD theory incorporates a self-interacting potential, as well as a dynamically varying Brans-Dicke(BD) parameter. Previously, Montenegro and Carneiro have investigated some cosmological models leading to Big Rip kinds of solutions within the Brans-Dicke theory in the presence of decaying vacuum density. In that work, they considered a time-independent negative BD parameter $\omega = -1$ [22]. They obtained cosmological solutions with a time-varying deceleration parameter which may lead to negative energy density. In general, the BD theory is a most popular modified gravity theory proposed as an alternative to GR, where the gravity is mediated by a scalar field. The BD theory has already been used emphatically in addressing many issues in cosmology and astrophysics including the explanation for the inflationary scenario [23]. Over a period of time, the BD theory has been well-studied through different tests, including the gravitational radiation from gravitational wave bursts [24–27].

The manuscript is presented as follows: In Section 2, the basic field equations for the GBD theory are estimated for an anisotropic LRS Bianchi I (LRSBI) metric. Additionally, the dynamics of the GBD theory when incorporating a cosmological constant is appraised. Section 3 is devoted to a Big Rip scenario through a scale factor that diverges at a specific time, and discusses the time evolution of the BD scalar field, BD parameter, and the self-interacting potential under the Big Rip scenario. Lastly, the conclusion is drawn, and a concise summary of the present investigation is given in Section 4. Throughout this work, we chose the natural unit system: $8\pi G_0 = c = 1$, where G_0 symbolises the Newtonian gravitational constant at the present epoch and c represents the speed of light in vacuum.

2. Basic Formalism

In a Jordan frame within the GBD theory with a self-interacting potential $V(\phi)$ and a time-dependent BD parameter $\omega(\phi)$, we have the action as [28,29]

$$S = \int d^4x\sqrt{-g}\left[\phi R - \frac{\omega(\phi)}{\phi}\phi^{,\mu}\phi_{,\mu} - V(\phi) + \mathcal{L}_m\right], \tag{1}$$

where R defines the curvature and $\mathcal{L}_m$ is the matter Lagrangian. The time-dependence aspect of the BD parameter emerges naturally in the Kaluza-Klein theory, string theory, or in the supergravity theory [30,31]. Different issues in cosmology have been investigated in recent times in the GBD framework with a time-dependent BD parameter [32–36]. Because of the GBD field equations may be obtained as [36,37]

$$G_{\mu\nu} = \frac{T_{\mu\nu}}{\phi} + \frac{\omega(\phi)}{\phi^2}\left[\phi_{,\mu}\phi_{,\nu} - \frac{1}{2}g_{\mu\nu}\phi_{,\alpha}\phi^{,\alpha}\right] + \frac{1}{\phi}\left[\phi_{,\mu;\nu} - g_{\mu\nu}\Box\phi\right] - \frac{V(\phi)}{2\phi}g_{\mu\nu}, \quad (2)$$

$$\Box\phi = \frac{T}{2\omega(\phi)+3} - \frac{2V(\phi) - \phi\frac{\partial V(\phi)}{\partial\phi}}{2\omega(\phi)+3} - \frac{\frac{\partial\omega(\phi)}{\partial\phi}\phi_{,\mu}\phi^{,\mu}}{2\omega(\phi)+3}, \quad (3)$$

where $\Box$ represents the d'Alembert operator. A perfect fluid distribution with the energy-momentum tensor $T_{\mu\nu} = (\rho + p)u_\mu u_\nu + pg_{\mu\nu}$ is considered, so that the $T = g_{\mu\nu}T^{\mu\nu}$ is the trace. In order to model the universe, we consider an LRSBI universe [38]

$$ds^2 = -dt^2 + A^2dx^2 + B^2(dy^2 + dz^2). \quad (4)$$

Even though the present observable Universe is mostly isotropic and homogeneous and can be mostly described as an FRW metric, some of the observations obviously hint of a possible departure from isotropy [39–42]. The amount of cosmic anisotropy present may be very small, but we cannot simply rule out its possibility. Additionally, the LRSBI model resembles the flat FRW model, but allowed us to incorporate a small but finite anistropy in the model. The GBD field equations for the anisotropic model become [32,36]

$$(2k+1)\xi^2H^2 = \frac{\rho}{\phi} + \frac{\omega(\phi)}{2}\left(\frac{\dot\phi}{\phi}\right)^2 - 3H\left(\frac{\dot\phi}{\phi}\right) + \frac{V(\phi)}{2\phi}, \quad (5)$$

$$2\xi\dot H + 3\xi^2H^2 = -\frac{p}{\rho} - \frac{\omega(\phi)}{2}\left(\frac{\dot\phi}{\phi}\right)^2 - 2\xi H\left(\frac{\dot\phi}{\phi}\right) - \frac{\ddot\phi}{\phi} + \frac{V(\phi)}{2\phi}, \quad (6)$$

$$(k+1)\xi\dot H + (k^2+k+1)\xi^2H^2 = -\frac{p}{\phi} - \frac{\omega(\phi)}{2}\left(\frac{\dot\phi}{\phi}\right)^2 - (k+1)\xi H\left(\frac{\dot\phi}{\phi}\right) - \frac{\ddot\phi}{\phi} + \frac{V(\phi)}{2\phi}. \quad (7)$$

In the above equations, $\xi = \frac{3}{k+2}$ is an anisotropic parameter where k is a positive constant that decides the relationship among the directional expansion rates: $\frac{\dot A}{A} = k\frac{\dot B}{B}$. One should note that the isotropic behaviour of the model may be obtained for $\xi = 1$. $H = \frac{1}{3}\left(\frac{\dot A}{A} + 2\frac{\dot B}{B}\right) = \frac{1}{\xi}\frac{\dot B}{B}$ is the Hubble parameter.

The BD scalar field satisfies the Klein-Gordon equation

$$\frac{\ddot\phi}{\phi} + 3H\frac{\dot\phi}{\phi} = \frac{\rho - 3p}{2\omega(\phi)+3} - \frac{\frac{\partial\omega(\phi)}{\partial\phi}\dot\phi^2}{2\omega(\phi)+3} - \frac{2V(\phi) - \phi\frac{\partial V(\phi)}{\partial\phi}}{2\omega(\phi)+3}. \quad (8)$$

The GBD theory may be recast as an effective GR picture by incorporating a cosmological constant Λ. In such a case, the total energy density and the total pressure respectively become $\rho_T = \rho + \Lambda$ and $p_T = p - \Lambda$, and the GBD field equations reduce to

$$(2k+1)\xi^2H^2 = \rho_T + \rho_\phi, \quad (9)$$

$$\left(\frac{k+3}{2}\right)\xi\dot H + \left(\frac{k^2+k+4}{2}\right)\xi^2H^2 = -(p_T + p_\phi), \quad (10)$$

where,

$$\rho_\phi = (2k+1)\xi^2H^2\Delta\phi - 3H\dot\phi + \frac{\omega(\phi)}{2}\left(\frac{\dot\phi^2}{\phi}\right), \quad (11)$$

$$p_\phi = -\left(\frac{k^2+k+4}{2}\right)\xi^2H^2\Delta\phi - \left(\frac{k+3}{2}\right)\xi\dot H\Delta\phi + \ddot\phi + \left(\frac{k+3}{2}\right)\xi H\dot\phi + \frac{\omega(\phi)}{2}\left(\frac{\dot\phi^2}{\phi}\right), \quad (12)$$

where $\triangle\phi = 1 - \phi$, which ensures the reduction to GR behaviour for $\triangle\phi = 0$. It is interesting to note here that the GBD theory provides an extra cosmic fluid which may shoulder the burden of late-time acceleration. The corresponding effective EoS parameter becomes

$$\omega_{eff} = \frac{p_\phi - \Lambda}{\rho_\phi + \Lambda} = -1 + \frac{-\left(\frac{k+3}{2}\right)\xi\dot{H}\triangle\phi + \left[(2k+1) - \left(\frac{k^2+k+4}{2}\right)\right]\xi^2H^2\triangle\phi + f(\phi,\dot{\phi},\ddot{\phi})}{\Lambda + (2k+1)\xi^2H^2\triangle\phi + g(\phi,\dot{\phi})}, \tag{13}$$

with $f(\phi,\dot{\phi},\ddot{\phi}) = \ddot{\phi} + \left[\left(\frac{k+3}{2}\right)\xi - 3\right]H\dot{\phi} + \omega(\phi)\left(\frac{\dot{\phi}^2}{\phi}\right)$ and $g(\phi,\dot{\phi}) = -3H\dot{\phi} + \frac{\omega(\phi)}{2}\left(\frac{\dot{\phi}^2}{\phi}\right)$.

In the low redshift epoch, there can be small values of $|\triangle\phi|$, and consequently, Λ becomes the dominant term in the denominator of the second term in Equation (13). Additionally, in the numerator of Equation (13), the contribution of $f(\phi,\dot{\phi},\ddot{\phi})$ may be neglected compared to other terms. In the limit of $|\triangle\phi| \to 0$, the effective EoS may be expressed as

$$\omega_{eff} \simeq -1 - \frac{\triangle\phi}{\Lambda}\left[-\left(\frac{k+3}{2}\right)\xi\dot{H} + \left[(2k+1) - \left(\frac{k^2+k+4}{2}\right)\right]\xi^2H^2\right], \tag{14}$$

which may be reduced to

$$\omega_{eff} \simeq -1 + 2\frac{\triangle\phi}{\Lambda}\xi\dot{H}. \tag{15}$$

for a small departure from cosmic anisotropy. The above equation tells us that we get a quintessence-like phase for $\triangle\phi > 0$ and a phantom-dominated phase for $\triangle\phi < 0$, at least in the low redshift epochs. In an earlier work, it has been shown from the reconstruction of the BD scalar field from observational $H(z)$ data that $\triangle\phi < 0$. This shows phantom-like behaviour in the GBD theory [38].

3. A Big Rip Scenario

An explanation of the late-time cosmic speed-up issue with dark energy requires that the EoS parameter should be $\omega_D < -1/3$. The cosmological constant corresponds to $\omega_D = -1$. For dark energy, cosmological models favouring $\omega_D < -1$ are usually dominated with phantom energy for which the energy density goes up with time violating the dominant energy condition. The energy density in a phantom dominated dark energy model is proportional to the scale factor. As a consequence, the scale factor blows up at a finite time $t_{BR} - t_0 \simeq \frac{2}{3|1+\omega_D|}\frac{1}{H_0\sqrt{1-\Omega_m}}$, where H_0 represents the present value of the Hubble parameter and $\Omega \simeq 0.3$ is the matter density parameter [43]. Such a scenario is termed as the Big Rip, whose occurrence dissolutes the bounded system [43–45]. The Big Rip scenario in Phantom models leads to a unique singularity in the Universe and can be associated with the fundamental quantum gravity formalism [46].

We consider a Big Rip scenario with the scale factor evolving as

$$a(t) \simeq (t_{BR} - t)^\alpha, \tag{16}$$

where t_{BR} is the epoch where the scale factor blows up. α is a constant parameter related to the EOS parameter as

$$\alpha = \frac{2}{3(1+\omega_D)}. \tag{17}$$

In a phantom field-dominated Universe, the EoS parameter is usually less than unity, that is, $\omega_D < -1$, which requires that the constant parameter α appearing in the scale factor should be negative, that is, $\alpha < 0$. For the given Big Rip scenario, the Hubble parameter and the deceleration parameter (DP) are, respectively, $H(t) = -\frac{\alpha}{t_{BR}-t}$ and $q = -1 - \frac{\dot{H}}{H^2} = -1 + \frac{1}{\alpha}$. Additionally, we have $\dot{H} = -\frac{H^2}{\alpha}$ and $\ddot{H} = \frac{2}{\alpha^2}H^3$. One should note that, while the Hubble parameter contains two adjustable parameters, the deceleration parameter contains only one parameter, α. The deceleration parameter in the present Big Rip scenario comes out to be a constant quantity which should be negative to provide an accelerating model. Because of this, we may constrain the parameter α from some recent observational constraint on the deceleration parameter. In a recent work, Camarena and Marra used the observational data from supernovae in a redshift range to constrain the DP as $q_0 = -1.08 \pm 0.29$ [47]. The central value of the deceleration parameter immediately fixes up the scale factor parameter as $\alpha = -12.49$. Consequently, the EoS parameter may be constrained as $\omega_D = -1.0533$. This value is in close agreement with some recent measurements, as mentioned earlier. Once α is fixed, the other parameter t_{BR} may be obtained from the present value of the Hubble parameter. In Figure 1, we show the evolution of the Hubble parameter for the constrained value of α that predicts a finite-time future singularity. The Hubble parameter increases with the cosmic expansion and blows up at a finite future. Assuming $H_0 = 74.3$ km s^{-1} Mpc^{-1}, the present model predicts a Big Rip occurring at a cosmic time $t_{BR} \simeq 16.14$ Gyr.

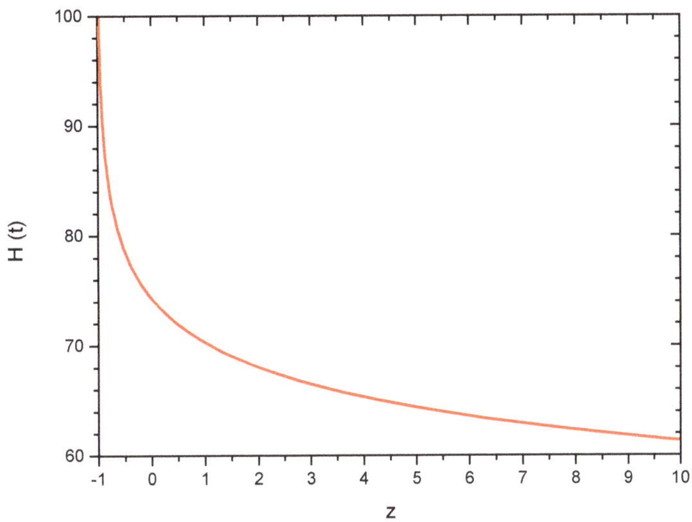

Figure 1. Hubble parameter as a function redshift. The Hubble parameter is in km s^{-1} Mpc^{-1} units.

3.1. Brans-Dicke Scalar Field

Considering the Big Rip scenario, the evolutionary aspect of the Brans-Dicke scalar field may be studied within the framework of the GBD theory. Algebraic simplification of the field Equations (6) and (7) leads to the evolution equation for the BD scalar field as

$$-\frac{\dot{H}}{H} - 3H = \frac{\dot{\phi}}{\phi}. \tag{18}$$

In terms of DP, we may express the evolution equation as

$$(q-2)H = \frac{\dot{\phi}}{\phi}. \tag{19}$$

In our model, we obtained the deceleration parameter to be a constant quantity, and consequently, the BD scalar field may be obtained by integrating Equation (19) as

$$\phi = \frac{\phi_0}{a_0^n} a^n, \tag{20}$$

where $n = q - 2 = \frac{1}{\alpha} - 3$. a_0 and ϕ_0 are respectively the value of the scale factor and the BD scalar field at the present epoch. This relation clearly articulates a power-law behaviour of the BD scalar field with respect to the scale factor. It is worth mentioning here that the use of power-law functional behaviour of the scalar field is quite common in the literature. Since the deceleration parameter is a negative quantity in our model, the BD scalar field should decrease with the cosmic expansion.

The scale factor may be expressed as $a = \left(-\frac{\alpha}{H}\right)^{\alpha}$, and consequently, the BD scalar field becomes

$$\frac{\phi}{\phi_0} = \left(\frac{H_0}{H}\right)^{n\alpha}, \tag{21}$$

which ultimately leads to

$$\frac{\ddot{\phi}}{\phi} = -3nH^2. \tag{22}$$

In terms of the redshift defined as $z = \frac{a_0}{a} - 1$, the BD scalar field reduces to

$$\frac{\phi}{\phi_0} = (1+z)^{-n}. \tag{23}$$

In Figure 2, the evolutionary aspect of the BD field is shown. The BD scalar field shows a decreasing trend from large positive values to vanishingly small values at late cosmic times. One should note from the figure that the BD scalar field behaves like $\sim(1+z)^3$ in low redshift epochs.

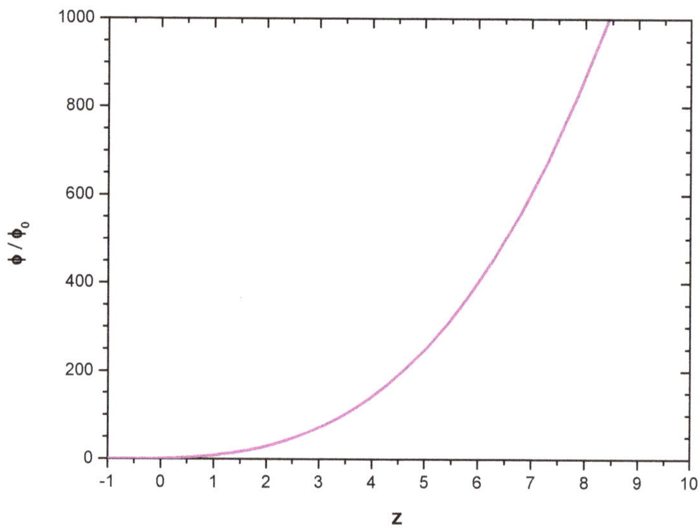

Figure 2. Evolutionary behaviour of the BD scalar field.

3.2. Brans-Dicke Parameter

The BD parameter is a dynamical quantity in the GBD theory, and its behaviour depends on the dynamical behaviour of the BD scalar field. The BD parameter may be obtained from the GBD field Equations (5)–(7) as

$$\omega(\phi) = \frac{1}{\left(\frac{\dot{\phi}}{\phi}\right)^2}\left[-\frac{\rho+p}{\phi} - \frac{\ddot{\phi}}{\phi} + k\xi H\frac{\dot{\phi}}{\phi} - 2\xi\dot{H} + 2(k-1)\xi^2 H^2\right]. \tag{24}$$

It should be mentioned here that, for a given Big Rip scenario as specified by a Hubble parameter H and a given cosmic anisotropy ξ, the dynamics of the BD scalar field is suitably obtained. Once the BD scalar field is obtained, the time-dependent aspect of the BD parameter now requires an equation of state $p = \omega_D\rho$, a relationship between the pressure p and the energy density ρ. Replacing $\rho + p$ by $(1+\omega_D)\rho$ and using the fact that $\frac{\dot{\phi}}{\phi} = nH$ in Equation (24), we get

$$\omega(\phi) = n^{-2}H^{-2}\left[-\left(\frac{1+\omega_D}{\phi}\right)\rho + \left((k\xi+3)n + 2(k-1)\xi^2\right)H^2\right]. \tag{25}$$

For the given Big Rip scenario, the conservation equation for the cosmic fluid

$$\dot{\rho} + 3H(\rho + p) = 0 \tag{26}$$

can be reduced to

$$\frac{\dot{\rho}}{\rho} = -\frac{2}{\alpha}H. \tag{27}$$

The energy density may be obtained from the integration of the conservation equation as

$$\rho = \rho_0\left(\frac{H}{H_0}\right)^2, \tag{28}$$

where ρ_0 is the present value of the energy density. Here, we used the fact that $\frac{a}{a_0} = \left(\frac{H_0}{H}\right)^\alpha$. It is now straight-forward to obtain the pressure as

$$p = \left(\frac{2}{3\alpha} - 1\right)\rho_0\left(\frac{H}{H_0}\right)^2, \tag{29}$$

so that $p + \rho = \frac{2}{3\alpha}\rho_0\left(\frac{H}{H_0}\right)^2$. The Brans-Dicke parameter may now be expressed in terms of the Hubble parameter as

$$\omega(H,\xi,\alpha) = \omega_0(H_0,\xi,\alpha) - \chi\left[\left(\frac{H}{H_0}\right)^{(n\alpha-2)} - 1\right], \tag{30}$$

where $\chi = \frac{2\rho_0 H_0^{((3+n)\alpha-1)}}{3\alpha\phi_0 n^2}$ and $\omega_0(H_0,\xi,\alpha) = \frac{k\xi+3}{n} + 2(k-1)\left(\frac{\xi}{n}\right)^2 - \chi$ is the present value of the BD parameter.

One may note that the anisotropy affects the Brans-Dicke parameter, but it does not participate in its evolution. Only the non-evolving part of the Brans-Dicke parameter is modified in the presence of cosmic anisotropy. In fact, as is evident from Equation (30), the cosmic anisotropy brings about a change in the required present value of the BD parameter. Similar results have already been observed in an earlier work [32], where it was shown that for a power-law expansion and an exponential expansion law of the scale factor, the cosmic anisotropy affects only that part of the BD parameter which does not evolve with time. The similarity between this Big Rip model and that of the power-law behaviour and the exponential expansion model is that the deceleration parameter is non-evolving. Because of this, we may infer that, for the time-independent deceleration parameter, the cosmic anisotropy will not contribute to the evolution of the Brans-Dicke parameter. However, for models with a time-dependent deceleration parameter, the cosmic anisotropy affects the BD parameter as a whole [36,38]. In Figure 3, the BD parameter (normalized to its value at the present epoch) is shown for a representative value of the cosmic anisotropy ξ. It is observed that the BD parameter increases with the cosmic expansion. Since the Hubble parameter blows up at a time $t \simeq t_{BR}$, the BD parameter also blows up at that epoch. In Figure 4, the evolutionary aspect of the BD parameter with respect to the BD scalar field is shown. With an increase in the BD scalar field, $\omega(H,\xi,\alpha)$ decreases from a much higher value to low positive values. However, the decrement in $\omega(H,\xi,\alpha)$ slows down for higher values of the scalar field. One may estimate the BD parameter at the present epoch within the purview of the present formalism to get a rip scenario at a finite future. In fact, for a finite time future singularity such as the one discussed in the present work within GBD theory, we require the BD parameter to be approximately $\omega(z=0) \simeq -3.6 \times 10^9$ at the present epoch.

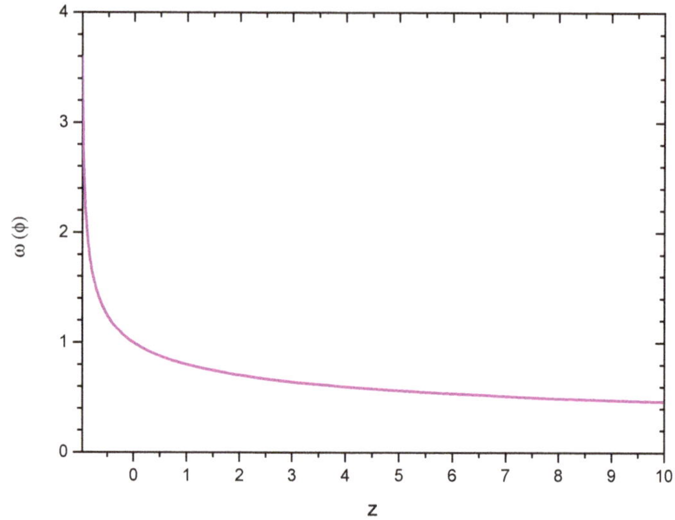

Figure 3. Brans-Dicke parameter as a function of redshift.

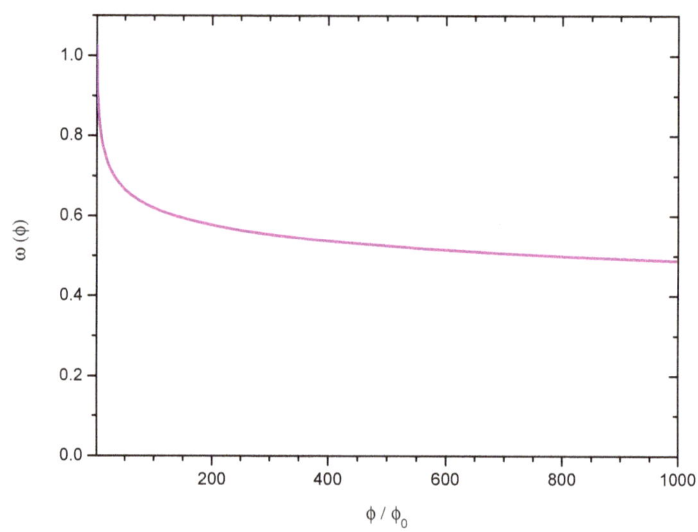

Figure 4. Evolutionary behaviour of the Brans-Dicke parameter with respect to the BD field.

3.3. *Self-Interacting Potential*

Within the GBD formalism, the time-dependence aspect of the self-interacting potential may be obtained as

$$V(\phi) = 2\phi\left[\left\{(2k+1)\xi^2 + 3n - \frac{\omega(\phi)}{2}n^2\right\}H^2 - \frac{\rho}{\phi}\right]. \tag{31}$$

In order to assess the effect of the cosmic anisotropy, we may express the self-interacting potential as

$$\frac{V(\phi)}{2\phi} = \left\{(2k+1)\xi^2 - \frac{\omega_0}{2}n^2\right\}H^2 + \left[3n + \frac{\chi n^2}{2}\left(\left(\frac{H}{H_0}\right)^{(n\alpha-2)} - 1\right)\right]H^2 - \frac{\rho}{\phi}. \quad (32)$$

In the above Equation (32), the first term in the right-hand side bears the contribution of the cosmic anisotropy. In Figure 5, we show the evolutionary behaviour of the self-interacting potential for a given cosmic anisotropy. Since in the Big Rip scenario, we discussed in the present work a large negative value of $\omega_0 = \omega(z=0)$ is being required, the effect of the cosmic anisotropy becomes negligibly small. However, in other scenarios such as a bouncing one, we may get a substantial effect of the cosmic anisotropy on the self-interacting potential [36]. Because of this, we chose a representative value $k = 1.0001$ or a corresponding $\xi = 0.999967$ to plot the figure. The value of the self-interacting potential in the figure is normalized to its present value. One should note that, with the growth of cosmic expansion, $V(\phi)$ decreases from a higher value to small values at late times. This behaviour may be translated in terms of the scalar field to infer that the self-interacting potential shows an increasing trend with the BD scalar field.

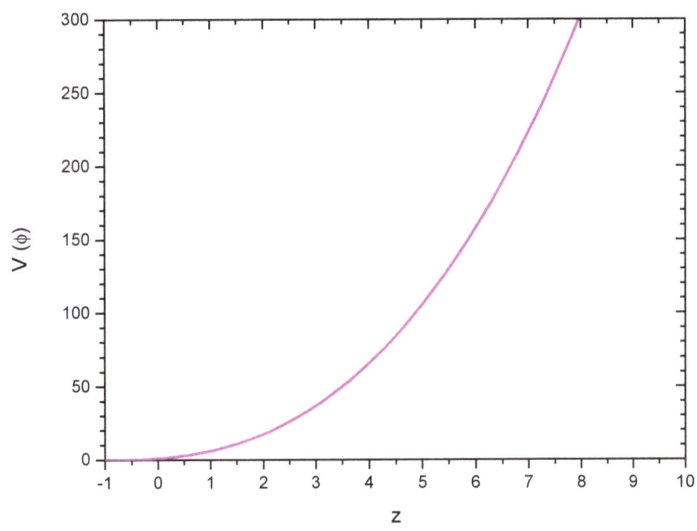

Figure 5. Evolutionary behaviour of the self-interacting potential.

4. Summary and Conclusions

In the present work, we studied a Big Rip scenario within the framework of a generalized Brans-Dicke theory. An LRSBI Universe is considered to incorporate directional anisotropy in the expansion rates. Such a model provides a more general approach compared to the FRW model. The generalized Brans-Dicke theory having a cosmological constant can be recast as a GR-like theory with cosmic fluid dominated by dark energy. The effective cosmological constant for such effective cosmic fluid dominated by dark energy may provide a quintessence-like or phantom-like behaviour depending on the nature of the scalar field. On the basis of the nature of the BD scalar field which has been reconstructed from the observational $H(z)$ data, we showed that the present model favours a phantom dark energy model. In Phantom models, the energy density and the scale factor may grow sharply within a finite time, leading to a Big Rip situation. We considered a Big Rip scale factor with an arbitrary parameter, that was fixed from a recent observational value of the deceleration parameter, which ultimately fixes the effective EoS parameter as $\omega_D = -1.05$

in close conformity with recent observational estimates. Additionally, the present model witnesses a Big Rip scenario at a time $t_{BR} \simeq 16.14$ Gyr. We studied the dynamical aspects of the scalar field, Brans-Dicke parameter, and the self-interacting potential. In conformity with other scalar field models, the BD scalar field shows a decreasing behaviour with cosmic time. The Brans-Dicke parameter, on the other hand, increases with the cosmic expansion. To the BD scalar field, the BD parameter (as normalized to its present value) decreases from a high value to almost constant values for higher values of the scalar field. In some recent tests concerning the gravitational radiation from gravitational wave bursts on the BD theory, the bounds on the BD parameter may be $\omega \geq 10^4 \sim 10^5$ [27] or $\omega \geq 10^6$ [24]. Montenegro et al. used $\omega = -1$ to obtain Big Rip kinds of solutions in the BD theory [22]. However, our present model requires a high negative value of the BD parameter at the present epoch of the order of $\sim -10^9$ to witness a Big Rip scenario in the finite future.

In our model, the anisotropy parameter does not affect the BD scalar field. It affects the BD parameter, but does not contribute to its time evolution aspect. In principle, only the non-evolving part of the BD parameter is affected by the anisotropy parameter. This is a feature usually observed for BD gravity models with a constant deceleration parameter [32]. However, the anisotropy affects the self-interacting potential. Since we require quite a large magnitude of the present epoch value of the BD parameter to get a viable Big Rip scenario, a small variation of the cosmic anisotropy does not substantially bring about a change in its numerical value.

Author Contributions: Conceptualization, S.K.T. and Z.N.; Methodology, S.K.T. and S.K.P.; Software, S.K.P.; Formal analysis, S.K.P., S.K.T. and D.B.; Investigation S.K.P., Z.N. and M.B.; Writing—original draft preparation, S.K.P. and S.K.T.; Writing—review and editing, M.B. and Z.N.; Supervision, S.K.T. and M.B.; Project administration, S.K.T. and M.B.; Funding acquisition, M.B. All authors have read and agreed to the published version of the manuscript.

Funding: This research was funded by FOSTECT Project No. FOSTECT.2019B.04, FAPESP Project No. 2017/05660-0, and the CNPq—Brasil.

Institutional Review Board Statement: Not applicable.

Informed Consent Statement: Not applicable.

Data Availability Statement: The Data are contained within the article.

Acknowledgments: SKT thanks IUCAA, Pune (India) for providing support through the visiting Associateship programme and MB thanks T. M. Joshua for the discussion throughout the work.

Conflicts of Interest: The authors declare no conflict of interest.

Abbreviations

The following abbreviations are used in this manuscript:

GR	General Relativity
BD	Brans-Dicke
GBD	Generalized Brans-Dicke
FRW	Friedmann-Robertson-Walker
LRSBI	Locally Rotationally Symmetric Bianchi type I
CMB	Cosmic Microwave Background
Λ CDM	Λ dominated Cold Dark Matter
WMAP	Wilkinson Microwave Anisotropy Probe

References

1. Ratra, B.; Peebles, P.J. Cosmological consequences of a rolling homogeneous scalar field. *Phys. Rev. D* **1988**, *37*, 3406. [CrossRef]
2. Sahni, V.; Starobinsky, A. The Case for a Positive Cosmological Lambda-term *Int. J. Mod. Phys. D* **2000**, *9*, 373. [CrossRef]
3. Sahni, V.; Starobinsky, A. Reconstructing dark energy. *Int. J. Mod. Phys. D* **2006**, *15*, 2105. [CrossRef]
4. Caldwell, R.R. A phantom menace? Cosmological consequences of a dark energy component with super-negative equation of state. *Phys. Lett. B* **2002**, *545*, 23–29. [CrossRef]

5. Tripathi, A.; Sangwan, A.; Jassal, H.K. Dark energy equation of state parameter and its evolution at low redshift. *J. Cosmol. Astropart. Phys.* **2017**, *2017*, 012. [CrossRef]
6. Hinshaw, G.; Larson, D.; Komatsu, E.; Spergel, D.N.; Bennett, C.; Dunkley, J.; Nolta, M.R.; Halpern, M.; Hill, R.S.; Odegard, N.; et al. Nine-year Wilkinson Microwave Anisotropy Probe (WMAP) observations: cosmological parameter results. *Astrophys. J. Suppl. Ser.* **2013**, *208*, 19. [CrossRef]
7. Amanullah, R.; Lidman, C.; Rubin, D.; Aldering, G.; Astier, P.; Barbary, K.; Burns, M.S.; Conley, A.; Dawson, K.S.; Deustua, S.E.; et al. Supernova Cosmology Project. Spectra and Hubble Space Telescope light curves of six type Ia supernovae at $0.511 < z < 1.12$ and the Union2 compilation. *Astrophys. J.* **2010**, *716*, 712.
8. Kumar, S.; Xu, L. Observational constraints on variable equation of state parameters of dark matter and dark energy after Planck. *Phys. Lett. B* **2014**, *737*, 244–247. [CrossRef]
9. Aghanim, N.; Akrami, Y.; Ashdown, M.; Aumont, J.; Baccigalupi, C.; Ballardini, M.; Banday, A.J.; Barreiro, R.B.; Bartolo, N.; Basak, S.; et al. Planck 2018 results-VI. Cosmological parameters. *Astron. Astrophys.* **2018**, *641*, A6.
10. Goswami, G.K.; Yadav, A.K.; Mishra, B.; Tripathy, S.K. Modeling of Accelerating Universe with Bulk Viscous Fluid in Bianchi V Space-Time. *Fortscr. Phys.* **2021**, *69*, 2100007. [CrossRef]
11. Stefancic, H. Expansion around the vacuum equation of state: Sudden future singularities and asymptotic behavior. *Phys. Rev. D* **2005**, *71*, 084024. [CrossRef]
12. Frampton, P.H.; Ludwick, K.J.; Scherrer, R.J. The little rip. *Phys. Rev. D* **2011**, *84*, 063003. [CrossRef]
13. Nojiri, S.I.; Odintsov, S.D. Unified cosmic history in modified gravity: From $F(R)$ theory to Lorentz non-invariant models. *Phys. Rep.* **2011**, *505*, 59–144. [CrossRef]
14. Frampton, P.H.; Ludwick, K.J.; Nojiri, S.; Odintsov, S.D.; Scherrer, R.J. Models for little rip dark energy. *Phys. Lett. B* **2012**, *708*, 204–211. [CrossRef]
15. Dabrowski, M.P. Future state of the universe. *Ann. Phys.* **2006**, *15*, 352–363. [CrossRef]
16. Balcerzak, A.; Dabrowski, M.P. Strings at future singularities. *Phys. Rev. D* **2006**, *73*, 101301. [CrossRef]
17. Granda, L.N.; Loaiza, E. Big Rip and Little Rip Solutions in Scalar Model with Kinetic and Gauss-Bonnet Couplings. *Int. J. Mod. Phys. D* **2012**, *21*, 1250002. [CrossRef]
18. Vasilev, T.B.; Bouhmadi-López, M.; Martín-Moruno, P. Classical and quantum fate of the little sibling of the big rip in $f(R)$ cosmology. *Phys. Rev. D* **2019**, *100*, 084016. [CrossRef]
19. Hanafy, W.E.; Saridakis, E.N. $f(T)$ cosmology: From Pseudo-Bang to Pseudo-Rip. *arXiv* **2020**, arXiv:2011.15070.
20. Ray, P.P.; Tarai, S.; Mishra, B.; Tripathy, S.K. Cosmological models with Big rip and Pseudo rip Scenarios in extended theory of gravity. *Forschr. Phys.* **2021**, *69*, 2100086. [CrossRef]
21. Pati, L.; Kadam, S.A.; Tripathy, S.K.; Mishra, B. Rip cosmological models in extended symmetric teleparallel gravity. *Phys. Dark Univ.* **2022**, *35*, 100925. [CrossRef]
22. Montenegro, A.E., Jr.; Carneiro, S. Exact solutions of Brans-Dicke cosmology with decaying vacuum density. *Class. Quant. Grav.* **2007**, *24*, 313–327. [CrossRef]
23. Tirandari, M.; Saaidi, K. Anisotropic inflation in Brans-Dicke gravity. *Nucl. Phys. B* **2017**, *925*, 403–414. [CrossRef]
24. Zhang, X.; Yu, J.; Liu, T.; Zhao, W.; Wang, A. Testing Brans-Dicke gravity using the Einstein Telescope. *Phys. Rev. D* **2017**, *95*, 124008. [CrossRef]
25. Koyama, K. Testing Brans-Dicke gravity with screening by scalar gravitational wave memory. *Phys. Rev. D* **2020**, *102*, 021502(R). [CrossRef]
26. Bonino, A.; Camera, S.; Fatibene, L.; Orizzonte, A. Solar system tests in Brans-Dicke and palatini $f(R)$-theories. *Eur. Phys. J. Plus* **2020**, *135*, 951. [CrossRef]
27. Arun, K.G.; Pai, A. Tests of General Relativity and alternative theories of gravity using gravitational wave observations. *Int. J. Mod. Phys. D* **2013**, *22*, 1341012. [CrossRef]
28. Nordtvedt, K., Jr. Post-Newtonian Metric for a General Class of Scalar-Tensor Gravitational Theories and Observational Consequences. *Astrophys. J.* **1970**, *161*, 1059. [CrossRef]
29. Wagoner, R.V. Scalar-tensor theory and gravitational waves. *Phys. Rev. D* **1970**, *1*, 3209. [CrossRef]
30. Freund, P.G. Kaluza-Klein cosmologies. *Nucl. Phys. B* **1982**, *209*, 146–156. [CrossRef]
31. Green, M.B.; Schwarz, J.H.; Witten, E. *Superstring Theory*; Two Volumes; Cambridge Univerity Press: Cambridge, UK, 1987.
32. Tripathy, S.K.; Behera, D.; Mishra, B. Unified dark fluid in Brans-Dicke theory. *Eur. Phys. J. C* **2015**, *75*, 1–11. [CrossRef]
33. Sahoo, B.K.; Singh, L.P. time-dependence of Brans-Dicke Parameter ω for an Expanding Universe. *Mod. Phys. Lett. A* **2002**, *17*, 2409–2415. [CrossRef]
34. Tahmasebzadeh, B.; Karami, K. Generalized Brans-Dicke inflation with a quartic potential. *Nucl. Phys. B* **2017**, *918*, 1–10. [CrossRef]
35. De Felice, A.; Tsujikawa, S. Conditions for the cosmological viability of the most general scalar-tensor theories and their applications to extended Galileon dark energy models. *J. Cosmol. Astropart. Phys.* **2012**, *2012*, 007. [CrossRef]
36. Tripathy, S.K.; Pandey, S.; Sendha, A.P.; Behera, D. Bouncing scenario in Brans-Dicke theory. *Int. J. Geom. Methods Phys.* **2020**, *17*, 2050056. [CrossRef]
37. Sharif, M.; Waheed, S. Cosmic acceleration and Brans-Dicke theory. *J. Exptl. Theor. Phys.* **2012**, *115*, 599–613. [CrossRef]

38. Tripathy, S.K.; Pradhan, S.K.; Naik, Z.; Behera, D.; Mishra, B. Unified dark fluid and cosmic transit models in Brans-Dicke theory. *Phys. Dark Univ.* **2020**, *30*, 100722. [CrossRef]
39. Buiny, R.V.; Berera, A.; Kephart, T.W. Asymmetric inflation: Exact solutions. *Phys. Rev. D* **2006**, *73*, 063529. [CrossRef]
40. Watanabe, M.; Kanno, S.; Soda, J. Inflationary Universe with anisotropic Hair. *Phys. Rev. Lett.* **2009**, *102*, 191302. [CrossRef]
41. Antoniu, I.; Perivolaropoulos. Searching for a cosmological preferred axis: Union 2 data analysis and comparison with other probes. *J. Cosmol. Astropart. Phys.* **2012**, *12*, 1012. [CrossRef]
42. Tripathy, S.K. Late-time cosmic acceleration and role of skewness in anisotropic models. *Astrophys. Space Sci.* **2014**, *350*, 367–374. [CrossRef]
43. Caldwell, R.R.; Kamionkowski, M.; Weinberg, N.N. Phantom energy: Dark energy with $w < -1$ causes a cosmic doomsday. *Phys. Rev. Lett.* **2003**, *91*, 071301. [PubMed]
44. Frampton, P.H.; Takahashi, T. The fate of dark energy. *Phys. Lett. B* **2003**, *557*, 135–138. [CrossRef]
45. Nesseris, S.; Perivolaropoulos, L. Fate of bound systems in phantom and quintessence cosmologies. *Phys. Rev. D* **2004**, *70*, 123529. [CrossRef]
46. McInnes, B. The phantom divide in string gas cosmology. *Nucl. Phys. B* **2005**, *718*, 55–82. [CrossRef]
47. Camarena, D.; Marra, V. Local determination of the Hubble constant and the deceleration parameter. *Phys. Rev. Res.* **2020**, *2*, 013028. [CrossRef]

foundations

MDPI

Article

Divergence in the Relativistic Mean Field Formalism: A Case Study of the Ground State Properties of the Decay Chain of ^{214,216,218}U Isotopes

Tolulope Majekodunmi Joshua [1,†], Nishu Jain [2], Raj Kumar [2,†], Khairul Anwar [1], Nooraihan Abdullah [1] and Mrutunjaya Bhuyan [3,4,*,†]

1 Institute of Engineering Mathematics, Faculty of Applied and Human Sciences, University Malaysia Perlis, Arau 02600, Malaysia; majekjoe1@gmail.com (T.M.J.); khairulanwar@unimap.edu.my (K.A.); raihan@unimap.edu.my (N.A.)
2 School of Physics and Materials Science, Thapar Institute of Engineering and Technology, Patiala 147004, India; nishujain1003@gmail.com (N.J.); rajkumar@thapar.edu (R.K.)
3 Center for Theoretical and Computational Physics, Department of Physics, Faculty of Science, University of Malaya, Kuala Lumpur 50603, Malaysia
4 Institute of Research and Development, Duy Tan University, Da Nang 550000, Vietnam
* Correspondence: bunuphy@um.edu.my; Tel.: +60-1137057605
† These authors contributed equally to this work.

Abstract: A new α-emitting ^{214}U has been recently observed experimentally. This opens the window to theoretically investigate the ground-state properties of the lightest known even–even neutron deficient ^{214,216,218}U isotopes and to examine α-particle clustering around the shell closure. The decay half-lives are calculated within the preformed cluster-decay model (PCM). To obtain the α-daughter interaction potential, the RMF densities are folded with the newly developed R3Y and the well-known M3Y NN potentials for comparison. The alpha preformation probability (P_α) is calculated from the analytic formula of Deng and Zhang. The WKB approximation is employed for the calculation of the transmission probability. The individual binding energies (BE) for the participating nuclei are estimated from the relativistic mean-field (RMF) formalism and those from the finite range droplet model (FRDM) as well as WS3 mass tables. In addition to $Z = 84$, the so-called abnormal enhancement region, i.e., $84 \leq Z \leq 90$ and $N < 126$, is normalised by an appropriately fitted neck-parameter ΔR. On the other hand, the discrepancy sets in due to the shell effect at (and around) the proton magic number $Z = 82$ and 84, and thus a higher scaling factor ranging from 10^{-8}–10^{-5} is required. Additionally, in contrast with the experimental binding energy data, large deviations of about 5–10 MeV are evident in the RMF formalism despite the use of different parameter sets. An accurate prediction of α-decay half-lives requires a Q-value that is in proximity with the experimental data. In addition, other microscopic frameworks besides RMF could be more reliable for the mass region under study. α-particle clustering is largely influenced by the shell effect.

Keywords: preformed cluster decay; relativistic mean-field; alpha-particle clustering; neck-length

Citation: Joshua, T.M.; Jain, N.; Kumar, R.; Anwar, K.; Abdullah, N.; Bhuyan, M. Divergence in the Relativistic Mean Field Formalism: A Case Study of the Ground State Properties of the Decay Chain of ^{214,216,218}U Isotopes. *Foundations* **2022**, 2, 85–104. https://doi.org/10.3390/foundations2010004

Academic Editor: Eugene Oks

Received: 13 December 2021
Accepted: 4 January 2022
Published: 9 January 2022

1. Introduction

Instability is highly prevalent in heavy nuclei due to the Coulomb repulsion of several protons involved. To gain stability, heavy nuclei naturally dispose of the excess positive charges via alpha (α)-decay. An α-particle itself is characterised by high stability and a tightly bound structure. The discovery of α-decay dates back to 1896 when it was observed as natural radioactivity and later confirmed in Rutherford's experiment in 1908. This was given credence in the theoretical interpretation of α-decay explained as a quantum tunnelling effect. Thereafter, the relationship between the decay energy of α-particles and half-lives was empirically deduced by Geiger and Nuttal [1,2]. Alpha decay has been proven to be a powerful tool that incorporates nuclear structure information [3,4], such as ground-state half-life and energy [5,6], stellar nucleosynthesis [7,8], cluster decay [9], exotic nuclei in the superheavy mass region [8], closed-shell region [10,11], proton-neutron drip line region [12,13], and the shell structure [11,14].

The development of radioactive beams across the globe since they were first used a few decades ago has extended the frontiers of the knowledge of the nuclear shell structure, especially in light nuclei. Yet, from the experimental front, little is presently known of the structure evolution in the heavy nuclei around and below the neutron shell closure at $N = 126$ [12,15,16]. Particularly, some neutron-deficient nuclei around the region $Z = 82$ in the heavy-ion reaction are often excluded in the in-beam γ-ray experiments [17]. On the other hand, all theoretical probes on this shell closure emanate from either the nuclear fission [18–20] or Gamow theory of α-decay [21–25]. The former relegates the idea of preformation; i.e., it is assumed that clusters are formed during the separation/deformation process of the parent nucleus while penetrating the confining interaction barrier. In Gamow's prescription [26], the α-particle decay is considered as a quantum tunnelling process of a preformed α-particle penetrating the potential barrier. This presupposes that α-preformation probability could be incorporated into α-decay theories as α-cluster preformation. One of the models that embraces this theory is the preformed cluster-decay model (PCM), which stems from the well-known quantum mechanical fragmentation theory (QMFT). In the PCM, an α-particle is assumed to be preborn within the parent nucleus and thereafter tunnels through the potential barrier generated by the superposition of the nuclear and Coulomb potentials [26–28]. Obtaining the Coulomb potential is straightforward, whereas the nuclear potential can be deduced via phenomenological [29,30] and microscopic approaches [31,32]. Thus, to gain insight into the decay phenomena, the choice of nuclear potential is crucial [30]. In addition to the fundamental approaches [33,34], the newly developed R3Y nucleon-nucleon (NN) potential [35,36], which is analogous to the phenomenological M3Y [37], is derived from the relativistic mean-field (RMF) Lagrangian and is applied in the present work for the study of α emission using the NL3* parameter set, which has been successfully employed in the study of various ground- and excited-state properties [38–40].

Recently, a new neutron-deficient α-emitting isotope ^{214}U produced by the fusion-evaporation reaction $^{182}\mathrm{W}(^{36}\mathrm{Ar}, 4\mathrm{n})^{214}\mathrm{U}$ has been observed [15]. Its α-decay energy and half-life were measured to be $Q_\alpha = 8533(18)$ KeV and $T_{1/2} = 0.52^{+0.95}_{-0.21}$ ms, respectively. Hence, ^{214}U is the lightest known even-A uranium isotope until now. This necessitates a detailed investigation of the ground-state properties of ^{214}U. The authors [15] also reported that an abnormal enhancement was observed in the reduced width of other re-measured light even-even ^{216,218}U isotopes around the proton closed-shell $84 \leq Z \leq 90$ and neutron closed-shell $N < 126$. This constitutes our motivation to theoretically examine this phenomenon in the α-decay chain of these lighter uranium isotopes using the PCM within the RMF framework [41–43], which is cognate with the energy density functional formalism and gives an accurate description of both ground and excited-state properties across the entire nuclear landscape [44]. However, previous theoretical studies [28,45] hold that a constant scaling factor (CSF) of 10^{-4} is required for the calculation of the α-decay half-lives of nuclei in the superheavy region within the PCM framework at the ground state (at temperature, $T = 0$). Thus, the present study is also aimed at answering two pertinent questions: (1) Does the α-decay of lighter uranium follow a similar trend as those from other superheavy nuclei? More precisely, does the constant scaling factor that appears in PCM calculations for the superheavy region apply to other regions in the nuclear chart? (2) If yes, what will be the course (of the CSF) for the lightest ^{214,216,218}U isotopes (around $84 \leq Z \leq 90$ and $N < 126$ shell closures) where an abnormal enhancement has just been observed?

The only variable parameter in the PCM is the neck-length parameter ΔR, which assimilates the neck formation effect between two nuclei and determines the first turning point of the barrier penetration. In this context, ΔR is always fitted to the experimental data within the proximity potential limit of up to 2 fm [28,46]. The inputs of PCM include the preformation probability (P_α), calculated from the analytic formula of Deng and Zhang [47,48], and penetration probability (P), using the WKB approximation [49–51]. The binding energy (BE) estimated from the microscopic RMF formalism and those from the FRDM [52] and WS3 [53] are utilised to calculate the Q-values of the α-decay. Thus, the dynamics of α-decay are analysed in detail.

The paper is organised as follows: In Section 2, the relativistic mean-field formalism and the nucleus–nucleus potential obtained from the double-folding procedure for R3Y and M3Y NN-potential using the densities of the daughter and cluster. This section concludes with a brief overview of the PCM. Section 3 is assigned to the discussion of the results obtained from our calculation. The summary of our findings and a brief conclusion is given in Section 4.

2. Theoretical Framework

The isotopes of uranium, namely ^{214,216,218}U, are studied here microscopically within the relativistic mean-field formalism in which the interaction between the many-body system of nucleons and mesons is expressed via the non-linear effective Lagrangian [41–43,54–58],

$$\begin{aligned}
\mathcal{L} &= \overline{\psi}_i\{i\gamma^\mu\partial_\mu - M\}\psi_i + \frac{1}{2}\partial^\mu\sigma\partial_\mu\sigma \\
&- \frac{1}{2}m_\sigma^2\sigma^2 - \frac{1}{3}g_2\sigma^3 - \frac{1}{4}g_3\sigma^4 - g_s\overline{\psi}_i\psi_i\sigma \\
&- \frac{1}{4}\Omega^{\mu\nu}\Omega_{\mu\nu} + \frac{1}{2}m_\omega^2 V^\mu V_\mu - g_\omega\overline{\psi}_i\gamma^\mu\psi_i V_\mu \\
&- \frac{1}{4}\vec{B}^{\mu\nu}.\vec{B}_{\mu\nu} + \frac{1}{2}m_\rho^2\vec{R}^\mu.\vec{R}_\mu - g_\rho\overline{\psi}_i\gamma^\mu\vec{\tau}\psi_i.\vec{R}^\mu \\
&- \frac{1}{4}F^{\mu\nu}F_{\mu\nu} - e\overline{\psi}_i\gamma^\mu(\frac{1-\tau_{3i}}{2})\psi_i A_\mu. \qquad (1)
\end{aligned}$$

The scalar meson σ and vector meson V^μ account for the medium-range attraction and the short-range repulsion between the nucleons, respectively. The isovector-vector meson $\vec{R}^\mu$ describes the isospin-dependent effects in the nuclei. Their respective masses are m_σ, m_ω and m_ρ with the coupling constants g_s, g_ω and g_ρ. The Dirac Spinor, isospin and its third component are denoted as ψ_i, τ and τ_3, respectively. Parameters g_2, g_3 and $\frac{e^2}{4\pi}$ are the coupling constants of the non-linear terms. M is the mass of nucleon and A_μ stands for the electromagnetic field. The π-meson is not considered since its contribution is negligible owing to its pseudoscalar nature [42,57]. From Equation (1), the classical variation principle is employed to deduce the Dirac equation,

$$[-i\alpha.\nabla + \beta(M^* + g_\sigma\sigma) + g_\omega\omega + g_\rho\tau_3\rho_3]\psi_i = \epsilon_i\psi_i \qquad (2)$$

to obtain the nuclear spinors and the Klein–Gordon equations

$$\begin{aligned}
(-\nabla^2 + m_\sigma^2)\sigma(r) &= -g_\sigma\rho_s(r) - g_2\sigma^2(r) - g_3\sigma^3(r), \\
(-\nabla^2 + m_\omega^2)V(r) &= g_\omega\rho(r), \\
(-\nabla^2 + m_\rho^2)\rho(r) &= g_\rho\rho_3(r). \qquad (3)
\end{aligned}$$

A numerical solution is then carried out self-consistently using an iterative approach with NL3* parameter set [59]. Taking the limit of a single-meson exchange for static baryonic medium, the scalar, σ- and vector (ω,ρ)-fields are expressed in terms of the nucleon–nucleon potentials as

$$V_\sigma = -\frac{g_\sigma^2}{4\pi}\frac{e^{-m_\sigma r}}{r} + \frac{g_2^2}{4\pi} r e^{-2m_\sigma r} + \frac{g_3^2}{4\pi}\frac{e^{-3m_\sigma r}}{r},$$
$$V_\omega(r) = \frac{g_\omega^2}{4\pi}\frac{e^{-m_\omega r}}{r}, \qquad V_\rho(r) = +\frac{g_\rho^2}{4\pi}\frac{e^{-m_\rho r}}{r}. \tag{4}$$

The contribution of δ-meson is subsumed in the ρ-field [57] and, hence, V_δ is negligible. The sum of the expressed NN interactions in Equations (4) gives the RMF-based-R3Y NN effective interactions plus a single-nucleon exchange effect [60]

$$\begin{aligned} V_{\text{eff}}^{R3Y}(r) &= \frac{g_\omega^2}{4\pi}\frac{e^{-m_\omega r}}{r} + \frac{g_\rho^2}{4\pi}\frac{e^{-m_\rho r}}{r} - \frac{g_\sigma^2}{4\pi}\frac{e^{-m_\sigma r}}{r} \\ &\quad + \frac{g_2^2}{4\pi} r e^{-2m_\sigma r} + \frac{g_3^2}{4\pi}\frac{e^{-3m_\sigma r}}{r} + J_{00}(E)\delta(s), \end{aligned} \tag{5}$$

where $J_{00}(E) = -276(1 - 0.005E_\alpha/A_\alpha)$ MeV fm^3. A_α represents the mass of the α-particle and E_α symbolises the energy measured in the centre of mass of the decay fragments (α-daughter) system whose magnitude is equal to energy released for the decay process (Q_α-value). Unlike the energies required in high-energy α-scattering, $J_{00}(E)$ could be independent of the Q_α-value and is often used in its approximated form, as seen in Ref. [61].

The M3Y (Michigan-3-Yukawa) is composed of a 0.25 fm medium-range attractive part, 0.4 fm short-range repulsive part and 1.414 fm long-range tail of one-pion exchange potential (OPEP), which proceeds from the fitting of G-matrix elements predicated on Reid-Elliott soft-core NN interaction [37] on an oscillator basis. The M3Y plus exchange term takes the form

$$V_{\text{eff}}^{M3Y}(r) = 7999\frac{e^{-4r}}{4r} - 2134\frac{e^{-2.5r}}{2.5r} + J_{00}(E)\delta(r), \tag{6}$$

where the unit of the ranges are in fm and the strength is in MeV. The nuclear interaction potential $V_n(R)$ is calculated here within the double-folding approach [37] and is given as

$$V_n(R) = \int \rho_\alpha(\vec{r}_\alpha)\rho_d(\vec{r}_d)V_{\text{eff}}(|\vec{r}_\alpha - \vec{r}_d + \vec{R}| \equiv r)d^3r_\alpha d^3r_d. \tag{7}$$

Here, ρ_α and ρ_d are the nuclear matter density distributions of the alpha particle (α) and the daughter nucleus (d), respectively. To obtain the alpha–daughter interaction potential, the nuclear potential $V_n(R)$ is added to the Coulomb potential $V_C(R)$ $(= \frac{Z_c Z_d}{R}e^2)$ and is given as

$$V(R) = V_n(R) + V_C(R). \tag{8}$$

This potential is used in the calculation of the WKB penetration probability in the PCM.

Preformed Cluster-Decay Model (PCM)

The alpha-decay half-life in the preformed cluster-decay model (PCM) can be estimated as [62,63]

$$T_\alpha^{1/2} = \frac{\ln 2}{\lambda}, \qquad \lambda = \nu_0 P_0 P. \tag{9}$$

The decay constant λ denotes the probability per unit time for each nucleus to decay. It is assumed that clusters are preborn within the parent nucleus with certain preformation P_α and hits the potential barrier with an assault frequency ν_0, given as

$$\nu_0 = \frac{\text{velocity}}{R_0} = \frac{\sqrt{2E_\alpha/\mu}}{R_0}, \tag{10}$$

and thereafter tunnels with a probability P. R_0 represents the radius of the parent nucleus. A necessary condition for the energetically favoured spontaneous emission of the α-particle is a positive Q-value. This is the total energy available for the decay process. The Q-values are obtained from the ground-state binding energies from the expression

$$Q = BE_p - (BE_d + BE_\alpha), \tag{11}$$

where BE_p, BE_d and BE_α are the binding energies of the parent and daughter nuclei and the emitted α-particle, respectively. The Q values are rationed between both fragments such that α-particle $E_\alpha = \frac{A_d}{A}Q$ and the recoil energy of the daughter $E_d = Q - E_\alpha$ since $Q = E_\alpha + E_d$.

The first turning point R_a shown in Figure 1, which illustrates the penetration path of the decaying compound nucleus $^{214}U \rightarrow {}^{210}Th + \alpha$, is given as

$$R = R_a = R_1(\alpha_1, T) + R_2(\alpha_2, T) + \Delta R. \tag{12}$$

Here, ΔR is the relative separation distance between two outcoming nuclei, which incorporates the neck formation effects between them and hence is referred to as the neck-length parameter. ΔR is introduced similarly as those of the scission point [64] and saddle point [65,66] in statistical fission models. In the present context, the neck parameter is fitted to predict the experimentally measured half-lives. It is pertinent to note that the Q-value of the reaction influences the choice of the neck length. As such, it is required that the potential at the first turning point $V(R_a)$ should be higher than the Q-value.

In the PCM framework, the preformation probability P_α (otherwise called the spectroscopic factor) encapsulates the structural information of the decaying parent nucleus. From a microscopic perspective, it is difficult to obtain the exact value of P_α due to the complexities associated with the nuclear many-body problem. Nonetheless, P_α could be several orders of magnitude below unity [22,62]. Here, the α-particle preformation is calculated from the analytic formula of Deng and Zhang [47,48], who successfully employed it in the investigation of some neutron-deficient nuclei. The authors also reported that this formula gives an accurate prediction of α-decay half-lives for known and unsynthesised superheavy nuclei and sheds light on some microscopic nuclear structure information such as odd-even staggering and the shell effect. It takes the expression

$$\log_{10} P_\alpha = a + bA^{1/6}\sqrt{Z} + c\frac{Z}{\sqrt{Q_\alpha}} - k\chi' - e\rho' + f\sqrt{l(l+1)}, \tag{13}$$

where

$$\chi' = Z_\alpha Z_d \sqrt{\frac{A_\alpha A_d}{(A_\alpha + A_d)Q_\alpha}} \quad \text{and}$$

$$\rho' = \sqrt{\frac{A_\alpha A_d}{(A_\alpha + A_d)} Z_\alpha Z_d (A_\alpha^{1/3} + A_d^{1/3})}.$$

The mass and proton numbers of the decaying parent nucleus are denoted as A, Z, respectively. l is the angular momentum carried by the α-particle. In this study, $l = 0$ since all nuclei are considered to be in the ground state. The adjustable parameters a, b, c, e, f and k have been fitted with the experimental data as given in Ref. [47] and their respective values are the same for the $N \leq 126$ region, as mentioned in Ref. [48]. It is worth noting that this analytical P_0 formula from Deng et al. is based on the GLDM with an elliptic lemniscatoid geometry. The family of the elliptic lemniscatoid geometries are obtained by the inversion of spheroids [67], in which the daughter nucleus is assumed to be almost spherical. Moreover, it has been established [68] that the shape evolves continuously from one spherical nucleus to two touching spherical nuclei and naturally results in the formation of a conspicuous neck. This geometry may not be the best compromise in α-

emission since its touching-point configuration is characterised by the existence of cusp, leading to an abrupt reversal in motion of the relative distance (see Ref. [69] for elaborate details). However, our goal in the present work is to investigate the behaviour of the RMF formalism. A more suitable preformation formula is being developed and will be communicated shortly.

The α-particle tunnels through the interaction potential $V(R)$, starting from the first turning point $R = R_a$, and terminates at the second turning point $R = R_b$, whose corresponding potential $V(R_b) = Q$ for ground-state decays (illustrated in the inset of Figure 1). On the other hand, $V(R_a) = Q + E_i$, where E_i (as adopted by Malik et al. [62]) is the energy with which the α-particle or daughter nucleus decays into an excited state. At the radius of the parent nucleus $R = R_0$, the potential of the system is equal to its Q value. The shape of the parent nucleus changes as instability sets in, leading to the separation of the α-particle and neck formation. To deduce the barrier penetration probability P, using the WKB approximation, three basic steps are involved [62]: (a) the penetrability P_i from R_a to R_i, (b) the (inner) de-excitation probability W_i at R_i, taken as one [70], and (c) the penetrability P_b from R_i to R_b, which leads to the penetration probability

$$P = P_i P_b, \tag{14}$$

where P_i and P_b are the integrals in the WKB approximation and are given as:

$$P_i = \exp\left(-\frac{2}{\hbar}\int_{R_a}^{R_i}\{2\mu[V(R) - V(R_i)]\}^{1/2}dR\right), \tag{15}$$

and

$$P_b = \exp\left(-\frac{2}{\hbar}\int_{R_i}^{R_b}\{2\mu[V(R_i) - Q]\}^{1/2}dR\right), \tag{16}$$

where μ is the reduced mass given by $\mu = A_d A_\alpha/(A_d + A_\alpha)$. The above integrals in Equations (15) and (16) are solved numerically to obtain the penetration probability.

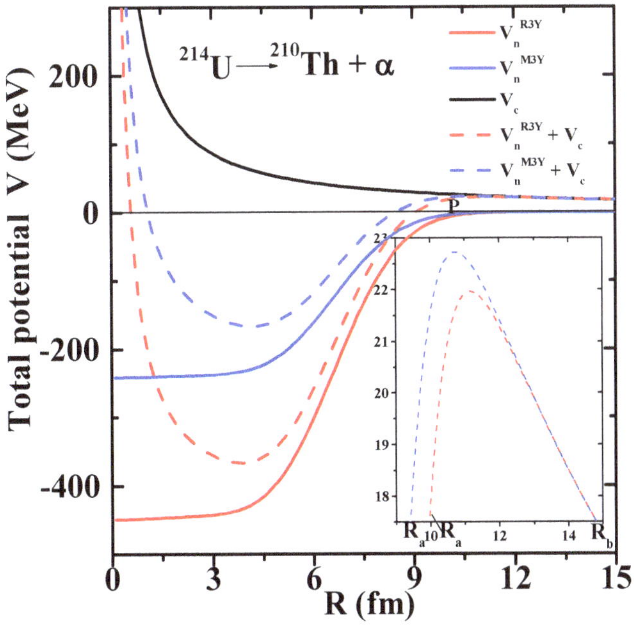

Figure 1. The total nucleus-nucleus interaction potential V (MeV) and its components, namely, nuclear and Coulomb potentials as a function of radial separation R (fm) for R3Y (NL3*) and M3Y NN potentials as a representative case of $^{214}U \rightarrow ^{210}Th + \alpha$. The inset shows a magnified view of the barrier height and position.

3. Calculations and Discussions

The present work aims to theoretically investigate the α-decay properties of the newly measured ^{214}U as well as the even-even $^{216,218}U$ isotopes within the PCM framework. The nuclear interaction potential (shown in Figure 1) is deduced from the RMF approach using the recently developed R3Y (with NL3* parameter set) and the well-known phenomenological M3Y NN interactions. Three different sets of Q-values calculated from the binding energy data FRDM [52], WS3 [53] and those from the RMF formalism (NL3*) were employed.

As a representative case, the total radial density distribution of the fragments (daughter nucleus ^{210}Th (red line) and α-particle (blue line)) in Figure 1 is conspicuously shown in Figure 2 as a function of the radius. A similar figure can be obtained for all the participating nuclei under study (but not shown here for the sake of clarity). From the figure, the density of ^{210}Th displays a lower magnitude around the central position and increases towards the surface region while the reverse is observed for the α-particle due to Coulomb repulsion and the difference in their respective mass. It is imperative to note that in Ref. [71], the density-dependent M3Y (DDM3Y) NN interactions were introduced to reproduce the saturation properties of nuclear matter. The inclusion of the DDM3Y NN interaction in the double-folding approach leads to the modification of nuclear potential, especially at small separation distances where the density overlap is larger and hence the density dependence of NN interaction becomes important. However, the present study aims to compare the results of widely adopted M3Y and recently proposed relativistic R3Y NN interactions. Therefore, to avoid complexities, we have only considered the simple density-independent M3Y NN interaction. A more comprehensive study involving the density dependence of both M3Y and R3Y is under process and will be communicated shortly. The respective neck parameter is optimised for each reaction system about the experimental half-life for both the cases of M3Y and R3Y NN interactions.

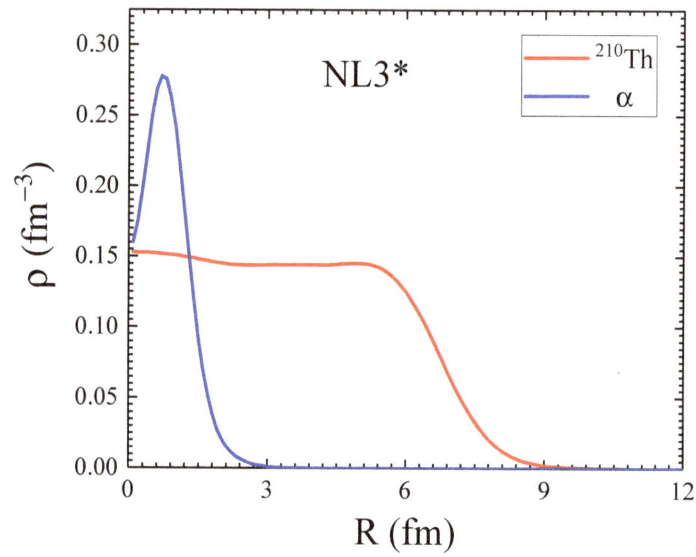

Figure 2. The total radial density distribution of ^{210}Th obtained from the RMF (NL3*) parameter set and those of α-particle, deduced from the experimental data [72] (following the illustration in Figure 1). See text for details.

Several parameters are involved in estimating the ground-state properties of nuclei using the relativistic mean-field Lagrangian. Examples of such parameters which include the binding energy (BE), charge radius r_c, quadrupole deformation parameter β_2, pairing energies E_{pair} and other bulk properties, are replete in the literature [40,73,74]. These mentioned parameters are obtained from RMF predictions for all the participating nuclei in the first decay chain ($^{214}U \rightarrow {}^{210}Th \rightarrow {}^{206}Ra \rightarrow {}^{202}Rn \rightarrow {}^{198}Po \rightarrow {}^{194}Pb$), second decay chain ($^{216}U \rightarrow {}^{212}Th \rightarrow {}^{208}Ra \rightarrow {}^{204}Rn \rightarrow {}^{200}Po \rightarrow {}^{196}Pb$) and the third decay chain ($^{218}U \rightarrow {}^{214}Th \rightarrow {}^{210}Ra \rightarrow {}^{206}Rn \rightarrow {}^{202}Po \rightarrow {}^{198}Pb$) and are compared with those of the FRDM ones given in the Table 1. As such, with an appropriate choice of parameter set, the predicted results are nearly force independent [6]. Note that, in Table 1, the RMF (NL3*) results are displayed in columns 2–10 and their respective FRDM and WS3 predictions are given in columns 11–13 and 14–16, respectively.

Table 1. The RMF (NL3*) predictions for the binding energy (BE), pairing energy E_{pair}, deformation β_2, root-mean-square radii (rms), charge radii r_c, proton radii r_p and neutron radii r_n in comparison with their results from FRDM [52] and WS3 [53]. The energy is in MeV and radii in fm.

Nuclei	RMF									FRDM			WS3		
	E_{pair}	$E_{c.m.}$	B.E	B.E/A	r_c	r_n	r_p	rms	β_2	B.E	B.E/A	β_2	B.E	B.E/A	β_2
^{214}U	16.49	−5.141	1640.51	7.67	5.66	5.73	5.60	5.68	0.001	1630.22	7.62	−0.115	1629.70	7.62	−0.106
^{210}Th	16.59	−5.173	1618.15	7.71	5.62	5.70	5.56	5.64	0.045	1610.37	7.67	−0.135	1610.53	7.67	−0.124
^{206}Ra	16.36	−5.207	1596.11	7.75	5.58	5.67	5.52	5.61	0.095	1589.89	7.72	−0.125	1590.06	7.72	−0.123
^{202}Rn	16.35	−5.241	1573.09	7.79	5.54	5.64	5.49	5.58	0.109	1568.87	7.77	−0.115	1569.16	7.77	0.096
^{198}Po	16.40	−5.276	1549.31	7.83	5.50	5.61	5.44	5.54	0.111	1547.43	7.82	0.075	1547.97	7.82	0.074
^{194}Pb	16.31	−5.312	1524.98	7.86	5.47	5.58	5.41	5.51	0.124	1526.18	7.87	0.000	1526.38	7.87	0.083
^{216}U	15.81	−5.125	1659.59	7.68	5.66	5.76	5.61	5.69	0.000	1648.69	7.63	−0.073	1648.16	7.63	−0.084
^{212}Th	16.00	−5.157	1637.16	7.72	5.62	5.72	5.57	5.66	0.003	1628.74	7.68	−0.094	1628.38	7.68	−0.101
^{208}Ra	15.89	−5.190	1614.15	7.76	5.59	5.69	5.53	5.62	0.059	1608.11	7.73	−0.125	1608.51	7.73	−0.116
^{204}Rn	15.84	−5.224	1590.80	7.80	5.55	5.66	5.489	5.59	0.083	1586.94	7.78	−0.115	1587.23	7.78	−0.101
^{200}Po	15.73	−5.258	1566.81	7.83	5.51	5.63	5.45	5.55	0.089	1565.52	7.83	−0.063	1565.66	7.83	−0.065
^{196}Pb	15.43	−5.294	1542.13	7.87	5.47	5.60	5.41	5.52	0.096	1543.48	7.88	0.000	1543.96	7.88	0.008
^{218}U	15.10	−5.109	1677.21	7.69	5.67	5.78	5.62	5.71	0.001	1666.17	7.64	0.000	1665.89	7.64	−0.004
^{214}Th	15.29	−5.141	1655.61	7.74	5.63	5.75	5.58	5.68	0.000	1646.39	7.69	−0.063	1645.97	7.69	−0.072
^{210}Ra	15.46	−5.173	1632.02	7.77	5.59	5.71	5.56	5.64	0.019	1625.80	7.74	−0.084	1625.57	7.74	−0.096
^{206}Rn	15.38	−5.207	1608.23	7.81	5.55	5.68	5.49	5.60	0.047	1604.44	7.79	−0.094	1604.86	7.79	−0.097
^{202}Po	15.22	−5.241	1583.85	7.84	5.51	5.65	5.45	5.57	0.056	1582.71	7.84	−0.063	1582.92	7.84	−0.070
^{198}Pb	14.66	−5.276	1558.98	7.87	5.47	5.62	5.42	5.54	0.075	1560.26	7.88	0.000	1560.75	7.88	0.022

The newly measured α-emitting isotope ^{214}U [15], produced via a fusion evaporation reaction and the re-measured even-even ^{216,218}U, furnishes us with helpful data with which the RMF (NL3*) is tested. In addition to the calculation of the α-decay energies, binding energies can be employed to probe the stability of nuclei as well as test the reliability of the model adopted if it can quantitatively replicate the experimental binding energies. Figure 3 displays the variation of the binding energy (B.E) and the binding energy per particle (B.E/A) with the mass number of their respective parent nuclei. In each case, the values of the binding energies increase proportionately with the increasing mass of the parent nuclei until notable peaks are formed at $A = 214, 216$ and 218 (near and) at magic numbers $N = 122, 124$ and 126, respectively. Unlike the former, the B.E/A values decrease, leading to a higher mass until a minimum is observed around the same closed shell. In both varied parameters, similar trends are observed. It is seen that the FRDM and WS3 predicted binding energies agree closely, while the RMF (NL3*) predictions display a considerable deviation, gradually decreasing from uranium to radon isotopes with about 5–10 MeV (comparing the corresponding values in columns 4, 11 and 14 of Table 1). As a necessary input for the calculation of the α-decay energies, this deviation would lead to an inaccurate estimation of the decay properties.

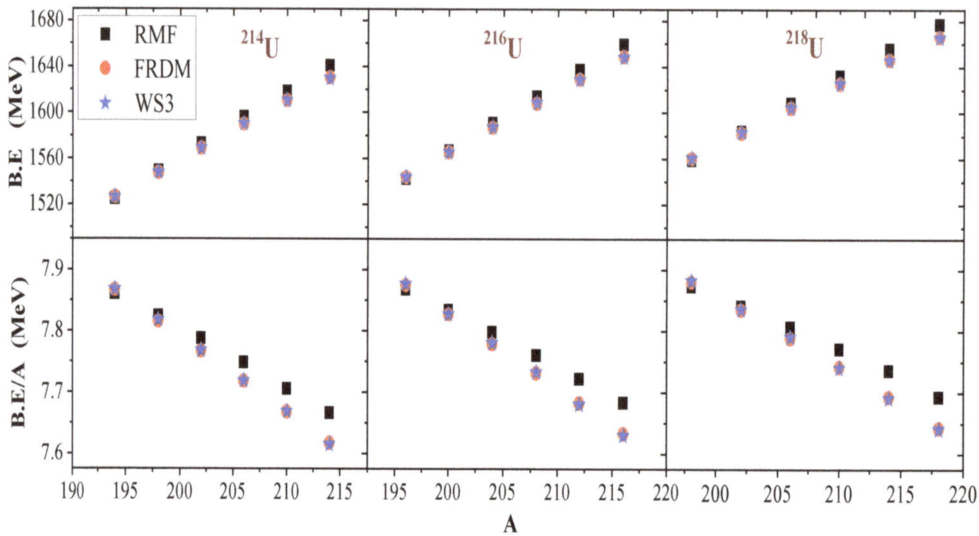

Figure 3. The binding energy B.E (upper panel) and the binding energy per particle B.E/A (lower panel) for the decay chains of even-even $^{214,216,218}U$ isotopes, estimated from the RMF (NL3*) in comparison with the results from FRDM [52] and WS3 [53].

The decay energy (Q-value) plays an indispensable role in the calculation of the half-lives of nuclei [75]. From Figure 4 and Table 2 (columns 5–7), it is observed that the estimated alpha-decay energies from FRDM and WS3 are in good agreement with the experimental data and notably, the FRDM gives a perfect fit. Meanwhile, the RMF predictions are found to be at least 2 MeV less as compared with the recent experimental measurement, although it accurately reproduces a similar trend of Q_α with the neutron number [17]. We have also compared the RMF Q-values using NL3, NL3* and DD-ME2 parameter sets with the experimental Q-values and those of WS3 (not shown in the present analysis for the sake of brevity), yet, a discrepancy of about 2 MeV is still present. This underestimated RMF prediction of the α-decay energies is obvious in all the comparisons made in this study. It is well known that little deviation in the Q values alters the resulting half-lives by a few orders of magnitude [5]. This suggests that the RMF formalism is not the best compromise to accurately estimate the Q_α values (for the region under study) and hence certain improvements in the interaction term of the Lagrangian density are needed to ensure its reliability in the prediction of the α-decay half-lives. The parameters of the macroscopic–microscopic FRDM have been fitted to the ground-state masses of about 1654 nuclei (Z, N $\geq$ 8), while the RMF approach is predicated on a Lagrangian describing the interactions of nucleons through the exchange of mesons and photons [76].

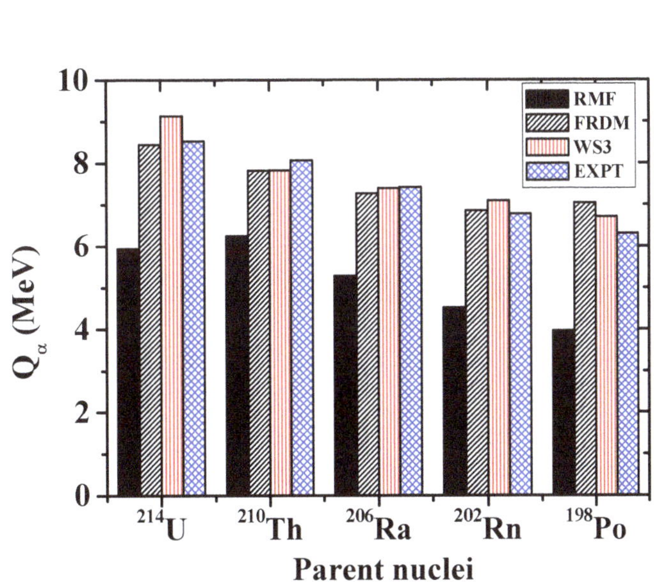

Figure 4. The Q_α energies for the α-decay chain of ^{214}U isotopes obtained from the RMF formalism (NL3*) in comparison with those from the FRDM [52], WS3 [53] and the recent experimental measurement of Zhang et al. [15].

Table 2. The R3Y and M3Y predictions of the α-decay half-lives $T_{1/2}$ within the PCM ($T = 0$) and the calculation details for the decay chains of the neutron-deficient even–even ^{214,216,218}U isotopes from RMF (NL3*) in comparison with the experimental data [15,77,78]. The Q-values also are calculated using the binding energies from FRDM [52] and WS3 [53] for comparison.

α-Transition		Experiment		Q Values (MeV)			R3Y					M3Y				
							ΔR	Scaling Factor	$\log_{10} T_{1/2}$			ΔR	Scaling Factor	$\log_{10} T_{1/2}$		
Parent	Daughter	Q_α^{expt}	$\log_{10} T_{1/2}$	RMF	FRDM	WS3			RMF	FRDM	WS3			RMF	FRDM	WS3
^{214}U	^{210}Th	8.533	−3.284	5.942	8.446	9.126	1.270	10^{-4}	5.672	−3.285	−4.950	0.540	10^{-4}	5.291	−3.197	−4.721
^{210}Th	^{206}Ra	8.069	−1.790	6.247	7.816	7.831	1.537	10^{-4}	3.913	−1.789	−1.835	0.590	10^{-4}	3.654	−1.846	−1.889
^{206}Ra	^{202}Rn	7.415	−0.620	5.280	7.276	7.394	1.737	10^{-4}	8.001	−0.633	−1.020	0.836	10^{-4}	7.622	−0.833	−1.209
^{202}Rn	^{198}Po	6.774	1.093	4.518	6.856	7.099	0.728	10^{-4}	9.098	1.094	0.717	1.487	10^{-4}	12.858	1.088	0.230
^{198}Po	^{194}Pb	6.309	2.270	3.966	7.046	6.710	1.469	10^{-8} [a]	18.299	2.269	3.404	1.302	10^{-7} [a]	18.525	2.271	3.433
^{216}U	^{212}Th	8.384	−2.326	5.863	8.346	8.513	1.810	10^{-4}	7.004	−2.319	−2.779	0.914	10^{-4}	6.796	−2.387	−2.837
^{212}Th	^{208}Ra	7.958	−1.499	5.287	7.666	8.424	1.393	10^{-4}	8.513	−1.496	−3.643	0.610	10^{-4}	8.219	−1.505	−3.553
^{208}Ra	^{204}Rn	7.273	0.137	4.951	7.126	7.019	1.913	10^{-4}	10.236	0.119	0.492	1.110	10^{-4}	10.291	0.214	0.584
^{204}Rn	^{200}Po	6.546	2.013	4.303	6.876	6.727	0.700	10^{-4}	10.568	2.014	2.205	0.956	10^{-6} [a]	15.236	2.014	2.543
^{200}Po	^{196}Pb	5.981	3.794	3.614	6.256	6.589	0.797	10^{-6} [a]	16.290	3.795	3.123	1.103	10^{-6} [a]	20.254	3.794	2.513
^{218}U	^{214}Th	8.775	−3.292	6.692	8.516	8.366	1.318	10^{-4}	2.609	−3.293	−2.896	0.592	10^{-4}	2.316	−3.293	−2.920
^{214}Th	^{210}Ra	7.827	−1.060	4.708	7.706	7.896	1.812	10^{-4}	12.896	−1.060	−1.641	0.872	10^{-4}	−9.977	−1.060	−1.631
^{210}Ra	^{206}Rn	7.151	0.585	4.500	6.936	7.591	1.622	10^{-4}	13.039	0.585	−1.599	0.828	10^{-4}	12.905	0.585	−1.558
^{206}Rn	^{202}Po	6.384	2.740	3.924	6.566	6.353	0.964	10^{-7} [a]	16.978	2.741	3.486	1.188	10^{-5} [a]	18.083	2.740	2.735
^{202}Po	^{198}Pb	5.701	5.143	3.422	5.846	6.130	1.655	10^{-6} [a]	21.800	5.142	3.913	0.840	10^{-6} [a]	21.656	5.144	3.934

[a] A higher scaling factor between the range 10^{-8}–10^{-5} is required to calculate the α-decay half-lives with daughter nuclei near (and at) proton shell closure Z = 82.

In summary, the difference in the Q-values estimated from the considered mass tables can be attributed to their respective parameterisations [76] and the mass correlation distance between surrounding nuclei, which is usually $r \leq 3$, but ranges up to $r \sim 10$ for RMF, being characterised with a larger rms deviation [79]. The α-decay half-life is a viable tool used in nuclear structure physics to reveal the shell effect of both parent and daughter nuclei. Figure 5 shows the profile of the logarithmic half-lives $\log_{10} T_{1/2}$ as a function of different α-emitting parent nuclei for the three decay chains, starting from the lightest uranium isotope ^{214}U. Again, the FRDM predictions appear to be the most consistent with the measured data [77,80]. Likewise, the WS3 is in fair agreement with the experimental data with a tolerable difference. On the other hand, the RMF half-life predictions were all-time higher with a wide difference, traceable to its relatively low alpha-decay energy. These discrepancies in the RMF predictions are reconciled by a random fitting (not following the systematic trend of FRDM and WS3) of the scaling factor, as well as the neck-length parameter ΔR, as shown in Table 3. From the table, it is clear that an appropriate scaling and fitting can compensate for the divergent RMF values. Nonetheless, the refitting process appears to play a small effect on the penetration probability P. This is because the decay energy directly imparts and determines the magnitude of cluster penetration in the decay process. As reported by Kumar and collaborators [81–83], parent nuclei with stabilised shells are marked with high half-life values and such a situation in the daughter nuclei results in relatively lower values of the half-life. The shell structure effect is prominent in all cases for the α-emitting parent ^{198,200,202}Po ($Z = 84$) having higher $\log_{10} T_{1/2}$ values. On the other hand, from Figure 5, the decay energy Q_α of Polonium isotopes, whose daughter Pb has a proton number at (or near) the magic shell closure $Z = 82$, which assumes the deepest minima as a result of its stability. This stability can be linked to the magicity of protons at (or near) ($Z = 82, 84$) or of the neutrons magic numbers $N = 214, 126, 218$. This phenomenon was also reported by Phookan [22] and Manjunatha and Sowmya [84]. Interestingly, in Figure 5, the variation in Q_α seems to be the inverse of those of the logarithmic half-lives $\log T_{1/2}$. Yet, a similar profile with the experimental data is maintained from the calculated values of RMF, FRDM and WS3. It is also noticed that Q_α increases to the size of the (neutron number of) the parent nuclei.

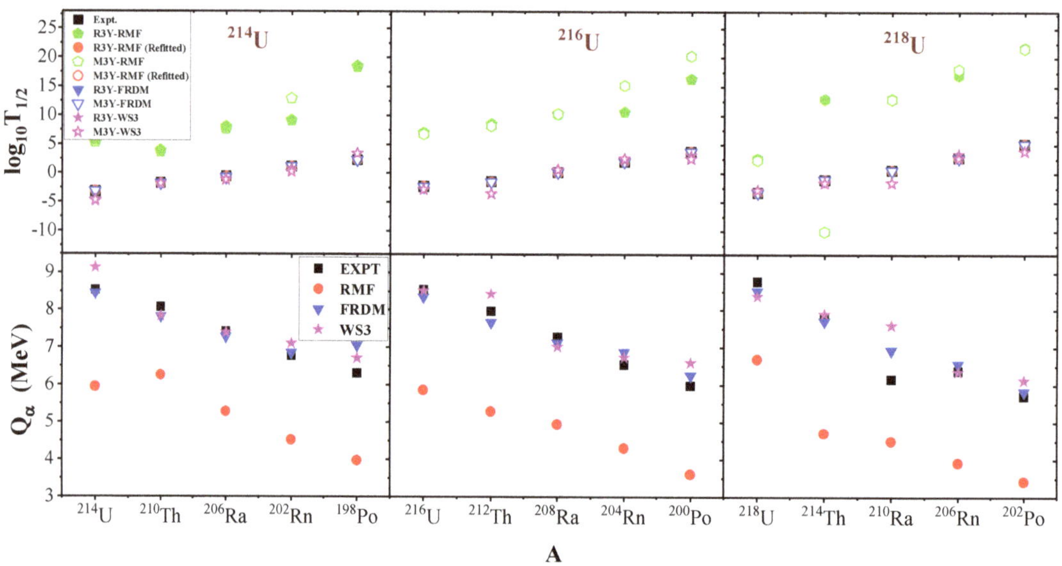

Figure 5. The logarithmic half-lives (upper panel) and the Q_α for the α-decay chain of ^{214}U, ^{216}U and ^{218}U, obtained of from the RMF formalism (NL3*) in contrast with those from the FRDM [52], WS3 [53] and the recent experimental measurement of Zhang et al. [15].

In the PCM framework, it is assumed that the penetration probability P_0 will be less than unity and manifests an abrupt reduction with an increasing mass number A of the parent nuclei. Here, our calculated P_0 values are in tune with the shell model, with notable minima around the magic numbers. This conforms with the recent findings of [24,25], in which it was demonstrated that P_0 can be influenced by the isospin asymmetry of the parents, the deformation of the daughter nuclei, and pairing and shell effects and that the minima of P_0 can be ascribed to the presence of proton, neutron shells and sub-shell closures. A careful evaluation of Figure 6, which portrays the relationship between the preformation probability P_0 and the penetrability P, where both parameters are shown as a function of the mass number, reveals that lower P_0 values correspond to a higher P and vice versa, such that their products P_0P are near to the same order for all the decay chains under study.

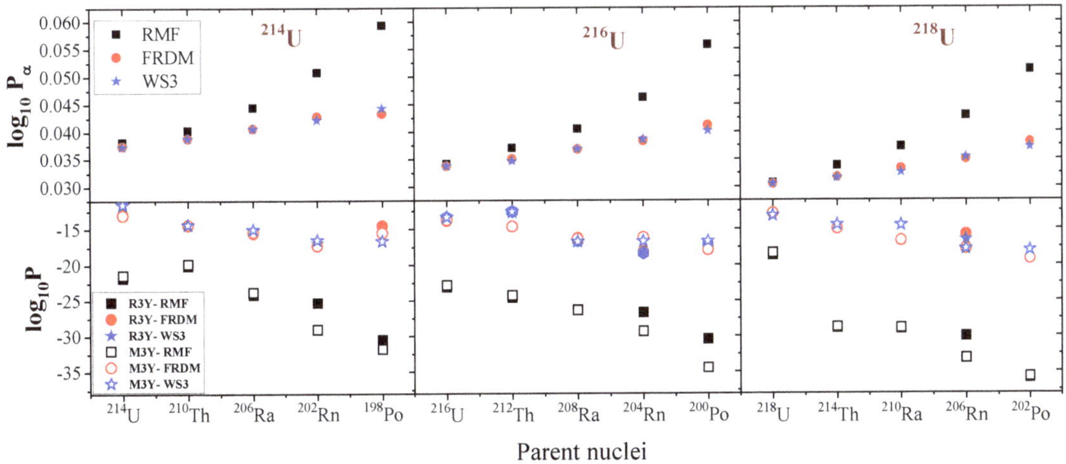

Figure 6. Preformation probability P_0 (calculated from Equation (13)) and the penetration probability P using the Q-values estimated from RMF (NL3*), FRDM [52] and WS3 [53] for ^{214}U, ^{216}U and ^{218}U decay chains.

Pairing is a vital quantity used in determining the nuclear properties of open-shell nuclei. Here, the pairing correlation is taken into account using the BCS approach. There exists a marginal relationship between the pairing energy E_{pair} and the deformation parameter β_2 [6]. In other words, the variation of β_2 yield a meagre change in E_{pair}. Figure 7 (upper panel) outlines the variation of the pairing energy E_{pair} with the mass number A of the parent nuclei. In our context, the pairing is found to increase until a peak is reached at ^{210}Th, ^{212}Th and ^{210}Ra, corresponding to $N = 120, 122, 122$, respectively, just before the neutron magic shell closure. However, in the first decay chain, the sudden surge at ^{198}Po $(Z = 84)$ indicates the presence of shape coexistence [14,85–87]. Shape coexistence is a habitual hallmark of neutron-deficient nuclei [14,87,88]. The RMF formulation has been successfully employed in the investigation of the quadrupole moment and found to be in harmony with the experimental data [86,89]. Figure 7 (lower panel) shows the changes in the quadrupole deformation parameter β_2 obtained from the RMF (NL3*), FRDM and WS3 results as a function of the mass number A. The RMF predicts far from the others. In the first chain, shape changes are observed with increasing mass numbers. Except for ^{194}Pb from the FRDM results, a shape change is noticed, from prolate to oblate, at $A = 198$ and thereafter descends steeply to a highly deformed oblate at $A = 202$. Most of these isotopes are nearly spherical in their ground-state configurations [90]. Additionally, a closer examination of the FRDM predictions (which is the most consistent with the experimental data in our study) of β_2 values in Table 1 shows that the deformation at (and near) the shell closures $Z = 82$ and $Z = 126$ are approximately zero, indicating a state of stability.

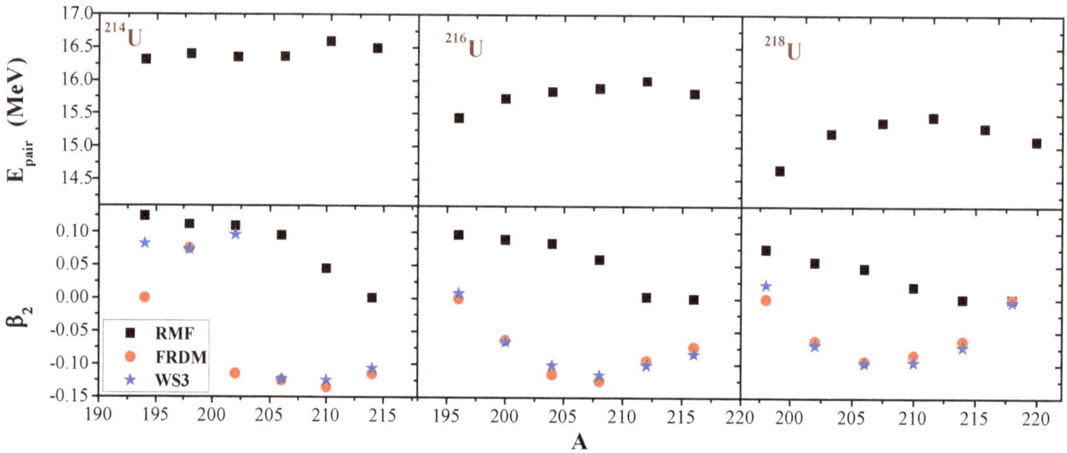

Figure 7. Pairing energies E_{pair} from RMF prediction and the quadrupole deformation parameter β_2 obtained from the RMF (NL3*), FRDM [52] and WS3 [53] for the three decay chains.

In the present study, we have also extended the theoretical investigation of Sawhney et al. [28] to neutron-deficient nuclei using the preformed cluster-decay model (PCM) within the RMF framework at temperature $T = 0$ for generalisation. The details of our calculations are shown in Table 2, highlighting the α-transitions (Columns 1 and 2), the experimental and calculated Q values (Columns 3 to 7), and details of R3Y calculations (Columns 8 to 12), as well as those of the M3Y (Columns 13 to 17) of the alpha-decay chains. The large deviation of the RMF-calculated Q values seen in their corresponding logarithmic half-life predictions have been extensively discussed. Columns 8 and 13 of Table 2 show the trend in the fitted neck-length parameter participating in the three decay chains for R3Y and M3Y, respectively. A close difference can be observed in the ΔR values, except in the case of Polonium ($Z = 84$), in which a smaller ΔR is required for its fitting. One very clear observation is that, in addition to the Po, almost all the ΔR values in M3Y were smaller than those of R3Y. This alludes to the fact that the R3Y and M3Y NN interactions have different barrier characteristics [91].

The difference in the effective R3Y and M3Y interactions is conspicuously shown in the driving potential ($V(R_a) - Q_{FRDM}$). Figure 8 displays the driving potential as a function of the mass number of the parent nuclei for R3Y (in black squares) and M3Y (in red circles) NN interactions for the three decay chains. The FRDM-calculated Q-values were chosen for this comparison. Despite the well-fitted neck-parameter ΔR, the influence of the different barrier characteristics of the R3Y and M3Y NN potential is visible. Mostly, the R3Y driving potential was found to be higher than M3Y, except for Rn and (in the second decay chain) Po isotopes. This anomaly occurs because these nuclei are found in the vicinity of the proton shell closure $Z = 82$. Indeed, α-decay studies of neutron-deficient isotopes provide information on nuclei near the proton drip line [92].

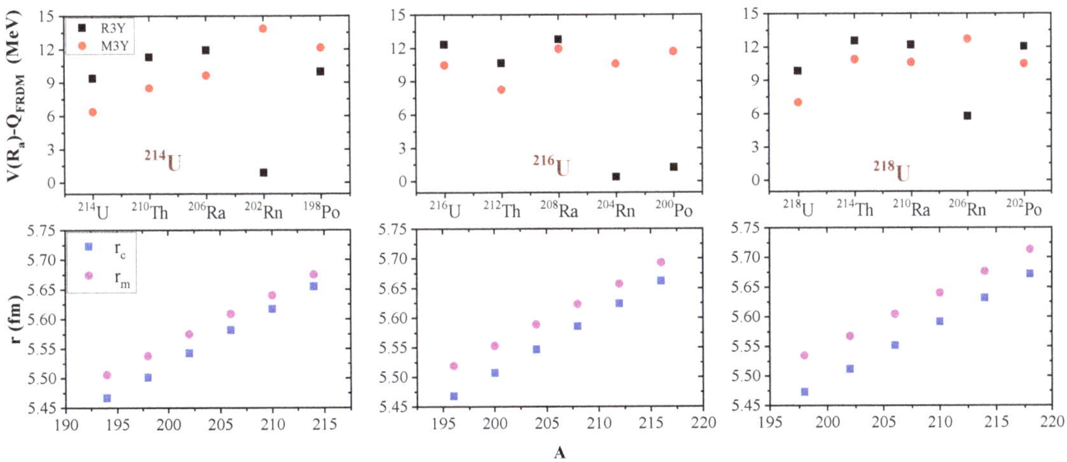

Figure 8. Upper panel: The driving potential ($V(R_a) - Q_{FRDM}$) for the decay chains of ^{214,216,218}U isotopes for R3Y (in black squares) and M3Y (in red circles) NN potentials as a function of the parent nuclei A. The lower panel illustrates the rms radii for matter distribution r_m and charge distribution r_c for the ^{214,216,218}U isotopes using the relativistic mean-field formalism RMF (NL3*).

The scaling factors for the decay chains of the considered Uranium isotopes are given in columns 9 and 14 of Table 2. The previous findings [28,45] at PCM ($T = 0$), indicating ground-state α-decays, are confirmed for the first three α-transitions in each of the chains. However, higher scaling factors in the range 10^{-8}–10^{-5} are required to normalise the penetrability P of the decay channels around the proton shell closure $Z = 82$. This is obvious for $Z = 82$ and 84, whose scaling factors are given in the footnote "a". The deviation in the RMF Q-values necessitates the use of random scaling to attain the measured experimental half-lives, as shown in Table 3. Despite this fact, the need for higher scaling factors around the proton magic number $Z = 82$ is confirmed, as highlighted with the footnote "b". Thus, the previous knowledge is now broadened around the proton magic number 82. The root mean square (rms) r_m and the charge radii r_c are deduced from the RMF formalism with the NL3* force parameter set and the corresponding values are given in the lower panel of Figure 8 as well as Columns 6 and 9 of Table 1, respectively. A similar behaviour and trend are noticed in the variation of both r_m and r_c. Specifically, both rms and charge radii increase monotonously with the increase in the mass of the parent nuclei. A close examination of the three decay chains reveals that the difference between r_m and r_c becomes larger with the increase in the neutron number. Currently, there are no available data for comparison.

Table 3. Details of the random fitting of the neck-length ΔR and scaling factor due to the deviations found in RMF (only).

Parent Nuclei	Expt.	R3Y			M3Y		
	$\log_{10} T_{1/2}$	ΔR (fm)	Scaling Factor	$\log_{10} T_{1/2}$	ΔR (fm)	Scaling Factor	$\log_{10} T_{1/2}$
^{214}U	−3.284	1.249	10^5	−3.285	0.544	10^5	−3.200
^{210}Th	−1.790	1.208	10^1	−1.791	1.390	10^3	−1.850
^{206}Ra	−0.620	1.070	10^3	−0.636	0.560	10^4	−0.833
^{202}Rn	1.093	0.728	10^4	1.098	1.010	10^7	1.104
^{198}Po	2.270	1.480	$10^{8\ b}$	2.276	1.050	$10^{9\ b}$	2.267
^{216}U	−2.326	1.156	10^4	−2.319	0.492	10^4	−2.326
^{212}Th	−1.499	1.399	10^6	−1.499	0.730	10^6	−1.501
^{208}Ra	0.137	1.911	10^6	0.189	0.980	10^6	0.127
^{204}Rn	2.013	1.030	10^5	2.008	0.961	10^7	2.225
^{200}Po	3.794	1.710	$10^{10\ b}$	3.816	1.400	$10^{11\ b}$	3.796
^{218}U	−3.292	0.793	10^1	−3.280	0.625	10^2	−3.296
^{214}Th	−1.060	0.591	10^9	−1.061	1.020	10^5	−1.129
^{210}Ra	0.585	1.090	10^7	0.549	0.730	10^8	0.581
^{206}Rn	2.740	1.364	10^9	2.734	1.200	10^8	2.742
^{202}Po	5.143	0.690	10^7	5.144	0.780	$10^{10\ b}$	5.194

^{b} The need for a higher 'random' scaling factor is evident to estimate the α-decay half-lives with daughter nuclei near (and at) proton shell closure $Z = 82$.

4. Summary and Conclusions

The RMF framework is employed for the calculation of the ground-state properties (binding energies, quadrupole deformation, rms charge and matter radii and Q-values) of the newly measured ^{214}U and remeasured ^{216,218}U together with the decay product in their respective α-decay chains and compared with those from the FRDM and WS3 mass tables. The α-decay process was treated as a quantum tunnelling effect across a potential barrier using the WKB approximation and the half-lives are deduced within the PCM framework. The phenomenological M3Y and the recently developed R3Y NN potentials are used to obtain the nuclear interaction potential following the double-folding technique. In all cases, the FRDM data with the new R3Y NN potential are found to be in close agreement with the experimental half-lives. Moreover, from our calculations, we observed that in the so-called region of *abnormal enhancement*, the scaling factor 10^{-4} is maintained and normalised with the optimised neck-parameter ΔR. Nonetheless, a discrepancy is noticed around the proton shell closure with $Z = 82$ and 84, and this is consistently found in all three decay chains. As such, the required scaling factor to estimate the experimental half-lives may range from 10^{-8}–10^{-5} at Z = 82, 84. However, it has been lately demonstrated in another study that the need for scaling factor is not necessary when the temperature effect $T \neq 0$ is considered. This assertion will be subsequently verified around the magic number $Z = 82$. Moreover, a number of studies have demonstrated that the quantisation, deformation and orientation degrees of freedom can influence the alpha-decay half-lives quantitatively. These effects will also be incorporated in our future study. In contrast with the FRDM- and WS3-estimated binding energies for the considered neutron-deficient nuclei, an unusually large difference (gradually decreasing from uranium to radon isotopes with about 10.288 MeV up to 4.219 MeV) is also observed in the those of RMF despite various parameter sets employed. This contrast can be attributed to the difference in their respective parameterisations as well as the mass correlation distance between surrounding nuclei, which has a larger rms deviation for RMF. This indicates that certain improvements are needed in the RMF framework in this mass region.

Author Contributions: Conceptualisation, T.M.J., M.B. and R.K.; methodology, T.M.J.; software, T.M.J. and R.K.; validation, T.M.J., M.B. and R.K.; formal analysis, T.M.J.; investigation, T.M.J.; resources, K.A and N.A.; data curation, T.M.J., M.B. and R.K.; writing—original draft preparation, T.M.J. and N.J.; writing—review and editing, T.M.J. and N.J.; visualisation, T.M.J. and N.J.; supervision, M.B., R.K., K.A. and N.A.; project administration, T.M.J., M.B. and R.K.; funding acquisition, K.A. and N.A. All authors have read and agreed to the published version of the manuscript.

Funding: This research was funded by the Ministry of Education Malaysia (under the Grant No: FRGS/1/2019/STG02/UNIMAP/02/2). This work was supported by DAE-BRNS Project Sanction No. 58/14/12/2019-BRNS, FOSTECT Project Code: FOSTECT.2019B.04, and FAPESP Project Nos. 2017/05660-0.

Institutional Review Board Statement: Not applicable.

Informed Consent Statement: Not applicable.

Data Availability Statement: The data are contained within the article.

Acknowledgments: One of the authors (TMJ) take this opportunity to convey special thanks to Shilpa Rana for fruitful discussions and suggestions towards the improvement of this manuscript. TMJ is also thankful to the Institute of Engineering Mathematics (IMK), UniMAP for providing computer facilities during the work.

Conflicts of Interest: The authors declare no conflict of interest.

Abbreviations

The following abbreviations are used in this manuscript:

CSF	Common Scaling Factor
PCM	Preformed Cluster-decay Model
WKB	Wentzel–Kramers–Brillouin
FRDM	Finite-Range-Droplet Model
RMF	Relativistic Mean-Field
M3Y	Michigan-3-Yukawa
R3Y	Relativistic (Mean-Field)-3-Yukawa
OPEP	One-Pion Exchange Potential
NN	Nucleon–Nucleon

References

1. Geiger, H.; Nuttall, J.M. The ranges of the α particles from various radioactive substances and a relation between range and period of transformation. *Philos. Mag.* **1911**, *22*, 613. [CrossRef]
2. Geiger, H. Reichweitemessungen an α-Strahlen. *Z. Phys.* **1922**, *8*, 45. [CrossRef]
3. Van Duppen, P.; Andreyev, A.N. Alpha decay and beta-delayed fission: Tools for nuclear physics studies. In *The Euroschool on Exotic Beams*; Scheidenberger, C., Pfutzner, M., Eds.; Springer International Publishing: Cham, Switzerland, 2018; Volume 5, pp. 65–116.
4. Qi, C.; Liotta, R.; Wyss, R. The single-particle unit for alpha decay. *Prog. Part. Nucl. Phys.* **2019**, *105*, 214. [CrossRef]
5. Gupta, M.; Heinz, S.; Munzenberg, G. Half lives and Q values of nuclei appearing in the α-decay chains of recently reported new isotopes. *Rom. J. Phys.* **2018**, *63*, 304.
6. Bhuyan, M.; Patra, S. K.; Gupta, R.K. Relativistic mean-field study of the properties of Z = 117 nuclei and the decay chains of the $^{293,294}117$ isotopes. *Phys. Rev. C* **2011**, *84*, 014317. [CrossRef]
7. Fynbo, H.O.; Diget, C.A.; Bergmann, U.C.; Borge, M.J.; Cederkäll, J.; Dendooven, P.; Fraile, L.M.; Franchoo, S.; Fedosseev, V.N.; Fulton, B.R.; et al. Revised rates for the stellar triple-α process from measurement of ^{12}C nuclear resonances. *Nat. Lond.* **2005**, *433*, 136. [CrossRef]
8. Delion, D.S.; Sandulescu, A.; Greiner, W. Evidence for α clustering in heavy and superheavy nuclei. *Phys. Rev. C* **2004**, *69*, 044318. [CrossRef]
9. Lovas, R.G.; Liotta, R.J.; Insolia, A.; Varga, K.; Delion, D.S. Microscopic theory of cluster radioactivity. *Phys. Rep.* **1998**, *294*, 265. [CrossRef]
10. Seweryniak, D.; Starosta, K.; Davids, C.N.; Gros, S.; Hecht, A.A.; Hoteling, N.; Khoo, T.L.; Lagergren, K.; Lotay, G.; Peterson, D.; et al. α decay of ^{105}Te. *Phys. Rev. C* **2006**, *73*, 061301(R). [CrossRef]

11. Andreyev, A.N.; Huyse, M.; Van Duppen, P.; Weissman, L.; Ackermann, D.; Gerl, J.; Hessberger, F.P.; Hofmann, S.; Kleinböhl, A.; Münzenberg, G.; et al. A triplet of differently shaped spin-zero states in the atomic nucleus ^{186}Pb. *Nat. Lond.* **2000**, *405*, 430. [CrossRef]
12. Andreyev, A.N.; Huyse, M.; Van Duppen, P.; Qi, C.; Liotta, R.J.; Antalic, S.; Ackermann, D.; Franchoo, S.; Hessberger, F.P.; Hofmann, S.; et al. Signatures of the Z = 82 shell closure in α-decay process. *Phys. Rev. Lett.* **2013**, *110*, 242502. [CrossRef]
13. Auranen, K.; Seweryniak, D.; Albers, M.; Ayangeakaa, A.D.; Bottoni, S.; Carpenter, M.P.; Chiara, C.J.; Copp, P.; David, H.M.; Doherty, D.T.; et al. Superallowed α Decay to Doubly Magic ^{100}Sn. *Phys. Rev. Lett.* **2018**, *121*, 182501. [CrossRef]
14. Wauters, J.; Bijnens, N.; Dendooven, P.; Huyse, M.; Hwang, H.Y.; Reusen, G.; von Schwarzenberg, J.; Van Duppen, P.; Kirchner, R.; Roeckl, E. Fine structure in the alpha decay of even-even nuclei as an experimental proof for the stability of the Z = 82 magic shell at the very neutron-deficient side. *Phys. Rev. Lett.* **1994**, *72*, 1329. [CrossRef]
15. Zhang, Z.Y.; Yang, H.B.; Huang, M.H.; Gan, Z.G.; Yuan, C.X.; Qi, C.; Andreyev, A.N.; Liu, M.L.; Ma, L.; Zhang, M.M.; et al. New α-Emitting Isotope ^{214}U and Abnormal Enhancement of α-Particle Clustering in Lightest Uranium Isotopes. *Phys. Rev. Lett.* **2021**, *126*, 152502. [CrossRef]
16. Sorlin, O.; Porquet, M.G. Nuclear magic numbers: New features far from stability. *Prog. Part. Nucl. Phys.* **2008**, *61*, 602. [CrossRef]
17. Zong-Qiang, S.; Jian-You, G. Study of the Alpha-Decay Chain for ^{194}Rn with Relativistic Mean-Field Theory. *Commun. Theor. Phys.* **2008**, *49*, 1583. [CrossRef]
18. Santhosh, K.P.; Sukumaran, I. Alpha decay studies on Po isotopes using different versions of nuclear potentials. *Eur. Phys. J. A* **2017**, *53*, 1. [CrossRef]
19. Poenaru, D.N.; Ivascu, M.; Greiner, W. Half lives for spontaneous emission of heavy ions from atomic nuclei. International Journal of Radiation Applications and Instrumentation. *Part D Nucl. Tracks Radiat. Meas.* **1986**, *12*, 313. [CrossRef]
20. Cui, J.P.; Xiao, Y.; Gao, Y.H.; ; Wang, Y.Z. α-decay half-lives of neutron-deficient nuclei. *Nucl. Phys. A* **2019**, *987*, 99–111. [CrossRef]
21. Adel, A.; Abdulghany, A.R. Proton radioactivity and α-decay of neutron-deficient nuclei. *Phys. Scr.* **2021**, *96*, 125314. [CrossRef]
22. Phookan, C.K. α-particle preformation of heavy nuclei using proximity potential. *Chin. J. Phys.* **2017**, *55*, 176. [CrossRef]
23. He, Y.; Yu, X.; Zhang, H.F. Improved empirical formula for α particle preformation factor. *Chin. Phys. C* **2021**, *45*, 014110. [CrossRef]
24. Seif, W.M. The α decay spectroscopic factor of heavy and superheavy nuclei. *J. Phys. G Nucl. Part. Phys.* **2013**, *40*, 105102. [CrossRef]
25. Seif, W.M.; Botros, M.M.; Refaie, A.I. Preformation probability inside α emitters having different ground state spin-parity than their daughters. *Phys. Rev. C* **2015**, *92*, 044302. [CrossRef]
26. Gamow, G. Zur quantentheorie des atomkernes. *Z. Phys.* **1928**, *51*, 204. [CrossRef]
27. Gurney, R.W.; Condon, E.U. Wave mechanics and radioactive disintegration. *Nature* **1928**, *122*, 439. [CrossRef]
28. Niyti, G.; Sawhney, M.K.; Sharma, R.K. Gupta, α-decay chains of recoiled superheavy nuclei: A theoretical study. *Phys. Rev. C* **2015**, *91*, 054606. [CrossRef]
29. Quentin, P.; Flocard, H. Self-consistent calculations of nuclear properties with phenomenological effective forces. *Annu. Rev. Nucl. Part. Sci.* **1978**, *28*, 523. [CrossRef]
30. Hornyak, W. *Nuclear Structure*; Elsevier: Amsterdam, The Netherlands, 2012.
31. Schunck, N.; Robledo, L.M. Microscopic theory of nuclear fission: A review. *Rep. Prog. Phys.* **2016**, *79*, 116301. [CrossRef]
32. Vautherin, D.; Brink, D.T. Hartree-Fock calculations with Skyrme's interaction. I. Spherical nuclei. *Phys. Rev. C* **1972**, *5*, 626. [CrossRef]
33. Epelbaum, E.; Hammer, H.W.; Meiner, U.G. Modern theory of nuclear forces. *Rev. Mod. Phys.* **2009**, *81*, 1773. [CrossRef]
34. Ekström, A.; Baardsen, G.; Forssén, C.; Hagen, G.; Hjorth-Jensen, M.; Jansen, G.R.; Machleidt, R.; Nazarewicz, W.; Papenbrock, T.; Sarich, J.; et al. Optimized chiral nucleon-nucleon interaction at next-to-next-to-leading order. *Phys. Rev. Lett.* **2013**, *110*, 192502. [CrossRef]
35. Singh, B.; Bhuyan, M.; Patra, S.K.; Gupta, R.K. Optical potential obtained from relativistic-mean-field theory-based microscopic nucleon–nucleon interaction: Applied to cluster radioactive decays. *J. Phys. G Nucl. Part. Phys.* **2012**, *39*, 069501. [CrossRef]
36. Singh, B.; Bhuyan, M.; Patra, S.K.; Gupta, R.K. A new microscopic nucleon-nucleon interaction derived from relativistic mean field theory. *arXiv* **2010**, arXiv:1011.5732.
37. Satchler, G.R.; Love, W.G. Folding model potentials from realistic interactions for heavy-ion scattering. *Phys. Rep.* **1979**, *55*, 183. [CrossRef]
38. Biswal, S.K.; El Sheikh, M.A.; Biswal, N.; Yusof, N.; Kassim, H.A.; Patra, S.K.; Bhuyan, M. Nuclear matter properties of finite nuclei using relativistic mean field formalism. *Nucl. Phys. A* **2020**, *1004*, 122042. [CrossRef]
39. Itagaki, N.; Afanasjev, A.V.; Ray, D. Possibility of ^{14}C cluster as a building block of medium-mass nuclei. *Phys. Rev. C* **2020**, *101*, 034304. [CrossRef]
40. Taninah, A.; Agbemava, S.E.; Afanasjev, A.V. Covariant density functional theory: An estimation of systematic uncertainties. *Bull. Am. Phys. Soc.* **2020**, *65*, 19.
41. Horowitz, C.J.; Serot, B.D. Self-consistent hartree description of finite nuclei in a relativistic quantum field theory. *Nucl. Phys. A* **1981**, *368*, 503. [CrossRef]
42. Serot, B.D.; Walecka, J.D. Relativistic nuclear many-body theory. *Recent Prog. Many-Body Theor.* **1992**, *49*, 49–92.
43. Reinhard, P.G. The relativistic mean-field description of nuclei and nuclear dynamics. *Rep. Prog. Phys.* **1989**, *52*, 439. [CrossRef]

44. Ebran, J.P.; Khan, E.; Lasseri, R.; Marevic, P.; Niksic, T.; Sandulescu, N.; Vretenar, D. Investigation of nuclear cluster phenomenology with the relativistic EDF approach. *Il Nuovo C. C-Colloq. Phys.* **2019**, *42*, 2–3.
45. Kumar, R.; Sandhu, K.; Sharma, M.K.; Gupta, R.K. Fusion-evaporation residues and α-decay chains of the superheavy element Z = 115 formed in the ^{243}Am+ ^{48}Ca reaction using the dynamical cluster-decay model. *Phys. Rev. C* **2013**, *87*, 054610. [CrossRef]
46. Sharma, K.; Sawhney, G.; Sharma, M.K.; Gupta, R.K. Decay of Plutonium isotopes via spontaneous and heavy-ion induced fission paths. *Nucl. Phys. A* **2018**, *972*, 1. [CrossRef]
47. Deng, J.G.; Zhang, H.F. Analytic formula for estimating the α-particle preformation factor. *Phys. Rev. C* **2020**, *102*, 044314. [CrossRef]
48. Deng, J.G.; Zhang, H.F. Systematic study of α decay half-lives within the Generalized Liquid Drop Model with various versions of proximity energies. *Chin. Phys. C* **2021**, *45*, 024104. [CrossRef]
49. Wentzel, G. Eine verallgemeinerung der quantenbedingungen für die zwecke der wellenmechanik. *Z. Physik* **1926**, *38*, 518. [CrossRef]
50. Kramers, H.A. Wellenmechanik und halbzahlige Quantisierung. *Z. Physik* **1926**, *39*, 828. [CrossRef]
51. Brillouin, L. Schr "o dinger's undulatory method; a general method of r 'e solution by successive approximations. *C. R. Acad. Sci.* **1926**, *183*, 24.
52. Möller, P.; Sierk, A.J.; Ichikawa, T.; Sagawa, H. Nuclear ground-state masses and deformations: FRDM. *At. Data Nucl. Data Tables* **2016**, *109*, 1. [CrossRef]
53. Liu, M.; Wang, N.; Deng, Y.; Wu, X. Further improvements on a global nuclear mass model. *Phys. Rev. C* **2011**, *84*, 014333. [CrossRef]
54. Singh, B.; Patra, S.K.; Gupta, R.K. Cluster radioactive decay within the preformed cluster model using relativistic mean-field theory densities. *Phys. Rev. C* **2010**, *82*, 014607. [CrossRef]
55. Sahu, B.B.; Singh, S.K.; Bhuyan, M.; Biswal, S.K.; Patra, S.K. Importance of nonlinearity in the NN potential. *Phys. Rev. C* **2014**, *89*, 034614. [CrossRef]
56. Bhuyan, M.; Patra, S.K.; Arumugam, P.; Gupta, R.K. Nuclear sub-structure in $^{112-122}$Ba nuclei within relativistic mean field theory. *Int. J. Mod. Phys. E* **2011**, *20*, 1227. [CrossRef]
57. Ring, P. Relativistic mean field theory in finite nuclei. *Prog. Part. Nucl. Phys.* **1996**, *37*, 193. [CrossRef]
58. Bhuyan, M. Structural evolution in transitional nuclei of mass $82 \leq A \leq 132$. *Phys. Rev. C* **2015**, *92*, 034323. [CrossRef]
59. Lalazissis, G.A.; Karatzikos, S.; Fossion, R.; Arteaga, D.P.; Afanasjev, A.V.; Ring, P. The effective force NL3 revisited. *Phys. Lett. B* **2009**, *671*, 36. [CrossRef]
60. Gupta, R.K. Collective clusterization in nuclei and excited compound systems: The dynamical cluster-decay model. In *Clusters in Nuclei*; Springer: Berlin/Heidelberg, Germany, 2010; pp. 223–265.
61. Basu, D.N. Folding model analysis of alpha radioactivity. *J. Phys. G Nucl. Part. Phys.* **2003**, 29, 2079. [CrossRef]
62. Malik, S.S.; Gupta, R.K. Theory of cluster radioactive decay and of cluster formation in nuclei. *Phys. Rev. C* **1989**, *39*, 1992. [CrossRef]
63. Gupta, R.K.; Greiner, W. Cluster radioactivity. *Int. J. Mod. Phys. E* **1994**, *3*, 335. [CrossRef]
64. Matsuse, T.; Beck, C.; Nouicer, R.; Mahboub, D. Extended Hauser-Feshbach method for statistical binary decay of light-mass systems. *Phys. Rev. C* **1997**, *55*, 1380. [CrossRef]
65. Sanders, S.J.; Kovar, D.G.; Back, B.B.; Beck, C.; Henderson, D.J.; Janssens, R.V.F.; Wang, T.F.; Wilkins, B.D. Binary decay of ^{56}Ni formed in the $^{32+24}$Mg reaction. *Phys. Rev. C* **1989**, *40*, 2091. [CrossRef]
66. Sanders, S.J. Fusion-fission in nuclear systems with $40 \leq A_{CN} \leq 80$. *Phys. Rev. C* **1991**, *44*, 2676. [CrossRef]
67. Royer, G., Normand, C., Druet, E. Analytic description of the fusion and fission processes through compact quasi-molecular shapes. *Nucl. Phys. A* **1998**, *634*, 267. [CrossRef]
68. Gao, J.; Bao, X.; Zhang, H.; Li, J.; Zhang, H. New numerical method for fission half-lives of heavy and superheavy nuclei at ground and excited states. *Nucl. Phys. A* **2014**, *929*, 246. [CrossRef]
69. Poenaru, D.N. *Particle Emission From Nuclei: Volume I: Nuclear Deformation Energy*; CRC Press: Boca Raton, FL, USA, 2018.
70. Greiner, M.; Scheid, W. Radioactive decay into excited states via heavy ion emission. *J. Phys. G Nucl. Part. Phys.* **1986**, *12*, L229. [CrossRef]
71. Khoa, D.T.; Von Oertzen, W. Refractive alpha-nucleus scattering: A probe for the incompressibility of cold nuclear matter. *Phys. Lett. B* **1995**, *342*, 6. [CrossRef]
72. De Vries, H.; De Jager, C.W.; De Vries, C. Nuclear charge-density-distribution parameters from elastic electron scattering. *At. Data Nucl. Data Tables* **1987**, *36*, 495. [CrossRef]
73. Sahoo, T.; Patra, S.K. Search for the stable isotopes for Z = 119 and 121 superheavy elements using relativistic mean field model. *Phys. Scr.* **2020**, *95*, 085302. [CrossRef]
74. Pattnaik, J.A.; Panda, R.N.; Bhuyan, M.; Patra, S.K. Isotopic shift and search of magic number in the superheavy region. *Phys. Scr.* **2021**, *96*, 125319.
75. Hong-Fei, Z.; Jun-Qing, L.; Wei, Z.; Xiao-Hong, Z.; Zai-Guo, G. Systematic study on alpha decay half-lives of superheavy nuclei. *Commun. Theor. Phys.* **2007**, *48*, 545. [CrossRef]
76. Patyk, Z.; Baran, A.; Berger, J.F.; Dechargé, J.; Dobaczewski, J.; Ring, P.; Sobiczewski, A. Masses and radii of spherical nuclei calculated in various microscopic approaches. *Phys. Rev. C* **1999**, *59*, 704. [CrossRef]

77. Ma, L.; Zhang, Z.Y.; Gan, Z.G.; Yang, H.B.; Yu, L.; Jiang, J.; Wang, J.G.; Tian, Y.L.; Wang, Y.S.; Guo, S.; Ding, B. α-decay properties of the new isotope ^{216}U. *Phys. Rev. C* **2015**, *91*, 051302. [CrossRef]
78. NNDC. *National Nuclear Data Center, Chart of Nuclides*; Brookhaven National Lab. (BNL): Upton, NY, USA, 2020.
79. Niu, Z.M.; Zhu, Z.L.; Niu, Y.F.; Sun, B.H.; Heng, T.H.; Guo, J.Y. Radial basis function approach in nuclear mass predictions. *Phys. Rev. C* **2013**, *88*, 024325. [CrossRef]
80. Leppänen, A.P.; Uusitalo, J.; Leino, M.; Eeckhaudt, S.; Grahn, T.; Greenlees, P.T.; Jones, P.; Julin, R.; Juutinen, S.; Kettunen, H.; et al. α decay studies of the nuclides ^{218}U and ^{219}U. *Phys. Rev. C* **2007**, *75*, 054307.
81. Kumar, S.; Balasubramaniam, M.; Gupta, R.K.; Münzenberg, G.; Scheid, W. The formation and decay of superheavy nuclei produced in ^{48}Ca-induced reactions. *J. Phys. G Nucl. Part. Phys.* **2003**, *29*, 625. [CrossRef]
82. Kumar, S.; Thakur, S.; Kumar, R. Decay studies of ^{288}-287115 alpha-decay chains. *J. Phys. G Nucl. Part. Phys.* **2009**, *36*, 105104. [CrossRef]
83. Kumar, S. α decay chains study for the recently observed superheavy element Z = 117 within the isospin cluster model. *Phys. Rev. C* **2012**, *85*, 024320. [CrossRef]
84. Manjunatha, H.C.; Sowmya, N. Competition between spontaneous fission ternary fission cluster decay and alpha decay in the super heavy nuclei of Z = 126. *Nucl. Phys. A* **2018**, *969*, 68. [CrossRef]
85. Van de Vel, K. Shape Coexistence in Neutron-Deficient Polonium and Lead Nuclei. Ph.D. Thesis, Katholieke Universiteit Leuven, Leuven, Belgium, 2003.
86. Bender, M.; Cornelius, T.; Lalazissis, G.A.; Maruhn, J.A.; Nazarewicz, W.; Reinhard, P.G. The Z = 82 shell closure in neutron-deficient Pb isotopes. *Eur. Phys. J. A-Hadron. Nucl.* **2002**, *14*, 23. [CrossRef]
87. Sun, X.D.; Zhang, H.F. α decay preformation probabilities across the N = 126 shell closure based on the single particle energy spectra. *J. Phys. G Nucl. Part. Phys.* **2018**, *45*, 075106. [CrossRef]
88. Van Duppen, P.; Huyse, M. Shape coexistence around the Z = 82 closed shell probed by α-decay. *Hyperfine Interact.* **2000**, *129*, 149. [CrossRef]
89. Swain, R.; Patra, S.K.; Sahu, B.B. Nuclear structure and decay modes of Ra isotopes within an axially deformed relativistic mean field model. *Chin. Phys. C* **2018**, *42*, 084102. [CrossRef]
90. Julin, R.; Helariutta, K.; Muikku, M. Intruder states in very neutron-deficient Hg, Pb and Po nuclei. *J. Phys. G Nucl. Part. Phys.* **2001**, *27*, R109. [CrossRef]
91. Sahu, B.; Agarwalla, S.K.; Patra, S.K. Half-lives of proton emitters using relativistic mean field theory. *Phys. Rev. C* **2011**, *84*, 054604. [CrossRef]
92. Leino, M.; Uusitalo, J.; Allatt, R.G.; Armbruster, P.; Enqvist, T.; Eskola, K.; Hofmann, S.; Hurskanen, S.; Jokinen, A.; Ninov, V.; et al. Alpha decay studies of neutron-deficient radium isotopes. *Phys. A Hadron. Nucl.* **1996**, *355*, 157. [CrossRef]

foundations

MDPI

Communication

Peculiar Features of Molecular Hydrogen Ions Formed by Proton Collisions with Hydrogen Atoms of the Second Flavor

Eugene Oks

Physics Department, Auburn University, 380 Duncan Drive, Auburn, AL 36849, USA; goks@physics.auburn.edu

Abstract: We analyze Molecular Hydrogen Ions (MHIs) formed by collisions of low-energy protons with the Second Flavor of Hydrogen Atoms SFHA, whose existence was previously proven by two kinds of atomic experiments and also evidenced by two kinds of astrophysical observations. We find that the resulting MHIs would lack a significant number of terms compared to the MHIs formed by collisions of low-energy protons with the usual hydrogen atoms. We show that, in this situation, the radiative transition between the terms of such MHIs of the lowest quantum numbers would be between the terms 5fσ and 4dσ. We calculate the position of the edge of the corresponding molecular band and find it to be at the frequency 14,700 cm^{-1} or equivalently at the wavelength of 680 nm, which belongs to the visible range. It should be easier to observe this band compared to the spectral bands that are completely beyond the visible range. We emphasize that these results open up another avenue for finding an additional experimental proof of the existence of the SFHA. Namely, if the SFHA is present in gas (in addition to the usual hydrogen atoms), on which a beam of low-energy protons is incident, then the relative intensity of the band, corresponding to the radiative transitions between the terms 5fσ and 4dσ of the MHIs, would be enhanced compared to the absence of the SFHA.

Keywords: second flavor of hydrogen atoms; molecular hydrogen ion; proton collisions; molecular spectral bands

Citation: Oks, E. Peculiar Features of Molecular Hydrogen Ions Formed by Proton Collisions with Hydrogen Atoms of the Second Flavor. *Foundations* **2022**, *2*, 228–233. https://doi.org/10.3390/foundations2010016

Academic Editor: Florentino Borondo

Received: 9 January 2022
Accepted: 9 February 2022
Published: 14 February 2022

1. Introduction

The Second Flavor of Hydrogen Atoms (SFHA) has been discovered theoretically and proven experimentally to exist (by analyzing atomic experiments related to the distribution of the linear momentum in the ground state of hydrogen atoms) first in 2001 [1]. The gist of the theoretical discovery was that, for the states of zero angular momentum, the so-called "singular" solution of the Dirac equation outside the atomic proton, which was usually disregarded, can be tailored without any problem with the regular solution inside the proton. Therefore, for this second solution, the wave function did not have a singularity at the origin and thus there was no reason to disregard this solution. Since then, there was also found evidence of the existence of the SFHA from atomic experiments: this time from the experiments on the charge exchange of hydrogen atoms with incoming protons [2].

There are also two kinds of astrophysical evidence for the existence of the SFHA. The first one is related to the puzzling observation of the redshifted 21 cm spectral line from the early universe by Bowman et al. [3]. They found that the absorption in this spectral line was about two times stronger than predicted by standard cosmology. This meant that the primordial hydrogen gas was significantly cooler than predicted by standard cosmology. In paper [4], it was shown that this big discrepancy between the observations by Bowman et al. [3] and standard cosmology can be eliminated—qualitatively and quantitatively—in the case where the additional cooling was due to collisions with the SFHA. These results made the SFHA a candidate for dark matter.

The second astrophysical evidence of the existence of the SFHA is the following. Jeffrey et al. [5] recently found through observations that the distribution of dark matter in the universe is smoother than predicted by Einstein's gravitation, which prompted calls for a non-Einsteinian gravity, i.e., for new physical laws. However, in paper [6] it was demonstrated that this perplexing observation can be also explained—qualitatively and quantitatively—by using the SFHA.

The SFHA-based explanations of the puzzling astrophysical observations by Bowman et al. [3] and by Jeffrey et al. [5] did not require any change of physical laws (and thus were favored by the Occam razor principle). The theoretical discovery of the SFHA in paper [1] was based on the standard Dirac equation of quantum mechanics.

In the present paper, we offer an experimental possibility for an additional proof of the existence of the SFHA; it is about collisions of low-energy protons with the SFHA. One of the possible outcomes of such collisions is charge exchange: for the SFHA it has a slightly larger cross-section than for the usual hydrogen atoms—as shown in paper [2] where the comparison with experiments confirmed this theoretical result. In the present paper, we focus on another outcome of such collisions: the capture of the incoming proton and the formation of the Molecular Hydrogen Ions (MHIs). We reveal the peculiar features of the MHIs formed in this way. This opens up another avenue for finding an additional experimental proof of the existence of the SFHA.

2. Allowed Molecular Terms and Radiative Transitions

There have been lots of theoretical studies of the MHI—see, e.g., reviews [7–9] and references therein, as well as papers [10,11] and references therein. There are several reasons for this. From the theoretical point of view, the MHI is the simplest stable molecule and thus represents the test-bench for molecular quantum mechanics. From the applied point of view, the MHI is important in astrophysics because it is involved in reaction chains leading to the formation of polyatomic molecules.

From the theoretical point of view, at fixed nuclei (which in this situation are protons), the MHI is a particular case of the two-Coulomb center system, the latter consisting of two fixed nuclei of charges Z_1 and Z_2 separated by a distance R and one electron. The two-Coulomb center system allows the complete separation of variables (in elliptical coordinates)—see, e.g., the textbook [12]. This is the consequence of the higher than geometric symmetry of this system, manifested by the existence of an additional conserved quantity: the projection of the super-generalized Runge–Lenz vector on the internuclear axis [13].

As a result of the complete separation of variables, the states of the system are described by sets of elliptical quantum numbers {k, q, m}. Here, k is the radial quantum number, q is the angular quantum number, and m is the azimuthal quantum number. They are equal to the number of nodes of the corresponding parts of the wave function. In particular, the azimuthal quantum number m is the projection of the orbital momentum (in atomic units) on the internuclear axis.

Usually, instead of (k, q, m), there are introduced the following linear combinations—i.e., the orbital quantum number,

$$l = q + m \tag{1}$$

and the principal quantum number,

$$N = k + q + m + 1 = k + l + 1. \tag{2}$$

Then the states are described by the sets (N, l, m). Finally, the numerical values $l = 0, 1, 2, 3, \ldots$ are substituted by letters s, p, d, f, ... , respectively, and the numerical values of $m = 0, 1, 2, \ldots$ are substituted by Greek letters σ, π, δ, ... For example, the state (1, 0, 0) becomes denoted as $1s\sigma$, the state (3, 2, 1) becomes denoted as $3d\pi$, and so on.

When the incoming slow proton is relatively far from the hydrogen atom, i.e., for a relatively large internuclear separation R, the states are described by the sets of the parabolic quantum numbers [n_1, n_2, m]—see, e.g., the textbook [12]. For the purpose of our paper, the most important is the correspondence between these parabolic quantum numbers (relevant for relatively large R) and the elliptical quantum numbers (relevant for relatively small R). According to book [14], the correspondence is the following.

For any Z_1 and Z_2:

$$k = n_1. \tag{3}$$

As for the correspondence between the other quantum numbers, for $Z_1 = Z_2$ (i.e., for the situation we are interested in), there are two subcases. For even q, one has,

$$q = 2n_2, \tag{4}$$

while for odd q, one has,

$$q = 2n_2 + 1. \tag{5}$$

Here we come to the central point. The primary distinctive feature of the SFHA is that it only has states of the zero-orbital momentum [1,4,15]: the S-states, which are spherically symmetric. Only for S-states, the highest probability of the location of the atomic electron is at or near the origin, so that only the S-states are sensitive to the finite size of the atomic proton and to the behavior of the solution of the Dirac equation in and around the proton. In terms of the parabolic quantum numbers, the spherically symmetric states can correspond only to the situation where,

$$n_1 = n_2, m = 0. \tag{6}$$

Therefore, in this situation, Equation (1) becomes,

$$l = q \tag{7}$$

and Equation (3) becomes,

$$k = n_2, \tag{8}$$

so that Equation (2) becomes,

$$N = l + n_2 + 1. \tag{9}$$

On substituting $q = l$ (according to Equation (7)) in Equations (4) and (5), we arrive to the following results (taking into account also Equation (9)). For even n_2, the possible states have the quantum numbers,

$$l = 2n_2, N = 3n_2 + 1. \tag{10}$$

For odd n_2, the possible states have the quantum numbers,

$$l = 2n_2 + 1, N = 3n_2 + 2. \tag{11}$$

(We remind that $n_2 = 0, 1, 2, \ldots$)

In other words, the MHI formed by collisions with low-energy protons with the SFHA would have a significantly smaller number of energy terms compared to the usual MHI. Namely, the even terms are,

$$1s\sigma, 4d\sigma, 7f\sigma, \ldots \tag{12}$$

The odd terms are,

$$2p\sigma, 5f\sigma, 8g\sigma, \ldots \tag{13}$$

Now, let us consider the consequences of these peculiar features of the energy terms (of this kind of the MHI) in the radiation spectrum. All allowed radiative transitions are between the terms of $m = 0$, so that the radiation would have the π-polarization only ($\Delta m = 0$). Further, taking into account the selection rule $|\Delta l| = 1$, we find the following.

The radiative transitions between the terms of the lowest quantum numbers might have been between the terms $2p\sigma$ and $1s\sigma$. The term $1s\sigma$ corresponds to the stable state of the molecule, the term $2p\sigma$ does not.

During the collisions of low-energy protons with hydrogen atoms, the stable theoretical terms of the MHI indicate the possibility of capturing the proton and forming a stable state for the NHI. As for the unstable theoretical terms of the MHI, they should correspond to the formation of transient, quasi-molecular states. Radiative transitions between the meta-stable and stable molecular states in principle should be possible to observe.

However, for the radiative transition between the terms $2p\sigma$ and $1s\sigma$ there is a problem, as follows. Figure 1 shows the frequency F (in units of 10^5 cm^{-1}) of the radiative transitions between these two terms versus the internuclear distance R (in atomic units). It would seem that the corresponding spectral band does not have an edge (theoretically it occupies the entire continuous spectrum of frequencies). Therefore, for this band, its intensity per unit frequency range would be very low, thus making it virtually impossible to observe.

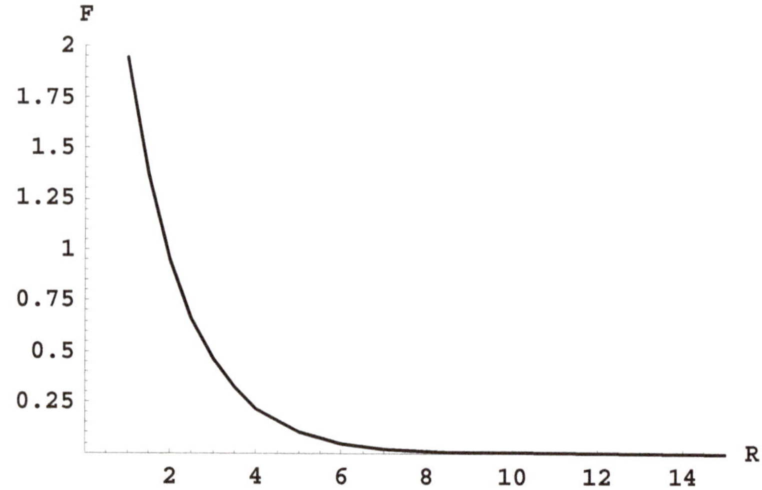

Figure 1. The frequency F (in units of 10^5 cm^{-1}) of the radiative transitions between the terms 2pσ and 1sσ versus the internuclear distance R (in atomic units).

As we proceed to the terms of higher quantum numbers, the next possibility is the radiative transition between the terms 5fσ and 4dσ. The term 4dσ corresponds to the stable state of the molecule, the term 5fσ does not. Fortunately, the corresponding spectral band has an edge.

Figure 2, which we created using the tables from the Madsen and Peek paper [16], shows the frequency F (in units of 10^5 cm^{-1}) of the radiative transitions between these two terms versus the internuclear distance R (in atomic units). At the maximum (located at R = 5), the frequency is F = 14,700 cm^{-1}. This means that this spectral band indeed has an edge, and it is in the vicinity of which the intensity per unit frequency range should be heightened. Moreover, the edge of this spectral band is at a wavelength of 680 nm. Thus, it should be easier to observe compared to the spectral bands that are completely beyond the visible range.

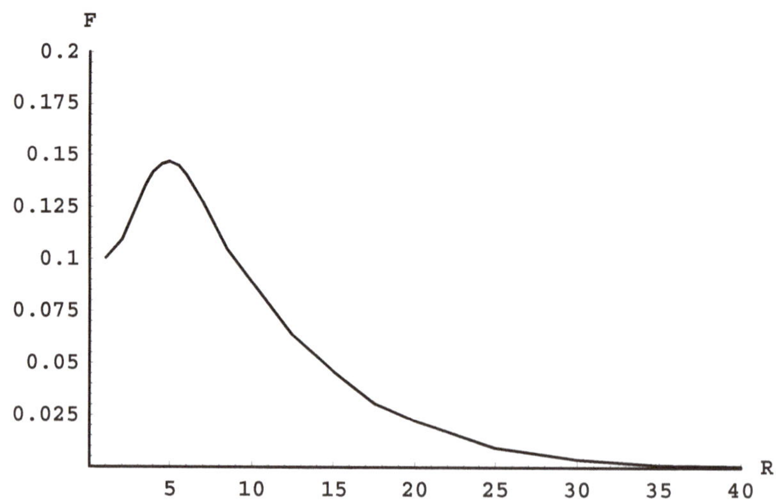

Figure 2. The frequency F (in units of 10^5 cm^{-1}) of the radiative transitions between the terms 5fσ and 4dσ versus the internuclear distance R (in atomic units).

These results open up another avenue for finding an additional experimental proof of the existence of the SFHA. Indeed, let us consider an experiment where a beam of low-energy protons is incident on a gas of hydrogen atoms. If the SFHA is present in the gas (in addition to the usual hydrogen atoms), then the relative intensity of the band, corresponding to the radiative transitions between the terms 5fσ and 4dσ, would be enhanced compared to the absence of the SFHA. This is because the SFHA would not contribute to the usually observed bands, corresponding to the radiative transitions between the terms of lower quantum numbers.

In closing this section, we make the following comments. First, some terms of the MHI, formed by collisions of the usual hydrogen atoms with low-energy protons, can exhibit an additional (typically shallow) potential well caused by the van der Waals interaction. Generally, it corresponds to the dipole–dipole interaction (see, e.g., the textbook [12]). It has the form,

$$V = [-\mathbf{d}_1\mathbf{d}_2 + 3(\mathbf{d}_1\mathbf{n})(\mathbf{d}_2\mathbf{n})]/R^3 \tag{14}$$

where $\mathbf{n}$ is the unit vector along the internuclear direction. Since generally the non-diagonal matrix elements of the dipole moment operators are non-zero, then in the second order of the perturbation theory one obtains a non-zero result proportional to $1/R^6$ (because the result is quadratic with respect to V). In particular, for the ground state, the second order correction is always negative so that the corresponding interaction energy $U(R) = -\text{const}/R^6 < 0$ (where const > 0). This corresponds to the attraction and might manifest in molecular terms as a shallow potential well located at relatively large R. The above is true for the usual hydrogen atoms. However, the SFHA has only the S-states, so that, due to the selection rules, all matrix elements of the dipole moment operator are zeros. Therefore, the dipole–dipole interaction vanishes, as well as the charge–dipole interaction. Thus, in the case of the SFHA, there is no van der Waals interaction.

Second, in paper [15] we discussed the consequences of the fact that the ground state of hydrogen atoms turns out to be double-degenerate (due to the SFHA) despite these two states having the same values of all the known conserved quantities. Actually, this additional double-degeneracy is related to all S-states of hydrogen atoms. According to the fundamentals of quantum mechanics, this means that there should be an additional, new conserved quantity by eigenvalues of which any S-state of the SFHA differs from the corresponding S-state of the usual hydrogen atoms. Consequently, hydrogen atoms have two flavors, differing by the eigenvalue of an additional, new conserved quantity that we called *isohydrogen spin* (abbreviated as *isohyspin*)—by analogy with the isotopic spin (isospin) distinguishing two flavors of quarks (the up-quark and down-quark). We also emphasized in paper [15] that, since the isospin does not couple to the electromagnetic force/interaction, there seems to be no reason for the isohyspin to couple to the electromagnetic force/interaction either. A hypothetical (though actually impossible) transition from a state of the SFHA (the state being the S-state) would require the "flip" of the isohyspin, thus transforming the state into the S-state (of the same energy) of the usual hydrogen atom. Obviously, this would not be a radiative transition because it would be the transition between states of the same energy. Finally, in paper [15] we emphasized that the idea of the isohyspin is not necessary for considering the second kind of hydrogen atoms as the candidate for dark matter or for a part of dark matter. The idea of the isohyspin is also not necessary for the analysis of different types of atomic and molecular experiments involving the SFHA, such as the experiments analyzed in paper [1,2] and the experiments suggested in the present paper.

3. Conclusions

We considered the MHIs formed by collisions of low-energy protons with the SFHA. We found that the resulting MHIs would lack a significant number of terms compared to the MHIs formed by collisions of low-energy protons with the usual hydrogen atoms.

We showed that, in this situation, the radiative transition between the terms of such MHIs of the lowest quantum numbers would be between the terms 5fσ and 4dσ. We calculated the position of the edge of the corresponding molecular band and found it to be at the frequency 14,700 cm^{-1} or equivalently at the wavelength of 680 nm, which belongs to the visible range. So, it should be easier to observe this band compared to the spectral bands that are completely beyond the visible range.

We emphasized that these results open up another avenue for finding an additional experimental proof of the existence of the SFHA. Namely, if the SFHA is present in the gas (in addition to the usual hydrogen atoms) on which a beam of low-energy protons is incident, then the relative intensity of the band, corresponding to the radiative transitions between the terms 5fσ and 4dσ of the MHIs, would be enhanced compared to the absence of the SFHA.

Funding: This research received no external funding.

Institutional Review Board Statement: Not applicable.

Informed Consent Statement: Not applicable.

Data Availability Statement: All data is included in the manuscript.

Conflicts of Interest: The author declares no conflict of interest.

References

1. Oks, E. High-Energy Tail of the Linear Momentum Distribution in the Ground State of Hydrogen Atoms or Hydrogen-like Ions. *J. Phys. B At. Mol. Opt. Phys.* **2001**, *34*, 2235–2243. [CrossRef]
2. Oks, E. Analysis of Experimental Cross-Sections of Charge Exchange between Hydrogen Atoms and Protons Yields another Evidence of the Existence of the Second Flavor of Hydrogen Atoms. *Foundations* **2021**, *1*, 265–270. [CrossRef]
3. Bowman, J.D.; Rogers, A.E.E.; Monsalve, R.A.; Mozdzen, T.J.; Mahesh, N. An Absorption Profile Centred at 78 Megahertz in the Sky-Averaged Spectrum. *Nature* **2018**, *555*, 67–70. [CrossRef] [PubMed]
4. Oks, E. Alternative Kind of Hydrogen Atoms as a Possible Explanation of the Latest Puzzling Observation of the 21 cm Radio Line from the Early Universe. *Res. Astron. Astrophys.* **2020**, *20*, 109. [CrossRef]
5. Jeffrey, N.; Gatti, M.; Chang, C.; Whiteway, L.; Demirbozan, U.; Kovacs, A.; Pollina, G.; Bacon, D.; Hamaus, N.; Kacprzak, T. Dark Energy Survey Year 3 results: Curved-Sky Weak Lensing Mass Map Reconstruction. *Mon. Not. R. Astron. Soc.* **2021**, *505*, 4626–4645. [CrossRef]
6. Oks, E. DES Map Shows a Smoother Distribution of Matter than Expected: A Possible Explanation. *Res. Astron. Astrophys.* **2021**, *21*, 241–245. [CrossRef]
7. Roth, B.; Koelemeij, J.; Schiller, S.; Hilico, L.; Karr, J.-P.; Korobov, V.; Bakalov, D. Precision Spectroscopy of Molecular Hydrogen Ions: Towards Frequency Metrology of Particle Masses. In *Precision Physics of Simple Atoms and Molecules*; Karshenboim, S.G., Ed.; Lecture Notes in Physics; Springer: Berlin/Heidelberg, Germany, 2008; Volume 745.
8. Leach, C.A.; Moss, R.E. Spectroscopy and Quantum Mechanics of the Hydrogen Molecular Cation: A Test of Molecular Quantum Mechanics. *Annu. Rev. Phys. Chem.* **1995**, *46*, 46–82. [CrossRef] [PubMed]
9. Carrington, A.; McNab, I.R.; Montgomerie, C.A. Spectroscopy of the Hydrogen Molecular Ion. *J. Phys. B At. Mol. Opt. Phys.* **1989**, *22*, 3551–3586. [CrossRef]
10. Scott, T.C.; Aubert-Frecon, M.; Grotendorst, J. New Approach for the Electronic Energies of the Hydrogen Molecular Ion. *Chem. Phys.* **2006**, *324*, 323–338. [CrossRef]
11. Carrington, A.; McNab, I.R. Spectroscopy of the H … H^+ and D … D^+ Charge/Induced-Dipole Complexes. *J. Chem. Soc. Faraday Trans.* **1990**, *86*, 1957–1962. [CrossRef]
12. Landau, L.D.; Lifshitz, E.M. *Quantum Mechanics*; Pergamon: Oxford, UK, 1965.
13. Kryukov, N.; Oks, E. Super-generalized Runge-Lenz Vector in the Problem of Two Coulomb or Newton Centers. *Phys. Rev. A* **2012**, *85*, 054503. [CrossRef]
14. Komarov, I.V.; Ponomarev, L.I.; Slavjanov, S.Y. *Spheroidal and Coulomb Spheroidal Functions*; Nauka: Moscow, Russia, 1976; Section II.3. (In Russian)
15. Oks, E. Two Flavors of Hydrogen Atoms: A Possible Explanation of Dark Matter. *Atoms* **2020**, *8*, 33. [CrossRef]
16. Madsen, M.M.; Peek, J.M. Eigenparameters for the Lowest Twenty Electronic States of the Hydrogen Molecule Ion. *At. Data* **1971**, *2*, 171–204. [CrossRef]

foundations

Article

On the Topological Structure of Nonlocal Continuum Field Theories

Said Mikki [1,2]

[1] International Campus, Zhejiang University/University of Illinois at Urbana-Champaign (ZJU-UIUC) Institute, Zhejiang University, 718 East Haizhou Road, Haining 314400, China; said.m.mikki@gmail.com
[2] Electrical and Computer Engineering Department, University of Illinois at Urbana-Champaign, Engineering Hall MC 266, 1308 West Green Street, Champaign, IL 61820, USA

Abstract: An alternative to conventional spacetime is proposed and rigorously formulated for nonlocal continuum field theories through the deployment of a fiber bundle-based superspace extension method. We develop, in increasing complexity, the concept of nonlocality starting from general considerations, going through spatial dispersion, and ending up with a broad formulation that unveils the link between general topology and nonlocality in generic material media. It is shown that nonlocality naturally leads to a Banach (vector) bundle structure serving as an enlarged space (superspace) inside which physical processes, such as the electromagnetic ones, take place. The added structures, essentially fibered spaces, model the topological microdomains of physics-based nonlocality and provide a fine-grained geometrical picture of field–matter interactions in nonlocal metamaterials. We utilize standard techniques in the theory of smooth manifolds to construct the Banach bundle structure by paying careful attention to the relevant physics. The electromagnetic response tensor is then reformulated as a superspace bundle homomorphism and the various tools needed to proceed from the local topology of microdomains to global domains are developed. For concreteness and simplicity, our presentations of both the fundamental theory and the examples given to illustrate the mathematics all emphasize the case of electromagnetic field theory, but the superspace formalism developed here is quite general and can be easily extended to other types of nonlocal continuum field theories. An application to fundamental theory is given, which consists of utilizing the proposed superspace theory of nonlocal metamaterials in order to explain why nonlocal electromagnetic materials often require additional boundary conditions or extra input from microscopic theory relative to local electromagnetism, where in the latter case such extra input is not needed. Real-life case studies quantitatively illustrating the microdomain structure in nonlocal semiconductors are provided. Moreover, in a series of connected appendices, we outline a new broad view of the emerging field of nonlocal electromagnetism in material domains, which, together with the main superspace formalism introduced in the main text, may be considered a new unified general introduction to the physics and methods of nonlocal metamaterials.

Keywords: nonlocal metamaterials; multiscale structures; fiber bundles; superspace; mathematical methods; mathematical physics; nonlocal continuum field theory; semiconductor materials

Citation: Mikki, S. On the Topological Structure of Nonlocal Continuum Field Theories. *Foundations* **2022**, *2*, 20–84. https://doi.org/10.3390/foundations2010003

Academic Editor: Eugene Oks

Received: 21 October 2021
Accepted: 13 December 2021
Published: 31 December 2021

1. Introduction

Numerous research studies point toward a basic fact: topology and physics are destined to come closer to each other in the following decades [1–4]. This in itself is not totally new because several authors, for example, Henri Poincare, E. Cartan, and Hermann Weyl, had already advocated topological thinking in physics [5–7]. However, a salient feature of this convergence is the focus on material engineering applications, for example, metamaterials and topology-based devices. In this paper, we look into the general and rigorous foundations of the discipline behind these applications, namely the framework of *nonlocal continuum field theories* [8,9], with focus on explicating the generic multiscale topological structure of continua studied by such theories. We propose that in addition to the now mainstream approach to topological materials [10,11], where the emphasis is often laid on exploiting the global dependence of the wave function on momentum (Fourier) space, there is a need to consider how materials can be assigned an indirect structure indexed by parameters taken directly from the *spatial* side of the configuration space, i.e., either space–time or space–frequency.

Our key observation is that arriving at an adequate understanding and characterization of nonlocality in generic scenarios would naturally require gathering information at the microtopological level of what we dub *nonlocal microdomains* (the topological level of small regions around every point where the response is nonlocal), then collectively aggregating these microdomains in order to obtain the global topological structure (the macro-topological level). The fundamental insight coming from topology is precisely how this process of "moving from the local to the global" can be enacted. We have found that a very efficient method to do this is the natural formulation of the entire problem in terms of a fiber bundle superspace, where conventional spacetime or space–frequency are here understood as nothing but "index spaces" embedded into a larger (in our opinion more fundamental) fibered superspace characteristic of nonlocal continuum field theories. In other words, and in contrast to existing approaches to local field theories and topological materials, our strategy is not to first solve Maxwell's equations in order to find the state function as expressed within the Fourier **k**-space, after which one proceeds to study topology over momentum space; instead, we start in spacetime (or space–frequency), and then formulate the *extended* or superspace structure of a topology over a fiber bundle where the conventional position space of the nonlocal continuum, e.g., Euclidean space, would manifest itself merely as the index space of the fiber bundle superspace.

The principal conceptual and philosophical message behind this work is that spacetime (or space–frequency) is not adequate for formulating nonlocal continuum field theories, and that a more appropriate natural approach is the superspace formalism proposed below, which, in our case, is based on a specific fiber bundle construction taking into account the intricate physics-based microdomain structure of the generic nonlocal continuum. It is the hope of the author that by helping scientists generate new insights into their physics and models, this formalism may provide a rigorous approach complementing some of the exciting theories and researches currently addressing various topics in continuum field theories, nonlocal metamaterials, and topological materials, while possibly stimulating the creation of novel algorithms for the computation of suitable topological invariant characterizing complex material domains. Due to the wide scope and complexity of this work, we first provide in Section 2 a relatively lengthy overview on the our contribution, where high-level information about this work, in addition to a guide to the literature and how to read the present paper, are outlined before moving to the more technical treatments of the subsequent sections and appendices.

2. Preliminary Considerations

While the essential idea of the superspace formalism introduced here will be valid for a generic nonlocal continuum field theory, it is much easier sometimes to work with a concrete example, especially in explaining what nonlocality is for someone who is coming to the subject for the first time. Therefore, in this preliminary section, we emphasize the special but very important case of *electromagnetic* nonlocality.

2.1. What Is Nonlocality?

In classical electromagnetic (EM) theory, it is currently widely held that there are no nonlocal interactions or phenomena in vacuum because Maxwell's equations, which capture the ultimate content of the physics of electromagnetic fields, are essentially local differential equations [12]. In other words, an effect applied at point $\mathbf{r}$ in space will first be felt at the same location but then spread or propagate slowly into the *infinitesimally* immediate neighborhood. Long-term disturbances, such as electromagnetic waves, propagate through both vacuum and material media by cascading these infinitesimal perturbations in outward directions (rays or propagation paths) emanating from the time-varying point source that originated the whole process. However, if we leave behind vacuum electromagnetism and move into electromagnetically-responsive matter-filled space, then we note that *nonlocal* interactions in material domains differ fundamentally from the de facto local vacuum-like picture in allowing fields applied at position $\mathbf{r}'$ to influence the medium at *different* location $\mathbf{r}$ [13]. That is, in the nonlocal material system, a location not *infinitesimally* close to the source position $\mathbf{r}'$ can experience a nonvanishing effect emanating from the source location. While the "nonlocality scale" $|\mathbf{r} - \mathbf{r}'|$ tends to be quite small in most natural media (and certainly zero in vacuum), in some types of materials, the so-called *nonlocal media*, observable response can be found such that this "radius of nonlocality" $|\mathbf{r} - \mathbf{r}'|$ becomes appreciably different from zero [14–16].

The existence of multiple scales in the fundamental physics of nature is not really new. The scaling properties are important in Yang–Mills fields, the non-abelian field theory, and it has been recently used to propose the presence of fractal structures in the dynamical evolution of the fields. For example, one may consider the fractal structure of Yang–Mills fields [17] as an example of a multiple-scale effect in fundamental field theories.[1] In a more familiar setting, it is generally accepted that Aharonov–Bohm-type effects, which lead to observable nonlocal electrodynamic effects [18], have their origin in quantum physics. Bringing quantum physics into field theory can be shown to lead to intrinsically nonlocal effects since quantum field theory may be considered a fundamentally nonlocal theory due to, for example, entanglement effects [19,20]. However, in this paper, we focus on *classical* field theory realized through *phenomenological* models of the electromagnetic response of the material domain. The phenomenological model itself (the constitutive relations [9]) may have as its ultimate origin a purely quantum effect. For example, the main example considered in this paper, the nonlocal semiconductor material domain, has as its "origin of nonlocality" the essentially *quantum* process of exciton polariton coupling in solids (Section 7). It should be noted that in recent years some authors suggested that *classical* electromagnetism, under certain conditions, *may* induce nonlocal effects [21–23]; nevertheless, such scenarios are *outside* the scope of the physical paradigm treated in the present paper.

On the other hand, and interestingly enough for our purposes, Cvijanovich proposed several decades ago a theoretical model in which *vacuum* itself is modeled as a nonlocal constitutive *non*-material domain, where the standard Lorentzian spacetime manifold of general relativity is assumed here to play the role of the "medium" transmitting nonlocal actions [24]. Such proposal might be linked to field–matter interaction regimes where there is a strong coupling between gravitational fields and electromagnetic degrees of freedom. For flat spacetime, however, we already know from experiments that classical electromagnetism is strictly local. Nevertheless, it was discovered recently that classical electromagnetism can be made nonlocal if the photon mass is nonzero. More precisely, classical massive electromagnetism can be shown to arise in certain nonlocal (spatially dispersive) homogeneous domains [25]. Therefore, the statement that "classical electromagnetism is strictly local" should be qualified by allowing for the possibility that the photon mass might be proved experimentally to be non-vanishing, say in a future empirical research. In spite of all these interesting proposals on how to modify classical electromagnetic theory in order to make it compatible with nonlocality at the very fundamental level, the system of field theory treated in this paper is mainly classical, and the underlying spacetime structure is flat (the gravitational degrees of freedom are ignored).

The research field concerned with the study of the classical electromagnetism of nonlocal material domains is called *nonlocal electromagnetism/electromagnetics/electrodynamics*. This paper introduces a comprehensive general approach to this emerging discipline together with a series of selected applications. An extensive literature survey on past researches into nonlocal electromagnetism is given in Appendixes A.1 and A.2. The subject of nonlocal electromagnetism, here understood as the electromagnetism of nonlocal material domains, is presently treated as a subdomain of the science of metamaterials. Historically, it has not been a well-defined direction of research, with researchers working on nonlocal structures often coming from very diverse and distinct fields, such as plasma physics, crystal optics, periodic structures, metasurfaces, and so on. One of the objectives of this article is to propose a coherent view of the inherently cross-disciplinary nonlocal materials research program, encompassing contributions coming from theoretical physics, applied physics, chemistry, engineering, with mathematical physics as the unifying framework of our inquiry.

2.2. *Key Contributions and Motivations in the Present Work*

Currently, there is an interest within applied physics and engineering in harnessing nonlocal media as a new generation of metamaterials for use in various settings, e.g., optical devices, energy control, antennas, circuit systems, etc., see Appendixes A.1 and A.3. The main goal of the present work is to explore, at a very general level, the conceptual and mathematical foundations of nonlocality in connection with applied electromagnetic metamaterials (MTMs). Our approach is conceptual and theoretical, with the main emphasis being laid on understanding the mathematical foundations of the subject and how they relate to the underlying physical bases of some illustrative examples. Indeed, while a massive amount of numerical and experimental data on all types of nonlocal materials abound in a literature that goes back to as early as the 1950s, the purpose of the present paper is attaining some clear understanding of the essentials of the subject, particularly in connection with the ability to build a very general superspace formalism for nonlocal continuum field theory without restricting the formalism first to particular classes of materials such as metals, plasma, or semiconductors.

The central theoretical idea in this work is the introduction of the superspace concept into the process of constructing a general formalism suitable for understanding, analyzing, and designing nonlocal material systems in classical field theory. The superspace formalism has a long history in physics, mathematical physics, and mathematics (see Appendix A.4). It will be shown below that nonlocal continuum field theory appears to lead very naturally to a reformulation of its essential configuration space by upgrading the conventional space–time or frequency space to a larger superspace in which the former spaces serve as base spaces for the new (larger) superspace. Such reconsideration of the fundamental structure of the problem may help foster future numerical methods and potential applications as will be discussed later, e.g., see Appendix A.11.2.

The key motivation behind the proposed superspace approach is explicating a subtle, but often overlooked, difference between two fundamental scales of interactions in nature:

1. *Infinitesimal interactions*: this characterizes local field theories, e.g., local electromagnetism, where all operators are differential operators.
2. Non-infinitesimal but local interactions: here, nonlocal operators, such as integral operators, may be present. In this type of theory, interactions are extended into small *topological neighborhoods* around the source/observation point.

We believe that this topological difference has not received the attention it deserves in the growing theoretical and methodological literature on nonlocal media. In particular, the author believes that a majority of present approaches to nonlocal metamaterials conflate the topologically local (but EM nonlocal) domain of small neighborhoods and global domains. However, general topology and much of modern mathematical physics is based on clearly distinguishing the last two topological levels. Explicating these subtle conceptual differences emanating from the existence of distinct types of spatial scales in field–matter interactions, while aided by a precise, rigorous, and powerful mathematical language, is one of the principal aims of this work. In fact, we believe that a complete understanding of material nonlocality in nature cannot be attained without relying on a fairy advanced mathematical apparatus such as the theory of smooth fiber bundles and infinite-dimensional manifolds developed below.

Let us give a brief summary of the main conceptual findings of this research. First, we highlight that the main idea of the superspace formalism is not restricted to electromagnetic theory, but applies to all types of nonlocal continuum theories, i.e., field theories in nonlocal continuous media. However, for concreteness, and in order to reduce the complexity of the mathematical formalism, we chose to work with a specific type of field theories, namely the classical field paradigm based on Maxwell's equations. As will be seen below, It turns out that the standard formalism of local field theory, which is based on spacetime points and their *differential* (but not topological) neighborhoods, viewed as the basic configuration space of the problem, is *not* the most natural or convenient framework for formulating field theory in nonlocal materials. This is mainly because the *physics*-based domain of nonlocality (to be defined precisely below), which captures the effective region of field–matter nonlocal interactions, is found to not always be naturally transportable into the mathematical formalism of boundary-value problems characteristic of classical field theory, as practiced in several domains, such as applied electromagnetism, heat transfer, hydrodynamics, etc. By investigating the subject from an alternative but enlarged and intrinsically broader perspective, it will be shown that a natural space for conducting nonlocal metamaterials research is the *vector bundle* structure, more specifically, a Banach bundle [26] where every element in the fiber superspace is a vector field on the entire domain of nonlocality.

The main result of this paper is that every generic nonlocal domain can be topologically described by a superspace comprised of a Banach (infinite-dimensional) vector bundle $\mathcal{M}$. If two materials described by their corresponding vector bundles $\mathcal{M}_1$ and $\mathcal{M}_2$ are juxtaposed, then one may use topological methods to combine them and to compare their topologies. The present paper's focus is mainly on the first part, i.e., how to construct the material bundle $\mathcal{M}$. That is, the derivation of the various vector bundle structures starting from a generic phenomenological model of electromagnetic nonlocality is the main contribution of the present work. It is hoped by the author that the superspace theory developed below will stimulate new approaches to computational field theories by adopting methods borrowed from or inspired by computational topology and differential topology to help supporting ongoing efforts to solve challenging problems in complex material domains as in nanoscale hydrodynamics, nonlocal optical materials, topological insulators, topological photonic devices, and other areas where nonlocality is currently important or expected to play an increasingly dominant role in the future.

2.3. An Outline of the Present Work

Because of the considerable complexity of the present article, which is unavoidable in treatment of the subject of nonlocality in the continuum field theory at this broad theoretical level, and in order to help make our contribution accessible to a wider audience involving, for instance, physicists, engineers, and mathematicians, we have divided the argument into different stages with different flavors, as follows. First, Section 3 provides a general mathematical description of nonlocality in the continuum field theory, emphasizing the settings of the electromagnetic case. The key ingredients of nonlocal metamaterials/materials are illustrated in Section 3.1 using an abstract excitation-response model. This is followed in Section 3.2 by a more detailed description of the special but important case of spatial dispersion, which tends to arise naturally in many investigations of nonlocal metamaterials. In Section 4, we begin the elucidation of the main topological ideas behind electromagnetic nonlocality, most importantly, the concept of EM *nonlocality microdomains*, which provides the key link between physics, material engineering, and topology in this paper. The various physical and mathematical structures are spelled out explicitly, followed in Section 5 by a more careful construction of a natural fiber bundle superspace structure that appears to satisfy simultaneously both the physical and mathematical requirements of EM nonlocality (Sections 5.1 and 5.2). We then provide a key computational application of the proposed theory in Section 5.3, where it is shown that the material response function is representable as a special fiber bundle homomorphism over the metamaterial base space. In this way, a more general map than linear operators in local field theory is derived, providing solid mathematical foundations for possible future computational topological methods where, for example, the bundle homomorphism itself might be discretized instead of the original spacetime-based linear operator. The fiber bundle superspace algorithm is summarized in Section 6, where it is highlighted that the main data needed are the physics-based (e.g., electromagnetic) nonlocality microdomains, which do not arise solely from purely mathematical considerations, but require some empirical input, for example the microscopic theory of materials, which ultimately would involve both electromagnetism and quantum mechanics. In this manner, the entire construction of the nonlocal metamaterial superspace may proceed as per the procedure outlined there. In order to illustrate how the above mentioned microdomain structure can be actually estimated in practice, in Section 7 we present a fairly detailed computational example based on nonlocal semiconductors, where we also explore in depth the physical origin of nonlocality in this particular setting. Insights into the lack of general EM boundary conditions in nonlocal EM are provided in Section 8 based on the superspace formalism.

This paper provides a series of technical appendices designed to provide necessary information to expand the scope of the treatment found in the main text. In Appendix A.1, we back up our major formulation as developed by introducing a general review of electromagnetic nonlocality targeting a wide audience of mathematicians, physicists, engineers, and applied scientists. This review does not restrict itself to specific types of materials, such as plasma, metals, and semiconductors, but aims at integrating the author's own understanding of the vast literature on the subject in a tentative and necessarily provisional, but somehow more coherent view. Because of the extreme importance of the special case of spatial dispersion for understanding nonlocality, we provide some brief historical remarks on this subject in a separate Appendix A.2. Some technical and historical explications of the concept of superspace, as needed and used in the main text, is given in Appendix A.4, which is not meant as a complete rigorous introduction to the concept of superspace in mathematics and theoretical physics, a topic far from being well-defined and focused. Instead, the goal of this appendix is to fix the very specific meaning we have in mind in this paper whenever we speak about superspace structures in order to avoid confusing our concept with other usages found in physics, such as in supersymmetry.

The Appendixes A.6–A.9 supply important technical information needed in order to fully comprehend the specific main example developed in this paper, to illustrate the use of the superspace formalism in actual real-life scenarios (the inhomogeneous nonlocal excitonic semiconductor material system of Section 7). We opted to separate the content of these appendices from the main text in order to simplify the presentation. The subject of nonlocal semiconductor metamaterials is already well-known in the specialized literature, but is also highly technical. In order to help keep the flow of the various ideas treated in the main text tightly focused on the conceptual and mathematical aspects of our proposed superspace theory, we relegated some background material, especially detailed derivations and explanations more related to semiconductor physics than the superspace formalism, to the three appendices mentioned above.

Some basic familiarity with vector bundles and Banach spaces is assumed, but essential definitions and concepts will be reviewed briefly within the main formulation and references where more background on vector bundles can be found will be pointed out. The paper intentionally avoids the strict theorem-proof format to make it accessible to a wider audience. Most of the time we give only proof sketches and leave out straightforward but lengthy computations. In general, just the very basic definitions of smooth manifolds, vector bundles, Banach spaces, etc., are needed to comprehend this theory (also see Appendix A.5 for a guide to the mathematical background.) The only place where the treatment is mildly more technical is in Section 5.3 when the bundle homomorphism is constructed using partition of unity technique as a detailed computational application of the superspace theory.

In Appendixes A.3 and A.11, various additional current and future applications to fundamental methods, applied physics, and engineering are outlined in brief form. Some of the applications mentioned there, for instance numerical methods and topological devices, appear to us to be directly relevant to the scope of a superspace extension of conventional nonlocal electromagnetic field continuum theory, such as the one attempted below within the main text. On the other hand, some of the other applications discussed there, e.g., digital communications and energy, are of a more general nature and belong to our broader tentative global review of the subject of nonlocality in nature and engineering attempted in the Appendix A sections of this paper. Finally we end with the conclusion.

3. The Nonlocal Continuum Response Model

3.1. A Generic Nonlocal Response Model in Inhomogeneous Continua

In order to introduce the concept of nonlocality in the simplest way possible, let us first start with a scalar field theory setting. As mentioned in the introduction, vacuum classical fields cannot exhibit nonlocality, so in order to attain this phenomenon, one must consider fields in specialized domains. We therefore kick-start the technical mathematical treatment by reviewing the broad theory of such media. The goal is to outline the main ingredients of the spacetime-based *configuration space* on which such theories are often founded in literature. To further simplify the presentation, we work in the regime of *linear response theory*: i.e., all material media considered throughout this paper are assumed to be linear with respect to field excitation.

In detail, if the medium response and excitation fields are captured by the spacetime functions $R(\mathbf{r},t)$ and $F(\mathbf{r},t)$, respectively, then the most general response is given by an operator equation of the form [9]

$$R(\mathbf{r},t) = \mathcal{L}\{F(\mathbf{r},t)\}, \tag{1}$$

where $\mathcal{L}$ is a linear operator describing the medium, and is ultimately determined by the laws of physics relevant to the structure under consideration [27–29].

Now, the entire physical process will occur in a spacetime domain. In a nonrelativistic formulation (like the one in the present work), we intentionally separate and distinguish space from time. Therefore, let us consider a process of field–matter interactions where $t \in \mathbb{R}$, while we spatially restrict to a "small" region spanned by the position

$$\mathbf{r} \in D \subset \mathbb{R}^3, \tag{2}$$

where D is an open set containing $\mathbf{r}$. (Throughout this paper, we assume the normal Euclidean topology on $\mathbb{R}^3$ for all spatial domains.) Since the operator $\mathcal{L}$ is linear, one may argue (informally) that its associated *Green's function* or *kernel function*

$$K(\mathbf{r},\mathbf{r}';t,t') \tag{3}$$

must exist. Strictly speaking, this is not correct in general and one needs to prove the existence of the Green's function for every given linear operator on a case by case basis by actually constructing one [30,31].[2] However, we will follow (for now) the common trend in physics and engineering by assuming that linearity alone is enough to justify the construction of Green's function. If this is accepted, then we can immediately infer from the very definition of the Green's function itself that [12,32]

$$R(\mathbf{r},t) = \int_D \int_{\mathbb{R}} \mathrm{d}^3r'\, \mathrm{d}t'\, K(\mathbf{r},\mathbf{r}';t,t') F(\mathbf{r}',t'). \tag{4}$$

The relation (4) represents the most general response function of a (scalar) material medium valid for linear field–matter interaction regimes [37,38]. The kernel (Green) function $K(\mathbf{r},\mathbf{r}';t,t')$ is often called the *medium response function* [9,32,37].

If we further assume that all of the material constituents of the medium are time invariant (the medium is not changing with time), then the relation (4) maybe replaced by

$$R(\mathbf{r},t) = \int_D \int_{\mathbb{R}} \mathrm{d}^3r'\,\mathrm{d}t'\,K(\mathbf{r},\mathbf{r}';t-t')F(\mathbf{r}',t'), \tag{5}$$

where the only difference is that the kernel function's temporal dependence is replaced by $t - t'$ instead of two separated arguments. Such superficially small difference has nevertheless considerable consequences. Most importantly, by working with (5) instead of (4), it becomes possible to apply the Fourier transform method to simplify the time-dependent formulation of the problem [39]. Indeed, taking the temporal Fourier transform of both sides of (5) leads to

$$R(\mathbf{r},\omega) = \int_D \mathrm{d}^3r' K(\mathbf{r},\mathbf{r}';\omega)F(\mathbf{r}';\omega), \tag{6}$$

where the Fourier spectra of the fields are defined by

$$\begin{aligned} F(\mathbf{r};\omega) &:= \int_{\mathbb{R}} \mathrm{d}t F(\mathbf{r};t)e^{-\mathrm{i}\omega t}, \\ R(\mathbf{r};\omega) &:= \int_{\mathbb{R}} \mathrm{d}t R(\mathbf{r};t)e^{-\mathrm{i}\omega t}. \end{aligned} \tag{7}$$

On the other hand, the medium response function's Fourier transform is given by the essentially equivalent formula

$$K(\mathbf{r},\mathbf{r}';\omega) := \int_{\mathbb{R}} \mathrm{d}(t-t')K(\mathbf{r},\mathbf{r}';t-t')e^{-\mathrm{i}\omega(t-t')}. \tag{8}$$

In this paper, we focus on time invariant material media and, hence, work exclusively with frequency domain expressions, such as (6), (7), and (8), though we often suppress the frequency dependence on ω in order to simplify the notation whenever no confusion would arise.

The generalization to the three-dimensional (full-wave) electromagnetic picture is straightforward when the dyadic formalism is employed [28,40]. The relation corresponding to (4) is

$$\mathbf{R}(\mathbf{r},t) = \int_D \mathrm{d}^3r' \int_{\mathbb{R}} \mathrm{d}t'\,\overline{\mathbf{K}}(\mathbf{r},\mathbf{r}';t-t')\cdot\mathbf{F}(\mathbf{r}',t'), \tag{9}$$

where we replaced the scalar fields $F(\mathbf{r})$ and $R(\mathbf{r})$ by vector fields $\mathbf{F}(\mathbf{r}),\mathbf{R}(\mathbf{r}) \in \mathbb{R}^3$. The kernel function K, however, must be transformed into a *dyadic* function (tensor of second rank) [14,28,41,42]:

$$\overline{\mathbf{K}}(\mathbf{r},\mathbf{r}';t-t'). \tag{10}$$

In the (temporal) Fourier domain, (9) becomes

$$\mathbf{R}(\mathbf{r},\omega) = \int_D \mathrm{d}^3r'\,\overline{\mathbf{K}}(\mathbf{r},\mathbf{r}';\omega)\cdot\mathbf{F}(\mathbf{r}';\omega), \tag{11}$$

where

$$\overline{\mathbf{K}}(\mathbf{r},\mathbf{r}';\omega) := \int_{\mathbb{R}} \mathrm{d}(t-t')\,\overline{\mathbf{K}}(\mathbf{r},\mathbf{r}';t-t')e^{-\mathrm{i}\omega(t-t')} \tag{12}$$

is the frequency domain response kernel, while

$$\begin{aligned} \mathbf{F}(\mathbf{r};\omega) &:= \int_{\mathbb{R}} \mathrm{d}t\,\mathbf{F}(\mathbf{r};t)e^{-\mathrm{i}\omega t}, \\ \mathbf{R}(\mathbf{r};\omega) &:= \int_{\mathbb{R}} \mathrm{d}t\,\mathbf{R}(\mathbf{r};t)e^{-\mathrm{i}\omega t}, \end{aligned} \tag{13}$$

are the corresponding frequency domain excitation and response fields, respectively.

The essence of electromagnetic nonlocality can be neatly captured by the mathematical structure of the basic relation (9). It says that the field response $\mathbf{R}(\mathbf{r})$ is determined not only by the excitation field $\mathbf{F}(\mathbf{r}')$ applied at location $\mathbf{r}'$, but at all points $\mathbf{r}' \in D$. Consequently, here we find that the following is true:

> *In nonlocal continuum field theories, knowledge of the field response at a specific point* $\mathbf{r}$ *requires knowledge of the cause (excitation field) on an entire topological neighborhood set* $D \ni \mathbf{r}$.

On the other hand, if the medium is local, then the material response function can be written as

$$\overline{\mathbf{K}}(\mathbf{r}, \mathbf{r}'; \omega) = \overline{\mathbf{K}}_0(\omega)\delta(\mathbf{r} - \mathbf{r}'), \tag{14}$$

where $\overline{\mathbf{K}}_0$ is a spatially constant tensor and $\delta(\mathbf{r} - \mathbf{r}')$ is the three-dimensional Dirac delta function. In this case, (11) reduces to [13]

$$\mathbf{R}(\mathbf{r}; \omega) = \overline{\mathbf{K}}_0(\omega) \cdot \mathbf{F}(\mathbf{r}; \omega), \tag{15}$$

which is the standard constitutive relation of linear electromagnetic materials. Clearly, (15) says that only the exciting field $\mathbf{F}(\mathbf{r})$ data at $\mathbf{r}$ is needed in order to induce a response at the same location. In a nutshell, locality implies that the natural configuration space of the electromagnetic problem is just the point-like spacetime manifold $D \subset \mathbb{R}^3$ or the entire Euclidean space $\mathbb{R}^3$.

Remark 1 (Infinitesimal domains). One may use the "infinitesimally immediate vicinity" of a given point $\mathbf{r}$, where a response is sought, for computing that response itself, yet while still remaining within the *local* regime of continuum field theory. Indeed, for the case of electromagnetic theory, we note that, according to the constitutive relation (15), while only the exciting field at $\mathbf{r}$ is required for computing the response, Maxwell's equations themselves, on the other hand, *still* must be coupled with the local constitutive relation model of the problem. Now, the fact that Maxwell's equations are *differential* equations implies that the "largest" domain beside the point $\mathbf{r}$ needed for carrying out the mathematical description of the details of the relevant field–matter interaction physics is just the region *infinitesimally* close to $\mathbf{r}$. In other words, in continuum field theories, infinitesimal domains should be treated as neither topological domains nor neighborhoods. The infinitesimal belong to *any* type of continuum field theory built on the differential calculus and, hence, is not a criterion for distinguishing local and nonlocal theoretical structures.

Conventional boundary-value problems in applied electromagnetism are formulated in this manner, i.e., with a three-differential manifold as the main problem space on which spatial fields live [28,29,32,40,41,43–46]. Note that, strictly speaking, the full configuration space in local electromagnetism (also called *normal optics* [16]) is the four-dimensional manifolds $D \times \mathbb{R}$ or $\mathbb{R}^4$ since either time t or the (temporal) circular frequency ω must be included to engender a full description of electromagnetic fields. However, nonlocal materials are most fundamentally a *spatial* type of materials/metamaterials where it is the spatial structure of the field what carries most of the physics involved [32,47]. For that reason, *throughout this paper, we investigate the required configuration spaces with focus mainly on the spatial degrees of freedom.* This will naturally lead to the discovery of the fiber bundle structure of nonlocality, the main topic of the present work.

3.2. Spatial Dispersion in Homogeneous Nonlocal Material Domains

Spatial dispersion is considered by some researchers as one of the most promising routes toward nonlocal metamaterials, e.g., see [16,47–49]. It is by large the most intensely investigated class of nonlocal media, receiving both theoretical and experimental treatments by various research groups since the early 1960s.[3] The basic idea is to restrict electromagnetism to the special, but important case of media possessing *translational symmetry*, an important special scenario of material nonlocality that holds when the medium is *homogeneous*. In such situation, the material tensor function satisfies

$$\overline{\mathbf{K}}(\mathbf{r},\mathbf{r}';\omega) = \overline{\mathbf{K}}(\mathbf{r}-\mathbf{r}';\omega). \tag{16}$$

The spatial Fourier transforms are defined by

$$\overline{\mathbf{K}}(\mathbf{k},\omega) := \int_{\mathbb{R}^3} \mathrm{d}^3(r-r')\, \overline{\mathbf{K}}(\mathbf{r}-\mathbf{r}';\omega) e^{\mathrm{i}\mathbf{k}\cdot(\mathbf{r}-\mathbf{r}')}, \tag{17}$$

with

$$\begin{aligned} \mathbf{F}(\mathbf{k},\omega) &:= \int_{\mathbb{R}^3} \mathrm{d}^3 r\, \mathbf{F}(\mathbf{r};t) e^{\mathrm{i}\mathbf{k}\cdot\mathbf{r}}, \\ \mathbf{R}(\mathbf{k},\omega) &:= \int_{\mathbb{R}^3} \mathrm{d}^3 r\, \mathbf{R}(\mathbf{r};t) e^{\mathrm{i}\mathbf{k}\cdot\mathbf{r}}. \end{aligned} \tag{18}$$

After inserting (16) into (11), taking the spatial (three-dimensional) Fourier transform of both sides, the following equation is obtained:

$$\mathbf{R}(\mathbf{k},\omega) = \overline{\mathbf{K}}(\mathbf{k},\omega) \cdot \mathbf{F}(\mathbf{k},\omega). \tag{19}$$

The dependence of $\overline{\mathbf{K}}(\mathbf{k},\omega)$ on the wave vector ("spatial frequency") $\mathbf{k}$, here added to the already existing temporal frequency ω dependence, is *the* signature of spatial dispersion. As a spectral transfer function of *a homogeneous* medium, $\overline{\mathbf{K}}(\mathbf{k},\omega)$ includes *all* the information needed to compute the nonlocal material domain's response to arbitrary spacetime field excitation functions $\mathbf{F}(\mathbf{r},t)$ (through the application of inverse four-dimensional Fourier transform [16]).

Remark 2. In several treatments of the subject within electromagnetic theory, the excitation field is taken as the electric field $\mathbf{E}(\mathbf{r},t)$, while the response function is $\mathbf{D}(\mathbf{r},t)$. In such formulation, the material tensor function $\overline{\mathbf{K}}(\mathbf{k},\omega)$ takes into account *both* electric and magnetic effects [14–16,37–39,50–54]. This is different from the permittivity tensor often invoked in local electromagnetism [28], which is ultimately based on the popular *multipole* model [43] of electromagnetic interactions in material media. A comparison between the two material response formalisms, the one based on $\overline{\mathbf{K}}(\mathbf{k},\omega)$ and the multipole model, is given in [32,47,53].

Complex heterogeneous arrangements of various nonlocal materials can be realized by juxtaposing several subdomains where each subunit is homogeneous, hence can be described by a spatial dispersion profile of the form $\overline{\mathbf{K}}(\mathbf{k},\omega)$ discussed above. The idea is that even materials that are inhomogeneous at a given spatial scale may become homogeneous at a different (less refined) spatial level, leading to a "grid-like" spatially dispersive cellular building blocks at the lower level. In Figure 1, we show a nonlocal metamaterial system with various multiscale structures. A large nonlocal domain, e.g., $\overline{\mathbf{K}}_3(\mathbf{r},\mathbf{r}')$ in the figure, acts like a "substrate" holding together several other smaller material constituents, such as $\overline{\mathbf{K}}_n(\mathbf{r},\mathbf{r}'), n = 1,2,4$. We envision that each nonlocal subdomain may possess its own specially tailored nonlocal response function profile serving one or several applications.[4] By concatenating multiple regions, interfaces between subdomains with different material constitutive relations are created. We here show subdomains $\mathcal{D}_n, n = 1,2,3,4$, while some of the possible intermaterial interfaces include $\mathcal{D}_1/\mathcal{D}_2$, $\mathcal{D}_1/\mathcal{D}_3$, $\mathcal{D}_2/\mathcal{D}_3$, $\mathcal{D}_3/\mathcal{D}_4$. More complex geometrical and topological interfaces than those shown in Figure 1 are possible where the topological type of the interface manifold can be controlled by introducing handles, holes, gluing, cutting, and so on.

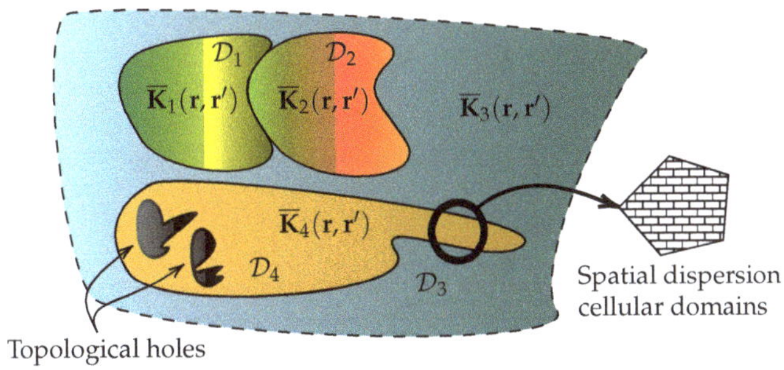

Figure 1. A generic depiction of an electromagnetic nonlocal metamaterial system. Each of the domains $\mathcal{D}_n$ is captured by a general linear nonlocal response function $\overline{\mathbf{K}}_n(\mathbf{r},\mathbf{r}')$.

Recall that in *local* electromagnetism each intermaterial interface should be assigned a special electromagnetic boundary condition in order to ensure the existence of a unique solution to the problem [9,41]. This, however, is not possible in nonlocal electromagnetism. Indeed, and as already mentioned earlier, nonlocal electromagnetism introduces several subtle issues that are absent in the local case: *additional boundary conditions* are often invoked to handle the transition of fields along barriers separating different domains, such as between two nonlocal domains, or even one nonlocal and another local domain [16,55,56]. The topological fiber bundle theory to be developed in Section 5 will provide a clarification of why this is so since it turns out that the traditional spacetime approach often employed in local electromagnetism is not necessarily the most natural one (see also Section 8). There is a need, then, to examine in a more in-depth fashion the detailed structural phenomena associated with the presence of *multiple topological scales* in nonlocal metamaterials. This paper will provide some new insights into these issues.

3.3. Preliminary Remarks on the Existence of Multiple Topological Scales in Nonlocal Continuum Field-Theoretic Structures

For completeness and maximal clarity, we discuss here some of the directly observable topological scales in nonlocal continuum systems whose preliminary understanding at this stage of our presentation does not require the use of the quite elaborate mathematical apparatus to be carefully constructed in the remaining parts of this paper. We list the most important of these topological levels as follows:

1. The first is the geometrical separation between different nonlocal domains, such as $\mathcal{D}_1$ and $\mathcal{D}_2$ discussed in Section 3.2 and illustrated by Figure 1.
2. The second is the case captured by the inset in the right hand side in Figure 1. Fine "microscopic" cells, each homogeneous and, hence, describable by a response function of the form $\overline{\mathbf{K}}(\mathbf{k},\omega)$, can be combined to build up a complex effective nonlocal response tensor $\overline{\mathbf{K}}_n(\mathbf{r},\mathbf{r}')$ over its topologically global domain $\mathcal{D}_n$. Such juxtaposition at the microscopically local level that effectively leads to the emergence of a global behavior is a classic example of multiscale physics. However, note that it even acquires a higher importance in the present context due to the fact that *both* of the constituent cell level (rectangular "bricks" in the inset of Figure 1) and the global domain level $\mathcal{D}_n$ already belong to the *physically*, e.g., electromagnetically, nonlocal dimension of the relevant nonlocal continuum field theory.
3. Finally, the third directly observable topological scale is that connected to what we termed "topological holes" in Figure 1. These are arbitrarily-shaped gaps, such as holes, vias, etchings, etc., which are intentionally introduced in order to influence the electromagnetic response by modifying the topology of the three-dimensional material manifolds $\mathcal{D}_n$.

The above topological levels are called "directly observable" because their determination does not require the use of abstract and advanced concepts from continuum field theory. This is in contrast to the more subtle distinction that will be discussed next.

In Remark 3, we discuss the very important conceptual distinction between topology-based and physics-based nonlocal domains, a demarcation between two concepts that has already been invoked several times above, and will also figure up repeatedly throughout the remaining parts of this article.

Remark 3 (Distinction between physics- and topology-based locality/nonlocality). The terms *local* and *global* possess two different senses, one physics-based, e.g., electromagnetic theory; the other is spatio-geometric in essence, belonging to the purely formal and mathematical dimensions of the structure of the nonlocal continuum theory of the material system. Elucidating this subtle interconnection between the two senses will be one of the main objectives of the present work but we will first need to introduce the various relevant microscale topological concepts to be given in Section 4 (see also Remark 17) For the time being, let the following be known:

1. *Physics-based local/non-local distinction*: this is where basically physical considerations are at stakes. We distinguish between:
   (a) *Physics-based non-local level*: this includes how the response of the material continuum depends on locations $\mathbf{r}'$ *not* infinitesimally close to the point $\mathbf{r}$ where the excitation field is applied. That is, $\mathbf{r}-\mathbf{r}'$ is nonzero but it is also not a differential. (On infinitesimal domains, see Remark 1.)
   (b) *Physics-based local level*: this is the physical regime whose essence is captured by local constitutive relations of the form (15).
2. *Topology-based local/non-local distinction*: mathematical considerations dominate at this level. We have:
   (a) *Topology-based non-local level*: this is the topologically global level, e.g., the entire topological manifold in contrast to the local description applicable only to a coordinate patch [57], and so on. At this level, the non-local-as-global is an emerging structure based on gluing together "smaller pieces" of the total manifold. We will see examples of processes occurring basically at this level when we use partition of unity methods.
   (b) *Topology-based local level*: this is the topological layer associated with structures, such as open sets, topological neighborhoods, closed sets, and so on. A topological space is defined as a collection of all such local sets [58,59].

The two concepts outlined above interact with each. There is a subtle relation between physics and topology. This paper will address some of these delicate interrelations in subsequent sections.

Remark 4 (Electromagnetic Domains). For simplicity, in what follows we will occasionally use 'electromagnetic (EM) domain' and 'physics-based nonlocal domain' as interchangeable terms. It should be kept–in min–that the concept of physics-based nonlocality is broader than EM nonlocality. The former refers to a characteristic structural trait enjoyed by all nonlocal continuum field theories, while the latter is restricted to the realm of just one such theory, that of the electromagnetism of continuous media.

4. The Microscopic Topological Structure of Physics-Based Nonlocal Domains

4.1. Introduction

In this section, we begin our careful examination of the mode of interrelation between the physics- and topology-based types of nonlocality introduced and discussed above.[5] Let the nonlocality domain of the electromagnetic medium, the region $D \subset \mathbb{R}^3$ in (11), be bounded. Corresponding to (1), a similar operator equation in the frequency domain representing the most general form of a nonlocal electromagnetic medium can be posited, namely

$$\mathbf{R}(\mathbf{r};\omega) = \mathcal{L}_\omega\{\mathbf{F}(\mathbf{r};\omega)\}, \tag{20}$$

where the nonlocal medium linear operator is itself frequency dependent. For simplicity, and as stated before, whenever it is understood from the context that the material response operator is formulated in the frequency domain, all dependencies on ω appearing in its formal expression will be removed.

We are going to propose a change in the mathematical framework inside which electromagnetic nonlocality is usually defined. This will be done in two stages:

- Initially, in the present Section, we introduce the rudiments of the main physics-based microtopological structure associated with nonlocality in continuum field theories, but without delving into considerable mathematical details. The aim is to familiarize ourselves with the minimal necessary physical setting and how it naturally gives rise to a more refined picture of the nonlocal material domain compared with the traditional (and much simpler) topological structure of local electromagnetism based on spacetime points.
- In the second stage, treated in Section 5, a more careful mathematical picture is developed using the theory of topological fiber bundles. We eventually show (Section 5.3) that the physics-based (in this case the electromagnetic) nonlocal operator (20) can be reformulated as a Banach bundle map (homomorphism) over the three-dimensional space of the material domain under consideration. Some computational examples and applications are provided in the later Sections, e.g., see Section 7.

The key conceptual idea behind the entire theory presented here is that of the *topological microdomains* associated with the field theory of nonlocal continua, e.g., the electromagnetism of continuous media, which we first develop thematically in the next Section 4.2 before moving subsequently to the more rigorous and exact topological formulation of Section 5.

4.2. The Concept of Topological Microdomains in Nonlocal Continuum Field Theories

In conventional frequency domain local electromagnetism, the boundary-value problem of multiple domains is formulated as a set of coupled partial differential equations or integro-differential equations interwoven with each other via the appropriate intermaterial interface boundary conditions dictating how fields change while crossing the various spatial regions inside which the equations hold [28,40,41]. This has been traditionally achieved by taking up the electromagnetic response function $\overline{\mathbf{K}}(\mathbf{r},\mathbf{r}';\omega)$ as an essential key ingredient of the problem description, which traditionally has been exploited in two stages: First, the constitutive relations would enter into the governing equations in each separate solution domain. Second, the constitutive relations themselves are used in order to construct the proper electromagnetic boundary conditions prescribing the continuity/discontinuity behavior of the sought field solutions as they move across the various interfaces separating domains with different material properties.

Unfortunately, it has been well known for a long time that it is not possible to formulate a universal electromagnetic boundary condition for nonlocal media, especially for the case of spatial dispersion. This will be discussed later with more details in Section 8, but also see the discussion around additional boundary conditions (ABC) in Appendix A.1. For now, we concentrate on gaining a deeper understanding of the generic structure of spatial nonlocality in continuum field theories.

Consider the microdomain structure depicted in Figure 2. A key starting observation is how nonlocality forces us to associate with every spacetime point $(\mathbf{r},t)$, or frequency space point $(\mathbf{r};\omega)$, a *topological neighborhood* of $\mathbf{r}$, say $V_\mathbf{r}$, such that $\mathbf{r}\in V_\mathbf{r}$. For now, let us assume that the spatial material domain $D \ni \mathbf{r}$ is just an *open set* in the technical sense of the topology of the Euclidean space $\mathbb{R}^3$ inherited from the standard Euclidean metric [59]. By restricting D to be open, we avoid the notorious problem of dealing with boundaries or interfaces between such (possibly overlapping) open sets. That is, the topological closure of D, denoted by $\text{cl}(D)$, is excluded from the domain of nonlocality. Let D be the maximal such topological neighbored for the problem under consideration.[6] We now associate with each point $\mathbf{r}$ a "smaller" open set $V_{\mathbf{r}}$ where the following holds:

$$\boxed{\forall\, \mathbf{r} \in D,\ \exists\, V_{\mathbf{r}} \subset D \text{ such that } \mathbf{r} \in V_{\mathbf{r}}, \text{ and } V_{\mathbf{r}} \text{ is open.}} \tag{21}$$

Note that the assumed openness of D makes the above construction technically possible. We will call the proposition (21) the *principle of nonlocal microdomain generation*. It formally captures the main content of the structure of nonlocality at the microscopic level. In Section 7, a practical example taken from nonlocal semiconductor metamaterials will be investigated in depth in order to illustrate the applicability of (21).

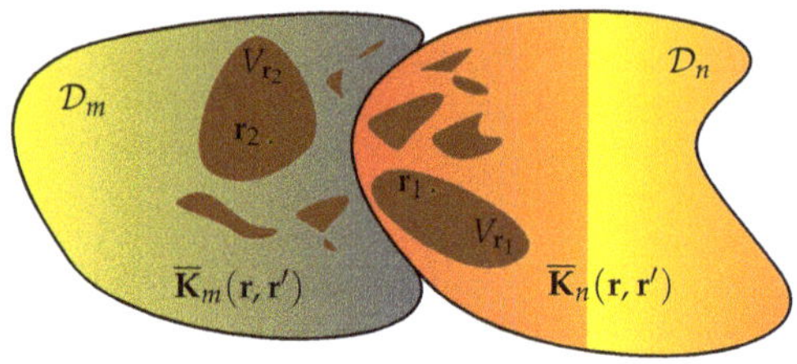

Figure 2. The microtopological structure of nonlocal metamaterial systems includes more than just the three-dimensional spatial domains $\mathcal{D}_n, n = 1, 2, \ldots$ It is best captured by classes $\mathcal{V}(\mathcal{D}_n)$ composed of various open sets $V_{\mathbf{r}} \subset \mathcal{D}_n$ based at each point $\mathbf{r} \in \mathcal{D}_n$. On every such subset a vector field is defined, representing the external field excitation field. The collection of all vector fields on a given set $V_{\mathbf{r}}$ gives rise to a linear topological function space $\mathcal{F}(V_{\mathbf{r}})$. The topologies consisting of the base spaces $\mathcal{D}_n$, the nonlocal microdomains $V_{\mathbf{r}}$, and the function spaces $\mathcal{F}(V_{\mathbf{r}})$, collectively give rise to a total "macroscopic" topological structure (superspace) that is considerably more complex than the base spaces $\mathcal{D}_n$.

Now, instead of considering fields like $\mathbf{R}(\mathbf{r})$ and $\mathbf{F}(\mathbf{r})$ defined on the entire maximal domain of nonlocality D (which can grow "very large") we propose to reformulate the problem of nonlocal continua as a *topologically local*[7] structure by exploiting the fact that the physics of field–matter interactions gives the field response at location $\mathbf{r}$ due to independent excitation fields essentially confined within a "smaller domain" around $\mathbf{r}$, namely the open set $V_{\mathbf{r}}$.[8]

Furthermore, if the response at another *different* point $\mathbf{r} \neq \mathbf{r}'$ is needed, then a *new*, generally different, "small" open set $V_{\mathbf{r}'}$ will be required. That is, in general we allow that

$$V_{\mathbf{r}} \neq V_{\mathbf{r}'} \tag{22}$$

even though it is expected that typically there should be some overlap between these two small local domains of electromagnetic nonlocality in the sense that

$$V_{\mathbf{r}} \cap V_{\mathbf{r}'} \neq \emptyset, \tag{23}$$

especially if the nonlocality radius $|\mathbf{r} - \mathbf{r}'|$ is small.

The following fundamental collection of "smaller" sets, where a metric scale characterizing "smallness" is not implied, written as

$$\{V_{\mathbf{r}}, \mathbf{r} \in D\}, \tag{24}$$

will be dubbed *nonlocal microdomains*, or just *microdomains* in short. A possible precise definition is given next.

Definition 1 (Nonlocal microdomains: the physics-based scenario). Consider a material domain D with the associated nonlocal response function $\overline{\mathbf{K}}(\mathbf{r}', \mathbf{r})$. We define the (physics-based) nonlocal microdomain $V_{\mathbf{r}} \subset D$, labeled by $\mathbf{r} \in D$, as the interior of the compact[9] support of $\overline{\mathbf{K}}(\mathbf{r}', \mathbf{r})$. The support itself is defined by the standard formula

$$\operatorname{supp} \overline{\mathbf{K}}(\mathbf{r}', \mathbf{r}) := \mathrm{cl}_D\{\mathbf{r}' \in D, \ \|\overline{\mathbf{K}}(\mathbf{r}', \mathbf{r})\| \neq 0\}, \tag{25}$$

where $\|\cdot\|$ is a suitable tensor norm, for example the matrix norm.[10] The topological closure operator cl_D is here taken with respect to the total material space D where the latter is viewed as a topological space on its own.

Remark 5 (Microdomain topology). By Definition 1 above, the nonlocal microdomain $V_{\mathbf{r}}$ is always open. It can be shown that the collection of open sets $\{V_{\mathbf{r}}, \mathbf{r} \in D\}$ induces a topology on the total space occupied by the nonlocal material (the details are omitted since they are lengthy though straightforward.) In what follows, this topology will be referred to by the term *microdomain topology*. The set of physics-based nonlocality microdomains (microdomains for short), as constructed in Definition 1, explicate the fine microtopological structure of nonlocal electromagnetic domains at a spatial scale different from that of the (topologically "larger") material domain D itself and are fundamental for the theory developed in this paper.

Remark 6 (Discrete topology in local continua). In local media, the microdomains topology reduces to the trivial discrete topology

$$\{\{\mathbf{r}\}, \mathbf{r} \in D\} \tag{26}$$

since the external field interacts only with the point $\mathbf{r}$ at which it is applied and hence

$$V_{\mathbf{r}} = \{\mathbf{r}\} \tag{27}$$

holds as the "smallest" possible topological microdomain in that rather special case. Therefore, the microdomain topology is interesting only for the case of physics-based nonlocality, e.g., the scenario of EM microdomains discussed in more details in the examples and applications below. In particular, from the point of view of this article, *local* metamaterials are not topologically interesting.

4.3. Construction of Excitation Field Function Spaces on the Topological Microdomains of Nonlocal Media

After enriching the MTM domain D with the finer topology of nonlocality microdomains $V_{\mathbf{r}}, \mathbf{r} \in D$, we wish to equip this total medium with additional mathematical structure based on the physics of field–matter interaction. Consider the set of all sufficiently differentiable vector fields $\mathbf{F}(\mathbf{r})$ defined on $V_{\mathbf{r}}$, $\mathbf{r} \in D$. This set possesses an obvious complex *vector space structure*: for any two complex numbers $a_1, a_2 \in \mathbb{C}$, the sum

$$a_1 \mathbf{F}_1(\mathbf{r}) + a_2 \mathbf{F}_2(\mathbf{r})$$

is defined on $V_{\mathbf{r}}$ whenever $\mathbf{F}_1(\mathbf{r})$ and $\mathbf{F}_2(\mathbf{r})$ are, while the null field plays the role of the origin. In what follows, we will denote such function spaces by $\mathcal{F}(V_{\mathbf{r}})$ or just $\mathcal{F}$ if it is understood from the context on which material spatial domains the fields are defined.

Remark 7 (The excitation field function space and Sobolev spaces). It is possible to equip $\mathcal{F}(V_{\mathbf{r}})$ with a suitable topology in order to measure how "near" to each other are any two fields defined on $V_{\mathbf{r}}$, e.g., see [59,62,63]. Therefore, in this manner $\mathcal{F}(V_{\mathbf{r}})$ acquires the structure of a *topological vector space* [59]. In particular, it can be made a *Sobolev space*, where the latter is not only a Banach space (normed space), but also a Hilbert space (inner product space) [64–66].

The detailed construction of a Sobolev space on a given microdomain is not needed for what follows in this paper, but can be found in the literature, including the references quoted in this Remark 7.

4.4. *The Global Topological Structure of Nonlocal Electromagnetic Material Domains: First Look*

In light of the analysis above, each microdomain $V_{\mathbf{r}}$ induces an infinite-dimensional linear function space (Sobolev space) $\mathcal{F}(V_{\mathbf{r}})$ indexed by the position $\mathbf{r} \in D$, with the corresponding topology being essentially determined by the geometry of $V_{\mathbf{r}}$. On the other hand, this latter geometry is obtained from the *physics* of field–matter interaction in nonlocal media. Consequently, the physical content of nonlocal materials is encoded at the level of the topological microstructure encapsulated by the following formal scheme:

$$\boxed{D \ni \mathbf{r} \xrightarrow[\text{Physics-Based Nonlocality}]{\text{Physical Data}} V_{\mathbf{r}} \xrightarrow[\text{Sobolev Space}]{\text{Mathematical Data}} \mathcal{F}(V_{\mathbf{r}})} \tag{28}$$

Let us first identify the main relevant collections of subsets needed in order to understand the formal set-theoretic structure of the problem. We begin by

$$\mathcal{V}(D) := \{V_{\mathbf{r}} \subset D \big| \, \mathbf{r} \in D, V_{\mathbf{r}} \text{ is open}\}, \tag{29}$$

as the class of physics-based nonlocal microdomains (Definition 1). On the other hand, it is also possible to introduce the useful construction

$$\mathcal{G}[\mathcal{V}(D)] := \{\mathcal{F}(V_{\mathbf{r}}) \big| \, \mathbf{r} \in D, \mathcal{F}(V_{\mathbf{r}}) \text{ is a Sobolev function space}\}, \tag{30}$$

as a convenient class into which we collect all the function spaces of excitation fields on each nonlocality microdomain $V_{\mathbf{r}}$ as spanned by the position index $\mathbf{r} \in D$ (see Remark 7 for the construction of each such function space.) It follows then that (28) can be neatly captured by the ordered triplet

$$D \times \mathcal{V}(D) \times \mathcal{G}[\mathcal{V}(D)]. \tag{31}$$

We wish now to unpack this compact structure in a careful, step-by-step manner, proceeding as follows:

1. Each open domain in $D \subseteq \mathbb{R}^3$ will by assigned a distribution $\mathcal{V}(D)$ of open sets $V_{\mathbf{r}}$, i.e., the physics-based nonlocality microdomains topology defined in Section 4.2, see in particular Definition 1 and Remark 5. Physically, it expresses the fine microtopological structure of nonlocal continua, e.g., electromagnetic material nonlocality.
2. The structure $\mathcal{V}(D)$ is solely determined by the physics of field–matter interaction. A concrete example explicitly illustrating how the detailed physical content of the underlying process contributes to the construction of $\mathcal{V}(D)$ will be given in Section 7.
3. We further emphasize that the various sets $V_{\mathbf{r}} \in \mathcal{V}(D)$ constitute an open cover of D, that is, we have
$$D = \bigcup_{\mathbf{r} \in D} V_{\mathbf{r}}. \tag{32}$$
In this way, the model can accommodate excitation fields $\mathbf{F}(\mathbf{r})$ applied at every point in $\mathbf{r} \in D$.
4. The decomposition of the material domain D into smaller building blocks exemplified by (32) is fundamental for computational topological models of nonlocal MTMs. For example, in Section 7 we will exploit this expansion in order to construct a topological coarse-grained model for inhomogeneous nonlocal semiconductor metamaterials.
5. Finally, the topology $\mathcal{V}(D)$ induces the "function superspace" $\mathcal{G}[\mathcal{V}(D)]$ (30) defined as a class of function spaces $\mathcal{F}(V_{\mathbf{r}}), \mathbf{r} \in D$,, where each vector field acts on one microdomain element $V_{\mathbf{r}}$ chosen from the topology $\mathcal{V}(D)$.

Remark 8 (Topology, physics, and multiple scales). It is interesting to observe how, within the framework proposed above, some sort of delicate constructive "division of labor" is seen to emerge into the picture, where a fruitful interaction between physics and mathematics generates the various required multiscale topological microstructures characteristic of nonlocality in continuum field theories. This is also the source of some potential difficulties hidden in the formal set-theoretic structure (31). Indeed, we will next try to smooth out the differences between the two main substructures $\mathcal{V}(D)$, which is principally controlled by physics, on one side, and $\mathcal{G}[\mathcal{V}(D)]$, which is dominated by purely mathematical considerations. One way to achieve a resolution of this philosophical tension between the physical and mathematical is by developing the entire theory of the set-theoretic structure (31) in a form that can encode all of its main substructures within a single, rich enough "metastructure": the Banach vector bundle *superspace* (see Section 5 for the detailed construction).

As can be seen from Remark 8, there is indeed some strong motivation to search for alternative formulations of physical theory in complex and rich systems such as nonlocal material continua, where there exists multiple spatial topological scales. It will be seen that the superspace theory appears to provide some form of rare direct and transparent unity between physics and topology in this regard. In order to reach there, gradual, step-by-step changes in the conventional formulation of continuum field theory will be introduced. We now begin to look into such a reformulation, starting with a straightforward one.

4.5. A Reformulation of the Nonlocal Continuum Response Function

It is now possible to provisionally construct the nonlocal continuum response function by working on the fundamental topological domain structure (31) instead of the global domain D, the later being the favored arena of conventional continuum field theory that we would like to ultimately move beyond. Again, for concrete expressions, the special case of electromagnetic theory will be presupposed but it should always be kept in mind that the mathematical structure of the theory is quite general and applies to all nonlocal continuum field theories governed by an abstract material response function model, such as the one discussed in Section 3.

We start by noting that the response field $\mathbf{R}(\mathbf{r})$ can be re-expressed by the map

$$\mathbf{R} : D \times \mathcal{V}(D) \to \mathbb{C}^3, \tag{33}$$

where the codomain is taken to be $\mathbb{C}^3$ because the electric or magnetic response functions $\mathbf{D}$ or $\mathbf{B}$, respectively, are complex vector fields in the frequency domain.[11] The value of the EM nonlocal response field due to excitation field $\mathbf{F}(\mathbf{r})$ applied at a microdomain $V_{\mathbf{r}}$ can be computed by means of

$$\mathbf{R}(\mathbf{r};\omega) = \int_{V_{\mathbf{r}}} \mathrm{d}^3 r' \, \overline{\mathbf{K}}(\mathbf{r},\mathbf{r}';\omega) \cdot \mathbf{F}(\mathbf{r}';\omega). \tag{34}$$

Although (34) may appear at first sight to be only slightly different from (11), the underlying difference between the two formulas is significant. In essence, the construction of the EM response field $\mathbf{R}(\mathbf{r})$ via the map (33) amounts to *topological localization* of electromagnetic nonlocality, since in the latter case, the EM response function $\overline{\mathbf{K}}(\mathbf{r},\mathbf{r}';\omega)$ is no longer allowed to extend globally onto "large and complicated material domains." Indeed, with the recipe (34) only the response to "small"–or more rigorously *topologically local*[12]—domains, namely the microdomains $V_{\mathbf{r}}$, is admitted. On the other hand, in order to find the response field $\mathbf{R}(\mathbf{r})$ everywhere in D, one needs to use sophisticated topological techniques to extend the response from one point to another until it covers the entirety of D. This local-to-global extension application of differential topology is discussed in detail in Section 5.3 and again briefly in Appendix A.11.

In such a manner, it becomes possible to provide an alternative, more detailed explication of the behavior of the medium at topological interfaces (boundary conditions in nonlocal metamaterials are treated–provisionally–in Section 8) and also explore the effect of the topology of the bulk medium itself on the allowable response functions and the production of non-trivial edge state, with obvious applications to emerging areas such as nonlocal metamaterials.[13]

5. The Fiber Bundle Superspace Formalism in the Field Theory of Generic Nonlocal Continua

Here, an outline of the direct construction of a fiber (Banach) bundle over an entire (global) nonlocal generic material domain is given, where our purpose is to attach to every point $\mathbf{r} \in U_i$ a *fiber* $\mathcal{F}_i$, actually a vector space in our case. The contents of this section are the most technically advanced in this paper. Readers interested in applications may skim through Sections 5.1 and 5.2, skip Section 5.3, then move directly to Section 6 for a general summary of the fiber bundle algorithm. Concrete computational models are outlined in Section 7 using a practical nonlocal model, while additional remarks and discussions about current and future uses of the theory are provided in Appendixes A.3 and A.11. However, even readers not fully familiar with the differential manifold theory will benefit from reading the present technical section, because we strive to illustrate the physical intuition behind the various mathematical computations and steps therein.

5.1. Preparatory Step: Promoting the Material Domain D to a Manifold $\mathcal{D}$

In order to investigate in depth the fundamental physico-mathematical constraints imposed on nonlocal continua, the domain D, which we have working with so far as the main total spatial space of the material, should be upgraded in complexity to the higher level of a *differential manifold*, the latter which posses a quite rich and sophisticated structure that allows performing calculus and geometrical reasoning simultaneously [26,57,62,63,68]. There are several reasons why this is highly desirable:

1. It provides a natural and obvious generalization of the basic structure (31) from the mathematical perspective.
2. Engineers often need to insert metamaterials into specific device settings, hence the shape of the material becomes highly restricted. It is therefore important to develop efficient tools to deal with variations of geometric and topological degrees of freedom and how they could possibly impact the design process.
3. Applied scientists and engineers are often interested in deriving fundamental limitation on metamaterials, e.g., what are the ultimate allowable response–excitation relations or constitutive response functions possible given this material domain topology?
4. Sophisticated full-wave electromagnetic numerical solvers prefer working with local coordinates in order to handle complicated shapes, even if a global coordinate system is sometimes available, making the deployment of the three-manifold structures for describing the material domain D useful.
5. In topological photonics and materials [11], most applications seem to focus on lower-dimensional states of matter like those associated with quantum Hall effects and edge states (surface waves).[14] There, new phenomena appear at material structures where the base space (material domain D) is a two-surface, which is best described mathematically as a differential two-manifold.

For all these reasons, it is desirable to strive to furnish the domain D with the most general and flexible mathematical apparatuses available to us, which, in this case, amounts to equipping the material/metamaterial spatial domain with a *smooth manifold structure*.

We quickly illustrate how this can be accomplished. If we denote by $\mathcal{D}$ a three-manifold (three-dimensional smooth manifold), then, since $D \subset \mathbb{R}^3$, there is a natural differential structure defined on D, inherited from the ambient three-dimensional Euclidean space itself. (Throughout this paper, such differential three-manifold structure will be presupposed as the de facto space for the *total*, i.e., largest, material space.)

Following the standard theory of smooth manifolds, let

$$(U_i, \phi_i) \tag{35}$$

be a countable collection of charts (an atlas), labeled by

$$i \in I \subset \mathbb{N}, \tag{36}$$

where I is an index set. Together, the devices (35) and (36) can equip $\mathcal{D} \subset \mathbb{R}^3$ with a differential three-manifold structure. For simplicity, we will refer to the points of the manifold $\mathcal{D}$ by $\mathbf{r}$, i.e., using the language of the global (ambient) Euclidean space $\mathbb{R}^3$. Symbolically, by adding a differential manifold structure, we effected the transformation

$$D \xrightarrow[\text{Introduce a differential atlas}]{\text{Insert a smooth manifold structure}} \mathcal{D} \tag{37}$$

This well-known construction [26,57,63] constitutes the differential atlas on $\mathcal{D}$, which will be used in what follows.

5.2. Attaching Fibers to Generic Points in the Nonlocal Material Manifold D

Our current goal is to attach a *vector fiber* (a linear function space in this case) at every point $\mathbf{r} \in \mathcal{D}$, namely the function space $\mathcal{F}(V_{\mathbf{r}})$ introduced in Section 4.3. It turns out that accomplishing this requires finding suitable "compatibility laws" dictating how coordinates change when two intersecting charts U_i and U_j interact with each other, which is typical in such types of constructions [26]. In particular, we will need to later find the law of mutual transformation of vectors in the fibers $\mathcal{F}(V_{\phi_i(\mathbf{r})})$ and $\mathcal{F}(V_{\phi_j(\mathbf{r})})$. Here, the expression

$$\mathcal{F}\left(V_{\phi_i(\mathbf{r})}\right) \tag{38}$$

means the fiber space attached to the point whose coordinates are $\phi_i(\mathbf{r})$, i.e., the *function* space where all functions are expressed in terms of the language of the ith chart $(U_i, \phi_i(\mathbf{r}))$.

In this connection, the major technical problem facing us is a mathematical one induced by the physics of the situation. We first isolate and describe the main problem by the following brief technical resume:

Since the differential structure associated with charts

$$\{(U_i, \phi_i(\boldsymbol{r})), i \in I, \}$$

can be fixed by essentially mathematical considerations alone, while the collection of microdomains

$$\mathcal{V}(\mathcal{D}) = \{V_{\boldsymbol{r}}, \boldsymbol{r} \in \mathcal{D}\}$$

is solely determined by the physics of electromagnetic nonlocality (See Remark 3 and Section 4), there is no direct and simple way to determine and express the vector transformation

$$\mathcal{F}(V_{\phi_i(\boldsymbol{r})}) \longrightarrow \mathcal{F}(V_{\phi_j(\boldsymbol{r})}),$$

because several different coordinate patches other than U_i and U_j, belonging to the differential three-manifold $\mathcal{D}$ atlas, might be involved in geometrically building up the microdomain $V_{\boldsymbol{r}}$.

The above technical problem will be solved in Section 5.3 by using the technique of *partition of unity* borrowed from differential topology [26,57,62]. It will allow us to split up each full microdomain $V_{\mathbf{r}}$ into several suitable sub-microdomains (details below), which can be later joined up together in order to give back the original EM nonlocality microdomain $V_{\mathbf{r}}$.

For now, we start by recalling that the microdomain structure represented by the collection $\mathcal{V}(\mathcal{D}) = \{V_{\mathbf{r}}, \mathbf{r} \in \mathcal{D}\}$ is an *open cover* of the manifold $\mathcal{D}$. Therefore, and since the material domain manifold $\mathcal{D}$ possesses a *countable* topological base [59], it contains a locally finite open cover subordinated to $\mathcal{V}(\mathcal{D})$ [26,57].[15] This implies that an atlas $(U_i, \phi_i), i \in I$, with diffeomorphisms

$$\phi_i : U_i \to \mathbb{R}^3, \tag{39}$$

describing the differential structure of the manifold $\mathcal{D}$ exists such that the elements $\{U_i, i \in I\}$ constitute the above mentioned locally finite subcover *subordinated* to the microdomains collection $\mathcal{V}(\mathcal{D})$. Moreover, the images $\phi_i(U_i)$ are *open balls* centered around 0 in $\mathbb{R}^3$ with finite radius $a > 0$ (henceforth, such balls will be denoted by B_a) [26].

In this way, the physics-based open cover set $\mathcal{V}(\mathcal{D})$ provides a first step toward the construction of a complete *topological* description of the *physics-based* nonlocal microdomain structure. The reason is that the coordinate patches $(U_i, \phi_i), i \in I$, are *subordinated* to the microdomains $\{V_{\mathbf{r}}, \mathbf{r} \in \mathcal{D}\}$ [26].

It is also known that there exists a *partition of unity* associated with the $\mathcal{D}$-atlas $(U_i, \phi_i), i \in I$, constructed above summarized by the following lemma [26,57,62,63,68]:

Lemma 1 (Partition of Unity). There is a collection of functions

$$\psi_i : U_i \subset \mathcal{D} \to \mathbb{R} \tag{40}$$

satisfying the following requirements:

1. $\psi_i(\mathbf{r}) \geq 0$ and each function is $C^p, p \geq 1$.[16]
2. The *support* of $\psi_i(\mathbf{r})$, denoted by supp ψ_i, is contained within U_i, that is, the condition

$$\text{supp}\, \psi_i \subset U_i, \tag{41}$$

holds. Recall that the support is defined as the (topological) closure of the set

$$\{\mathbf{r} \in \mathcal{D} |\, \psi_i(\mathbf{r}) \neq 0\}. \tag{42}$$

See for example [30,68,69].

3. Since the open cover $U_i, i \in I$, is *locally finite*, at each point $\mathbf{r} \in \mathcal{D}$, only a *finite* number of U_i will intersect $\mathbf{r}$.
4. Let the set of indices of those intersecting U_is be $I_{\mathbf{r}}$. Then we require that

$$\sum_{i \in I_{\mathbf{r}}} \psi_i(\mathbf{r}) = 1, \tag{43}$$

where the sum is always convergent because the set $I_{\mathbf{r}}$ is finite.

Remark 9. It can be shown that the sets

$$\phi_i^{-1}(B_{a/3}), i \in I, \tag{44}$$

where $B_{a/3}$ is a standard Euclidean ball centered at the origin with radius $a/3$, already cover $\mathcal{D}$ [57]. Moreover, the closure

$$\text{cl}\{\phi_i^{-1}(B_{a/3})\} \tag{45}$$

may be taken to constitute the support of $\psi_i(\mathbf{r})$, while [26,57,70]

$$\mathbf{r} \notin \text{supp}\{\psi_i(\mathbf{r})\} \implies \psi_i(\mathbf{r}) = 0. \tag{46}$$

The partition of unity functions ψ_i can be computationally constructed using standard methods, most prominently the bump functions, see [57,71] for details.

The motivation behind the deployment of the partition of unity technique and how it immediately arises in connection with our fundamental EM nonlocal structure should now be clear. We have found that the following three-step process is natural:

1. Initially, the *physics*-based collection of sets
$$\mathcal{V}(\mathcal{D}) = \{V_{\mathbf{r}}, \mathbf{r} \in \mathcal{D}\},$$
for example, the EM nonlocal microdomain structure based on each point $\mathbf{r}$ in the nonlocal metamaterial $\mathcal{D}$, is obtained using a suitable physical microscopic theory or some other procedure.[17]
2. Introduce a differential atlas
$$(U_i, \phi_i(\mathbf{r})), \ i \in I,$$
on the smooth manifold $\mathcal{D}$ subordinated to $\mathcal{V}(\mathcal{D})$ and representing the nonlocal material domain under consideration.
3. Finally, the same atlas is linked to a set of functions $\psi_i(\mathbf{r})$ (partition of unity) that can be recruited as "topological bases" in order to expand any differentiable field excitation function into sum of individual sub-fields defined on open subsets of the material domain $\mathcal{D}$ (see Section 5.3).

The three-step process outlined above is summarized in Figure 3, illustrating how to progressively construct micro-coordinate systems allowing one to see through increasingly smaller spatial scales in the fundamental characterization of electromagnetic material nonlocality.

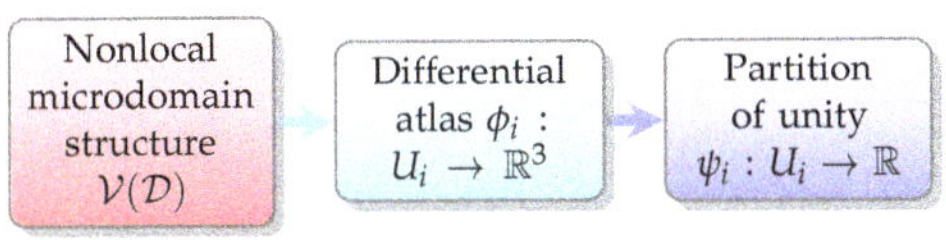

Figure 3. The three-step process of constructing micro-coordinate representations of material nonlocality starting from the nonlocal microdomain set and ending with the partition of unity on the material continuum's superspace.

The key idea to be developed next is that both the base manifold $\mathcal{D}$ and the nonlocal physics-based microdomains $V_{\mathbf{r}}$ are described *locally* (in the topological sense[18]) by the *same* collection of charts, namely $(U_i, \phi_i(\mathbf{r})), i \in I$. This will permit us to construct a direct unified description of *both* the base manifold $\mathcal{D}$ *and* its fibers, i.e., the linear topological function spaces $\mathcal{F}(V_{\mathbf{r}})$, the latter being the model of the physical electromagnetic fields exciting the nonlocal material $\mathcal{D}$.

The construction of a fiber bundle superspace for nonlocal electromagnetic materials will be completed in two steps:

- Step I: Construct a tailored fiber bundle based on the partition of unity charts $(U_i, \phi_i(\mathbf{r}))$ introduced above.
- Step II: the original physical structure (31) is recovered by gluing together various sub-microdomain $U_i \subseteq V_{\mathbf{r}}$ of each EM nonlocal microdomain $V_{\mathbf{r}}$.

We start with Step I, while we leave the more complicated Step II to Section 5.3.

Consider the $(U_i, \phi_i(\mathbf{r})), i \in I$, as our atlas on the three-manifold $\mathcal{D}$ introduced in Section 5.1. At each point $\mathbf{r} \in U_i$, we attach a linear topological space $\mathcal{F}(U_i)$ defined as the Sobolev space
$$W^{p,2}(U_i), \ p \geq 1, \tag{47}$$
of functions on the open set U_i, i.e., we write
$$\forall\, i \in I, \ \mathcal{F}(U_i) := \{\psi_i(\mathbf{r})\mathbf{F}(\mathbf{r}), \mathbf{r} \in U_i, \text{ is in the Sobolev space } W^{p,2}(U_i)\}, \tag{48}$$
where $\mathbf{F}(\mathbf{r})$ is a suitable $C^{p,2}$ vector field.

Remark 10 (Sobolev Spaces). For the precise technical definition of the infinite-dimensional Sobolev function space $W^{p,2}(U_i)$, see [64,65]. Appendix A.5 provides some additional information on the literature. Section 4.3 gives a simplified intuitive definition of the physics-based function space $\mathcal{F}_i$, in particular see Remark 8. The intricate details of the theory of such Sobolev function spaces will not be needed for our immediate purposes in what follows (compare with Remark 11).

Physically, the multiplication of the global excitation field $\mathbf{F}(\mathbf{r})$ by $\psi_i(\mathbf{r})$ in constructions like (48) above and (50) below effectively "localizes" (in the topological sense) the field into a smaller compact subdomain, namely the support of the "topological localization basis function" $\psi_i(\mathbf{r})$ itself. Moreover, because the C^p-functions $\psi_i(\mathbf{r})$ have *compact* supports satisfying the inclusion restrictions

$$\text{supp}\{\psi_i\} \subset U_i, \ \ i \in I, \tag{49}$$

it follows that $\mathcal{F}(U_i)$ is effectively a *local* Sobolev space on U_i [66]. Alternatively, it is also possible to seek different constructions, such as the one captured by the following remark.

Remark 11. We may define a less complicated function space on U_i using the following construction:

$$\forall\, i \in I, \ \ \mathcal{F}'(U_i) := \{\psi_i(\mathbf{r})\mathbf{F}(\mathbf{r}), \mathbf{r} \in U_i, \text{ is an element of a } C^p \text{sup-norm function space}\}, \tag{50}$$

where the C^p-sup-norm is defined by

$$||\psi_i(\mathbf{r})\mathbf{F}(\mathbf{r})|| := \sup_{\mathbf{r}\in\text{supp}\{\psi_i\}}[\psi_i(\mathbf{r})\mathbf{F}(\mathbf{r})]. \tag{51}$$

In the case of $\mathcal{F}'(U_i)$, one may further consider only C^p-vector excitation fields $\mathbf{F}(\mathbf{r})$. A choice of which linear function space to work with depends on the particular application under consideration. In what follows, we further simplify our notation by writing $\mathcal{F}_i$ instead of $\mathcal{F}(U_i)$ whenever the partition of unity's differential atlas' coordinate patches U_i are used.

5.3. Direct Construction of Bundle Homomorphism as Generalization of Linear Operators in Electromagnetic Theory

We now demonstrate how the material constitutive relations in conventional (local) continuum theory may be absorbed into a new structure, the *bundle homomorphism*, which is the most natural generalization of linear operators in local electromagnetism taking us into the enlarged stage of the generic nonlocal medium's superspace formalism. In the future, these bundle homomorphisms may be discretized using topological numerical methods, e.g., see [72]. In what follows, we focus on the rigorous exact construction using the technique of partition of unity, which allows computations going from local to global domains.[19]

5.3.1. The Basic Definition of the Nonlocal Material, (or Continuum, Metamaterial (MTM), etc.), Banach (Fiber) Bundle Superspace

The initial step in formally defining the proposed nonlocal MTM bundle superspace is the following disjoint union construction:

Definition 2 (Preliminary Definition of the Bundle Superspace). Let the material continuum's superspace be denoted by $\mathcal{M}$, which is also called the *total bundle space*. We define this space as the disjoint union of all spaces $\mathcal{F}_i$ of the form:

$$\mathcal{M} := \{(\mathbf{r}, \mathcal{F}_i) | \, \forall i \in I, \mathbf{r} \in U_i \subset \mathcal{D}\}. \tag{52}$$

Associated with $\mathcal{M}$ is a surjective map

$$p : \mathcal{M} \to \mathcal{D}, \tag{53}$$

which "projects" the fiber onto its corresponding point in the base manifold $\mathcal{D}$, i.e., $p((\mathbf{r}, \mathcal{F})) := \mathbf{r}$.

Remark 12 (Other constructions of bundle spaces). In mainstream literature, the fiber bundle concept is often approached in a manner slightly differently from that of Definition 2. Indeed, the *fiber* of $\mathcal{M}$ at $\mathbf{r} \in \mathcal{D}$ is *defined* as the set $p^{-1}(\mathbf{r})$, but provided the map p is already given as part of the bundle's initial data. However, in this paper, we *construct* the bundle data starting with the physics-based topological structure (31).

Remark 13 (Fiber Projections and Local Isomorphisms). The map p is called the *projection* of the vector bundle $\mathcal{M}$ onto its *base space* $\mathcal{D}$. Moreover, from now on, we will also use the notation $\mathcal{F}_{\mathbf{r}}$ to denote the fiber $p^{-1}(\mathbf{r})$. By construction, it should be clear that

$$\forall i \in I : \quad p^{-1}(\mathbf{r}) = \mathcal{F}_i \iff \mathbf{r} \in U_i. \tag{54}$$

From the topological viewpoint, the material continuum superspace $\mathcal{M}$ manifests itself *locally* as a product space in the form

$$U_i \times \mathcal{F}_i. \tag{55}$$

In other words, the map p should behave locally as a conventional projection operator; i.e., in a local domain U_i, the material's total bundle space $\mathcal{M}$ is isomorphic to $U_i \times \mathcal{F}_i$, and $p(U_i \times \mathcal{F}_i)$ should be isomorphic to U_i. Symbolically, we have:

$$\mathcal{M} \cong_{\text{locally}} U_i \times \mathcal{F}_i, \quad p(U_i \times \mathcal{F}_i) \cong_{\text{locally}} U_i, \tag{56}$$

for all $i \in I$, and where $\cong_{\text{locally}}$ means local topological (in this case also smooth) isomorphism.[20]

In order to complete the specification of the nonlocal material continuum superspace, we next construct the linear function space X_i defined by

$$\forall i \in I, \ X_i := \left\{\psi_i\left[\phi_i^{-1}(\overline{x})\right]\mathbf{F}\left[\phi_i^{-1}(\overline{x})\right] \text{ is an element of a Sobolev space for all } \overline{x} \in B_a\right\}, \tag{57}$$

which is the Sobolev space of $W^{p,2}(B_a)$ functions on the Euclidean 3-ball B_a. Here, each function is defined with respect to the *local* coordinates

$$\overline{x} := \phi^{-1}(\mathbf{r}), \ \mathbf{r} \in U_i. \tag{58}$$

In fact, it should be straightforward to deduce from the above that there exists maps

$$\tau_i : p^{-1}(U_i) \to U_i \times X_i, \tag{59}$$

for all $i \in I$, that are *isomorphisms* (*diffeomorphism* in our case), where such diffeomorphism may be expressed by

$$\forall i \in I : \ p^{-1}(U_i) \cong U_i \times \mathcal{F}_i. \tag{60}$$

We also add that the fact of (59) actually playing the role of such an isomorphism would naturally follow from the respective definitions of the spaces $\mathcal{F}_i$ and X_i, as specified by (48) and (57), and from the proposition that each ϕ_i is a diffeomorphism from U_i into $\mathbb{R}^3$ (or, equivalently, to the unit 3-ball B_a with radius a instead of $\mathbb{R}^3$.) Furthermore, note that by construction the diffeomorphism τ_i satisfies

$$\mathrm{proj}_1 \circ \tau_i = p, \tag{61}$$

where proj_1 is the standard projection map defined by $\mathrm{proj}_1(x, y) := x$. Finally, if we restrict τ_i to $p^{-1}(\mathbf{r})$, the resulting map

$$\tau_i|_{p^{-1}(\mathbf{r})} : p^{-1}(\mathbf{r}) \to \{\mathbf{r}\} \times X_i \tag{62}$$

is a (linear) topological vector space isomorphism from $\mathcal{F}_\mathbf{r}$ to X_i; namely, we have

$$\forall i \in I,\ \mathbf{r} \in U_i : \quad \mathcal{F}_\mathbf{r} \cong X_i. \tag{63}$$

Remark 14. The charts (U_i, τ_i) are called *trivialization covering* of the vector bundle $\mathcal{M}$. They provide a coordinate representation of local patches of the vector bundle. (The global topology of the bundle, however, is rarely trivial [62].) Since here all maps are C^p smooth, τ_i are also called *smooth* trivialization maps. The complete derivations of the diffeomorphism (60) and the topological vector space isomorphism (63) are straightforward, but lengthy, and the full proofs are omitted.

Consider now two patches U_i and U_j with $U_i \cap U_j \neq \emptyset$. By restricting τ_i and τ_j to $U_i \cap U_j$, two diffeomorphisms

$$\begin{aligned} \tau_i &: p^{-1}(U_i \cap U_j) \to (U_i \cap U_j) \times X_i, \\ \tau_j &: p^{-1}(U_i \cap U_j) \to (U_i \cap U_j) \times X_j, \end{aligned} \tag{64}$$

are obtained, which together imply in turn that

$$(U_i \cap U_j) \times X_i \cong (U_i \cap U_j) \times X_j, \tag{65}$$

or, equivalently, the following expected Banach space isomorphism:

$$X_i \cong X_j, \tag{66}$$

In particular, it can be shown that the composition map

$$\tau_j \circ \tau_i^{-1} : (U_i \cap U_j) \times X_i \to (U_i \cap U_j) \times X_j \tag{67}$$

possesses the simple form

$$\tau_j \circ \tau_i^{-1}(\mathbf{r}, \mathbf{F}) = (\mathbf{r}, g(\mathbf{r})\mathbf{F}), \tag{68}$$

with the following formal structure:

$$\forall\, \mathbf{r} \in U_i \cap U_j,\ \mathbf{F} \in X_i,\ \exists g \in \mathrm{L}(X_i, X_j), \tag{69}$$

where the abstract vector linear space

$$\mathrm{L}(X_i, X_j) \tag{70}$$

is defined as the space of all linear operators [26]

$$g : X_i \to X_j \tag{71}$$

on Banach vector spaces. In particular, $g(\mathbf{r})$ is a C^p-Banach space isomorphism.

Remark 15. In the mathematical literature, the smooth maps $\tau_j \circ \tau_i^{-1}$ are called the *vector bundle transition maps*. They are essential technical tools for computing global data by starting from local data then gluing them together. For example, they will be used in Sections 5.3.2 and 5.3.3 as part of the toolbox needed in the process of generalizing local information into global domains.

We have now succeeded in directly constructing a specialized smooth Banach vector bundle $(\mathcal{M}, \mathcal{D}, \tau, p)$ consisting of the nonlocal material continuum's total fiber bundle space $\mathcal{M}$, the material domain's base three-manifold $\mathcal{D}$, a set of smooth trivialization charts $\tau_i, i \in I$, and a projection map p. The base manifolds $\mathcal{D}$ itself is described by a differential atlas (U_i, ϕ_i), also associated with the partition of unity $(U_i, \psi_i), i \in I$ as per our discussion in Section 5.2 above. This incredible increase in the complexity of the mathematical space of nonlocal continuum field theory; that is, the transition from spacetime (or space–frequency) as the configuration space to a a larger superspace, here the fiber bundle space (which might be time- or frequency-dependent), is a direct expression of the very significant complexity and richness of the *physics* of nonlocal field theory in general.

As will be seen in the next Section 5.3.2, it is possible to demonstrate yet another remarkable departure from conventional theory where the concept of *linear operator*, as such, a fundamental structural object in the mathematical and computational physics of local continuum field theories [65], is found to be generalizable to the concept of *homomorphism*, which is essentially topological in nature.

5.3.2. The Nonlocal Material Continuum Fiber Bundle Homomorphism

At this point, we need to describe how the *evaluation* process of the response field (33) may be formulated within the new enlarged framework of the fibered superspace $\mathcal{M}$. The most obvious method is to introduce a *new* vector bundle with the base space being the same base space $\mathcal{D}$, but with the fibers now taken as the complex Hilbert space $\mathbb{C}^3$. This is a well-known vector bundle, which we denote by $\mathcal{R}$, and dub the *range vector bundle*. Formally, the structure of this vector bundle is expressed by the ordered quadruple $(\mathcal{R}, \mathcal{D}, \tau', p')$, where τ' and p' are the range bundle $\mathcal{R}$'s smooth trivialization and projection maps, respectively. On the other hand, the *source vector bundle* is taken as $\mathcal{M}$.

As a preparation for introducing the concept of the nonlocal continuum homomorphism, let us recap and comment on the overall physical process of exciting a material nonlocal continuous domain $\mathcal{D}$ as follows:

1. The continuum itself is mathematically modeled as a Banach bundle superspace $\mathcal{M}$ instead of its conventional differential manifold representation $\mathcal{D}$. The response of the medium is to be sought at some point $\mathbf{r} \in \mathcal{D}$.
2. The bundle superspace $\mathcal{M}$ encompasses an *additional* structure compared to $\mathcal{D}$, namely a distinct copy of a linear function space attached at each point $\mathbf{r} \in \mathcal{D}$. This is nothing but the fiber $p^{-1}(\mathbf{r})$, which is a Banach space of functions defined on the region U_i. This function space can be intuitively understood as a rigorous and exact model of the excitation field $\mathbf{F}(\mathbf{r})$ when the latter is restricted to (topologically localized at) the physics-based nonlocal domain U_i.
3. It should be noted that in local continuum field theory, e.g., conventional electromagnetism in normal temporally dispersive media, each one of the subdomains $U_i, i \in I$, is essentially *one* point $\mathbf{r} \in \mathcal{D}$. Therefore, in the case of local continua, the excitation field $\mathbf{F}$ is there found to be preferably defined as acting on the conventional space $\mathcal{D}$ instead of being a section of a Banach bundle superspace $\mathcal{M}$.
4. A vector bundle homomorphism (to be formalized in Definition 3) will map one element of this fiber function space, namely, the particular excitation field $\mathbf{F}(\mathbf{r}), \mathbf{r} \in U_i$, to its value in the range vector bundle $\mathcal{R}$. For the case of electromagnetic field theory, the latter may be taken as a vector space fiber isomorphic to $\mathbb{C}^3$ with a copy of this fiber attached to each $\mathbf{r} \in \mathcal{D}$.

We turn now to a precise definition for convenient maps between bundle superspaces. Formally, we may directly use the standard concept of homomorphism in fiber bundle theory, adapted to our purposes in the following manner [57,68]:

Definition 3 (Bundle homomorphism). A smooth bundle homomorphism over a common base space $\mathcal{D}$ shared between the two vector bundles $\mathcal{M}$ and $\mathcal{R}$ is defined by the (smooth) map:

$$\mathcal{L} : \mathcal{M} \to \mathcal{R} \tag{72}$$

satisfying $p' \circ \mathcal{L} = p$. Moreover, the restriction of $\mathcal{L}$ to each fiber $p^{-1}(\mathbf{r})$ induces a linear operator on the corresponding vector space of that fiber. In effect, the following diagram

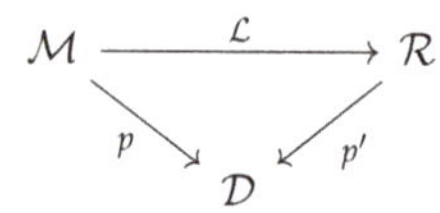

is commutative.

Remark 16. Because the nonlocal material continuum's superspace $\mathcal{M}$ and its range fiber bundle $\mathcal{R}$ both share an identical base manifold $\mathcal{D}$, the action of the homomorphism $\mathcal{L}$ as a bundle map is effectively reduced to how it interacts with each fiber $p^{-1}(\mathbf{r})$ by acting on the latter as a standard vector space linear operator. Therefore, a large portion of the conventional linear algebra and computational methods extensively deployed in the mathematical and numerical apparatus of local continuum field theory, such as nonlinear functional analysis [65], Hilbert space methods [64], and the Finite Element Method [75], may be reused as "sub-algorithms" within the larger, more general formalism of nonlocal continuum field theory proposed in this paper.

Now, since the Banach space X_i is isomorphic to $p^{-1}(\mathbf{r})$, we may assemble the homomorphism $\mathcal{L}$ by specifying its *local* expression in each topological subdomain $U_i \subset \mathcal{D}$ of the open cover $\{U_i, i \in I\}$. In particular, we define the local action using the source and range bundles' trivialization maps τ_i and τ_i' by the intuitively obvious formula:

$$\tau_i' \circ \mathcal{L}_\omega \circ \tau_i^{-1} : U_i \times X_i \to U_i \times \mathbb{C}^3, \tag{73}$$

with

$$\tau_i' \circ \mathcal{L}_\omega \circ \tau_i^{-1} := (\mathbf{r}, \mathcal{L}_{i,\omega}\mathbf{F}), \quad \mathbf{F} \in X_i, \tag{74}$$

where

$$\mathcal{L}_{i,\omega} : X_i \to \mathbb{C}^3 \tag{75}$$

is the linear operator defined by

$$\mathcal{L}_{i,\omega}(*) = \int_{U_i} \mathrm{d}^3r' \, \overline{\mathbf{K}}(\mathbf{r}, \mathbf{r}'; \omega) \cdot (*), \tag{76}$$

in which '*' stands for an element of the smooth Banach function space X_i.

Therefore, within the frequency domain formulation of this paper, the operator $\mathcal{L}$ will leave every point in the base space $\mathcal{D}$ unchanged while mapping each smooth function on U_i (component of the total electromagnetic excitation field, see below) into its complex vector value in $\mathbb{C}^3$ at $\mathbf{r} \in U_i$. Physically, $\mathcal{L}_i$ models a (topologically) localized "piece" of the global electromagnetic material operator mapping excitation fields $\mathbf{F}(\mathbf{r})$ to response fields $\mathbf{R}(\mathbf{r})$, where the entire physics here is restricted to the physics-based nonlocal subdomain U_i. The global operator itself is assembled by gluing together these small pieces using the partition of unity technique, as we endeavor to show next.

5.3.3. Computing Global Data Starting from Local Data

The final step is tying up together the fundamental source Banach bundle superspace $\mathcal{M}$, range bundle $\mathcal{R}$ and the nonlocal microdomain physics space (31). The essential ingredients of the *physics* of nonlocal field–matter interaction are encoded in the geometrical construction of the collection of microdomains $\mathcal{V}(\mathcal{D}) = \{V_{\mathbf{r}}, \mathbf{r} \in D\}$, and the excitation fields $\mathbf{F}(\mathbf{r})$ defined on them, i.e., the sets $\mathcal{V}(\mathcal{D})$ *and* the (excitation) function spaces $\mathcal{G}(\mathcal{D})$ combined together in one space, the superspace $\mathcal{M}$.

So far, the vector bundle homomorphism $\mathcal{L}$ introduced above (Definition 3) can handle excitation fields supported on the open sets $U_i, i \in I$. However, the latter sets are *mathematical* fundamental building blocks, or "set-theoretic atoms", deployed in order to formally construct the source vector bundle superspace $\mathcal{M}$. The question that will be addressed presently is the following one:

> *How can we extend the description of the nonlocal continuum's response operators starting from excitation fields defined locally to excitation fields applied on the entire physical cluster of nonlocal microdomains* $\{V_{\mathbf{r}}, \mathbf{r} \in \mathcal{D})$*?*

As mentioned before, it is the partition of unity $(U_i, \psi_i), i \in I$, what will make this expansion of the topological formulation technically feasible.

To see this, let us consider an electromagnetic field $\mathbf{F}(\mathbf{r})$ interacting with a nonlocal medium extended over the manifold $\mathcal{D}$. Our goal is to compute the response field $\mathbf{R}(\mathbf{r})$; that is, at point $\mathbf{r}$. Let us recall what the fundamental idea of EM nonlocality is: in order to compute the nonlocal material continuum's response at one point $\mathbf{r}$, one must know the excitation field in the *entirety* of an open set $V_{\mathbf{r}}$. This set $V_{\mathbf{r}}$ is one of these nonlocal microdomains composing $\mathcal{D}$ as per (29). Moreover, such $V_{\mathbf{r}}$ is also a topological neighborhood of its continuous index point $\mathbf{r} \in \mathcal{D}$ (cf. Section 4.2). However, in general this microdomain will change depending on the position $\mathbf{r}$. The goal now is to find $\mathbf{R}(\mathbf{r})$ using the vector bundle map $\mathcal{L}$ defined by (72) starting from the data:

1. Region $V_{\mathbf{r}}$;
2. Vector field $\mathbf{F}(\mathbf{r})$ acting on $V_{\mathbf{r}}$.

To accomplish this, we exploit the properties of the partition of unity functions ψ_i (Lemma 1) for expanding the excitation field $\mathbf{F}(\mathbf{r})$ over *all* patches U_i covering $V_{\mathbf{r}}$, resulting in

$$\mathbf{F}(\mathbf{r}';\omega) = \sum_{i \in I_{\mathbf{r}}} \psi_i(\mathbf{r}')\mathbf{F}_i(\mathbf{r}';\omega), \tag{77}$$

where (43) was used. The truncated function $\mathbf{F}_i$ is equal to $\mathbf{F}(\mathbf{r})$ only if $\mathbf{r} \in U_i$ and zero elsewhere, i.e., we have

$$\mathbf{F}_i(\mathbf{r}';\omega) := \begin{cases} \mathbf{F}(\mathbf{r}';\omega), & \mathbf{r} \in U_i, \\ 0, & \mathbf{r} \notin U_i. \end{cases} \tag{78}$$

Recall that according to Lemma 1, the set $I_{\mathbf{r}}$ is defined as the collection of indices $i \in I$ of all U_i having the point $\mathbf{r}$ in their common set intersection; by construction, this index set $I_{\mathbf{r}}$ is always finite.

The main idea behind our construction should now become clear: while each truncated sub-field $\mathbf{F}_i$ fails to be differentiable (it is not even continuous), the multiplication by $\psi_i(\mathbf{r})$ fixes this problem. In fact, each function

$$\psi_i(\mathbf{r}')\mathbf{F}_i(\mathbf{r}';\omega) \tag{79}$$

is a *smooth* component of the total excitation field $\mathbf{F}$ with support *fully* contained *inside* the coordinate patch U_i; that is, we have

$$\text{supp}\{\mathbf{F}_i\} \subset U_i. \tag{80}$$

Consequently, the vector bundle map constructed in (72) can be applied to each such component field. From (73)–(75) and (77), the following can be deduced:

$$\mathbf{R}(\mathbf{r};\omega) = \sum_{i\in I_{\mathbf{r}}} \mathcal{L}_{i,\omega}[\psi_i(\mathbf{r}')\mathbf{F}_i(\mathbf{r}';\omega)]. \tag{81}$$

Finally, using (76), we arrive at our main superspace map theorem:

Theorem 1 (global superspace bundle map). For the fiber bundle superspace $\mathcal{M}$ of the nonlocal continuum whose differential manifold representation is $\mathcal{D}$ and the nonlocal continuum response function (tensor) $\overline{\mathbf{K}}$, the response and excitation fields $\mathbf{R}$ and $\mathbf{F}$ can be related to other other via the global bundle (superspace) map:

$$\boxed{\mathbf{R}(\mathbf{r};\omega) = \sum_{i\in I_{\mathbf{r}}} \int_{U_i} \mathrm{d}^3r'\, \overline{\mathbf{K}}(\mathbf{r},\mathbf{r}';\omega)\cdot\psi_i(\mathbf{r}')\mathbf{F}_i(\mathbf{r}';\omega),} \tag{82}$$

where $\psi_i, i \in I$, are the partition of unity basis functions subordinated to the $\mathcal{D}$-atlas $(U_i, \phi_i), i \in I$.

Physically, Theorem 1 states that the nonlocal continuum's source bundle (superspace) $\mathcal{M}$, the range bundle $\mathcal{R}$, and the nonlocal response superspace map $\mathcal{L}$, together, supply the fundamental formal scaffold upon which the material domain's response to generic excitation field, when the latter field operates on arbitrary configurations of nonlocality microdomain, can be constructed. By aggregating all of those physics-based microdomains constituting the topological microstructure of nonlocal processes in material continua, the main field-theoretic structures of the medium may be couched, computed and reformulated in the richer language of this more general superspace framework belonging to the Banach fiber bundle $\mathcal{M}$ instead of the position space $\mathcal{D}$ of conventional spacetime extensively used in local field theories. At this stage of our formulation, the vector bundle formalism of nonlocality becomes essentially complete, where the connection between the purely mathematical fiber superspace and the physical microdomain structures is secured by Theorem 1, especially the Formula (82).

6. Interlude: The Nonlocal Continuum Fiber Bundle Superspace Algorithm—Summary and Transition to Applications

We review and summarize here the salient features of the fiber bundle superspace construction, carefully developed above, by explicitly outlining the algorithm implicit in the various detailed derivations of the previous sections. Our main objective in this short transitional section is to highlight again the fact, already discussed above, which is that our superspace formalism is based on estimating the *physics*-based nonlocality microdomain set $\mathcal{V}(\mathcal{D}) = \{V_{\mathbf{r}}, \mathbf{r} \in \mathcal{D}\}$ associated with the nonlocal continuum $\mathcal{D}$. These data can be obtained only through physical theory and/or measurement. However, once available, the construction of the fibered space proceeds in a computationally well-determined manner. We first summarize the algorithm then provide few additional preparatory remarks before moving to the more detailed and concrete computational examples of Section 7.

In Figure 4, we show two distinct points $\mathbf{r}_1, \mathbf{r}_2 \in \mathcal{D}$ and their associated microdomains $V_{\mathbf{r}_1}$ and $V_{\mathbf{r}_2}$, respectively. From the locally finite subcover $\{U_i\}_{i\in I}$ subordinated to $\mathcal{V}(\mathcal{D}) = \{V_{\mathbf{r}}, \mathbf{r} \in \mathcal{D}\}$ we highlight two sets

$$U_i \subseteq V_{\mathbf{r}_1}, \quad U_j \subseteq V_{\mathbf{r}_2}, \tag{83}$$

where in general it is allowed that

$$V_{\mathbf{r}_1} \cap V_{\mathbf{r}_2} \neq \emptyset, \quad U_i \cap U_j \neq \emptyset, \tag{84}$$

as could be inferred from a glance at the Figure itself. For the partition of unity $(U_i, \psi_i)_{i\in I}$, which is subordinated to the open cover $\{U_i\}_{i\in I}$, we also highlight the two compact sets

$$S_i := \text{supp}\{\psi_i(\mathbf{r})\}, \quad S_j := \text{supp}\{\psi_j(\mathbf{r})\}, \tag{85}$$

forming the support of the corresponding partition of unity functions.

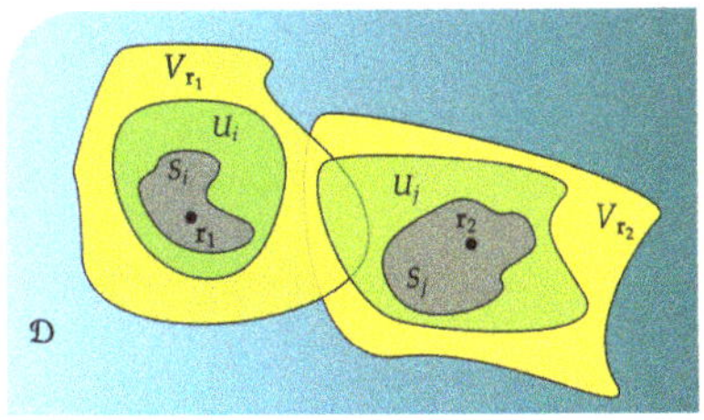

Figure 4. An example illustrating the various topological microstructures involved in modeling a generic nonlocal material. The microdomains $V_{\mathbf{r}_1}, V_{\mathbf{r}_2} \in \mathcal{V}(\mathcal{D})$ are open sets and belong to the nonlocal microstructure of the MTM $\mathcal{D}$. The open sets U_i and U_j are the corresponding coordinate sets and partition of unity functions $\{\psi_i\}_{i\in I}$'s domains subordinated to $V_{\mathbf{r}_1}$ and $V_{\mathbf{r}_2}$, respectively. The compact sets S_i and S_j are defined by $S_i := \text{supp}\{\psi_i(\mathbf{r})\}$ and $S_j := \text{supp}\{\psi_j(\mathbf{r})\}$.

The nonlocal material continuum's superspace algorithm itself is summarized in Algorithm 1. Once the microdomain dataset $\mathcal{V}(\mathcal{D})$ is given, the construction proceeds automatically using the partition of unity basis functions $(U_i, \psi_i)_{i\in I}$. The latter may be computed directly in terms of the standard bump functions, see [57,68,71], and also Remark 9.

Because of the fundamental importance of the physics-based nonlocality microdomain structure $\mathcal{V}(\mathcal{D})$, Section 7 will be entirely devoted to the explication of a quantitative practical example illustrating the origin of these microdomains in the concrete setting of a real-life advanced material system, including how the microdomain topology itself may be estimated in practice. In the subsequent sections Section 8 and Appendixes A.3 and A.11, we also explore the usefulness of the superspace homomorphism construction developed in Section 5 for reformulating boundary-value problems in the nonlocal continuum field theories of mathematical physics, besides also providing some hints and additional remarks on other current and future applications.

Algorithm 1 The nonlocal continuum fiber bundle algorithm.

1. Start with a physics-based microdomain structure $\mathcal{V}(\mathcal{D}) = \{V_\mathbf{r}, \mathbf{r} \in \mathcal{D}\}$.
2. The open cover $\mathcal{V}(\mathcal{D})$ of $\mathcal{D}$ induces a locally finite subcover $\{U_i\}_{i\in I}$ subordinated to $\mathcal{V}(\mathcal{D})$. It is then automatically equipped with the differential structure of the manifold $\mathcal{D}$, generating the differential atlas $(U_i, \phi_i)_{i\in I}$.
3. The subcover $\{U_i\}_{i\in I}$ is equipped with a partition of unity function set $\{\psi_i\}_{i\in I}$, producing the partition of unity $(U_i, \psi_i)_{i\in I}$.
4. Generate an appropriate Banach/Sobolev/Hilbert space X_i attached to each point $\mathbf{r} \in \mathcal{D}$ using constructions, such as (57).
5. Declare $\mathcal{D}$ the base manifold of the fiber bundle. Construct the bundle space $\mathcal{M}$ using

$$\mathcal{M} := \{(\mathbf{r}, X_i) | \forall i \in I, \mathbf{r} \in U_i \subset \mathcal{D}\}. \tag{86}$$

6. Construct the projection map $p : \mathcal{M} \to \mathcal{D}$ through the operation $(\mathbf{r}, X_i) \to \mathbf{r}$.
7. Use (68) to transform vector from one fiber (function) space to another.

7. Applications to Advanced Materials: Nonlocal Inhomogeneous Semiconductors

7.1. Introduction

A concrete example involving spatially-dispersive isotropic media is considered in this Section, where the intention is to provide an outline of how the intricate fiber bundle type topological fine structure (the topology of microdomains attached to each point explored above as developed in detail in Section 5 and summarized in Section 6) may be estimated in actual practice. The contents of the example given below are rather detailed, and that is for two main reasons. First, in spite of the fact that nonlocal metamaterials are not proposed here for the first time, the author's experience indicates that there is still a general lack of appreciation of the subject in the large community, where most research on "metamaterials" concentrate on *temporally*-dispersive media. Because of that, we provide a very detailed example, including reintroducing some of the well-known physics of semiconductors (in some of the appendices) in order to make the presentation complete and self sufficient. Second, the detailed example to be found below is itself novel. The estimation of the microdomain nonlocal structure in inhomogeneous semiconductors seems to be achieved here for the first time. Therefore, it is a topic that could be treated not merely as an example illustrating the more general and abstract superspace theory developed in the earlier sections, but possibly as a stand-alone contribution to semiconductor materials and their physics. However, the main intention behind the inclusion of this highly-technical physical example continues to be the illustration of the fundamental superspace formalism. More detailed examinations of nonlocality in semiconductor metamaterials belong to a more specialized literature than the current article, whose main topic is the mathematical physics of nonlocal continuum field theories.

7.2. A Topological Coarse-Grained Model for Inhomogeneous Nonlocal Material Domains

A review of the homogeneous medium model of spatial dispersion is provided in Appendix A.6. Below, we describe a method that can help transitioning from the generic form (A3), valid for homogeneous nonlocal domains, to the *inhomogeneous* medium situation developed throughout this paper where nonlocality cannot be captured by a simple global dependence of the dielectric function on $\mathbf{k}$. However, instead of working with the full nonlocal function $\overline{\mathbf{K}}(\mathbf{r},\mathbf{r}')$, an alternative simplified model is proposed which we entitle *the topological coarse-grained model.* The idea is as follows. Consider a global material domain $\mathcal{D}$, which is an open three-manifold, say an open subset of $\mathbb{R}^3$ that may be either simply connected or disconnected.[21] The material is nonlocal and inhomogeneous. At each point $\mathbf{r} \in \mathcal{D}$, a microdomain, i.e., and open set $V_{\mathbf{r}} \subset \mathcal{D}$, is assigned. The medium is *locally isotropic and homogeneous* in the sense that within each microdomain we can describe the response to an external field excitation $\mathbf{E}$ by means of a relation similar to (34), namely:

$$\mathbf{D}(\mathbf{r};\omega) = \varepsilon_0 \int_{V_{\mathbf{r}}} d^3r'\, \overline{\mathbf{K}}(\mathbf{r}-\mathbf{r}';\omega)\cdot \mathbf{E}(\mathbf{r}';\omega). \tag{87}$$

That is, the only difference between (87) and (34) is that, in the former, we use the correct form of *homogeneous* nonlocality $\overline{\mathbf{K}}(\mathbf{r}-\mathbf{r}';\omega)$ instead of $\overline{\mathbf{K}}(\mathbf{r},\mathbf{r}';\omega)$. Moreover, we have put the proper response and excitation fields $\mathbf{D}(\mathbf{r})$ and $\mathbf{E}(\mathbf{r})$ and inserted the free space permittivity ε_0.

Fundamentally speaking, each material microdomain is now described by a *spatially-dispersive* model of the form (87). The "topological atoms" of nonlocality, namely the sets $V_{\mathbf{r}}$, spanned by the continuous index $\mathbf{r} \in \mathcal{D}$, are each a spatially-dispersive "medium" on its own. As will be seen later in this section, the idea of the locally–spatially-dispersive nonlocal semiconductor system is to build an *inhomogeneous* metamaterial that goes *beyond* spatial dispersion by assembling a more general form of nonlocality using the spatially-dispersive material "atoms" $V_{\mathbf{r}}$. In such systems, the engineered metamaterial is only *locally* spatial dispersive. On the other hand, at a larger spatial scale it does not follow the standard spatial dispersion law, but rather appears to belong to a more complicated class of nonlocal continua which, we believe, are best mathematically described using the fiber bundle superspace formalism of Section 5.

It may be seen then that as a topological coarse-grained process, the original inhomogeneous nonlocal medium, ultimately described by the material tensor $\overline{\mathbf{K}}(\mathbf{r},\mathbf{r}';\omega)$, is sub-decomposable into "small topological cells", the microdomains $V_{\mathbf{r}}$, $\mathbf{r} \in \mathcal{D}$, such that each "topological cell" or "atom" would in itself behave like a homogeneous nonlocal isotropic subdomain, hence may be described by (87), where the material tensor in that case takes the (topologically) locally correct form (A3). This can be considered a *quasi-local* model (also sometimes called *locally spatially dispersive*), where the global domain, electromagnetically speaking, is nonlocal, while, on the other hand, seen at the scale of a small region (cells) it would more or less behave like a typical electromagnetic local medium, see, for example, the discussion of some special cases of complex nonlocal crystals in [76].

Remark 17. We remind the reader again about the subtle difference between *mathematical* nonlocality and *physics*-based nonlocality, a distinction at the conceptual level that will become quite visible throughout this section. The term *local* is used in this paper in two senses. The first sense is the physical one in which *local* is set against *physical nonlocality*, which includes spatial dispersion (EM local/nonlocal.) On the other hand, in topology, a *local* property is that which holds in a small open neighborhood of a given point, in our case the topological microdomain $V_{\mathbf{r}}$. The distinction between the two technical senses of the same term should always be clear from the context. In the few cases when there is a risk of confusion, we say *topologically local* to emphasize the second meaning above from *EM local*. (see also Remark 3 and Section 3.3).

Our key objective now is to first develop a simple estimation of the "size" of the nonlocal microdomains $V_{\mathbf{r}}$. To do so, some metric methods must be introduced. An attractive approach would be to approximate the topology of the nonlocal metamaterial system using arrays of various *spheres*, and then use this array in order to obtain the topological content of the microdomain structure described in Section 4.

Let us illustrate the main ideas with a simple example first. Consider a point $\mathbf{r}_1$, which provides a label for one of the micro cells, we may deploy for creating a coarse-grained model for the inhomogeneous medium. To be more specific, let us construct the topological open ball defined by

$$B(\mathbf{r}, a_{\mathbf{r}}) := \left\{ \mathbf{r} \in \mathbb{R}^3 \middle| \, d(\mathbf{r}, \mathbf{r}') < a_{\mathbf{r}} \right\}, \tag{88}$$

where $a_{\mathbf{r}} \in \mathbb{R}^+$ is a number quantifying the smallness of this "nonlocality ball" centered at $\mathbf{r}'$, while d is the distance metric. The number $a_{\mathbf{r}}$ will be determined later based on the actual physics of the problem.

Next, the fine-grained topological microdomain structure can be constructed by aggregating all these balls in order to produce a coarse-grained of the overall inhomogeneous nonlocal material domain $\mathcal{D}$. The choice of the *shape* of the microdomain $V_{\mathbf{r}}$ as a sphere $B(\mathbf{r}, a_{\mathbf{r}})$ defined by (88) is justified by our earlier assumption that the material is (topologically) *locally* isotropic. However, note that *globally* electromagnetic processes need *not* behave as they do in isotropic domains.

In Figure 5, a diagrammatic depiction of the two local and global processes is provided where we illustrate:

1. The proposed topological coarse-grained model utilizing the set of balls $V_{\mathbf{r}}$, $\mathbf{r} \in \mathcal{D}$ (left).
2. The conventional paradigm where the unit cells are non-overlapping (right).

As can be seen from the diagram, in the topological approach, there exists an open set (microdomain) $V_{\mathbf{r}}$ attached to each point $\mathbf{r} \in \mathcal{D}$ such that nearby microdomains may *overlap* with each other; i.e., in such case the set

$$\cup_{\mathbf{r}_1, \mathbf{r}_2 \in \mathcal{D}} \left[V_{\mathbf{r}_1} \cap V_{\mathbf{r}_2} \right] \tag{89}$$

is not necessarily empty. On the other hand, the conventional approach to coarse-grained, depicted in Figure 5 (right), involves subdomains like $V'_{\mathbf{r}_1}$ and $V'_{\mathbf{r}_2}$ that are *non-overlapping*, leading to a grid-like structures or "tile covering up" of the material domain $\mathcal{D}$ where in general no holes are left.

In both approaches, it should be noted, we find that each type of the two subdomains, whether $V_{\mathbf{r}}$ or $V'_{\mathbf{r}}$, was already assumed to be *homogeneous*. The disadvantage of the conventional approach is that any *abrupt* change in the electromagnetic properties of the material, experienced when transitioning between two neighboring subdomains through their interface region, often requires imposing a suitable "boundary condition" at this geometric interface in order to obtain an accurate computational assessment of the physics. On the other hand, this problem does *not* exist in the topological approach, illustrated in Figure 5 (left), because the microdomains are allowed to overlap, where common regions between overlapping microdomains are treated correctly using the partition of unity basis functions as described in Section 5.3.

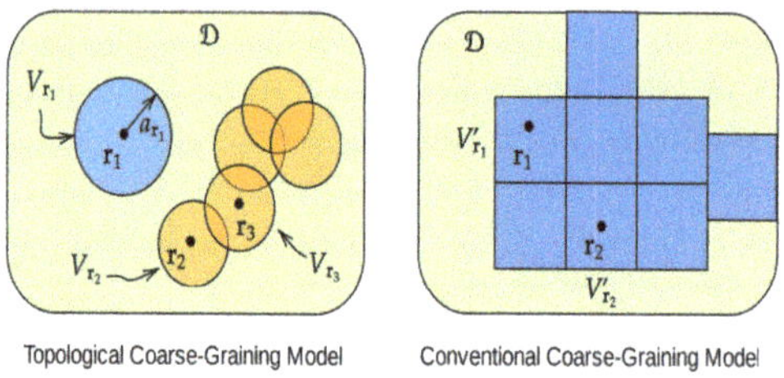

Figure 5. Topological coarse-grained model for an inhomogeneous nonlocal material domain $\mathcal{D}$ (**left**) in comparison with a conventional coarse-grained process (**right**). The topological microdomains constitute an open cover of the domain in the sense that $\mathcal{D} = \bigcup_{\mathbf{r}\in\mathcal{D}} V_{\mathbf{r}}$, which is the obvious generalization of (32). Note how the topological approach allows overlapping microdomains, e.g., between microdomains $V_{\mathbf{r}_2}$ and $V_{\mathbf{r}_3}$. The technique of the partition of unity will take care of electromagnetic data "repeated" in such regions of overlap by assigning proper weights that always sum to unity at each point in $\mathbf{r} \in \mathcal{D}$.

7.3. Resonant Nonlocal Semiconductor Domains and the Nonlocal Exciton-Polariton Model

A concrete application of the topological coarse-grained algorithm proposed in Section 7.2 is now in order. The specific nonlocal metamaterial is a semiconductor with dielectric function exhibiting a single strong resonant exciton transition at the frequency $\omega = \omega_{\mathrm{e}}$. We first examine in detail the nonlocal exciton–polariton model to be used below. For a review on the physics of exciton–polariton interactions in solids, see Appendix A.7.

A *polariton* is simply a "photon living inside a dielectric medium". The quantum of an electromagnetic wave inside a dielectric domain is often called polariton instead of photons (sometimes polaritons are called "dressed photons"). An *exciton–polariton* is a polariton coupled with a mechanical exciton, e.g., an electron-hole pair. The latter should be distinguished from other types of polaritons such as *phonon-polaritons* defined as polaritons coupled with *phonons*, the quantum of lattice vibrations [37].

It is well known from quantum theory that near resonance, the dielectric function of such semiconductor materials may be approximated by the formula [16,77–80]:

$$\varepsilon(\mathbf{k}, \omega) = \varepsilon_0 + \frac{\chi}{k^2 - \gamma^2(\omega)}, \tag{90}$$

where

$$\chi = 4\pi \frac{\alpha m_{\mathrm{e}}^{\star} \omega_{\mathrm{e}}}{\hbar}, \quad \gamma^2(\omega) = \frac{m_{\mathrm{e}}^{\star}}{\hbar \omega_{\mathrm{e}}} \left(\omega^2 - \omega_{\mathrm{e}}^2 + \mathrm{i}\omega\Gamma \right). \tag{91}$$

Here, $\hbar$ is the reduced Planck constant, while α serves as the oscillator strength.[22] The effective mass of the exciton is denoted by $m_{\mathrm{e}}^{\star}$.[23]. On the other hand, the *exciton lifetime* τ_{e} is defined by

$$\tau_{\mathrm{e}} := \frac{2\pi}{\Gamma}, \tag{92}$$

hence, Γ can be thought of as the *exciton decay* or *relaxation rate*. We emphasize that any dependence of Γ and the oscillator strength α on k is ignored in the excitonic model (90).

In Appendix A.7, the physical origin of nonlocality in the semiconductor is revisited, where it is traced to the quantum mechanical energy–momentum relations of exciton–polaritons. In order to actually see significant nonlocal physics taking place in the excitonic material system described by (90), the following sufficient condition may be imposed:

$$\Gamma \ll \frac{\hbar k^2}{2m_e^*}. \tag{93}$$

It can be shown that under such Γ-bound, the kinetic energy term in (A12) can induce significant nonlocal effects in (90). One way to realize nonlocal (spatially dispersive) semiconducting metamaterials is to operate with intrinsic semiconductors satisfying (93) by keeping the temperature low and the material pure (undoped) [83].

The model described by (90) and (91) can be viewed as a natural generalization of the local Lorentz model widely utilized to model temporal dispersion in solids and plasma [77,79]. It represents the simplest nonlocal resonant model with a single strong resonance at a characteristic frequency, here $\omega = \omega_e$. All other off-resonance excitonic transitions are gathered into the background dielectric constant ε_0 for simplicity. For frequencies well below $\omega \ll \omega_e$, the exciton–polariton behaves essentially like a photon propagating in a medium with background permittivity ε_0. For $\omega \gg \omega_e$, we again recover photons but usually with a background described by ε_∞, the high-frequency limit of permittivity. In general, the difference between the static and high-frequency permittivities is quite small in the sense that

$$|\varepsilon_0 - \varepsilon_\infty| \ll \varepsilon_0. \tag{94}$$

Hence, for simplicity, in this example the two permittivities are treated as identical ($\varepsilon_0 \simeq \varepsilon_\infty$) since we are interested in the EM response around a single excitonic resonance while in fact the oscillator strength α in (90) is small. One consequence of this assumption is that the splitting between longitudinal and transverse modes can be neglected. Indeed, since the longitudinal and transverse frequencies ω_L and ω_T are related to each other via the relation [16]

$$\frac{\omega_L^2}{\omega_T^2} = \frac{\varepsilon_0}{\varepsilon_\infty}, \tag{95}$$

then the assumption (94) is equivalent to neglecting the longitudinal–transverse splitting

$$\omega_{L,T} := |\omega_L - \omega_T| \tag{96}$$

in the sense that

$$\omega_{L,T} \ll \omega_T. \tag{97}$$

A consequence of this is the near equality of the longitudinal and transverse frequencies, which allows us to considerably simplify the mathematical treatment.[24] In addition, assuming that the oscillator strength α in (90) is nearly the same for both the longitudinal and transverse part of the response function, then it follows that we need only work with a single *scalar* response function, namely the form (90) itself instead of the more general tensorial Formula (A3).[25]

Nonlocal effects associated with the model (90) emerge from the quantum mechanical nature of exciton–polariton interactions and the need to enforce conservation of energy/momentum as discussed in Appendix A.7, leading to the strong dependence on k observed in (90). There is yet another physical explanation of nonlocality. Within the regime of the large exciton mass limit

$$m_e^* \to \infty, \tag{98}$$

the kinetic energy term in (A12) drops out and the excitonic dielectric function (90) becomes local. This is why spatial dispersion is sometimes referred to as the "finite-mass model", with some suggestions that the origin of nonlocality in this case is the inertial effects of the exciton [79].[26] In what follows, we assume that the effective mass of the exciton is always finite and positive:

$$0 < m_e^\star < \infty. \tag{99}$$

However, it should be noted that since excitons are *collective excitations* of solids [84,85], they may have negative mass [86]. While this will not be pursued here, the negativity of the excitonic mass may be exploited in order to further design and control the EM behavior of nonlocal MTMs constructed using excitonic semiconductors.

In order to gain a deeper insight into the various resonance structures of the exciton–polariton response function (90), we rewrite it in the equivalent form

$$\varepsilon(\mathbf{k},\omega) = \varepsilon_0 + \frac{\chi/k_e^2}{k^2/k_e^2 + 1 - \omega^2/\omega_e^2 - i\omega\Gamma/\omega_e^2}, \tag{100}$$

where

$$k_e := \frac{2\pi}{\lambda_e} = \sqrt{\frac{m_e^\star \omega_e}{\hbar}} \tag{101}$$

is called the *exciton wave number*. The wavelength λ_e is a fundamental resonance spatial scale, which we will refer to as the *exciton wavelength* and is given by

$$\lambda_e = \frac{1}{2\pi}\sqrt{\frac{\hbar}{m_e^\star \omega_e}}. \tag{102}$$

For example, with $\hbar\omega_e = 2.5$ eV and $m_e^\star = 0.9 m_{el}$, where m_{el} is the electron mass, the exciton wavelength λ_e is around 0.0293 nm, which is the same order of magnitude of interatomic spacing. The excitation field wavelength λ is at least one order of magnitude larger. Later we will show typical values for the topological microdomain radius a_r.

There are several fundamental spatial and temporal scales involved in the process of describing the generic nonlocal metamaterial domain $\mathcal{D}$. The excitation field $\mathbf{E}(\mathbf{r})$ itself introduces its own temporal excitation period

$$T := \frac{2\pi}{\omega}, \tag{103}$$

in addition to a purely spatial scale (wavelength) measured by the formula

$$\lambda := \frac{2\pi}{k}. \tag{104}$$

On the other hand, the excitonic transition as such is associated with the fundamental (temporal) transition period

$$T_e := \frac{2\pi}{\omega_e}, \tag{105}$$

while a fundamental spatial scale

$$\lambda_e := \frac{2\pi}{k_e} \tag{106}$$

can be unambiguously linked to the exciton at the same time. Table 1 gives a summary of all these parameters with their meaning explicitly stated. Moreover, it will be demonstrated later that the radius of a topological microdomain V_r, which is based at a generic position $\mathbf{r} \in \mathcal{D}$, can be given by a special Formula (120). Nonlocality arises from the delicate interplay between all these different spatial and temporal scales. In what follows, we will emphasize their relative roles in determining the rich nonlocal microstructure of the material domain, while introducing quantitative calculations.

Table 1. A summary of the various spatial and temporal scales involved in understanding and designing generic nonlocal metamaterials with exciton–polariton resonance-type of nonlocality.

Scale	Type	Meaning	Formula
λ	spatial	excitation field wavelength	$\lambda = 2\pi/k$
λ_e	spatial	exciton wavelength	$2\pi/k_e$
a_r	spatial	microdomain radius	$1/\lvert\gamma''\rvert$
T	temporal	excitation field period	$2\pi/\omega$
τ_e	temporal	exciton lifetime	$2\pi/\Gamma$
T_e	temporal	exciton period	$2\pi/\omega_e$

Armed with this typology of spatial and temporal scales, we are now better positioned to understand the resonance structure associated with the exciton–polariton nonlocal dielectric function (100). Figure 6 illustrates two cases of resonance where the value of the dielectric function is examined with respect to variations in the excitation field wave number k (or equivalently the wavelength λ). In order to focus on nonlocality, we only plot the nonlocal part of the total response, which is found here to be proportional to the dielectric residue

$$\varepsilon(\mathbf{k},\omega) - \varepsilon_0. \tag{107}$$

As we may infer from Figure 6, a strong resonance takes place when the ratio

$$\frac{k}{k_e} = \frac{\lambda_e}{\lambda} \tag{108}$$

becomes comparable in magnitude to the quantities remaining in the denominator of (100). That is, the spatial resonance condition is

$$\frac{k^2}{k_e^2} + 1 - \frac{\omega^2}{\omega_e^2} \sim \frac{\omega\Gamma}{\omega_e^2}. \tag{109}$$

However, the condition (109) holds only if the imaginary part of the denominator of (100), i.e., the quantity $\omega\Gamma/\omega_e^2$, is relatively small. Otherwise, since k and k_e are real, the ratio k/k_e can never lead to strong resonance when the relaxation rate Γ is sufficiently large. Another way to say the same thing is the following: strong *spatial* resonances, whose main origin is nonlocality, can take place either when dissipation is small, or when the exciton lifetime is long enough. The latter scenario of long exciton lifetime is characterized by the condition

$$\frac{\omega\Gamma}{\omega_e^2} \ll \left|1 - \frac{\omega^2}{\omega_e^2}\right|. \tag{110}$$

In such case, it is evident that the appropriate spatial and temporal sufficient conditions needed to secure nonlocal resonance are mutually related by the simple relation

$$\frac{k^2}{k_e^2} \approx \frac{\omega^2}{\omega_e^2} - 1. \tag{111}$$

From this, it can be inferred that nonlocal resonances generally occur only for $\omega/\omega_e > 1$. In Figure 6 (left), we can see that for the above-resonance condition of $\omega/\omega_e = 1.5$, the nonlocal domain possesses a spatial resonance at roughly $\lambda \approx \lambda_e$. On the other hand, if we operate the material at larger frequency $\omega/\omega_e = 2.5$, i.e., well above the exciton transition frequency, then spatial resonances may occur only at values of the excitation field wavelength λ that are considerably smaller than the exciton wavelength λ_e.

Finally, we add that when the nonlocal response is plotted as function of ω instead of k, resonance structures similar to Figure 6 are obtained under the condition (110) since in that case (111) approximately holds. In general, we would expect that for the best operation of the designed nonlocal MTM (maximal nonlocal response), the operating frequency should be selected to be as close as possible to the exciton transition frequency, i.e., we would like to maintain the material design condition

$$\frac{\omega}{\omega_e} \approx 1, \tag{112}$$

which is needed since, in general, the excitonic relaxation rate Γ is never exactly zero and, hence, the condition (110) seldom holds otherwise for all frequencies.

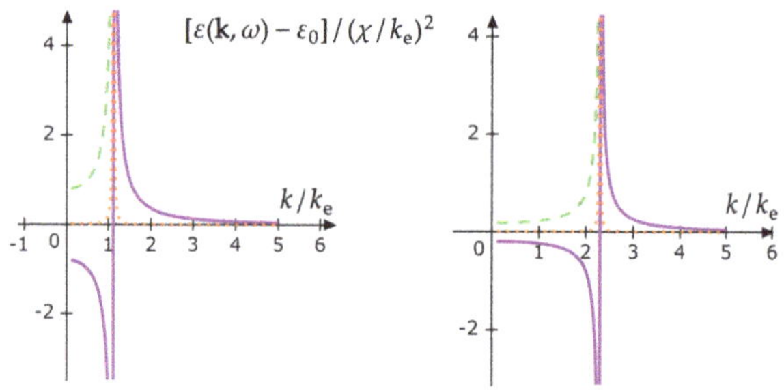

Figure 6. The nonlocal spatial resonance structure of the exciton–polariton dielectric response as a function of the excitation field wave number k. The normalized response function $(\varepsilon(\mathbf{k},\omega)-\varepsilon_0)/(\chi/k_e)^2$ is plotted, where the dashed line is the absolute value, the solid line represents the real part, while the dotted line is the imaginary part. For both figures, $\Gamma/\omega_e = 0.01$. **(Left)** $\omega/\omega_e = 1.5$. **(Right)** $\omega/\omega_e = 2.5$.

7.4. Quantitative Estimation of the Electromagnetic Nonlocality Microdomain Structure in the Exciton-Polariton Dielectric Model

In the spatial domain, the dielectric function can be obtained by computing the inverse Fourier transform

$$\varepsilon(\mathbf{r}-\mathbf{r}';\omega) = \mathcal{F}_{\mathbf{k}}^{-1}\{\varepsilon(\mathbf{k},\omega), \tag{113}$$

where $\mathcal{F}_{\mathbf{k}}^{-1}$ is the converse of the forward Fourier transformation defined by (17). We will need the following inverse Fourier transform relation (proved in Appendix A.9):

$$\mathcal{F}_{\mathbf{k}}^{-1}\left\{\frac{\chi}{k^2-\gamma^2(\omega)}\right\} = \frac{\chi}{4\pi}\frac{e^{\gamma''(\omega)|\mathbf{r}-\mathbf{r}'|}e^{-i\gamma'(\omega)|\mathbf{r}-\mathbf{r}'|}}{|\mathbf{r}-\mathbf{r}'|}, \tag{114}$$

where

$$\gamma' = -\sqrt{\frac{m_e^\star}{2\hbar\omega_e}}\sqrt{(\omega^2-\omega_e^2)+\sqrt{(\omega^2-\omega_e^2)^2+(\omega\Gamma)^2}}, \tag{115}$$

$$\gamma'' = -\sqrt{\frac{m_e^\star}{2\hbar\omega_e}}\sqrt{-(\omega^2-\omega_e^2)+\sqrt{(\omega^2-\omega_e^2)^2+(\omega\Gamma)^2}}. \tag{116}$$

Hence, by substituting (100) into (113) and using (114), we arrive at

$$\varepsilon(\mathbf{r}-\mathbf{r}';\omega) = \underbrace{\varepsilon_0\delta(\mathbf{r}-\mathbf{r}')}_{\text{local response}} + \underbrace{\varepsilon_{\rm NL}(\mathbf{r}-\mathbf{r}';\omega)}_{\text{nonlocal response}}. \tag{117}$$

The first terms in the RHS of (117) provides the background local response of the medium. On the other hand, all nonlocal effects are relegated to the second term in the RHS of (117):

$$\varepsilon_{\rm NL}(\mathbf{r}-\mathbf{r}';\omega) := \frac{\alpha m_e^\star\omega_e}{\hbar}\frac{e^{-i\gamma'(\omega)|\mathbf{r}-\mathbf{r}'|}}{|\mathbf{r}-\mathbf{r}'|}e^{\gamma''(\omega)|\mathbf{r}-\mathbf{r}'|}, \tag{118}$$

which is nothing but the Green's function of the electromagnetic semiconductor material system under investigation.

The Green's function (118) is the most fundamental physical quantity needed for the construction of the microdomain structure $\mathcal{D}$ of the nonlocal medium. It has some similarity with the scalar free-space Green function for radiation fields, i.e., spherical waves of the form:

$$\frac{\exp(ik|\mathbf{r}-\mathbf{r}'|)}{|\mathbf{r}-\mathbf{r}'|}. \tag{119}$$

However, there are notable differences:

1. First, we note that (118) exhibits strong dispersive behavior due to the dependence of γ' and γ'' on frequency per their Formulas (115) and (116).
2. Second, the presence of a spatially-decaying exponential factor of the form $\exp(\gamma'|\mathbf{r}-\mathbf{r}'|)$ makes the Green function $\varepsilon_{\mathrm{NL}}(\mathbf{r}-\mathbf{r}';\omega)$ highly attenuating in spite of the fact that this attenuation is *not* mainly due to thermodynamic losses.

Indeed, as can be seen from (90), dissipation is controlled by the exciton lifetime τ_e, or, equivalently, the decay rate Γ. Dissipation decreases as the lifetime increases, i.e., when Γ is small. Figure 7 illustrates some examples where we plot both γ' and γ'' as functions of frequency. The frequency-dependent behavior observable there strongly depends on Γ/ω_e, i.e., the ratio between the relaxation frequency and the excitonic transition frequency. For ratios as small as $\Gamma/\omega_e = 0.1$, the intensity of attenuation per unit length γ'' is nearly constant for $\omega > \omega_e$, while it assumes higher values for frequencies below the ω_e as can be seen from Figure 7a. This is consistent with a "high-pass filtering behavior" typical for this type of resonance phenomena, where waves are often excited with frequencies slightly larger than the cutoff threshold at ω_e. For the propagation constant γ' at the same relaxation-to-exciton transition ratio Γ/ω_e, Figure 7b shows that it becomes nearly straight line. Such behavior, when combined with nearly constant per-unit-length attenuation, represents negligible dispersion effects. On the other hand, when Γ/ω_e increases, we begin to see strong dispersion effects, manifested by non-constant per-unit-length attenuation and nonlinear phase-delay relations.

In fact, the attenuation process described by the per-unit-length rate γ'' is not merely an expression of dissipation, but is also *the* signature of nonlocality in exciton–polariton semiconductor materials. The medium response weakens as the distance from the source increases, while the characteristic length scale of this nonlocality radius is found to be solely controlled by γ''. Figure 8 illustrates the real part of the dielectric function Green's function (118). The ability of the excitonic semiconducting medium to respond to spatially distant sources is graphically illustrated by its dielectric profile's functional spread around the origin $|\mathbf{r}-\mathbf{r}'| = 0$. The size of the nonlocal domain is then directly reflected by the rapidity of the decay of the Green's function (118) as one moves away from $\mathbf{r}'$, which is the origin here.

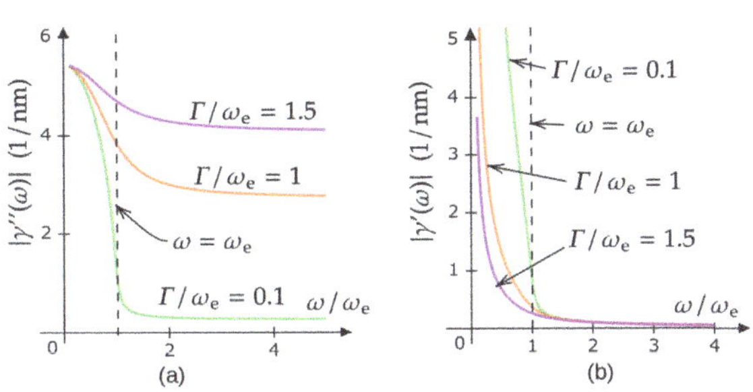

Figure 7. Frequency dependence of γ'' (**a**) and γ' (**b**) for several values of the exciton decay rate Γ. Here, $m_e^\star = 0.9m_e$, where m_e is the electron mass. The exciton transition frequency is $\omega_e = 3.7977 \times 10^{15}$ rad/s ($\hbar\omega_e = 2.5$ eV).

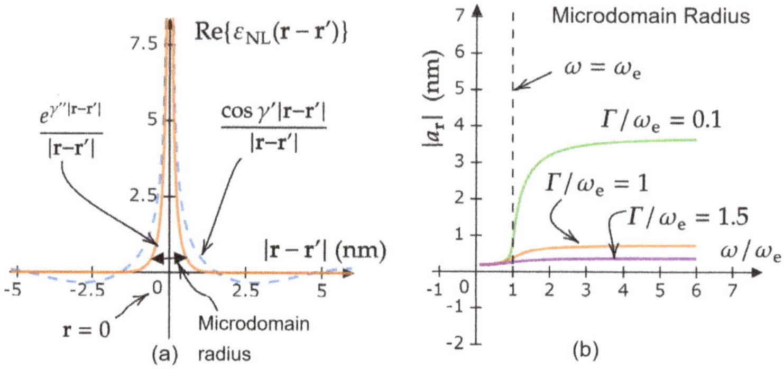

Figure 8. (**a**) Comparison between the real parts of the long-range decay of the excitonic nonlocal domain Green function $\varepsilon_{NL}(\mathbf{r}-\mathbf{r}')$ with and without the full spatial dependence, including the exponential short-range decay factor $\exp(\gamma''|\mathbf{r}-\mathbf{r}'|)$ for $\gamma' = 1\ \mathrm{nm}^{-1}$ and $\gamma'' = 2\ \mathrm{nm}^{-1}$. (**b**) Frequency dependence of $a_{\mathbf{r}}$, the radius of the topological microdomain $B(\mathbf{r}, a_{\mathbf{r}})$ centered at some generic point $\mathbf{r}$ in the nonlocal excitonic material domain $\mathcal{D}$ for several values of the exciton lifetime Γ^{-1}. Here, $m_e^\star = 0.9m_e$, where m_e is the electron mass. The exciton transition frequency is $\omega_e = 3.7977 \times 10^{15}$ rad/s ($\hbar\omega_e = 2.5$ eV).

7.5. The Locally-Homogeneous Model of Nonlocal Semiconducting Domains

Quasi-inhomogeneous, also known as smoothly-inhomogeneous or locally spatially dispersive nonlocal media, are some of the simplest possible prototypes of general (inhomogeneous) nonlocal materials where the spatial dispersion model $\varepsilon(\mathbf{k})$, with a dependence on only one spatial spectral variable $\mathbf{k}$, is found to be not adequate for the mathematical description of the physics of the nonlocal system [76,87]. In contrast, one would need the considerably more complex spectral functions of the form $\varepsilon(\mathbf{k}, \mathbf{k}')$, which are three-dimensional spatial Fourier transformations of generic nonlocal response functions like (3) or (10). In general, there has been quite few investigations aimed at going beyond spatial dispersion in homogeneous media. Examples include inhomogeneous plasma, such as those in controlled-fusion reactors [88], cold collisionless magnetoplasma [88], the electrodynamics of nanostructures [89–92], and incommensurately-modulated superstructures in insulators [76,93].

Here, we will analyze a simple inhomogeneous model of semiconductors experiencing exciton–polariton transitions as outlined above. The EM nonlocal model is *locally-homogeneous* in the sense that around each point $\mathbf{r} \in \mathcal{D}$ there exists a *topologically*-local neighborhood, namely the microdomain $V_{\mathbf{r}}$, inside which the medium can be modeled as a homogeneous and spatially dispersive domain for all $\mathbf{r} \in V_{\mathbf{r}}$ (i.e., the second mention of "locally" here means *topological* nonlocality, see Remarks 3 and 17). It should be noted though that for maximum generality, we allow for variations in the spatial dispersion model to take place from one microdomain $V_{\mathbf{r}}$ to another.

We now wish to estimate the size of each nonlocality microdomain with the help of the exponential law in (118). Let us first expand the homogeneous model treated in Section 7.3 to the inhomogeneous setting of the present discussion, where currently we need to allow that at each point $\mathbf{r} \in \mathcal{D}$, the parameters of the original exciton–polariton model (100) would all become generally functions of the position. That is, in this more general case, one should write $\gamma'(\mathbf{r})$, $\gamma''(\mathbf{r})$, $\omega_e(\mathbf{r})$, $\alpha(\mathbf{r})$, $m_e^\star(\mathbf{r})$, etc, where it is understood that the medium's microscopic composition may change from one position to another.

The main formula for computing the size (radius) of the topological microdomain balls $V_{\mathbf{r}} = B(\mathbf{r}, a_{\mathbf{r}})$ can be easily given by the following expression:

$$a_{\mathbf{r}} \simeq \frac{1}{|\gamma''(\mathbf{r})|}. \tag{120}$$

Roughly speaking, the radius given by (120) quantifies the spatial extension of that characteristic phenomenon of field localization entailed by the presence in the medium Green function (118) of exponential factors like $\exp(-|\gamma''|r')$. Using the formula (116), the relation (120) becomes:

$$a_{\mathbf{r}} = \sqrt{\frac{2\hbar/m_e^\star(\mathbf{r})\omega_e(\mathbf{r})}{1-\frac{\omega^2}{\omega_e^2(\mathbf{r})}+\sqrt{\left(\frac{\omega^2}{\omega_e^2(\mathbf{r})}-1\right)^2+\frac{\Gamma^2(\mathbf{r})}{\omega_e^2(\mathbf{r})}\frac{\omega^2}{\omega_e^2(\mathbf{r})}}}}. \tag{121}$$

This expression (121) is illustrated with some basic examples as given in Figure 8b for various values of the crucial parameter Γ/ω_e. When this ratio between the relaxation rate and the exciton transition frequency is small, the size of the EM nonlocality domain will increase due to the weakening of the corresponding nonlocality-based attenuation (field localization or confinement) processes. Conversely, one may control the size of each EM nonlocality microdomain $V_{\mathbf{r}}$ by modifying the ratio $\Gamma(\mathbf{r})/\omega_e(\mathbf{r})$ evaluated at that position. This may provide a path toward an experimental realization of generalized nonlocal MTMs with controlled microtopological structures. In order to give a view on the numerical values of this structure, Table 2 provides some relevant microdomain data computed by means of the expression (121).

Table 2. Topological microdomain data at a generic position $\mathbf{r} \in \mathcal{D}$. The exciton transition frequency is $f_e = 23{,}862$ THz ($\hbar\omega_e = 2.5$ eV), while $m_e^\star = 0.9m_e$. For the left table, $\Gamma/\omega_e = 2\times10^{-5}$.

f (THz)	ω/ω_e	$a_{\mathbf{r}}$ (μm)	Γ/ω_e	ω/ω_e	$a_{\mathbf{r}}$ (μm)
19,090	0.8	0.0003	0.00002	1.01	2.5834
21,476	0.9	0.0004	0.00020	1.01	0.2583
23,862	1.0	0.0582	0.00200	1.01	0.0259
26,248	1.1	7.6670	0.02000	1.01	0.0028
35,793	1.5	13.7174	0.02000	1.01	0.0028
47,724	2.0	15.9382	2.0000	1.01	0.0002
59,655	2.5	16.8674	20.000	1.01	0.0001

Remark 18. The approximation (120), strictly speaking, is not compatible with Definition 1 since the latter is based on assuming that the material response kernel possesses a compact support. However, for all practical purposes, a decaying exponential can be taken to approximate the behavior of a function with compact support. Nevertheless, in a more careful future treatment it is always possible to modify the exact Definition 1 in order to incorporate the decaying-exponential response kernel as another valid example of effective physics-based nonlocality mathematically realized by a topologically-localized function. An elementary discussion of some possible such modifications is given in Appendix A.10.

8. Application to Fundamental Theory: Electromagnetic Boundary Conditions in the Fiber Bundle Superspace Formalism

Armed with the general superspace formalism of nonlocal continua (Section 5) and the detailed practical example illustrating the theory (Section 7), we now turn to a brief reexamination of a topic in fundamental theory: the role of boundary conditions in nonlocal continuum field theories. The well-known tension between nonlocal electromagnetism and intermaterial interfaces has been already mentioned several times above. Here, we provide some application of the fiber bundle theory of Section 5, aiming at elucidating the nature of this tension, and we suggest some possible new formulation of the problem.

The natural starting point is Figure 2, where a zoomed-in topological picture based on the general structures explicated in Section 4 is given. The focus now is on the *interface* between two generic nonlocal domains $\mathcal{D}_n$ and $\mathcal{D}_m$. In traditional local electromagnetism, the constitutive relation material tensor $\overline{\mathbf{K}}_n$ is usually exploited to deduce conditions dictating how various electromagnetic field components behave as they cross the $\mathcal{D}_n/\mathcal{D}_m$

intermaterial interface. However, even if each response function $\overline{\mathbf{K}}_{n/m}(\mathbf{r}, \mathbf{r}')$ was to be treated as one belonging to a spatially dispersive domain, i.e., replacing it by $\overline{\mathbf{K}}_{n/m}(\mathbf{r} - \mathbf{r}')$, the presence of a boundary between two *distinct* material profiles completely destroys the *translational symmetry* of the structure on which the very rigorous derivation of the specific spatially-dispersive nonlocal response tensor $\overline{\mathbf{K}}_{n/m}(\mathbf{r} - \mathbf{r}')$ was originally based.

The breakdown of translational symmetry in inhomogeneous crystal configurations was very clearly identified and explained by Agranovich and Ginzburg [16], together with several proposals for a solution of such unusual electromagnetic problem. For example, because it is evident that close to the intermaterial interface the response tensor of each medium, when seen from its own side while approaching the boundary, must be reverted back to the most general nonlocal form, namely $\overline{\mathbf{K}}_{n/m}(\mathbf{r}, \mathbf{r}')$ instead of $\overline{\mathbf{K}}_{n/m}(\mathbf{r} - \mathbf{r}')$, it was then proposed that one may use the former, more general, functional form, but only within a "thin transitional layer" that includes the intermaterial interface, yet while additionally extending, along some necessarily "ambiguous distance", into the depths of the two material domains $\mathcal{D}_n$ and $\mathcal{D}_m$ on both sides of the boundary. Outside this fuzzy region, a gradual transition, or a continuously changing profile (a tapered channel), is introduced to proceed from the most general forms $\overline{\mathbf{K}}_{n/m}(\mathbf{r}, \mathbf{r}')$, valid in the vicinity of the intermaterial interface, to the special spatially dispersive forms $\overline{\mathbf{K}}_{n/m}(\mathbf{r} - \mathbf{r}')$, which are more accurate the further one goes away from the material boundary, where the latter response tensor functions are considered characteristic of "bulk" homogeneous material domains [16].

Another proposal is to keep using everywhere spatial dispersion profiles of the form $\overline{\mathbf{K}}_n(\mathbf{r} - \mathbf{r}')$, but introduce specialized *additional boundary conditions* (ABCs) at the intermaterial interface based on each particular problem under consideration. Although this latter approach is both mathematically and physically inconsistent (due to the breakdown of symmetry caused by the presence of intermaterial interfaces), it nevertheless remains popular because—at least in outline—nonlocal electromagnetism is thereby held up in a form as close as possible to familiar local electromagnetic theory methods, especially numerical techniques, such as finite element method (FEM) [44], method of moment (MoM) [46], and finite difference time-domain method (FDTD) [45], i.e., established full-wave algorithms where it is quite straightforward to replace one boundary condition by another without essentially changing much of the code.[27]

Nevertheless, both approaches discussed above require considerable input from the *microscopic* theory, mainly to determine the tapering transition region in the case of the first, and the ABCs themselves in the second. That motivated the third approach, called, the ABC-free formalism, where the relevant microscopic theory was utilized right from the beginning in order to formulate and solve Maxwell's equations. For example, in [50,53], a global Hamiltonian of the matter-field system is constructed and Maxwell's equations are derived accordingly. In [38], the rim zone (field attached to matter) is investigated using different physical assumptions to understand the transition from nonlocal material domains to vacuum going through the entire complex near-field zone. In [89], the symmetry group of carbon nanotubes was exploited to construct a set of Maxwell's equations in nonlocal nanoscale problems without using a homogenized electromagnetic field-based boundary condition.

We believe that the main common conclusion from all these different formulations is that in nonlocal electromagnetism *it is not possible in general to formulate the electromagnetic problem at a fully phenomenological level.* In other words, microscopic theory appears to be in demand more often than in the case of systems involving only local materials. However, since all existing solutions use the traditional spatial manifold $\mathcal{D}$ as the main configuration space, the question now is whether the alternative formulation proposed in this paper, the extended fiber bundle superspace formalism, may provide some additional insights into the problem of why nonlocal continuum field theory cannot be formulated in general for inhomogeneous domains as in the local version of that theory.

We provide a provisional elucidation of the topological nature of field theory across intermaterial interfaces by noting that, in Figure 2, it is not only the behavior of the fields

$\mathbf{F}(\mathbf{r})$ in the two domains that is mostly relevant, but also the *entire* set of local topological microdomains $V_{\mathbf{r}}$ clustered on both sides of the interface inside the material domains. More specifically, we attach a great importance to *how these microtopological domains, together with the corresponding set of excitation fields that are applied on them, would behave as they move across the boundary*. In *general (set-theoretic)* topology, *boundaries* are defined fully in terms of the behavior of *open sets* [58,59].

We now build on this key set-theoretic topological concept in order to illustrate how the problem of nonlocal inhomogeneous continuum field theory may be reformulated through the superspace formalism developed in Section 5. First, Figure 9 provides a finer or more structured picture of the topological content of Figure 2 based on replacing the spaces $\mathcal{D}_m$ and $\mathcal{D}_n$ by the corresponding Banach bundle superspaces $\mathcal{M}_m$ and $\mathcal{M}_n$, respectively. The thick horizontal curved lines represent the base spaces $\mathcal{D}_m$ and $\mathcal{D}_n$, while the wavy vertical lines stands for the fiber spaces X_m and X_n attached at each point $\mathbf{r} \in \mathcal{D}_{n/m}$ in the corresponding base manifolds. The double discontinuous lines at the "junction" of the two base spaces $\mathcal{D}_m$ and $\mathcal{D}_n$ indicate the joining together of the two vector bundles $\mathcal{M}_m$ and $\mathcal{M}_n$.

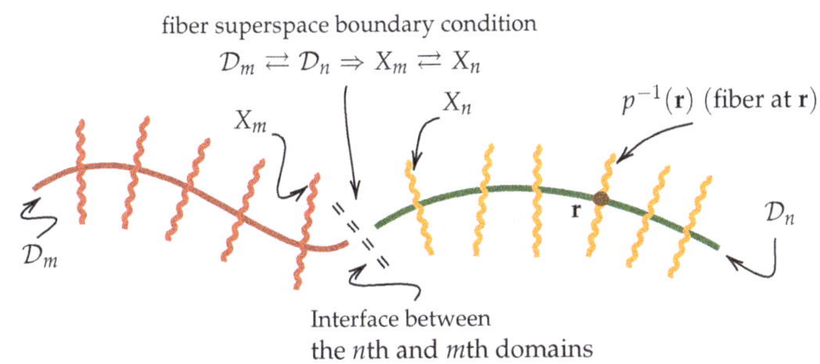

Figure 9. An abstract representation of the topological fiber bundle superspace structure behind Figure 2.

It should become clear now that since the two nonlocal material domains possess an *extra structure*, namely that of the individual copies of the fibers, each a linear vector Banach space attached to every point in the base space, we must also indicate how the various elements belonging to the Banach function spaces, i.e., the fields defined on the microdomains $V_{\mathbf{r}}$ in Figure 2,[28] would behave as they cross the boundary separating the two material domains $\mathcal{D}_m$ and $\mathcal{D}_n$. One obvious way to do this is to introduce a *bundle homomorphism* between the two vector bundles $\mathcal{M}_m$ and $\mathcal{M}_n$ over the interface submanifold $\partial\mathcal{D}_{mn}$ separating $\mathcal{D}_m$ and $\mathcal{D}_n$. This mathematical object is similar to the nonlocal response map $\mathcal{L}$ introduced by (72).

The motivation behind introducing this bundle homomorphism is to serve as a "boundary condition operator" acting on the fiber bundle *superspaces* $\mathcal{M}_m$ and $\mathcal{M}_n$ instead of the conventional spaces $\mathcal{D}_m$ and $\mathcal{D}_n$ always used in local continuum field theories. We will not go here into a detailed construction of such a new fiber bundle super-operator. Instead, we provide some additional remarks to illustrate the broad outline of the key idea behind our proposal. A more detailed investigation of the intermaterial interface homomorphism will be given somewhere else.

In continuum field theories, the formal expression of the traditional boundary condition applied to the two materials' $\mathcal{D}_n$ and $\mathcal{D}_m$ base spaces (spacetime, space–frequency, or space differential manifolds) will be summarized by the symbolic formula

$$\mathcal{D}_m \rightleftarrows \mathcal{D}_n \tag{122}$$

in order to highlight that in such traditional formulation, it is the direct geometric relations between the individual material manifolds that usually holds the center stage. For example, the electromagnetism of continuous media, is usually spelled out in the more specific space-limit form:

$$\lim_{\mathbf{r}\to\partial\mathcal{D}_{mn}} \{\mathbf{F}_m(\mathbf{r}) - \mathbf{F}_n(\mathbf{r})\} = \Gamma_{b_1}[\mathbf{F}_m(\mathbf{r}), \mathbf{F}_n(\mathbf{r})],$$
$$\lim_{\mathbf{r}\to\partial\mathcal{D}_{mn}} \{\mathbf{R}_m(\mathbf{r}) - \mathbf{R}_n(\mathbf{r})\} = \Gamma_{b_2}[\mathbf{R}_m(\mathbf{r}), \mathbf{R}_n(\mathbf{r})], \quad (123)$$

where $\partial\mathcal{D}_{mn}$ is the boundary between $\mathcal{D}_m$ and $\mathcal{D}_n$. Here, Γ_{b_1} and Γ_{b_2} are "base space boundary functions", which are not universal, but whose detailed expressions depends on the concrete content of the field theory and the material system under consideration.

On the other hand, in the superspace formalism of nonlocal metamaterials and continua, it can be seen that the various elements belonging to each fiber space $X_{n/m}$ attached at the point $\mathbf{r} \in \mathcal{D}_{n/m}$ of the base manifolds, i.e., the excitation field functions operating on the microdomains $V_{\mathbf{r}}$, $\mathbf{r} \in \mathcal{D}_{n/m}$, are to be mapped onto each other via an expression of the form:

$$X_m \rightleftarrows X_n : \quad \lim_{\mathbf{r}\to\partial\mathcal{D}_{mn}} (X_m - X_n) = \Gamma_f[X_m, X_n]. \quad (124)$$

Here, Γ_f is a new "fiber superspace boundary function". The full formulation of (124) is considerably more complex than the local field-theoretic case of (122) and (123) due to the fact that, additionally, the boundary condition quantity Γ_f must be also proven compatible with the detailed corresponding fiber bundle structures of the materials involved. Consequently, for the field theoretic treatment of complex nonlocal continuum systems, the *global* topology of the metamaterial superspaces $\mathcal{M}_m$ and $\mathcal{M}_n$ will have to be assessed and utilized in the process of formulating a generalized "superspace boundary condition" of the form (124).

We summarize our main provisional view on the status of boundary conditions in the nonlocal field theory of inhomogeneous continua as follows:

- The existence of *extra* or *additional* structures in the fiber bundle superspace approach to nonlocality in complex continua forces on us the need for introducing additional boundary conditions or information coming from the microscopic *topological* structure of the corresponding material superspaces.
- The fiber bundle superspace formalism of nonlocal metamaterials appears to be able to capture the intricate processes taking place inside and across various nonlocal material domains joined together through interfaces.
- This is achieved by providing an efficient apparatus to topologically encode some of rich and complex physics of field–matter interactions via the construction of appropriate infinite-dimensional function spaces (Banach space fibers) attached at each point of the materials' base manifold.
- It is suggested that the relations between those additional *fiber spaces* are in fact what should be mainly taken into account while formulating boundary conditions for nonlocal continuum field theories, hence not merely the conventional relations involving only spatial interfaces between the material base manifolds as has been usually the practice in local field theories.

However, despite the fact that the full mathematical formulation of the proposed fiber bundle boundary condition homomorphism (124) is beyond the scope of this paper, it is hopped that the initial insight provided in this section can at least clarify the subject and stimulate further researches into the fundamental theory of nonlocal continua and metamaterials. Additional possible applications are given in the Appendixes A.3 and A.11.

9. Conclusions

We provided a general theoretical and conceptual investigation of nonlocal continuum field theories that aimed to achieve several goals. First, the subject was revisited from a new perspective, with the intention of introducing it not only to mathematical and theoretical physicists, but also invite a wider audience, including engineers, material scientists, chemists, applied physicists, and applied mathematicians. The various essential ideas behind nonlocality in material continua were put under new light with the help of an abstract field-response model developed in three dimensions. Next, the fine-grained topological microstructure of nonlocal metamaterials was explicated in detail. We introduced the concept of physics-based nonlocality microdomains, demonstrating how the latter regions present an important structural topological feature of the physics of nonlocal media. Afterwards, it was proved using differential topology that a natural fiber bundle structure, serving as a "source (excitation) superspace", can be constructed. The material source fiber bundle superspace, or the material superspace in short, was shown to possess all of the required properties of a standard fiber bundle yet while faithfully reflecting the physics of nonlocal microdomains. Eventually, and using the technique of partition of unity, it was proved that the fiber bundle superspace can be deployed for the purpose of constructing and computing the complete nonlocal material response function over arbitrary microdomain configurations. This was accomplished by building a bundle homomorphism to replace the well-known, but now inadequate, material tensor linear operators commonly utilized in local continuum field theories, for instance, conventional electromagnetism. This new homomorphism can be viewed as a generalization of the linear operators of the various classical boundary-value problems of mathematical physics. It is hoped that in the future this homomorphism may be "topologically discretized" using suitable methods borrowed from other advanced fields such as algebraic topology, computational topology, and global analysis. The new fiber bundle superspace formulation suggested that nonlocal continuum field theories could be reformulated in an alternative way compared with the prevailing existing methods. Most importantly, nonlocality in material continua forces us to introduce an entire array of infinite-dimensional Banach spaces attached to every point in the conventional three-dimensional base space inside which the material is conventionally defined. This extra or additional fiber structure provides a natural explanation of why traditional boundary conditions often fail to account for the physics of nonlocal metamaterials. Moreover, the fiber bundle theory opens the door for several new applications, including the ability to understand the deep connection between topology and field theories, e.g., electromagnetism, in engineered artificial media. Overall, the author proposes that future research in metamaterials will gradually require more extensive collaboration between engineers and mathematicians in order to explore in full the deep consequences of this organic topology/electromagnetism relation.

Funding: This research received no external funding.

Conflicts of Interest: The author declares no conflict of interest.

Appendix A

Appendix A.1. Survey of the Literature on Nonlocal Metamaterials

Appendix A.1.1. Introduction

We first provide a non-exhaustive and selective review of the development of nonlocal electromagnetic materials research. More information and proposals regarding engineering applications are given in Appendix A.3, where additional references can be found. The main propose behind this literature overview is to suggest that the area of nonlocal metamaterials research might be approached as an approximately coherent field of investigation, i.e., more than just being merely a technical sub-discipline selected from within the sciences of metals, semiconductors, plasma, periodic structures, metasurfaces, etc. In fact, one of the main objectives of this paper is to demonstrate that a unified theoretical treatment of the entire subject is mathematically possible (the superspace formalism.) However, one needs to be convinced first of the presence of substantial past researches into this area. Hence, convincing readers not familiar with the topic about the long and very rich history of investigations into various nonlocal phenomena in material systems is one of the objectives of this Appendix.

Appendix A.1.2. Historically Important Examples

Some of the physical phenomena that cannot be understood using local electromagnetic theory include spatial dispersion effects [83], extreme negative group velocity and negative refraction [52,94], new diffraction behavior in optical beams [95], superconductivity [96], natural optical activity [16,97,98], non-Planck equilibrium radiation formulas in nonlocal plasma [99]. Outside electromagnetism but within wave phenomena, there also exists processes that cannot be fully accounted for through simple local material models, for instance, we mention phase transitions, Casimir force effects [100], and streaming birefringence [9]. By large, *spatial dispersion* has attracted most of the attention of the various research communities working on nonlocal electromagnetic materials. Indeed, few book-length researches on spatial dispersion already exist in literature, most notably [14–16,83]. We provide additional remarks on the history of spatial dispersion in Appendix A.2.

Appendix A.1.3. General Theories of Nonlocal Continua

The majority of the published research on nonlocal media and nonlocal electromagnetism tend to focus on applications and specialized materials (see the majority of the references quoted below). Few exceptions include investigations attempting to approach the subject at a more general level. For example, from the perspective of general thermodynamics, see [8,9]. A unified perspective inspired by condensed-matter physics, especially plasma physics, can be found in [47]. Within nanoscale electrodynamics, nonlocality was treated broadly as an essential feature of microscopic interactions at the nano- and mesoscopic scales [50,53]. Some of the topics reexamined within the framework of a general nonlocal field–matter interaction theory include the applicability of optical reciprocity theorems [101–104], energy/power balance [105], quantization [106–108], operator methods [87], extension of spatial dispersion to include inhomogeneous media [76], and alternative formulations of spatial dispersion in terms of the Jones calculus [109].

Appendix A.1.4. Semiconductors, Metals, Plasma, Periodic Structures

The bulk of the available literature on nonlocality is concentrated in the very large area of general field–matter interactions. There already exists a well-attested body of research on nonlocality in metals based on various phenomenological approaches, e.g., see [110] for a general review. Nonlocality has also been extensively investigated in dielectric media, for example semiconductors [83,111]. A comprehensive recent review of nonlocality in crystal structures is provided in [112], which updates the classic books [16,55]. Moreover, numerous researches conducted within condensed-matter physics and material science implicitly or explicitly assume that nonlocality is essentially based on microscopic (hence quantum) processes, and develop an extensive body of work where the spatially dispersive dielectric tensor is deployed as the representative constitutive material relation [37,38,50,53,54]. On the other hand, one can also treat nonlocality without resort to spatial dispersion by modeling certain classes of material media as periodic structures [113], e.g., photonic nanocrystals [114], where the susceptibility tensor is derived from the symmetry of the overall structure [37,89,115] or from the lattice dynamics approach [92,116].

Appendix A.1.5. Boundary Conditions in Nonlocal Metamaterials

For solving nonlocal problems, several methods have been proposed in order to deal with the notorious problem of the lack of exact universal nonlocal response models at the intermaterial interface between a nonlocal domain and other media. The so-called additional boundary condition (ABC) approach adjoins new boundary conditions to the standard Maxwell's equations in order to account for "additional waves" excited at the interface, which otherwise would not be explicable by the standard local theory alone [16]. However, it must be noted that without exception all ABC formulations are inherently model-specific since each boundary condition model presupposes a particular type of nonlocal media, or simply just postulates specific ABCs based on their ease of use in applications, e.g., see [55,56,80,117–119]. We note that such ABC formalisms are not inevitable since there exists several boundary-condition free formulations, e.g., see [50,53,89].[29]

Appendix A.1.6. Computational Techniques

For performing full-wave field analysis in the presence of nonlocal materials, a number of discretization strategies have been proposed. For example, an FDTD-based method was suggested to deal with metallic spatially dispersive objects [120]. The formulation, discretization, and solution of surface integral equations for nonlocal plasmonic materials were also attempted in [121,122], where the reduction of the electromagnetic problem to a finite-matrix form was achieved using the RWG basis functions. Moreover, specialized methods were proposed for various possible scenarios involving nonlocal field–matter interactions, such as nonlocal dielectric profile retrieval from measurable data [123], iterative solutions of nonlocal wave equations [124,125], applications of the derivative expansion method to nonlocal plasma analysis [126], application of Kramers–Kronig relation method [127], application of the Pade approximation to homogenization [128].

Appendix A.1.7. Novel Systems and Devices with New Electromagnetic Behavior

The idea of exploiting nonlocality to design and develop a new generation of metamaterials (MTMs) exhibiting novel EM behavior has also received a revival in recent years [32,47,129], though the basic concept in itself is not completely new, going back to at least the 1980s and possibly earlier [16]. Recent examples of research focused in explicating nonlocal behavior to harness the associated new physics include spatial dispersion in photonic crystals [130], wire media [131–134], semiconductor nanoparticles [135–138], optically nonlinear liquids [139], hyperbolic metamaterials [140], layered dielectric-metal structures [141,142] and thin films [143], plasma-based metamaterials [144–146], quantum wells [147], soliton interactions with matter [148–153], superconducting films [154] and circuits [155], plasmonic devices and structures [156–158], nanocubes [159], cloaking [160], Chern metamaterials [161] and superconductors [162], dispersion management profiles [52,163], biomedical applications in materials [164], nonlocal antennas [165,166], and nonlocal uniaxial metamaterials [167]. Due to the practical importance of this area of research, we provide additional information in Appendixes A.3 and A.11.

Appendix A.1.8. Homogenization

Numerous homogenization theories for nonlocal MTMs, where averaging operations are considered over multiple spatial scales, have been reported in the literature, e.g., see [133,168–171]. We note that the subject of estimating the effective electric and magnetic properties of electromagnetic metamaterials, with or without nonlocality, is enormous and it is beyond the scope of this paper to even summarize the main papers in the field. Nevertheless, it is curious to note that until fairly recently, most publications have tended to focus on *non*-spatially-dispersive media; hence, local scenarios are still dominant in the area of advanced artificial material systems. This situation has began to change in the last few years, and nowadays an increasing number of reports appear to move from the old opinion that "spatial dispersion is a bug" to the more positive and fruitful perspective that nonlocality may provide pathways to novel physical behavior that can be exploited for various applications in metamaterial system design. However, we also note that progress in this second direction, where nonlocality is embraced rather than being treated with suspicion, has been generally slow.

Appendix A.1.9. Topological Materials and Photonics

A particularly interesting direction of research in nonlocal media is the recent subject of *topological photonics*. The main idea was inspired by previous researches in Chern insulators and topological insulators [10], where the focus has been on electronic systems. There, it has already been observed that the nonlocal behavior of the fermion wave function may exhibit a rather interesting and nontrivial dependence on the entire configuration space of the system, in that case the momentum space (the wave vector **k** space). In addition to the already established role played by nonlocality in superconductors, quantum Hall effects are among the most intriguing physically observable phenomena that turned out to depend fundamentally on purely topological aspects of the electron wave function [96]. The major themes exhibited by electrons undergoing topological transition states include topological robustness of the excited edge (surface) states moving along a two-dimensional interface under the influence of an external magnetic fields. More recently, it was proposed that the same phenomenon may apply to photons (electromagnetism) [172], where the key idea is to use photonic crystals to emulate the periodic potential function experienced by electrons in fermion systems. However, since photons are bosons, transplanting the main theme of topological insulators into photonics is not trivial and is currently generating a great attention, see for example the extensive review article [11], which provides a literature survey of the field. One of the most important applications of topological photonics is the presence of "edge states", which are topologically robust unidirectional surface waves excited on the interface between two metamaterials with topologically distinct invariants. Since edge states are immune to perturbations on the surface, they have been advocated for major new applications where topology and physics become deeply intertwined [173]. Topology can also be exploited to devise non-resonant metamaterials [174] and to investigate bifurcation transitions in media [175]. Another different but related exciting subject illustrating the synergy between topology, physics, and engineering is non-Hermitian dynamics, especially in light of recent work related to the origin of surface waves [176,177], which is now being considered as essentially non-trivial topological effect. In Appendix A.11, the subject of topological photonics is taken up again but from the viewpoint of applications.

Appendix A.2. On the History of Spatial Dispersion in Crystal and Plasma Physics

Historically, spatial dispersion had been under the radar since the 1950s, especially in connection with researches on the optical spectra of material domains [77,78,178]. However, the first systematic and thorough treatment of the subject appeared in 1960s, prominently in the first edition of Ginzburg's book on plasma physics, which was dedicated to electromagnetic wave propagation in plasma media. The second edition of the book, published in 1970, contained a considerably extended treatment of the various mathematical and physical aspects of the electromagnetism of spatially dispersive media [14]. Spatial dispersion in crystals had been also investigated by Ginzburg and his coworkers during roughly the same time [179–181]. The book [83] contains good summaries on spatial dispersion research up to the end of the 1980s. More recently, media obtained by homogenizing arrays of wires, already very popular because of their connection with traditional (temporal) metamaterials, are known to exhibit spatial dispersion effects, though many researchers ignore that effect to focus on temporal dispersion [182–184]. Other types of periodic or large finite arrays composed of unit cells like spheres and desks also exhibit spatial dispersion effects [185]. Nonlinear materials with observable nonlocality have also been investigated in the optical regime [186]. More recently, much of the resurgence of interest in spatial dispersion can be traced back to the observation that nonlocal phenomena cannot be ignored at the nanoscale level [187], especially in problems of low-dimensional structures, such as carbon nanotubes [89,91,92,188] and graphene [189,190]. The subject was also introduced at a pedagogical level for applications involving current flow in spatially dispersive conductive materials, such as plasma and nanowires [191].

Appendix A.3. Some Further Engineering Applications of Nonlocal Metamaterials

The purpose of this Appendix is to provide a sample of some other current and future possible applications of nonlocal metamaterials based on the author's own experience, which may serve as a supplementary text to be read in conjunction with the general survey of Appendix A.1.

Appendix A.3.1. Communications Systems and Information Transmission

Nonlocal metamaterials offer a very wide range of potential applications in wireless communications and optical fibers. The basic idea is to introduce specially engineered nonlocal domains either as part of the communication channel (e.g., optical fibers, plasmonic circuits, microwave transmission lines) [189], or as a control structure integrated with existing antennas [129,192]. Spatial dispersion was also used as a method to engineer wave propagation characteristics in material domains, e.g., see [193] for applications to high-efficiency modulation of free-space EM waves. A general linear partial equation explicating how spatial and temporal dispersion can be jointly exploited to produce zero distortion (e.g., constant negative group velocity) was derived and solved in [52]. The main idea originated from the fact that one of the main sources of distortion in communication systems is that due to *non*-constant group velocity $\mathbf{v}_g := \nabla_{\mathbf{k}}\omega$ [194,195]. Since $\mathbf{v}_g$ is a strong function of the dependence of the material response tensor $\overline{\mathbf{K}}(\mathbf{k},\omega)$ on both $\mathbf{k}$ and ω, *dispersion management equations* can be derived for several applications. For example, it was proved in [52] that in simple isotropic spatially dispersive media with high-symmetry, one may obtain exact solutions where the group velocity is constant at an entire frequency band. This happens because while strong temporal dispersion is present (which alone causes strong distortion), incorporating optimized spatially dispersive profiles leads to *complete compensation* (cancellation) of distortion, resulting in essentially a distortion-free communication channel. There are enormous potentials of research into this new exciting area. The reason is that most practical realizations of nonlocal metamaterials involve complex material response tensors, where the relevant mathematics of dispersion engineering is still underdeveloped (and in fact underappreciated by researchers), which implies that, to the best of our knowledge, relatively very little has been done in this emerging field so far.

Appendix A.3.2. Electromagnetic Metamaterials

While this paper attempts to analyze and understand the general structure of nonlocality in generic field theories of continuous media, we have already mentioned above that artificial media, better known nowadays as metamaterials systems, could provide one of the most direct paths toward building new functional advanced materials and also providing models to further explore nonlocality both experimentally and numerically. As early as the 1960s, it was proposed that EM nonlocality can be exploited to produce materials with very unusual properties. For example, in [16], negative refraction materials were noted as one possible application of spatial dispersion where the path toward attaining this goes through controlling the direction of the group velocity vector. Since in nonlocal media, power does not flow along the Poynting vector [14], new (higher-order) effects were shown to be capable of generating arbitrary group velocity profiles by carefully controlling the spatial and temporal dispersion profiles. Overall, the ability of spatial dispersion to induce higher-order corrections to power flow is a unique advantage enjoyed by nonlocal metamaterials exhibiting weak or strong spatial dispersion in addition to normal dispersion. This extra spatial degrees of freedom provided by nonlocality was researched, reviewed and highlighted in many publications, including, for example, works such as [32,47,115,129,134,141,163,175,184,196].

Appendix A.3.3. Near-Field Engineering, Nonlocal Antennas, and Energy Applications

Another interesting application of nonlocality in electromagnetic media is near-field engineering, a subject that has not yet received the attention it deserves. It was observed in [129] that a source radiating in homogeneous, unbounded isotropic spatial dispersive medium may exhibit several unusual and interesting phenomena due to the emergence of extra poles in the radiation Green's function of such domains. Both longitudinal and transverse waves are possible (dispersion relations), and the dispersion engineering equations relevant to finding suitable modes capable of engineering desired radiation field patterns are relatively easy to set and solve. For example, by carefully controlling the modes of the radiated waves, it is possible to shape the near field profile, including total confinement of the field around the antenna even when losses is very small, opening the door for applications like energy harvesting, storage, and retrieval in such media [197]. The direct use of especially-engineered nonlocal metamaterials, however, has been explored only for simple materials so far and mainly at the theoretical level [32]. However, the increasing importance of energy localization [198] at both the level of numerical methods [199] and the device level applications [200], suggest the need to reconsider the role played by nonlocality in complex media.

On the other hand, away from the source region, the subject of *far*-field radiation by sources embedded into nonlocal media was investigated previously by some authors within the context of plasma domains [104]. Recently, it has been systematized into a general theory for nonlocal antennas with media possessing an arbitrary spatial dispersion profile [25,165,166,192]. However, no general theory exists for nonlocal media, which are inhomogeneous. The superspace formalism proposed in this paper may help stimulate research into this direction in order to overcome the limitations of the existing theory of nonlocal antenna systems.

Appendix A.4. On the Concept of Superspace

The concept of superspace is not new, and has been proposed several times in both physics and mathematics. For a brief but general view on the definition of superspaces, see [201]. For example applications, various superspaces have been proposed as fundamental structures in quantum gravity [202,203], which are frequently infinite-dimensional. Superspace concepts are also now extensively researched in quantum field theory and the standard model of particle physics, e.g., see [204–206]. In general, dealing with topics such as supergravity, supersymmetry, superfields, superstrings, and noncommutative geometry often requires the use of one superspace formalism or another [204]. In mathematics and mathematical physics, where the concept itself originated, a notable recent example of the superspace concept includes *sheaves*, which are used in differential and algebraic topology and algebraic geometry and have numerous applications in physics [30,207,208].

In this paper, the superspace concept has very little to do with applications to supersymmetry or supergravity, such as the examples mentioned above (and many others we do not mention.) Instead, our use of the concept is more aligned with the mathematical practice of *extending* one space by *embedding* it into a larger superspace as in the schema:

$$\text{Space} \xrightarrow[\text{injection}]{\text{embedding}} \text{Superspace}. \tag{A1}$$

In other words, the embedded space is injected as a substructure into the (larger) embedding superspace. The key interest behind the formula superspace-as-embedding (A1), of course, goes beyond mere set-theoretic inclusion. We are not here trying simply to say that Space $\subset$ Superspace, which would be devoid of mathematical substance. Instead, the main motivation behind the superspace construction (A1) is that the embedded Space becomes a *substructure* attached to or placed within the larger, embedding "container", which is here superspace.

The most important thing to note here is that the latter superspace acquires a more coherent and fundamental status than the former. Eventually, Space becomes nothing but a mere "substructure" or "index space" of the more originary mother space that we originally called superspace. Strangely, with time, superspaces tend to become so familiar and basic to the degree one begins to call them regular spaces, while the original Space fades into oblivion. This last observation regarding the ontological primacy of the superspace concept over space can be best seen from the *converse* generative schema:

$$\text{Superspace} \xrightarrow[\text{projection}]{\text{de-embedding}} \text{Space}. \tag{A2}$$

Here, we recover the original space through a *projection* operation by which a de-embedding of the superspace substructure, the interior placeholder occupied by space, is achieved by projecting the mother space, the superstructure, superspace, onto the substructure, space. It is really the purely formal *structural* relation dictating how sub- and super-structures are organized within a common unifying global schema what is at stake in such type of superspace theories, i.e., not just the simple set-theoretic inclusion of one space into another.

Both operations, the injection (A1) and projection (A2) are necessary to fully understand the idea of superspace in general. However, in practice, usually only one of them is emphasized on the expense of the other. It is rare to find in superspace theories that both projection and injection operations are allotted the same ontological status. For instance, in the fiber bundle approach adopted in the present paper, space is recovered (or generated) from the fundamental superspace through the projection map of the mother fiber bundle, which will send each fiber into its "representative point" in the base manifold. In this manner, regular space may be seen as if it was actually *generated* or "produced" by the more primordial superspace mother structure [18,209].

A specific example more related to the subject of nonlocal MTMs is the original superspace concept introduced earlier for the analysis of deformed crystal [210] and subsequently utilized for fundamental investigations of EM nonlocality in incommensurate (IC) superstructures in insulators [76]. Such modulated-structure materials possess spaces with dimensions greater than spacetime [211]. Nevertheless, for fairly concrete models one may exploit group theory to construct finite-dimensional (dimension > 4) approximations of them. The general theory of superspace formalisms in quasi-periodic crystals is presented in [212]. Other examples from condensed-matter physics where superspace methods where applied include mesoscopic superconductivity [213].

Appendix A.5. Guide to the Mathematical Background

We provide a brief overview on how to read the mathematical portions of this paper and where to find detailed references that might be needed in order to expand some of the technical proof sketches provided in the main text. We emphasize that in this paper only the *elementary* definitions of

1. Differential manifolds,
2. Banach and Sobolev spaces,
3. Vector bundles, and
4. Partition of unity

that are needed in order to understand the mathematical development. Here, we briefly go over the principal ideas behind each one of these four key mathematical topics listed above, providing also additional references for readers interested in learning more about the required background. The current Appendix is not intended as a complete review; some familiarity with all four elementary mathematical topics listed above is required for a complete understanding of the technical proofs and constructions found in Section 5.

Appendix A.5.1. Topology on Smooth Manifolds

A *differential manifold* is a collection of fundamental "topological atoms" each composed of an open set U_i and a chart $\phi_i(x)$, which serves as a *coordinate system*, basically an invertible differentiable map to the Euclidean space $\mathbb{R}^n$. That is, locally, every manifold looks like a Euclidean space with dimension n. When the differentiable map is smooth, the differential manifold is called *smooth manifold*. The collection of open sets $U_i, i \in I$, where I is an index set, covers this n-dimensional manifold. Since some of these open sets are allowed to overlap, the crucial idea underlying the concept of the differential manifold is that over the common intersection region $U_i \cap U_j$, there exists a smooth reversible coordinate transformation function mutually relating the two coordinates of the same abstract point when expressed in the two (generally different) languages belonging to the topological atoms U_i and U_j. *Note that the key concept of topology is how to propagate information from the local to the global levels.* In this sense, differential manifolds present elementary structure allowing us to rigorously conduct this process using the efficient apparatus of the differential calculus. Note that only the elementary definition of smooth manifolds is required in this paper, which can be found in virtually any book on differential or Riemannian geometry, e.g., see [26,30,57,62,65,70,204,214].

Appendix A.5.2. Banach and Hilbert Spaces

A *Banach space* is a vector space equipped with a norm satisfying the standard properties that a generic norm should have (namely, being positive, being zero only for the null vector, scale linearity, and the triangle inequality [215].) Most importantly, Banach spaces are also required to be *topologically complete* in the sense that every Cauchy sequence converges to an element in the space itself. In this way, no "holes" are left in the space thus defined, hence one may deploy a Banach space in order to do analysis on operators as in solving differential equations or the analysis of numerical methods. A *Hilbert space* is a Banach space equipped with an inner product. An important fact to remember about Banach and Hilbert spaces is that when they are employed to model *function* spaces (as in this paper), they most often lead to *intrinsically-infinite* dimensional vector spaces [30].

Appendix A.5.3. Banach and Hilbert Manifolds

A straightforward process of combining Banach or Hilbert spaces with differential manifolds leads to the concept of *Banach* or *Hilbert manifold*, which are prominent examples of *infinite-dimensional manifolds*. A Banach/Hilbert manifold is simply a differentiable/smooth manifold that is locally isomorphic to a Banach/Hilbert space instead of the regular n-dimensional Euclidean space $\mathbb{R}^3$ invoked in the basic definition of an n-dimensional manifold. The isomorphism itself can be either differentiable or smooth, where a suitable derivative operator, such as the Fréchet derivative, may be defined on Banach/Hilbert spaces, leading to the resulting Banach/Hilbert manifold itself being either a differentiable or smooth infinite-dimensional manifold. A Banach/Hilbert manifold is then an intrinsically infinite-dimensional manifold. An elegant formulation of the theory of Banach manifolds can be found in Lang's text [26]. Applications of this theory in the general fields of analysis and geometry can be found in textbooks on global analysis, e.g., see [70,216]. In general, much of the theory of n-dimensional manifolds carry over unchanged into the case of infinite-dimensional manifolds. However, there exists some subtle technical differences, which are carefully highlighted in [214].

Appendix A.5.4. Sobolev Spaces

The most economic approach to constructing *Sobolev spaces* is to define them as Hilbert spaces consisting of (Lebesgue) square integrable functions that posses "generalized derivative", a concept in itself technical but straightforward. For the basic definition of Sobolev spaces and their applications to partial differential equations in mathematical physics and finite-element method in engineering, we recommend [65]. The subject of Banach manifolds is less commonly treated in the literature on Sobolev spaces than finite-dimensional manifolds. For a very readable account on the functional analytic background to the use of Sobolev spaces, see [65], while [64] provides information on the applications of Sobolev spaces in the analysis of linear partial differential equations. The generalization of the theory of Sobolev spaces into the wider setting of functions defined on differential manifolds is tackled in [66] (with applications to nonlinear functional analysis).

Appendix A.5.5. Vector Bundles

The quite general structure known as *fiber bundles*, of which *vector bundles* are famous special cases, are now standard topics in both mathematics (topology, geometry, differential equations), theoretical physics (quantum field theory, cosmology, quantum gravity), and applied physics (condensed-matter physics, many-body problems). On the major importance of vector and fiber bundles within the overall area of modern fundamental physics, see [30,204,208]. In quantum field theory, gauge field theories use vector bundles as essential ingredients in the standard model of particle physics [18,30]. The increasing importance of methods based on quantum field theory in applications to condensed-matter physics has contributed into making knowledge of fiber bundle techniques useful and more widespread in physical and engineering research than originally anticipated; e.g., see the area of the Berry phase and the associated gauge connection [11,96]. The key idea behind the vector bundle is to attach an entire vector space to every point on a base manifold. To be more specific, consider a differential manifold $\mathcal{D}$ serving as the base manifold. Each copy of the vector space that is attached to a point in this base space will be called the *fiber* at that point. The standard *tangent space* of a smooth manifold is the most obvious example of such vector bundles. However, more complicated structures than finite-dimensional tangent spaces can also be captured by a suitable vector bundle concept. In this paper, we have shown that physics-based nonlocality in material continua can be modeled, very naturally in the mathematical sense, by considering the Banach space of all excitation fields acting on the microdomains indexed by a point in the material configuration space (base space). The fiber bundle superspace formalism may then be seen as a highly efficient and economic apparatus available for encoding, storing, and processing a large amount of topological and geometrical data pertinent to the problems of nonlocality in physics and engineering since fiber bundles lend themselves easily to complex calculations. Readable technical descriptions of vector bundles can be found in [30,57,63,70].

Appendix A.5.6. Additional Remarks on the Use of Sobolev Spaces in the Fiber Bundle Superspace Formalism

In Section 4, we introduced Sobolev space over the open domain D instead of simply operating with the more generic Banach space. The reason behind our decision to invoke the more specialized (and technical) structure of a Sobolev space was mainly to actually simplify the technical development and in anticipation of future work on the superspace formalism. Indeed, in this paper, the fiber bundle $\mathcal{M}$ is referred to just as Banach bundle, not Sobolev bundle for the reason that all our essential results and insights apply to the more general concept of Banach space, which contains Sobolev spaces as a special case. In fact, Sobolev spaces are easier to work with in problems involving integro-differential equations such as nonlocal continuum field theory. Nevertheless, we only used the elementary definition of Sobolev space itself in Section 5, not its advanced properties. In particular, none of the other technical properties of Sobolev spaces are needed in the paper. Nevertheless, since in the future the material bundle space $\mathcal{M}$ is expected to be employed in order to construct solutions of Maxwell's equations in new form (i.e., in superspace instead of conventional spacetime), Sobolev spaces are projected to play the most important role since they have proved very efficient in analysis and the theory of partial differential equations [64].

Appendix A.5.7. Partition of Unity Techniques

In analysis and differential topology, the title *the partition of unity lemma* refers to a somehow rather technical tool used by topologists and analysts in order to help propagate information from the local to the global setting. They were found to be quite handy and easy to apply. The main theorem (Lemma 1) permits us to move from one topological "atom" to another by "gluing" them together using smooth standard domain-division functions. The technique was stated and used only toward the end of Section 5 in order to justify expansions, such as (77) and can be skipped in first reading of the paper. Partition of unity is usually taught in all topology and some geometry textbooks, e.g., see [26,57,63,70].

Appendix A.6. The General Electromagnetic Model of Nonlocal (Spatially-Dispersive) Isotropic Domains

One of the simplest—yet still demanding and interesting—nonlocal media is the special case of isotropic, homogeneous, spatially-dispersive, but optically inactive domains [14]. In this case, very general principles force the generic expression of the material response tensor to acquire the concrete form [13,15,16]:

$$\overline{\mathbf{K}}(\mathbf{k},\omega) = K^{\mathrm{T}}(k,\omega)(\bar{\mathbf{I}} - \hat{k}\hat{k}) + K^{\mathrm{L}}(k,\omega)\hat{k}\hat{k}, \tag{A3}$$

where

$$k := |\mathbf{k}|,\ \hat{k} := \mathbf{k}/k, \tag{A4}$$

and $\mathbf{k}$ is the wave vector (spatial-frequency) of the field. The first term in the RHS of (A3) represents the *transverse* part of the response function, while the second term is clearly the *longitudinal* component, with behavior captured by the generic functions $K^{\mathrm{T}}(k,\omega)$ and $K^{\mathrm{L}}(k,\omega)$, respectively.[30] The tensorial forms involving the dyads $\hat{k}\hat{k}$, however, are imposed by the formal requirement of the need to satisfy the Onsager symmetry relations in the absence of external magnetic fields [16]

Using a proper microscopic theory, ultimately quantum theory, it is possible in general to derive fundamental expressions for the transverse and longitudinal components of the response functions in (A3) [14–16,37,38,50]. These forms are often obtained in the following way:

1. First, fundamental theory is deployed to derive analytical expressions for $K^{\mathrm{T}}(k,\omega;\mathbf{r}')$ and $K^{\mathrm{L}}(k,\omega;\mathbf{r}')$.
2. Afterwards, depending on the concrete values of the various physical parameters that enter into these expressions, e.g., frequency, temperature, molecular charge/mass/spin, density, etc., the obtained analytical expressions are expanded in power series with the proper number of terms.
3. The expression of the dielectric tensor function is then put in the form of either a polynomial or rational polynomial in $\mathbf{k}$.

A concrete example is given in Section 7.3 to illustrate the use of such physics-based dielectric functions for the case of exciton–polariton-based semiconductor materials.

Appendix A.7. Origin of Electromagnetic Nonlocality in Excitonic Semiconductors

Appendix A.7.1. Review of the Semiconductor Physics of Excitons

Very early in the history of condensed-matter physics, excitons were introduced by Frenkel [84,85], and further developed by other researchers, such as Wannier [217]. In the late 1950s, excitonic phenomena were transplanted into a central stage in the framework of light-matter interaction through the concept of *exciton–polariton* [178], which will be defined below. Pekar [178], Ginzburg [77], and others [79,119,218,219] affirmed the nonlocal approach to exciton–polariton materials by explicitly highlighting the strong impact of spatial dispersion near excitonic resonances. The subject of excitons is vast and multidisciplinary. For extensive treatments covering various applications in physics, chemistry, and technology, see [16,55,86,220–222].

In order to understand the particular nonlocal model to be presented in Section 7.3, let us first briefly explain the relevant physics of exciton–polariton interactions and why they can lead to strong nonlocal response. In a direct-band gap semiconductor the minimum of the conduction band is aligned along the maximum of the valance band, allowing electronic transitions from lower (unexcited) to excited bands upon interaction with external EM fields. For insulating semiconductors of the II-VI and III-V groups, exciton transitions occur in the visible or near-ultraviolet range of the electromagnetic spectrum. By engineering the material/metamaterial parameters, these transition frequencies can be shifted.

It should be noted that in contrast to metals and plasma, no free charged carriers are assumed to exist in the material. An electron exiting the valance band after the absorption of an external photon will leave behind a *hole*, which acts as an independent quasiparticle that can travel throughout the material in the form of a collective excitation [84,85,223,224]. The *exciton* is defined as a coupled pair composed of the two bound states of the electron and hole. Here, both electrons and holes must be understood as "dressed" particles (quasi-particles) with effective mass and charge different from those of the bare (noninteracting) particle [225]. We may apply the Bohr model to the exciton (electron-hole pair) with simple modifications that can be summarized by the following procedure:

1. The electron mass must be replaced by the reduced exciton mass

$$m_r := \frac{m_{el} m_h}{m_{el} + m_h}, \tag{A5}$$

   where m_{el} and m_h are the electron and hole masses, respectively.
2. The numerical values of m_{el} and m_h are determined by the curvature of the conduction and valance bands, respectively, and, hence, they follow from accurate quantum mechanical calculations of the band structure, see for example [55,226,227].
3. Due to the screening of Coulomb attraction by the dielectric medium, the effective electron charge $e^- = -e$ should be replaced by $e^- / \sqrt{\varepsilon_0}$, where ε_0 is the static dielectric constant.

From this, it follows that the exciton binding energy E_b is given by

$$E_b = \frac{m_r e^4}{2\hbar^2 \varepsilon_0^2}. \tag{A6}$$

Therefore, the total energy needed to create an exciton state is given by

$$\hbar\omega_e = E_g - E_b, \tag{A7}$$

where E_g is the semiconductor band gap energy. In most applications, the binding energy E_b is in the order of meV, while E_g is usually few eV. That is, the energy needed to create an exciton is slightly less than the band gap energy and typically we have

$$E_b \ll E_g. \tag{A8}$$

However, it is recommended to include binding energy in some applications for accurate calculations to help explaining the fine structure of measured excitonic transitions.

Appendix A.7.2. A Simple Explanation of How Nonlocality Emerges in the Excitonic Semiconductor

They key to the origin of nonlocality is the scenario when the excitation photon has an energy $\hbar\omega$ that is *greater* than the minimum exciton energy (A7). In the case where

$$\hbar\omega > \hbar\omega_e, \tag{A9}$$

the excess energy will be transformed into *kinetic energy*. Due to the conservation of momentum, the wave vector of the exciton is equal to the photon wave vector $\mathbf{k}$ and, hence, the exciton kinetic energy E_e^{kinetic} is given by

$$E_e^{\text{kinetic}} = \frac{\hbar\mathbf{k} \cdot \hbar\mathbf{k}}{2m_e}, \tag{A10}$$

where

$$m_e := m_{el} + m_h \tag{A11}$$

is the translational mass of the exciton in the effective-mass approximation [217]. Consequently, the total exciton energy E_e is given by [228]

$$E_e(\mathbf{k}) = \hbar\omega_e(\mathbf{k}) = \hbar\omega_e + \frac{\hbar^2 k^2}{2m_e}. \tag{A12}$$

Consequently, the exciton frequency $\omega_e(\mathbf{k})$ acquires a novel dependence on $\mathbf{k}$, which is mainly due to the kinetic energy term in expression (A12). It is *precisely* such dependence that eventually leads to the emergence of electromagnetic nonlocality in semiconductors around excitonic resonances when photons couple with excitons. In other words, away from the excitonic transition regime, the effective dielectric function of the semiconductor exhibits only the typical dependence on ω (normal or temporal dispersion.)

Appendix A.8. An Alternative Intuitive Derivation of the Dielectric Model (90) *and the Quantum Origin of Nonlocality in Excitonic Semiconductors*

The model (90) itself may be intuitively derived as follows. A generic oscillator model is the one having the following well-attested Lorentzian expression:

$$\frac{1}{\omega_e^2 - \omega^2 - \mathrm{i}\Gamma\omega}. \tag{A13}$$

This Lorentzian form models a large number of physical processes in nature, from lattice vibrations to electronic transitions and numerous many others [13,15,37,229]. Substituting the wave vector-dependent ω_e expression (A12) into the above Lorentzian form (A13), the dielectric function formula (90) can be immediately obtained when we keep only quadratic terms of $\mathbf{k}$. For a more careful quantum mechanical derivation, see [16,54,86].

Appendix A.9. Computation of the Inverse Fourier Transform (114)

We start from the standard Fourier transform pair

$$\mathcal{F}_{\mathbf{k}}^{-1}\left\{\frac{\chi}{k^2 - \gamma^2(\omega)}\right\} = \frac{\chi}{4\pi}\frac{e^{-\mathrm{i}\gamma(\omega)|\mathbf{r}-\mathbf{r}'|}}{|\mathbf{r}-\mathbf{r}'|}, \quad \mathrm{Im}\{\gamma(\omega)\} < 0, \tag{A14}$$

where the spatial Fourier transform is defined by (17). The condition

$$\mathrm{Im}\{\gamma(\omega)\} < 0 \tag{A15}$$

is due to the physical requirement that fields do not grow exponentially in passive domains [13]. We also have written $|\mathbf{r}-\mathbf{r}'|$ instead of $|\mathbf{r}|$ in anticipation of the fact that the inverse Fourier transform will produce a Green function.

Our main task now is to make a proper choice of the correct sign when performing the square root operation $\sqrt{\gamma^2}$. Let us write

$$\gamma^2 = \mathrm{Re}\left\{\gamma^2\right\} + \mathrm{i}\,\mathrm{Im}\left\{\gamma^2\right\}. \tag{A16}$$

From (91), we have

$$\mathrm{Re}\left\{\gamma^2\right\} = \frac{m_e^\star}{\hbar\omega_e}\left(\omega^2 - \omega_e^2\right), \quad \mathrm{Im}\left\{\gamma^2\right\} = \frac{m_e^\star}{\hbar\omega_e}\omega\Gamma. \tag{A17}$$

On the other hand, γ also can take the form

$$\gamma(\omega) = \gamma'(\omega) + \mathrm{i}\gamma''(\omega), \tag{A18}$$

where both γ' and γ'' are real. The goal now is to derive expressions for γ' and γ'' in terms of $\mathrm{Re}\{\gamma^2\}$ and $\mathrm{Im}\{\gamma^2\}$ with the correct sign since the square root is a many-one function.

To accomplish this, we use the following elementary theorem: let $x, y, a, b \in \mathbb{R}$. Then the square root of $x + iy$ is given by

$$\sqrt{x + iy} = \pm(a + ib), \tag{A19}$$

where the following expressions hold

$$a = \sqrt{\frac{x + \sqrt{x^2 + y^2}}{2}}, \quad b = \frac{y}{|y|}\sqrt{\frac{-x + \sqrt{x^2 + y^2}}{2}}. \tag{A20}$$

Substituting (A17) into (A20), the following is obtained

$$a = \sqrt{\frac{m_e^\star}{2\hbar\omega_e}}\sqrt{(\omega^2 - \omega_e^2) + \sqrt{(\omega^2 - \omega_e^2)^2 + (\omega\Gamma)^2}}, \tag{A21}$$

$$b = \sqrt{\frac{m_e^\star}{2\hbar\omega_e}}\sqrt{-(\omega^2 - \omega_e^2) + \sqrt{(\omega^2 - \omega_e^2)^2 + (\omega\Gamma)^2}}. \tag{A22}$$

Here, we used the calculation

$$y/|y| = \mathrm{Im}\left\{\gamma^2\right\}/|\mathrm{Im}\left\{\gamma^2\right\}| = \omega\Gamma/|\omega\Gamma| = 1, \tag{A23}$$

which follows from the fact that $\omega, \Gamma > 0$.

It remains now to find the correct signs. From (A14), the condition

$$\gamma'' = \mathrm{Im}\{\gamma(\omega)\} < 0 \tag{A24}$$

must be satisfied. Therefore, we choose the *negative* sign in (A19). The final expressions become $\gamma = -a, \gamma'' = -b$, and after inserting γ' and γ'' into (A14), the required relation (114) is obtained.

Appendix A.10. On Extending Definition 1 to Noncompact Regions

The localization of the physics-based nonlocality microdomains estimated using the formula (121) is based on approximating the exact mathematical definition of the topological microdomain (Definition 1) by response kernel functions possessing spatial decaying exponential behavior as in (114). It might be advisable then to provide a modification of Definition 1 taking into account the noncompact setting, which is the scenario more often encountered in physical applications.

Definition A1 (**Nonlocal Microdomains: The Noncompact Case**)**.** Consider a material domain D with the associated nonlocal response function $\overline{\mathbf{K}}(\mathbf{r}', \mathbf{r})$. Assume further that the following bound holds:

$$||\overline{\mathbf{K}}(\mathbf{r}', \mathbf{r})|| \leq A\exp(-a|\mathbf{r} - \mathbf{r}'|), \tag{A25}$$

for all $\mathbf{r}' \in D$, where a and A are some positive numbers. (Here, $||\cdot||$ is as in Definition 1). The form (A25) is called the *exponentially-decaying nonlocal kernel response function*. We define the effective (physics-based) nonlocal microdomain $V_{\mathbf{r}} \subset D$, labeled by $\mathbf{r} \in D$, as the open ball

$$\{\mathbf{r}' \in D, \ ||\overline{\mathbf{K}}(\mathbf{r}', \mathbf{r})|| < A\exp(-1)\}, \tag{A26}$$

where A is as defined in (A25). In other words, an effective onlocality microdomain of this type, such as the one in Definition 1, is still locally compact.

Remark A1. The physical meaning of the ball (A26) is that it effectively approximate the spatial region where most of the "energy" of the response is concentrated, hence providing a physical means to estimate practical microdomain physical problems, since the Coulomb interaction and other types of molecular forces are long range forces and, hence, cannot be described by an exact compact support such as the one originally introduced in the Definition 1.

Remark A2. It is straightforward to modify Definition A1 to include other forms of decay functions weaker than the exponential form. For example, we may replace the exponential in (A25) by a decaying function $1/r^n$, where n is a suitable positive integer. Obvious modifications of the ball (A26) can then be subsequently made to construct the nonlocality microdomain.

Appendix A.11. Possible Applications of the Superspace Formalism to Fundamental Methods in Metamaterials Research

Appendix A.11.1. Estimating Fundamental Limitations on Nonlocal Metamaterials

Fundamental continuum response maps, such as $\mathcal{L}$ (72), can be completely reformulated in a different setting, that of the space of *vector bundle sections* [57,62,70]. The latter topic, the theory of sections, is an extremely well-developed subject in mainstream differential topology. In fact, in some cases the electromagnetic response field function $\mathbf{R}(\mathbf{r})$ itself may be obtained by working directly with the source bundle superspace $\mathcal{M}$. For example, under some conditions, this can be achieved by replacing each fiber X_i by $X_i \times \mathbb{C}^3$. In this way, the entire nonlocal response problem reduces to understanding how vector bundle sections interact with the topology of the underlying base manifold $\mathcal{D}$. There is a large literature in differential topology and geometry focused on this latter technical mathematical problem, especially how local information can be transported from one place to another in order to extend local structures into global ones [26,57,63].

The author believes that by starting from local data in a given nonlocal metamaterial domain, e.g., the global shape of the device, the distribution of topological holes, etc, one may then use existing techniques borrowed from differential topology, e.g., the theory of characteristic classes, to determine allowable EM response functions that are in principle permissible at the global level. Engineers are typically interested in acquiring in advance the knowledge of what the best (or worst) performance measures obtainable from specific topologies are. Hence, reformulating the electromagnetism of nonlocal metamaterials in terms of vector bundles could be of help in this respect since it opens a pathway, within metamaterials research, toward a synergy between general topology, physics, and engineering.

Appendix A.11.2. Numerical Methods

Traditional full-wave numerical methods are sometimes deployed in order to deal with nonlocal EM materials, often using the additional boundary conditions framework, in spite of the latter's lack of complete generality.[31] At the heart of the traditional approach to numerical methods in local electromagnetism is the concept of operators between linear spaces. However, by reformulating the source space of field–matter interaction in terms of a Banach bundle, it should be possible to reformulate Maxwell's equations to act on this extended geometric superspace instead of the conventional spacetime framework. As an alternative to the concept of the linear operator of classical mathematical physics and numerical methods, we now have the much more general and richer concept of bundle homomorphism developed in Section 5. Some of the advantages anticipated from such reformulation include

1. The ability to resolve the issue of generalized boundary condition (already discussed in Section 8).
2. Since every point belonging to a fiber superspace is in itself a smooth function defined on an entire material sub-microdomain, by building a new system of discretized recursive equations approximating the behavior of electromagnetic solutions living in the enlarged superspaces $\mathcal{M}$ and $\mathcal{R}$ one may anticipate arriving at a deeper understanding of the physics of nonlocality. The reason is that the topology of the nonlocal

interaction regime is explicitly encoded into the *geometry* of the new expanded solution superspace $\mathcal{M}$ itself. Characterizing this geometry is then possible through a suitable discrete approximation of the *interior* microtopological content of the superspace (fiber bundle) structure itself; i.e., not just at the "exterior" parts often found in the boundary conditions of classical local field continuum theories, but also "going inside" the problem space as such.

3. It is also possible that such numerical methods may emerge as more computationally efficient and broader in applicability than the conventional methods rooted in local electromagnetism. One reason for this is that the Banach vector bundle formulation introduced in this paper is quite natural and appears to reflect the underlying physics of nonlocal metamaterials in a direct manner. From our general experience in numerical methods, "natural operations" tend to translate into numerical methods with better convergence, sensitivity, and robustness.

As directly related to the three possible advantages of the superspace formalism discussed above, we also add that in recent years the subject of *computational topology* has gained momentum, where some researchers are now building new numerical methods by exploiting the topological structure of the problems under considerations, e.g., see [72,230]. The fiber bundle superspace formalism of this paper might provide a way to link research done in electromagnetic and non-electromagnetic nonlocal materials with such advances in the computational and applied mathematical sciences.

Appendix A.11.3. Topological Photonics

One of the main applications of the proposed vector bundle formalism is that it opens the door for a new way to investigate the topological structure of materials. It has already been noted that the nonlocal EM response is essential in topological photonics, e.g., see [11,161] and also Appendix A.1. Indeed, since in topological photonics the wave function of bosons, usually the Bloch state, is examined over the entirety of momentum space (usually the Brillouin zone), then it is the dependence of the EM response on $\mathbf{k}$ what is at stake, which naturally brings in nonlocal issues. But now since by using our theory we can associate with every nonlocal material a concrete fiber bundle superspace reflecting the rich information about the multiscale topological microdomain structure and the global shape of the material plus the impact of the boundaries separating various material domains, it is natural to examine whether a topological classification of the corresponding fiber bundles may lead to a new way to characterize the topology of materials other than the Chern invariants used extensively in literature. The advantage of the superspace approach in this case is that the complicated topological and geometrical aspects of the boundaries and inhomogeneity in nonlocal media can be encoded very efficiently in the local structure of the material fiber bundle. Using standard techniques in differential topology [62], it should be possible to propagate this local information to the global domain (the entirety of the system), for example by computing suitable fiber bundle topological invariants like its homology groups [70]. Our approach is then a "dual" to the standard approach since we work on an enlarged configuration space (spacetime or space–frequency), while the mainstream approach operates in the momentum space of the wave function.

Notes

1 The author would like to thank one of the anonymous reviewers for suggesting this connection.

2 This is argued in detail in [32]. In particular, the recently-introduced current Green's function of electromagnetic devices was inspired by finding a Green's function structure similar to that corresponding to nonlocal media [33–36].

3 See Appendixes A.1 and A.2 for the literature review.

4 For a brief discussion of some possible engineering applications of metamaterials, see Section A.3.

5 Cf. Section 3.3.

6 If D asymptotically grows into an unbounded region, then the problem reduces to that of homogeneous unbounded domain (bulk media), well treated in the basic literature on spatial dispersion. Clearly, in this paper, we are not interested in such a topologically trivial problem.

7 Cf. Remark 3.

8 This is well known from the quantum theory of nonlocality [16,38,50], but concrete examples illustrating this behavior will be given in Section 7.

9 This compactness of the response kernel's support cannot be proved in general, but is very plausible on physical grounds (causality considerations). Therefore, we posit such compactness as an axiomatic feature of all physically-realizable causal nonlocal continuum field theories. However, see Appendix A.10.

10 For some possible definitions of matrix norms, see for example [60,61].

11 In this section and the one to follow, we do not worry much about the details of the electromagnetic model and for simplicity assume that only one vector field $\mathbf{F}$ acts as excitation and one response field $\mathbf{R}$ is induced. More complex media like bi-anisotropic domains and others [67] may also be treated within this formulation. For example, if two response fields are needed, the codomain in (33) can be simply changed to $\mathbb{C}^6$.

12 Cf. Remark 3.

13 See the discussion of nonlocal and topological metamaterials applications in Appendix A.3.

14 Cf. Appendix A.1.

15 See [59,68] for the full technical definition of *subordinated cover*. A collection of subsets of a topological space is said to be *locally finite*, if each point in the space has a neighborhood that intersects only finitely many of the sets in the collection. What we need here is that there exists some i and $\mathbf{r}$ such that U_i is inside $V_{\mathbf{r}}$, i.e., $U_i \subseteq V_{\mathbf{r}}$ where $\mathbf{r} \in U_i$.

16 The function space C^p is comprised of the set of real functions that are continuously differentiable p-times [62,69].

17 See Section 7 for one possible method and examples.

18 Cf. Remark 3.

19 The discretization of the nonlocal MTM bundle homomorphism itself is outside the scope of the present work and will be addressed elsewhere.

20 On the technical difference between local and global topological isomorphisms, especially in differential topology, see [62,73,74].

21 For instance, by introducing holes into a simply-connected domain in order to make the latter disconnected.

22 The numerical value of α may be different for transverse and longitudinal excitation fields.

23 In the effective-mass approximation, a simple way to estimate the exciton mass $m_e^\star$ is via the relation $m_e^\star = m_{el} + m_h$, i.e., the sum of the effective electron and hole masses introduced in Appendix A.7. However, it must be noted that this relation is far from being universal, e.g., it should be modified when there are strong interactions [81,82]

24 In typical crystal materials, the ratio $\omega_{L,T}/\omega_T$ is already about 10^{-3} [16,55].

25 It should be noted that there is no loss of generality here. The computational model to be presented shortly allows the estimation of the nonlocal microdomain topology based on a generic model of the form (90). If ε^L and ε^T are not identical, then the same procedure can be applied to each one of them separately.

26 There is a nice parallelism here with *temporal* dispersion where the latter is known to arise from the inertial effects of electrons in interaction with radiation fields [37].

27 This is more obvious in FEM or FDTD than MoM.

28 Cf. Section 5.2.

29 Cf. Section 8.

30 Even for isotropic materials, the response tensor $\overline{\mathbf{K}}$ remains a tensor. This is due to the manner in which the equivalent dielectric function is defined using the Fourier transform instead of the conventional multipole approach, e.g., see [16,32,37].

31 Cf. Section 8 and Appendix A.1.

References

1. Schwarz, A.S. *Topology for Physicists*; Springer: Berlin, Germany; New York, NY, USA, 1994.
2. Monastyrskiĭ, M.I. *Riemann, Topology and Physics*; Birkháuser: Boston, MA, USA, 1999.
3. Penrose, R. *Techniques of Differential Topology in Relativity*; Society for Industrial and Applied Mathematics: Philadelphia, PA, USA, 1972.
4. Mikki, S. Topological circuit theory: A Lie group perspective. *Prog. Electromagn. Res. B* **2020**, *89*, 133–156. [CrossRef]
5. Weyl, H. *The Theory of Groups and Quantum Mechanics*; Martino Publishing: Mansfield Centre, CT, USA, 2014.
6. Poincaré , H. *New Methods of Celestial Mechanics*; American Institute of Physics: Woodbury, NY, USA, 1993.
7. Poincare, H. Mémoire sur les courbes définies par une équation différentielle I-IV. *J. Mathématiques Pures Appliqués* **1881**, *7*, 375–422.
8. Eringen, A. *Nonlocal Continuum Field Theories*; Springer: New York, NY, USA, 2002.

9. Fabrizio, M.; Morro, A. *Electromagnetism of Continuous Media: Mathematical Modelling and Applications*; Oxford University Press: Oxford, UK, 2003.
10. Hasan, M.Z.; Kane, C.L. Colloquium: Topological insulators. *Rev. Mod. Phys.* **2010**, *82*, 3045–3067. [CrossRef]
11. Ozawa, T.; Price, H.M.; Amo, A.; Goldman, N.; Hafezi, M.; Lu, L.; Rechtsman, M.C.; Schuster, D.; Simon, J.; Zilberberg, O.; et al. Topological photonics. *Rev. Mod. Phys.* **2019**, *91*, 015006. [CrossRef]
12. Schwinger , J.; Deraad, L.L.; Milton, K.A.; Tsai, W.Y. *Classical Electrodynamics*; Perseus Books: New York, NY, USA, 1998.
13. Landau, L.D.; Lifshitz, E. *Electrodynamics of Continuous Media*; Butterworth-Heinemann: Oxford, UK, 1984.
14. Ginzburg, V.L. *The Propagation of Electromagnetic Waves in Plasmas*; Pergamon Press: Oxford, UK; New York, NY, USA, 1970.
15. Ginzburg, V.L. *Theoretical Physics and Astrophysics*; Pergamon Press: Oxford, UK; New York, NY, USA, 1979.
16. Agranovich, V.; Ginzburg, V. *Crystal Optics with Spatial Dispersion, and Excitons*; Springer: Berlin/Heidelberg, Germany, 1984.
17. Deppman, A.; Megías, E.; P. Menezes, D. Fractal Structures of Yang–Mills Fields and Non-Extensive Statistics: Applications to High Energy Physics. *Physics* **2020**, *2*, 455–480. [CrossRef]
18. Auyang, S. *How Is Quantum Field Theory Possible*; Oxford University Press: New York, NY, USA, 1995.
19. Coleman, S. *Quantum Field Theory: Lectures of Sidney Coleman*; World Scientific: Singapore, 2019.
20. Espagnat, B. *On Physics and philosophy*; Princeton University Press: Princeton, NJ, USA, 2013.
21. Frisch, M. Non-Locality in Classical Electrodynamics. *Br. J. Philos. Sci.* **2002**, *53*, 1–19. [CrossRef]
22. Bernabeu, J.; Navarro-Salas, J. A Non-Local Action for Electrodynamics: Duality Symmetry and the Aharonov-Bohm Effect, Revisited. *Symmetry* **2019**, *11*, 1191. [CrossRef]
23. Mashhoon, B. Toward Nonlocal Electrodynamics of Accelerated Systems. *Universe* **2020**, *6*, 229. [CrossRef]
24. Cvijanovich, G.B. Classical electrodynamics with nonlocal constitutive equations. *Found. Phys.* **1977**, *7*, 785–799. [CrossRef]
25. Mikki, S. Proca Metamaterials, Massive Electromagnetism, and Spatial Dispersion. *Ann. Der Phys.* **2021**, *533*, 2000625. [CrossRef]
26. Lang, S. *Introduction to Differentiable Manifolds*; Interscience: New York, NY, USA, 1962.
27. Colton, D.; Kress, R. *Inverse Acoustic and Electromagnetic Scattering Theory*; Springer: Cham, Switzerland, 2019.
28. Chew, W.C. *Waves and Fields in Inhomogenous Media*; Wiley-IEEE: Piscataway, NJ , USA, 1999.
29. Nedelec, J.C. *Acoustic and Electromagnetic Equations: Integral Representations for Harmonic Problems*; Springer: New York, NY, USA, 2001.
30. Zeidler, E. *Quantum Field Theory I: Basics in Mathematics and Physics*; Springer: New York, NY, USA, 2009.
31. Hassani, S. *Mathematical Physics: A Modern Introduction to Its Foundations*; Springer: Cham, Switzerland, 2013.
32. Mikki, S.; Antar, Y. *New Foundations for Applied Electromagnetics: The Spatial Structure of Fields*; Artech House: London, UK, 2016.
33. Mikki, S.; Antar, Y. The antenna current Green's function formalism–Part I. *IEEE Trans. Antennas Propagat* **2013**, *9*, 4493–4504. [CrossRef]
34. Mikki, S.; Antar, Y. The antenna current Green's function formalism—Part II. *IEEE Trans. Antennas Propagat* **2013**, *9*, 4505–4519. [CrossRef]
35. Mikki, S. The Antenna Spacetime System Theory of Wireless Communications. *Proc. R. Soc. Math. Phys. Eng. Sci.* **2019**, 475, 2224. [CrossRef]
36. Mikki, S.; Antar, Y. A rigorous approach to mutual coupling in general antenna systems through perturbation theory. *IEEE Antennas Wirel. Commun. Lett.* **2015**, *14*, 115–118. [CrossRef]
37. Ilinskii, Y.A.; Keldysh, L. *Electromagnetic Response of Material Media*; Springer Science Business Media: New York, NY, USA, 1994.
38. Keller, O. *Quantum Theory of Near-Field Electrodynamics*; Springer: Berlin, Germany; New York, NY, USA, 2011.
39. Melrose, D.B.; McPhedran, R.C. *Electromagnetic Processes in Dispersive Media: A Treatment Based on the Dielectric Tensor*; Cambridge University Press: Cambridge, UK; New York, NY, USA, 1991.
[illegible] L. *Radiation and Scattering of Waves*; IEEE Press: Piscataway, NJ, USA, 1994.
41. Geyi, W. *Foundations of Applied Electrodynamics*; Wiley: Hoboken, NJ, USA, 2010.
42. Melrose, D.B. *Instabilities in Space and Laboratory Plasmas*; Cambridge University Press: Cambridge, UK; New York, NY, USA, 1986.
43. Jackson, J. *Classical Electrodynamics*; Wiley: New York, NY, USA, 1999.
44. Jin, J. *The Finite Element Method in Electromagnetics*; John Wiley & Sons Inc: Hoboken, NJ, USA, 2014.
45. Taflove, A. *Computational Electrodynamics: The Finite-Difference Time-Domain Method*; Artech House: Boston, MA, USA, 2005.
46. Gibson, W.C. *The Method of Moments in Electromagnetics*; CRC Press: Boca Raton, FL, USA, 2015.
47. Mikki, S.; Kishk, A. Nonlocal Electromagnetic Media: A Paradigm for Material Engineering. In *Passive Microwave Components and Antennas*; InTech: Nappanee, IN, USA, 2010.
48. Agranovich, V.M.; Shen, Y.R.; Baughman, R.H.; Zakhidov, A.A. Linear and nonlinear wave propagation in negative refraction metamaterials. *Phys. Rev. B* **2004**, *69*, 165112. [CrossRef]
49. Agranovich, V.M.; Gartstein, Y.N.; Zakhidov, A.A. Negative refraction in gyrotropic media. *Phys. Rev. B* **2006**, *73*, 045114. [CrossRef]
50. Cho, K. *Optical Response of Nanostructures: Microscopic Nonlocal Theory*; Springer: Berlin, Germany; New York, NY, USA, 2003.
51. Cho, K. A single susceptibility scheme of macroscopic Maxwell equations: Beyond the 'E,D,B,H' approach. *J. Phys. Condens. Matter* **2008**, *20*, 175202. [CrossRef]
52. Mikki, S.; Kishk, A. Electromagnetic Wave Propagation in Nonlocal Media: Negative Group Velocity and Beyond. *Prog. Electromagn. Res. B* **2009**, *14*, 149–174. [CrossRef]

53. Cho, K. *Reconstruction of Macroscopic Maxwell Equations: A Single Susceptibility Theory*; Springer: Berlin, Germany, 2018.
54. Toyozawa, Y. *Optical Processes in Solids*; Cambridge University Press: Cambridge, UK; New York, NY, USA, 2003.
55. Pekar, S.I. *Crystal Optics and Additional Light Waves*; Benjamin/Cummings Pub. Co: Menlo Park, CA, USA, 1983.
56. Churchill, R.J.; Philbin, T.G. Electromagnetic reflection, transmission, and energy density at boundaries of nonlocal media. *Phys. Rev. B* **2016**, *94*, 235422. [CrossRef]
57. Lee, J. *Introduction to Smooth Manifolds*; Springer: New York, NY, USA; London, UK, 2012.
58. Hausdorff, F. *Set Theory*; American Mathematical Society: Providence, RI, USA, 2005.
59. Kelley, J. *General Topology*; Dover Publications, Inc.: New York, NY, USA, 2017.
60. Baker, A. *Matrix Groups: An Introduction to Lie Group THEORY*; Springer: London, UK; New York, NY, USA, 2002.
61. Horn, R.; Johnson, C.R. *Matrix Analysis*; Cambridge University Press: Cambridge, UK; New York, NY, USA, 2012.
62. Hirsch, M. *Differential Topology*; Springer: New York, NY, USA, 1976.
63. Dundas, B.I. *A Short Course in Differential Topology*; Cambridge University Press: Cambridge, UK; New York, NY, USA, 2018.
64. Showalter, R.E. *Hilbert space methods in partial differential equations*; Dover Publications: Mineola, NY, USA, 2010.
65. Zeidler, E. *Applied Functional Analysis: Applications to Mathematical Physics*; Springer: Berlin/Heidelberg, Germany, 1995.
66. Hebey, E. *Nonlinear Analysis on Manifolds: Sobolev Spaces and Inequalities*; Courant Institute of Mathematical Sciences American Mathematical Society: New York, NY, USA, 2000.
67. Mackay, T.; Lakhtakia, A. *Electromagnetic Anisotropy and Bianisotropy: A Field Guide*; World Scientific Publishing Co. Pte. Ltd: Singapore; Hackensack, NJ, USA, 2019.
68. Mukherjee, A. *Differential Topology*; Birkháuser: Cham, Switzerland, 2015.
69. Appel, W. *Mathematics for Physics and Physicists*; Princeton University Press: Princeton, NJ, USA, 2007.
70. Kahn, D. *Introduction to Global Analysis*; Dover Publications: Mineola, NY, USA, 2007.
71. Nestruev, J. *Smooth Manifolds and Observables*; Springer: Cham, Switzerland, 2020.
72. Edelsbrunner, H. *Computational Topology: An Introduction*; American Mathematical Society: Providence, RI, USA, 2010.
73. Guillemin, V.; Pollack, A. *Differential Topology*; AMS Chelsea Pub.: Providence, RI, USA, 2010.
74. Majthay, A. *Foundations of Catastrophe Theory*; Pitman Advanced Pub. Program: Boston, MA, USA, 1985.
75. Strang, G.; Fix, G. *An Analysis of the Finite Element Method*; Wellesley-Cambridge Press: Wellesley, MA, USA, 2008.
76. Kushnir, O.S. Spatial dispersion in incommensurately modulated insulators. *J. Physics: Condens. Matter* **2004**, *16*, 1245–1267. [CrossRef]
77. Ginzburg, V. Electromagnetic Waves in Isotropic and Crystalline Media Characterized by Dielectric Permittivity with Spatial Dispersion. *JETP* **1958**, *7*, 1096.
78. Hopfield, J.J. Theory of the Contribution of Excitons to the Complex Dielectric Constant of Crystals. *Phys. Rev.* **1958**, *112*, 1555–1567. [CrossRef]
79. Hopfield, J.J.; Thomas, D.G. Theoretical and Experimental Effects of Spatial Dispersion on the Optical Properties of Crystals. *Phys. Rev.* **1963**, *132*, 563–572. [CrossRef]
80. Forstmann, F. Optics in spatially dispersive media: The energy theorem and the boundary conditions. *Z. Phys. B Condens. Matter Quanta* **1979**, *32*, 385–393. [CrossRef]
81. Mattis, D.C.; Gallinar, J.P. What is the Mass of an Exciton? *Phys. Rev. Lett.* **1984**, *53*, 1391–1393. [CrossRef]
82. Davydov, A.; Enolskii, V. Effective mass of the Pekar polaron. *JETP* **1988**, *67*, 313.
83. Halevi, P. *Spatial Dispersion in Solids and Plasmas*; North-Holland: Amsterdam, The Netherlands; New York, NY, USA, 1992.
84. Frenkel, J. On the Transformation of light into Heat in Solids—Part I. *Phys. Rev.* **1931**, *37*, 17–44. [CrossRef]
85. Frenkel, J. On the Transformation of Light into Heat in Solids—Part II. *Phys. Rev.* **1931**, *37*, 1276–1294. [CrossRef]
86. Davydov, A.S. *Theory of Molecular Excitons*; Plenum Press: New York, NY, USA, 1971.
87. Balakin, A.A.; Gospodchikov, E.D. Operator formalism for permittivity tensor in smoothly inhomogeneous media with spatial dispersion. *J. Phys. B At. Mol. Opt. Phys.* **2015**, *48*, 215701. [CrossRef]
88. Balakin, A.A.; Balakina, M.A.; Permitin, G.V.; Smirnov, A.I. Wave beams in inhomogeneous anisotropic and gyrotropic media. *Radiophys. Quantum Electron.* **2007**, *50*, 955–972. [CrossRef]
89. Mikki, S.; Kishk, A. A Symmetry-Based Formalism for the Electrodynamics of Nanotubes. *Prog. Electromagn. Res.* **2008**, *86*, 111–134. [CrossRef]
90. Mikki, S.M.; Kishk, A.A. Exact derivation of the Dyadic Green's functions of carbon nanotubes using microscopic theory. In Proceedings of the 2007 IEEE Antennas and Propagation Society International Symposium, Honolulu, HI, USA, 9–15 June 2007; pp. 4332–4335. [CrossRef]
91. Mikki, S.M.; Kishk, A.A. Electrodynamics of low-dimensional structures: Lattice dynamics formalism for carbon nanotubes. In Proceedings of the 2008 IEEE Antennas and Propagation Society International Symposium, San Diego, CA, USA, 5–11 July 2008; pp. 1–4. [CrossRef]
92. Mikki, S.; Kishk, A. Derivation of the Carbon Nanotube Susceptibility Tensor Using Lattice Dynamics Formalism. *Prog. Electromagn. Res. B* **2008**, *9*, 1–26. [CrossRef]
93. Kushnir, O.S.; Lokot, L.O. Features of the optical response of dielectric crystals with incommensurate phases. *Phys. Solid State* **2001**, *43*, 818–822. [CrossRef]
94. Agranovich, V.M.; Gartstein, Y. Spatial dispersion and negative refraction of light. *Physics-Uspekhi* **2006**, *49*, 1029. [CrossRef]

95. Qi, G. Optical Beams in Media with Spatial Dispersion. *Chin. Phys. Lett.* **2002**, *20*, 64–67. [CrossRef]
96. Altland, A.; Simmons, B. *Condensed Matter Field Theory*; Cambridge University Press: Leiden, The Netherlands, 2010.
97. Morro, A. Optical activity and spatial dispersion. *Phys. Rev. E* **1997**, *56*, 1124–1128. [CrossRef]
98. Agranovich, V.; Yudson, V. On phenomenological electrodynamics of gyrotropic media. *Opt. Commun.* **1973**, *9*, 58–60. [CrossRef]
99. Trigger, S.A. Equilibrium radiation in a plasma medium with spatial and frequency dispersion. *Phys. Scr.* **2020**, *95*, 075504. [CrossRef]
100. Esquivel-Sirvent, R.; Villarreal, C.; Mochán, W.L.; Contreras-Reyes, A.M.; Svetovoy, V.B. Spatial dispersion in Casimir forces: A brief review. *J. Phys. A Math. Gen.* **2006**, *39*, 6323–6331. [CrossRef]
101. Leung, P.T.; Chang, R. Reciprocity in nonlocal nano-optics. *J. Opt. A Pure Appl. Opt.* **2008**, *10*, 075201. [CrossRef]
102. Xie, H.Y.; Leung, P.T.; Tsai, D.P. General proof of optical reciprocity for nonlocal electrodynamics. *J. Phys. A Math. Theor.* **2008**, *42*, 045402. [CrossRef]
103. Xie, H.Y.; Leung, P.T.; Tsai, D.P. Reciprocity theorem for nonlocal optics: Completion of proof and application to spectroscopic analysis. *J. Opt.* **2010**, *12*, 035006. [CrossRef]
104. Bareev, D.D.; Gavrilenko, V.G. On the Lorentz lemma for media with spatial dispersion and its application to calculation of the radiation fields of given sources. *Radiophys. Quantum Electron.* **2009**, *52*, 735–739. [CrossRef]
105. Tokman, M.; Westerhof, E.; Gavrilova, M. Wave power balance in resonant dissipative media with spatial and temporal dispersion. *Nucl. Fusion* **2003**, *43*, 1295–1304. [CrossRef]
106. Amooshahi, M.; Kheirandish, F. Electromagnetic field quantization in an anisotropic magnetodielectric medium with spatial–temporal dispersion. *J. Phys. A Math. Theor.* **2008**, *41*, 275402. [CrossRef]
107. Buhmann, S.Y.; Butcher, D.T.; Scheel, S. Macroscopic quantum electrodynamics in nonlocal and nonreciprocal media. *New J. Phys.* **2012**, *14*, 083034. [CrossRef]
108. Horsley, S.A.R.; Philbin, T.G. Canonical quantization of electromagnetism in spatially dispersive media. *New J. Phys.* **2014**, *16*, 013030. [CrossRef]
109. Nastyshyn, S.Y.; Bolesta, I.M.; Tsybulia, S.A.; Lychkovskyy, E.; Fedorovych, Z.Y.; Khaustov, D.Y.; Ryzhov, Y.; Vankevych, P.I.; Nastishin, Y.A. Optical spatial dispersion in terms of Jones calculus. *Phys. Rev. A* **2019**, *100*, 013806. [CrossRef]
110. Raza, S.; Bozhevolnyi, S.I.; Wubs, M.; Mortensen, N.A. Nonlocal optical response in metallic nanostructures. *J. Phys. Condens. Matter* **2015**, *27*, 183204. [CrossRef]
111. Adolph, B.; Gavrilenko, V.I.; Tenelsen, K.; Bechstedt, F.; Del Sole, R. Nonlocality and many-body effects in the optical properties of semiconductors. *Phys. Rev. B* **1996**, *53*, 9797–9808. [CrossRef]
112. Dommermuth, M.; Schopohl, N. On the theory of light propagation in crystalline dielectrics. *J. Phys. Commun.* **2018**, *2*, 075012. [CrossRef]
113. Brillouin, L. *Wave Propagation in Periodic Structures, Electric Filters and Crystal Lattices*; Dover Publications: New York, NY, USA, 1953.
114. Mikki, S.; Kishk, A. An Efficient Algorithm for the Analysis and Design of Carbon Nanotube Photonic Crystals. *Prog. Electromagn. Res. C* **2018**, *83*, 83–96. [CrossRef]
115. Mechelen, T.V.; Jacob, Z. Nonlocal topological electromagnetic phases of matter. *Phys. Rev. B* **2019**, *99*, 205146. [CrossRef]
116. Born, M.; Huang, K. *Dynamical theory of crystal lattices*; Clarendon Press Oxford University Press: Oxford, UK; New York, NY, USA, 1988.
117. Silveirinha, M.G. Additional boundary conditions for nonconnected wire media. *New J. Phys.* **2009**, *11*, 113016. [CrossRef]
118. Agarwal, G.S.; Pattanayak, D.N.; Wolf, E. Electromagnetic fields in spatially dispersive media. *Phys. Rev. B* **1974**, *10*, 1447–1475. [CrossRef]
119. Maradudin, A.A.; Mills, D.L. Effect of Spatial Dispersion on the Properties of a Semi-Infinite Dielectric. *Phys. Rev. B* **1973**, *7*, 2787–2810. [CrossRef]
120. Fang, M.; Huang, Z.X.; Sha, W.E.I.; Xiong, X.Y.Z.; Wu, X.L. Full hydrodynamic model of nonlinear electromagnetic response in metallic metamaterials (Invited Paper). *Prog. Electromagn. Res.* **2016**, *157*, 63–78. [CrossRef]
121. Galynsky, V.M.; Furs, A.N.; Barkovsky, L.M. Integral formalism for surface electromagnetic waves in bianisotropic media. *J. Phys. Math. Gen.* **2004**, *37*, 5083–5096. [CrossRef]
122. Forestiere, C.; Capretti, A.; Miano, G. Surface integral method for second harmonic generation in metal nanoparticles including both local-surface and nonlocal-bulk sources. *J. Opt. Soc. Am. B* **2013**, *30*, 2355–2364. [CrossRef]
123. Mnasri, K.; Khrabustovskyi, A.; Plum, M.; Rockstuhl, C. Retrieving effective material parameters of metamaterials characterized by nonlocal constitutive relations. *Phys. Rev. B* **2019**, *99*, 035442. [CrossRef]
124. Vallejos, P.; Johnson, T.; Hellsten, T. An iterative method to include spatial dispersion for waves in nonuniform plasmas using wavelet decomposition. *J. Phys. Conf. Ser.* **2016**, *775*, 012016. [CrossRef]
125. Vallejos, P.; Hellsten, T.; Jonsson, T. A numerical tool based on FEM and wavelets to account for spatial dispersion in ICRH simulations. *J. Phys. Conf. Ser.* **2018**, *1125*, 012020. [CrossRef]
126. JiSheng, C.; JiaRong, L. The Boundary Value Problem and the Nonlinear Spatial Dispersion in Collisionless Quark-Gluon Plasma. *Commun. Theor. Phys.* **2000**, *34*, 251–260. [CrossRef]
127. Musienko, T.; Rudakov, V.; Solov'ev, L. On the application of Kramers-Kronig relations to media with spatial dispersion. *J. Phys. Condens. Matter* **1989**, *1*, 6745–6753. [CrossRef]

128. Goffi, F.Z.; Mnasri, K.; Plum, M.; Rockstuhl, C.; Khrabustovskyi, A. Towards more general constitutive relations for metamaterials: A checklist for consistent formulations. *Phys. Rev. B* **2020**, *101*, 195411. [CrossRef]
129. Mikki, S.; Antar, Y. On electromagnetic radiation in nonlocal environments: Steps toward a theory of near field engineering. In Proceedings of the 2015 9th European Conference on Antennas and Propagation (EuCAP), Lisbon, Portugal, 13–17 April 2015; pp. 1–5.
130. Konovalenko, A.; Reyes-Avendaño, J.A.; Méndez-Blas, A.; Cervera, F.; Myslivets, E.; Radic, S.; Sánchez-Dehesa, J.; Pérez-Rodríguez, F. Nonlocal electrodynamics of homogenized metal-dielectric photonic crystals. *J. Opt.* **2019**, *21*, 085102. [CrossRef]
131. Zheng, L.; Qi-Huang, G. Surface Polaritons in a Wire Medium with Spatial Dispersion. *Chin. Phys. Lett.* **2008**, *25*, 298–301. [CrossRef]
132. Belov, P.A.; Marqués, R.; Maslovski, S.I.; Nefedov, I.S.; Silveirinha, M.; Simovski, C.R.; Tretyakov, S.A. Strong spatial dispersion in wire media in the very large wavelength limit. *Phys. Rev. B* **2003**, *67*, 113103. [CrossRef]
133. Silveirinha, M.G. Nonlocal homogenization model for a periodic array of ϵ-negative rods. *Phys. Rev. E* **2006**, *73*, 046612. [CrossRef] [PubMed]
134. Simovski, C.R.; Belov, P.A.; Atrashchenko, A.V.; Kivshar, Y.S. Wire Metamaterials: Physics and Applications. *Adv. Mater.* **2012**, *24*, 4229–4248. [CrossRef]
135. Moradi, A. Spatial nonlocality in the infrared absorption spectra of polar semiconductor nanospheres. *Semicond. Sci. Technol.* **2015**, *30*, 115003. [CrossRef]
136. Maack, J.R.; Mortensen, N.A.; Wubs, M. Size-dependent nonlocal effects in plasmonic semiconductor particles. *EPL (Europhys. Lett.)* **2017**, *119*, 17003. [CrossRef]
137. Dong, T.; Ma, X. Electromagnetic scattering by gyrotropic semiconductor spheres when considering spatial dispersion. *J. Phys. D Appl. Phys.* **2018**, *51*, 285302. [CrossRef]
138. Eremin, Y.A.; Mädler, L.; Wriedt, T. Influence of the Nonlocal Effect on the Optical Properties of Nonspherical Plasmonic Semiconductor Nanoparticles. *Comput. Math. Model.* **2020**, *31*, 58–74. [CrossRef]
139. Souza, R.F.; Alencar, M.A.R.C.; Meneghetti, M.R.; Dupont, J.; Hickmann, J.M. Nonlocal optical nonlinearity of ionic liquids. *J. Phys. Condens. Matter* **2008**, *20*, 155102. [CrossRef]
140. Poddubny, A.; Iorsh, I.; Belov, P.; Kivshar, Y. Hyperbolic metamaterials. *Nat. Photonics* **2013**, *7*, 948–957. [CrossRef]
141. Orlov, A.; Voroshilov, P.M.; Belov, P.A.; Kivshar, Y.S. Engineered optical nonlocality in nanostructured metamaterials. *Phys. Rev. B* **2011**, *84*, 045424. [CrossRef]
142. Becchi, M.; Oldano, C.; Ponti, S. Spatial dispersion and optics of crystals. *J. Opt. Pure Appl. Opt.* **1999**, *1*, 713–718. [CrossRef]
143. Shvartsburg, A.B.; Agranat, M.B.; Chefonov, O.V. Nanooptics of gradient dielectric films. *Quantum Electron.* **2009**, *39*, 948–952. [CrossRef]
144. Weissglas, P. Spatial dispersion and absorption of waves in bounded low-density plasma. *J. Nucl. Energy Part C Plasma Phys. Accel. Thermonucl. Res.* **1964**, *6*, 251–265. [CrossRef]
145. Brodin, G.; Marklund, M. On the possibility of metamaterial properties in spin plasmas. *New J. Phys.* **2008**, *10*, 115031. [CrossRef]
146. Sakai, O.; Tachibana, K. Plasmas as metamaterials: A review. *Plasma Sources Sci. Technol.* **2012**, *21*, 013001. [CrossRef]
147. Wang, C.Y.; Wang, G.H. Spatially nonlocal effects on optical absorption properties in coupled quantum wells with an applied electric field. *Chin. Phys. B* **2014**, *23*, 127103. [CrossRef]
148. Ikeda, K.; Akimoto, O. Steady propagation of a coherent light pulse in a dielectric medium – Part II. The effect of spatial dispersion. *J. Phys. A Math. Gen.* **1979**, *12*, 1105–1120. [CrossRef]
149. Alfassi, B.; Rotschild, C.; Manela, O.; Segev, M.; Christodoulides, D.N. Nonlocal Surface-Wave Solitons. *Phys. Rev. Lett.* **2007**, *98*, 213901. [CrossRef]
150. Zhong, W.P.; Yi, L.; Xie, R.H.; Belić, M.; Chen, G. Robust three-dimensional spatial soliton clusters in strongly nonlocal media. *J. Phys. B At. Mol. Opt. Phys.* **2008**, *41*, 025402. [CrossRef]
151. Zhang, H.; Xu, D.; Li, L. An approximate solution for describing a fundamental soliton in nonlocal nonlinear media. *J. Opt. A Pure Appl. Opt.* **2009**, *11*, 125203. [CrossRef]
152. Pu, S.; Hou, C.; Zhan, K.; Du, Y. Beam splitters in inhomogeneous nonlocal media. *Phys. Scr.* **2012**, *86*, 025404. [CrossRef]
153. Alberucci, A.; Peccianti, M.; Assanto, G.; Dyadyusha, A.; Kaczmarek, M. Two-Color Vector Solitons In Nonlocal Media. *Phys. Rev. Lett.* **2006**, *97*, 153903. [CrossRef]
154. Vestgården, J.I.; Mikheenko, P.; Galperin, Y.M.; Johansen, T.H. Nonlocal electrodynamics of normal and superconducting films. *New J. Phys.* **2013**, *15*, 093001. [CrossRef]
155. Abdumalikov, A.; Alfimov, G.L.; Malishevskii, A.S. Nonlocal electrodynamics of Josephson vortices in superconducting circuits. *Supercond. Sci. Technol.* **2009**, *22*, 023001.
156. Binkowski, F.; Zschiedrich, L.; Hammerschmidt, M.; Burger, S. Modal analysis for nanoplasmonics with nonlocal material properties. *Phys. Rev. B* **2019**, *100*, 155406. [CrossRef]
157. Luo, Y.; Fernandez-Dominguez, A.I.; Wiener, A.; Maier, S.A.; Pendry, J.B. Surface Plasmons and Nonlocality: A Simple Model. *Phys. Rev. Lett.* **2013**, *111*, 093901. [CrossRef] [PubMed]
158. Moaied, M.; Palomba, S.; Ostrikov, K. Quantum plasmonics: Longitudinal quantum plasmons in copper, gold, and silver. *J. Opt.* **2017**, *19*, 105402. [CrossRef]

159. Ramírez, E.V.G.; Hernández, S.A.S.; Rodríguez, J.J.M.; Díaz, G.; Carrasco, M.L.A.; Esqueda, J.A.R. Nonlocality in Ag nanocubes: Dependence on the concentration and incident wavelength. *Mater. Res. Express* **2020**, *6*, 1250c6. [CrossRef]
160. Gratus, J.; Kinsler, P.; McCall, M.W.; Thompson, R.T. On spacetime transformation optics: Temporal and spatial dispersion. *New J. Phys.* **2016**, *18*, 123010. [CrossRef]
161. Silveirinha, M.G. Chern invariants for continuous media. *Phys. Rev. B* **2015**, *92*, 125153. [CrossRef]
162. Banerjee, N.; Dutta, S.; Roychowdhury, D. Chern–Simons superconductor. *Class. Quantum Gravity* **2014**, *31*, 245005. [CrossRef]
163. Mikki, S.; Kishk, A. Electromagnetic wave propagation in dispersive negative group velocity media. In Proceedings of the 2008 IEEE MTT-S International Microwave Symposium Digest, Atlanta, GA, USA, 15–20 June 2008; pp. 205–208.
164. Hapuarachchi, H.; Gunapala, S.D.; Premaratne, M. Plasmonic metaresonances: Harnessing nonlocal effects for prospective biomedical applications. *J. Phys. Condens. Matter* **2019**, *31*, 325301. [CrossRef] [PubMed]
165. Mikki, S. Theory of Electromagnetic Radiation in Nonlocal Metamaterials—Part I: Foundations. *Prog. Electromagn. Res. B* **2020**, *89*, 63–86. [CrossRef]
166. Mikki, S. Theory of Electromagnetic Radiation in Nonlocal Metamaterials—Part II: Applications. *Prog. Electromagn. Res. B* **2020**, *89*, 87–109. [CrossRef]
167. Gorlach, M.A.; Belov, P.A. Nonlocality in uniaxially polarizable media. *Phys. Rev. B* **2015**, *92*, 085107. [CrossRef]
168. Skettrup, T. Effective optical constants in the treatment of spatial dispersion. *J. Phys. D Appl. Phys.* **1981**, *14*, 1343–1354. [CrossRef]
169. Pérez-Huerta, J.S.; Ortiz, G.P.; Mendoza, B.S.; Mochán, W.L. Macroscopic optical response and photonic bands. *New J. Phys.* **2013**, *15*, 043037. [CrossRef]
170. Ciattoni, A.; Rizza, C. Nonlocal homogenization theory in metamaterials: Effective electromagnetic spatial dispersion and artificial chirality. *Phys. Rev. B* **2015**, *91*, 184207. [CrossRef]
171. Mnasri, K.; Khrabustovskyi, A.; Stohrer, C.; Plum, M.; Rockstuhl, C. Beyond local effective material properties for metamaterials. *Phys. Rev. B* **2018**, *97*, 075439. [CrossRef]
172. Raghu, S.; Haldane, F.D.M. Analogs of quantum-Hall-effect edge states in photonic crystals. *Phys. Rev. A* **2008**, *78*, 033834. [CrossRef]
173. Lu, L.; Joannopoulos, J.D.; Soljačić, M. Topological photonics. *Nat. Photonics* **2014**, *8*, 821–829. [CrossRef]
174. Chen, W.J.; Hou, B.; Zhang, Z.Q.; Pendry, J.B.; Chan, C.T. Metamaterials with index ellipsoids at arbitrary k-points. *Nat. Commun.* **2018**, *9*, 2086. [CrossRef]
175. Gorlach, M.; Belov, P.A. Effect of spatial dispersion on the topological transition in metamaterials. *Phys. Rev. B* **2014**, *90*, 115136. [CrossRef]
176. Bliokh, K.Y.; Leykam, D.; Lein, M.; Nori, F. Topological non-Hermitian origin of surface Maxwell waves. *Nat. Commun.* **2019**, *10*, 1–7. [CrossRef]
177. Silveirinha, M.G. Topological theory of non-Hermitian photonic systems. *Phys. Rev. B* **2019**, *99*, 125155. [CrossRef]
178. Pekar, S. The Theory of Electromagnetic Waves in a Crystal in which Excitons Are Produced. *JETP* **1957**, *6*, 785.
179. Agranovich, V.M.; Ginzburg, V.L. Crystal optics with allowance for spatial dispersion: Exciton theory—Part I. *Sov. Phys. Uspekhi* **1962**, *5*, 323–346. [CrossRef]
180. Agranovich, V.M.; Ginzburg, V.L. Crystal optics with allowance for spatial dispersion: Exciton theory—Part II. *Sov. Phys. Uspekhi* **1963**, *5*, 675–710. [CrossRef]
181. Ginzburg, V.L. The Development of Crystal Optics with Allowance for Spatial Dispersion. *Sov. Phys. Uspekhi* **1973**, *15*, 839–841. [CrossRef]
182. Belov, P.A.; Simovski, C.R. On the low-frequency spatial dispersion in wire media. In Proceedings of the IEEE International Workshop on Antenna Technology: Small Antennas and Novel Metamaterials, Singapore, 7–9 March 2005; pp. 363–366. [CrossRef]
183. Wells, B.M.; Zayats, A.V.; Podolskiy, V.A. Nonlocal optics of plasmonic nanowire metamaterials. *Phys. Rev. B* **2014**, *89*, 035111. [CrossRef]
184. Mikki, S.; Kishk, A. Mean-Field Electrodynamic Theory of Aligned Carbon Nanotube Composites. *IEEE Trans. Antennas Propag.* **2009**, *57*, 1412–1419. [CrossRef]
185. Chebykin, A.V.; Gorlach, M.A.; Gorlach, A.; Belov, P.A. Spatial dispersion in metamaterials based on three-dimensional arrays of spheres and disks. In Proceedings of the 2015 Days on Diffraction (DD), St. Petersburg, Russia, 25–29 May 2015; pp. 1–5. [CrossRef]
186. Guan, Y.; Zhong, L.X.; Chew, K.H.; Chen, H.; Wu, Q.; Chen, R.P. Evolution of cos-gaussian beams in a strongly nonlocal nonlinear medium. *Prog. Electromagn. Res.* **2013**, *141*, 403–414. [CrossRef]
187. McMahon, J.M.; Gray, S.K.; Schatz, G.C. Nonlocal Optical Response of Metal Nanostructures with Arbitrary Shape. *Phys. Rev. Lett.* **2009**, *103*, 097403. [CrossRef]
188. Ciecierska, E.; Boczkowska, A.; Kurzydlowski, K.J. Quantitative description of the spatial dispersion of carbon nanotubes in polymeric matrix. *J. Mater. Sci.* **2010**, *45*, 2305–2310. [CrossRef]
189. Gomez-Diaz, J.S.; Mosig, J.R.; Perruisseau-Carrier, J. Effect of Spatial Dispersion on Surface Waves Propagating Along Graphene Sheets. *IEEE Trans. Antennas Propag.* **2013**, *61*, 3589–3596. [CrossRef]
190. Rodriguez-Lopez, P.; Kort-Kamp, W.J.M.; Dalvit, D.A.R.; Woods, L.M. Nonlocal optical response in topological phase transitions in the graphene family. *Phys. Rev. Mater.* **2018**, *2*, 014003. [CrossRef]

191. Hanson, G.W. Drift-Diffusion: A Model for Teaching Spatial-Dispersion Concepts and the Importance of Screening in Nanoscale Structures. *IEEE Antennas Propag. Mag.* **2010**, *52*, 198–207. [CrossRef]
192. Mikki, S. Exact Derivation of the Radiation Law of Antennas Embedded into Generic Nonlocal Metamaterials: A Momentum-Space Approach. In Proceedings of the 14th European Conference on Antennas and Propagation (EuCAP), Copenhagen, Denmark, 15–20 March 2020; pp. 1–5.
193. Jiafu Wang.; Shaobo Qu.; Jiequ Zhang.; Hua Ma. Spatial-temporal dispersion engineering of longitudinally coupled spoof surface plasmon polaritons for free-space EM wave modulation. In Proceedings of the 2016 Progress in Electromagnetic Research Symposium (PIERS), Shanghai, China, 8–11 August 2016; pp. 1571–1572. [CrossRef]
194. Lathi, B.P.; Ding, Z. *Modern Digital and Analog Communication Systems*; Oxford University Press: New York, NY, USA, 2019.
195. Mikki, S. Theory of Nonsinusoidal Antennas for Near-Field Communication System Analysis. *Prog. Electromagn. Res.* **2020**, *86*, 177–193. [CrossRef]
196. Mikki, S.; Kishk, A. Effective Medium Theory for Carbon Nanotube Composites and their Potential Applications as Metamaterials. In Proceedings of the 2007 IEEE/MTT-S International Microwave Symposium, Honolulu, HI, USA, 3–8 June 2007; pp. 1137–1140. [CrossRef]
197. Mikki, S.; Antar, Y. Reactive, localized, and stored energies: The fundamental differences and proposals for new experiments. In Proceedings of the 2015 IEEE AP-S Symposium on Antennas and Propagation and URSI CNC/USNC Joint Meeting, Vancouver, BC, Canada, 19–24 July 2015; p. 366.
198. Mikki, S.; Sarkar, D.; Antar, Y. On Localized Antenna Energy in Electromagnetic Radiation. *Prog. Electromagn. Res. M* **2019**, *79*, 1–10. [CrossRef]
199. Sarkar, D.; Mikki, S.; Srivastava, K.V.; Antar, Y. Dynamics of Antenna Reactive Energy Using Time-Domain IDM Method. *IEEE Trans. Antennas Propag.* **2019**, *67*, 1084–1093. [CrossRef]
200. Sarkar, D.; Mikki, S.M.; Antar, Y.M.M. Poynting Localized Energy: Method and Applications to Gain Enhancement in Coupled Antenna Systems. *IEEE Trans. Antennas Propag.* **2020**, *68*, 3978–3988. [CrossRef]
201. Shvarts, A.S. On the definition of superspace. *Theor. Math. Phys.* **1984**, *60*, 657–660. [CrossRef]
202. Barbour, J. *The End of Time: The Next Revolution in Physics*; Oxford University Press: Oxford, UK; New York, NY, USA, 2000.
203. Anderson, E. *The Problem of Time: Quantum Mechanics Versus General Relativity*; Springer: Cham, Switzerland, 2017.
204. Penrose, R. *The Road to Reality: A Complete Guide to the Laws of the Universe*; Vintage Books: New York, NY, USA, 2007.
205. McClain, B.; Niemi, A.; Taylor, C.; Wijewardhana, L.C.R. Superspace, Negative Dimensions, and Quantum Field Theories. *Phys. Rev. Lett.* **1982**, *49*, 252–255. [CrossRef]
206. Castagnino, M. Mathematical structure of quantum superspace as a consequence of time asymmetry. *Phys. Rev. D* **1998**, *57*, 750–767. [CrossRef]
207. Zeidler, E. *Quantum Field Theory II: Quantum Electrodynamics*; Springer: Berlin/Heidelberg, Germany, 2006.
208. Zeidler, E. *Quantum Field Theory III: Gauge Theory*; Springer: Berlin/Heidelberg, Germany, 2011.
209. Russell, B. *The Analysis of Matter*; Martino Fine Books: Mansfield Centre, CT, USA, 2014.
210. Janner, A.; Janssen, T. Symmetry of periodically distorted crystals. *Phys. Rev. B* **1977**, *15*, 643–658. [CrossRef]
211. Tsakalakos, T. *Modulated Structure Materials*; M. Nijhoff: Dordrecht, The Netherlands; Ner Boston, MA, USA, 1984.
212. Elcoro, L.; Perez-Mato, J.M. Superspace description of quasiperiodic structures and the nonuniqueness of superspace embedding. *Phys. Rev. B* **1996**, *54*, 12115–12124. [CrossRef]
213. Andreev, A.F. Mesoscopic superconductivity in superspace. *J. Exp. Theor. Phys. Lett.* **1998**, *68*, 673–678. [CrossRef]
214. Geroch, R. *Infinite-Dimensional Manifolds*; Minkowski Institute Press: Montreal, QC, USA, 2013.
215. Weyl, H. *Space, Time, Matter*; Dover Publications: New York, NY, USA, 1952.
216. Agricola, I. *Global Analysis: Differential Forms in Analysis, Geometry, and Physics*; American Mathematical Society: Providence, RI, USA, 2002.
217. Wannier, G.H. The Structure of Electronic Excitation Levels in Insulating Crystals. *Phys. Rev.* **1937**, *52*, 191–197. [CrossRef]
218. Quattropani, A.; Czaja, W. Quantum Theory of Exciton-Polaritons with Spatial Dispersion. *Phys. Scr.* **1989**, *T29*, 162–166. [CrossRef]
219. Bamba, M.; Ishihara, H. QED of excitons with nonlocal susceptibility in arbitrarily structured dielectrics. *Phys. Rev. B* **2008**, *78*, 085109. [CrossRef]
220. Cho, K., Ed.. *Excitons*; Springer: Berlin/Heidelberg, Germany, 1979.
221. Lagoudakis, K. *The Physics of Exciton-Polariton Condensates*; CRC Press: Lausanne, Switzerland; Boca Raton, FL, USA, 2013.
222. Kira, M.; Koch, S.W. *Semiconductor Quantum Optics*; Cambridge University Press: Cambridge, UK; New York, NY, USA, 2012.
223. Bohm, D.; Pines, D. A Collective Description of Electron Interactions–Part I: Magnetic Interactions. *Phys. Rev.* **1951**, *82*, 625–634. [CrossRef]
224. Pines, D.; Bohm, D. A Collective Description of Electron Interactions–Part II: Collective vs Individual Particle Aspects of the Interactions. *Phys. Rev.* **1952**, *85*, 338–353. [CrossRef]
225. Mattuck, R. *A Guide to Feynman Diagrams in the Many-Body Problem*; Dover Publications: New York, NY, USA, 1992.
226. Pines, D. *The Many-Body Problem*; Addison-Wesley: Reading, MA, USA, 1997.
227. Pines, D. *Elementary Excitations in Solids: Lectures on Protons, Electrons, and Plasmons*; Advanced Book Program, Perseus Books: Reading, MA, USA, 1999.

228. Dresselhaus, G. Effective mass approximation for excitons. *J. Phys. Chem. Solids* **1956**, *1*, 14–22. [CrossRef]
229. Prigogine, I. *The End of Certainty: Time, Chaos, and the New Laws of Nature*; Free Press: New York, NY, USA, 1997.
230. Zomorodian, A. *Topology for Computing*; Cambridge University Press: Cambridge, UK, 2009.

foundations

Article

Relativistic Effects for a Hydrogen Rydberg Atom in a High-Frequency Laser Field: Analytical Results

Nikolay Kryukov [1] and Eugene Oks [2,*]

[1] Universidad Nacional Autónoma de México, Av. Universidad 3000, col. Ciudad Universitaria, del. Coyoacán, México DF 04510, Mexico; kriukov@gmail.com
[2] Physics Department, Auburn University, 380 Duncan Drive, Auburn, AL 36849, USA
* Correspondence: goks@physics.auburn.edu

Abstract: Previously published analytical results for the effects of a high-frequency laser field on hydrogen Rydberg atoms demonstrated that the unperturbed elliptical orbit of the Rydberg electron, generally is engaged simultaneously in the precession of the orbital plane about the direction of the laser field and in the precession within the orbital plane. These results were obtained while disregarding relativistic effects. In the present paper, we analyze the relativistic effect for hydrogenic Rydberg atoms or ions in a high-frequency linearly- or circularly-polarized laser field, the effect being an additional precession of the electron orbit in its own plane. For the linearly-polarized laser field, the general case, where the electron orbit is not perpendicular to the direction of the laser field, we showed that the precession of the electron orbit within its plane can vanish at some critical polar angle θ_c of the orbital plane. We calculated analytically the dependence of the critical angle on the angular momentum of the electron and on the parameters of the laser field. Finally, for the particular situation, where the electron orbit is perpendicular to the direction of the laser field, we demonstrated that the relativistic precession and the precession due to the laser field occur in the opposite directions. As a result, the combined effect of these two kinds of the precession is smaller than the absolute value of each of them. We showed that by varying the ratio of the laser field strength F to the square of the laser field frequency ω, one can control the precession frequency of the electron orbit and even make the precession vanish, so that the elliptical orbit of the electron would become stationary. This is a counterintuitive result.

Keywords: hydrogenic atoms; high-frequency laser field; relativistic precession; laser-controlled precession

Citation: Kryukov, N.; Oks, E. Relativistic Effects for a Hydrogen Rydberg Atom in a High-Frequency Laser Field: Analytical Results. *Foundations* **2022**, 2, 105–113. https://doi.org/10.3390/foundations2010005

Academic Editor: Wiesław Leonski

Received: 12 December 2021
Accepted: 4 January 2022
Published: 10 January 2022

1. Introduction

Analytical studies of effects of a high-frequency laser field on various Rydberg atoms and ions—the studies using the method of separating rapid and slow subsystems—have been presented in the literature: see, e.g., book [1] and references therein. In particular, analytical results for hydrogen Rydberg atoms were presented in paper [2] for the case of the linear polarization of the high-frequency laser field and in paper [3] for the cases of the elliptical or circular polarization of the high-frequency laser field.

Specifically, in paper [2] it was shown that the unperturbed elliptical orbit of the Rydberg electron, generally is engaged simultaneously in the precession of the orbital plane about the direction of the laser field and in the precession within the orbital plane, the corresponding precession frequencies being calculated analytically. In paper [2] it was also pointed out that the situation has a celestial analogy: it is mathematically equivalent to the motion of a satellite around an oblate planet (such as, e.g., the Earth), the results for the latter system being presented, e.g., in book [4]. Later in paper [5] it was demonstrated that there is also another celestial analogy: it is mathematically equivalent also to the motion of a planet around a circular binary star.

As for paper [3], their authors showed that the case of the circular polarization of the high-frequency laser field is mathematically equivalent to the motion of a satellite around a (fictitious) prolate planet, the results for the latter system being presented, e.g., in book [6]. The orbit of the electron in this case is also engaged simultaneously in the precession of the orbital plane about the direction of the laser field and in the precession within the orbital plane, the corresponding precession frequencies being calculated analytically [3].

The authors of paper [3] also obtained analytical results for the situation where the high-frequency laser field is elliptically-polarized in the plane of the electron orbit. They demonstrated that this situation is mathematically equivalent to a problem of celestial mechanics, where a satellite moves in an equatorial orbit about a slightly non-spherical planet. For this case the plane of the orbit does not change its orientation over the course of time: the only precession that is the precession of the periapsis (and apoapsis) of the ellipse in the orbital plane.

All of the above analytical results were obtained while disregarding relativistic effects. In the present paper we study the role of the relativistic effect for hydrogenic Rydberg atoms or ions in a high-frequency linearly- or circularly-polarized laser field, the effect being an additional (relativistic) precession of the electron orbit in its own plane. For the linearly-polarized laser field, in the general case, where the electron orbit is not perpendicular to the direction of the laser field, there can exist a critical polar angle θ_c of the orbital plane, for which the precession within the plane vanishes and only the precession of the orbital plane remains. We study the dependence of the critical angle both on the angular momentum of the electron and on the laser field parameters.

For the particular situation, where the electron orbit is perpendicular to the direction of the linearly-polarized laser field, we show that the relativistic precession and the precession due to the laser field occur in the opposite directions, so that their combined effect is smaller than the absolute value of each of them. Moreover, we show the existence and calculate the specific value of the laser field parameters, for which the two precessions cancel each other out, so that the elliptical orbit of the electron becomes stationary. This is a counterintuitive result.

2. Analytical Calculations for the Linearly-Polarized Laser Field in the General Case

We study a hydrogen atom or a hydrogen-like ion of charge Z which is subjected to a high-frequency linearly-polarized laser field of amplitude F, directed along the z-axis, and frequency ω. The interaction of the laser field with Rydberg states can be described classically. Relativistic effects are taken into account. The Hamiltonian of the system is therefore

$$H = H_0 + zF\cos\omega t, H_0 = \sqrt{p^2c^2 + m^2c^4} - mc^2 - \frac{Ze^2}{r} \quad (1)$$

where m is the electron mass, e is the elementary charge, p is the momentum of the electron, r is the distance from the nucleus to the electron, and c is the speed of light. Atomic units ($m = e = \hbar = 1$) are used throughout this study.

In the absence of the laser field, we approximate the time-independent part of the Hamiltonian for the case $p << mc$:

$$H_0 = c^2\sqrt{1 + \frac{p^2}{c^2}} - c^2 - \frac{Z}{r} \approx \frac{p^2}{2} - \frac{p^4}{8c^2} - \frac{Z}{r} \quad (2)$$

From the non-relativistic Hamiltonian,

$$H_{NR} = \frac{p_0^2}{2} - \frac{Z}{r} = E_0 \quad (3)$$

where p_0 is the non-relativistic momentum of the electron and E_0 is its energy, we express p_0

$$p_0^2 = 2\left(E_0 + \frac{Z}{r}\right) \quad (4)$$

and substitute it in the second term in Equation (2), thus obtaining the further approximation:

$$H_0 \approx \frac{p^2}{2} - \frac{Z}{r}\left(1 + \frac{E_0}{c^2}\right) - \frac{Z^2}{2c^2r^2} - \frac{E_0}{2c^2} \equiv H_1 - \frac{Z^2}{2c^2r^2} - \frac{E_0}{2c^2} \tag{5}$$

where H_1 is the Hamiltonian of the system without the relativistic correction. Therefore, the penultimate term in Equation (5) is the perturbing term due to the relativistic effects. (The last term in Equation (5) is constant and thus does not affect the motion.) In book [7], a relativistic treatment of the Kepler problem is presented. The effect of the relativistic correction on the orbit dynamics is the precession of the orbit in its plane with the frequency (scaled by the Kepler frequency ω_K)

$$\frac{\Omega_c}{\omega_K} = \frac{1}{\sqrt{1 - \frac{Z^2}{L^2c^2}}} - 1 \approx \frac{Z^2}{2L^2c^2} \tag{6}$$

where L is the angular momentum of the electron (this result follows from Equation (10) in book [7]); we refer to the quantity (6) as the scaled relativistic precession. The precession is positive, i.e., its angular velocity has the same sign as the angular velocity of the Kepler motion.

Now we consider the above-mentioned system without the relativistic correction subjected to a linearly-polarized laser field of amplitude F and frequency ω which is much greater than the highest frequency of the unperturbed system. For such systems, it is appropriate to use the formalism of effective potentials [1,8–10]. As a result, the Hamiltonian H_1 in Equation (5) acquires a time-independent term. The zeroth-order effective potential,

$$U_0 = \frac{1}{4\omega^2}[V,[V,H_1]] = \frac{F^2}{4\omega^2} \tag{7}$$

where $V = zF$ and $[P, Q]$ are the Poisson brackets, is a coordinate-independent energy shift, so it does not affect the dynamics of the system. The first-order effective potential gives the first non-vanishing effect on the system:

$$U_1(r,\theta) = \frac{1}{4\omega^4}[[V,H_1],[[V,H_1],H_1]] = -\frac{a\left(1 + \frac{E_0}{c^2}\right)\left(3\cos^2\theta - 1\right)}{r^3} \approx -\frac{a\left(3\cos^2\theta - 1\right)}{r^3} \tag{8}$$

where $a = ZF^2/(4\omega^4)$; as $E_0 << c^2$, we can neglect the term E_0/c^2 in (8). The first term of U_1 is a perturbation of the Coulomb potential which makes the system mathematically equivalent to a satellite rotating around the oblate Earth [4], whose motion has the following property: the unperturbed elliptic orbit undergoes simultaneously two precessions, one of them being the precession of the orbit in its plane, and the other being the precession of the orbital plane about the vector **F**. Both precession frequencies are of the same order of magnitude and are much smaller than the Kepler frequency.

Without the relativistic correction, the first-order effective potential given in Equation (8) gives rise to two simultaneous effects on the Kepler orbit, as mentioned above. By using Equations (1.7.10) and (1.7.11) from book [4], we obtain the scaled frequencies of the precession of the orbit in its plane ("pip" stands for "precession in plane") and the precession of the plane about the direction of the laser field ("pop" stands for "precession of plane"):

$$\frac{\Omega_{pip}}{\omega_K} = \frac{3aZ}{2L^4}\left(1 - 5\sin^2\theta\right) \tag{9}$$

$$\frac{\Omega_{pop}}{\omega_K} = \frac{3aZ}{L^4}\sin\theta \tag{10}$$

where θ is the angle between the orbital plane and the laser field. The precession of the orbital plane is realized by the plane's rotation around the vector **F**, while its angle with the vector stays the same. For the case considered in the previous section,

$\theta = \pi/2$, the orbit plane precesses parallel to itself, therefore, the angular velocities from Equations (9) and (10) are both parallel to the laser field.

When the relativistic precession is taken into account, it creates an additional term for the precession in the plane. Thus, the plane of the orbit of the electron in this case undergoes the precession given by (10), while the orbit precesses in its own plane with the scaled frequency

$$\frac{\Omega_{pip}}{\omega_K} + \frac{\Omega_c}{\omega_K} = \frac{3aZ}{2L^4}\left(1 - 5\sin^2\theta\right) + \frac{Z^2}{2L^2c^2} \tag{11}$$

Without the relativistic effects, the critical angle θ_c at which there is no precession in the plane is given by arcsin $(1/5^{1/2}) \approx 26.6°$. The relativistic effects increase the value of this critical angle: its value is given by

$$\theta_c = \arcsin\sqrt{\frac{1}{5} + \frac{4\omega^4 L^2}{15F^2c^2}} \tag{12}$$

Figure 1 shows the value of the critical angle in degrees depending on the angular momentum of the electron, for selected values of the laser field strength and frequency.

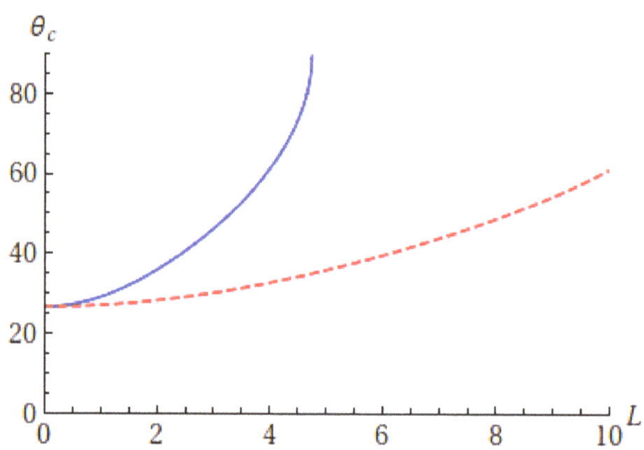

Figure 1. Dependence of the critical angle θ_c at which the precession in the orbital plane vanishes, on the angular momentum of the electron, for the laser field amplitude $F = 2$ (solid line) and $F = 5$ (dashed line) and the frequency $\omega = 10$.

From Equation (12), we see that the situation when the precession in the orbital plane vanishes is possible when

$$L < \frac{F}{\omega^2}c\sqrt{3} \tag{13}$$

i.e., the relativistic correction puts an upper limit on the value of the angular momentum of the electron when the vanishing of the precession in the orbital plane is possible, for the given values of the laser field strength and frequency. For example, when

$$\omega > \sqrt{Fc\sqrt{3}} \tag{14}$$

the precession in the orbital plane never vanishes for any $L \geq 1$; for example, if $F = 2$, then for the laser field frequency $\omega > 21.8$ the precession in the plane never vanishes for any $L \geq 1$.

3. The Case of the Electron Angular Momentum Collinear with the Linearly-Polarized Laser Field

Now we consider the situation when the angular momentum of the electron is collinear to the laser field, i.e., $\theta = \pi/2$. In this case, the perturbation takes the following form:

$$U_1(r) = \frac{a}{r^3} \tag{15}$$

The calculation of the $1/r^n$-perturbation for the Kepler orbit can be found in work [11] (the treatment for the cases $n = 2$ and $n = 3$ can be found also in the textbook [12]). For the Coulomb potential $-\alpha/r$ perturbed by the potential β/r^k, the orbit undergoes a precession with the perihelion advance

$$\delta\Phi = 2m\beta\frac{\partial}{\partial L}\left(\frac{1}{L}p^{2-k}\int_0^\pi (1+\varepsilon\cos\varphi)^{k-2}d\varphi\right) \tag{16}$$

with the substituted quantities

$$p = \frac{L^2}{m\alpha}, \varepsilon = \sqrt{1+\frac{2E_0L^2}{m\alpha^2}} \tag{17}$$

the first of which is the semi-latus rectum of the unperturbed elliptical orbit and the second is its eccentricity. The ratio of the precession frequency due to the perturbation to the Kepler frequency given by Equation (16) is therefore

$$\frac{\Omega_1}{\omega_K} = -\frac{3aZ}{L^4} = -\frac{3Z^2F^2}{4L^4\omega^4} \tag{18}$$

to which we refer as scaled high-frequency precession. The precession caused by the high-frequency laser field is negative (its angular velocity is of the opposite sign to that of the Kepler motion). The ratio of the magnitudes of the precessions is

$$\frac{\Omega_c}{\Omega_1} = \frac{2L^2\omega^4}{3c^2F^2} \tag{19}$$

For example, for the values of the laser field amplitude $F = 2$ and frequency $\omega = 10$, the ratio in Equation (19) is of the order of unity for L being in the approximate range between 3 and 6. Due to their opposite directions, the combined effect of the relativistic and high-frequency precessions is always less by absolute value than the greater precession by absolute value, and the two effects may cancel each other.

We note that the ratio of the frequencies in Equation (19) does not depend on the nuclear charge Z. However, if in the expansion in Equation (6) we would add higher order terms, then the ratio in Equation (19) would become weakly dependent on Z.

Figure 2 shows the dependence of the value of both corrections and of the combined effect of the two on the value of the angular momentum L of the electron for the nuclear charge $Z = 6$, the laser field amplitude $F = 2$ and frequency $\omega = 10$.

The high-frequency laser field cancels the relativistic effect when

$$\frac{F}{\omega^2} = \sqrt{\frac{2}{3}\frac{L}{c}} \tag{20}$$

For example, for $L = 3$, the laser field with $F = 2$ and $\omega = 10.5778$ will make the orbit's precession vanish. Figure 3 shows the critical value of the frequency of the laser field of selected amplitudes at which the precession of the electron orbit vanishes, depending on the angular momentum of the electron. As we see, the critical value of the laser field frequency stays much greater than the Kepler frequency of the electron $1/L^3$ and is therefore within the validity range of the method of effective potentials.

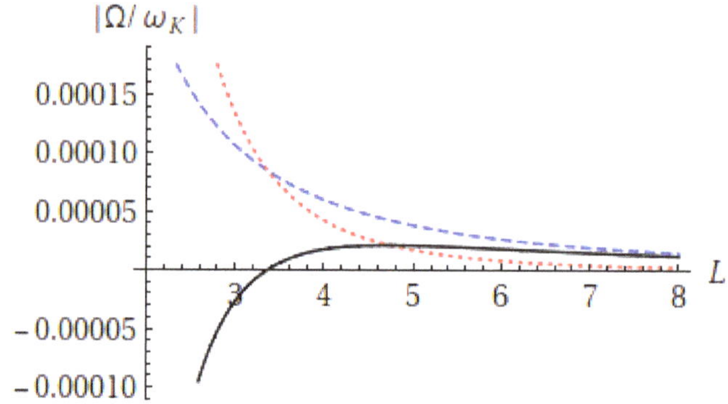

Figure 2. The scaled relativistic precession (dashed line), the absolute value of the scaled negative high-frequency precession (dotted line), and their combined effect (solid line) for $Z = 6$, $F = 2$ and $\omega = 10$.

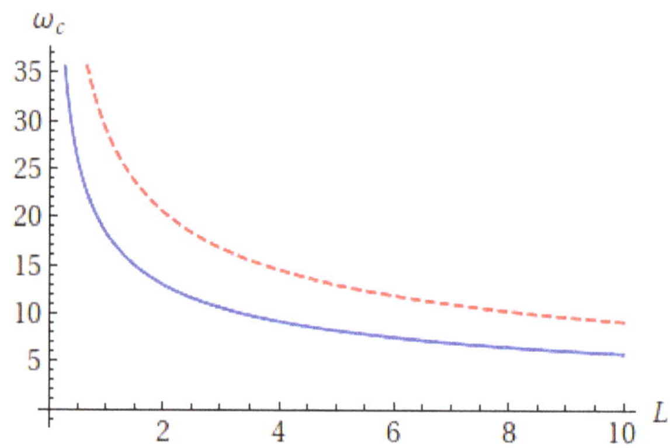

Figure 3. Dependence of the critical value of the laser field frequency, at which the precession of the electronic orbit vanishes, on the angular momentum of the electron, for the laser field amplitude $F = 2$ (solid line) and $F = 5$ (dashed line).

Thus, by varying the ratio of the laser field strength F to the square of the laser field frequency ω, one can control the precession frequency of the electron orbit and even make the precession vanish (according to Equation (20)), so that the elliptical orbit of the electron would become stationary. This is a counterintuitive result.

In general, any precession at some frequency Ω manifests in the radiation spectrum of a hydrogenic atom/ion as satellites separated from the unperturbed frequency of the spectral line by multiples of the precession frequency. Thus, the above situation where the elliptical orbit of the electron becomes stationary would manifest in the radiation spectrum as the disappearance of the satellites.

4. Analytical Calculations for the Circularly-Polarized Laser Field

In this section, we consider the case of circular polarization of the laser field. In this case, the field of amplitude F and frequency ω is perpendicular to the z-axis and varies as

$$\mathbf{F} = F\left(\mathbf{e}_x \cos \omega t + \mathbf{e}_y \sin \omega t\right) \tag{21}$$

where $\mathbf{e_x}$ and $\mathbf{e_y}$ are the Cartesian unit vectors. For our system, the relativistic correction stays the same as described in Section 2, while the effective potential due to the laser field is different. In this case, the Hamiltonian without the relativistic correction is

$$H_0 = H_1 + xF\cos\omega t + yF\sin\omega t \tag{22}$$

where H_1 is the Hamiltonian of the system without the relativistic correction. We denote

$$V = xF = Fr\sin\theta\cos\varphi, W = yF = Fr\sin\theta\sin\varphi \tag{23}$$

where (r, θ, φ) are the spherical coordinates. We apply the method of effective potentials [8–10] and obtain the zeroth-order effective potential

$$U_0 = \frac{1}{4\omega^2}([V,[V,H_1]] + [W,[W,H_1]]) = \frac{F^2}{2\omega^2} \tag{24}$$

which, as in the linear-polarization case, is a coordinate-independent energy shift, so it does not affect the dynamics of the system, and the first-order effective potential

$$\begin{aligned} U_1 = \tfrac{1}{4\omega^4}([[V,H_0],[[V,H_0],H_0]] + [[W,H_0],[[W,H_0],H_0]])+ \\ +\tfrac{-1}{2\omega^3}[[V,H_0],[W,H_0]] = \tfrac{a\left(3\cos^2\theta-1\right)}{r^3} \end{aligned} \tag{25}$$

which is the opposite of that in the linear-polarization case. Thus, the results for the circular-polarization case can be obtained by effectively replacing a with $-a$ in the linear-polarization case. In particular, in the case of the angular momentum collinear with the laser field, the precession due to the circularly-polarized laser field is positive, the same as the precession due to the relativistic correction, so these two effects cannot cancel each other. The ratio of the magnitudes of these two precessions is the same as in Equation (13) for the linear-polarization case. In the general case of the orientation of the angular momentum with respect to the direction of the laser field, the precession of the orbit in its plane and the precession of the orbit plane both have the direction opposite to that in the linear-polarization case and are expressed by Equations (15) and (16) multiplied by -1.

The precession in the orbital plane vanishes when the critical angle is

$$\theta_c = \arcsin\sqrt{\frac{1}{5} - \frac{4\omega^4L^2}{15F^2c^2}} \tag{26}$$

Figure 4 shows the dependence of the critical angle corresponding to zero precession on the angular momentum of the electron, for the cases of $F = 2$ and $F = 5$ and $\omega = 10$.

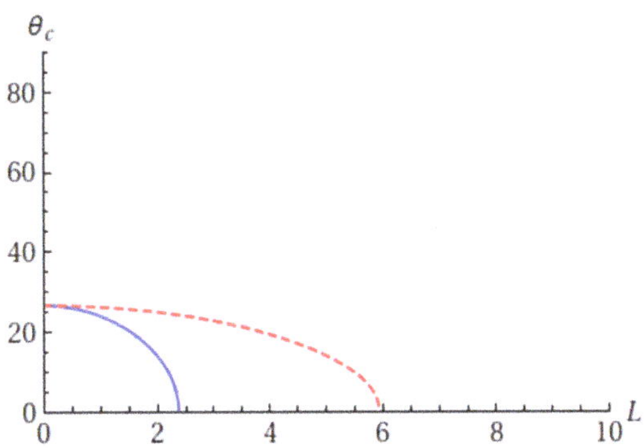

Figure 4. Dependence of the critical angle θ_c in degrees at which the precession in the orbital plane vanishes in the circular-polarization case, on the angular momentum of the electron, for the laser field amplitude $F = 2$ (solid line) and $F = 5$ (dashed line) and the frequency $\omega = 10$.

We see that the range of possible angles is more narrow than in the linear-polarization case, being $0 \leq \theta \leq \arcsin(1/5^{1/2})$, with the possibility of zero precession at polar orbits ($\theta = 0$) when

$$L = \frac{F}{\omega^2} c \frac{\sqrt{3}}{2} \quad (27)$$

So the upper limit of the angular momentum is twice as small as in the linear-polarization case (19).

5. Conclusions

We analyzed the relativistic effect for hydrogenic Rydberg atoms or ions in a high-frequency linearly- or circularly-polarized laser field. For the general case, where the electron orbit is not perpendicular to the direction of the laser field, we showed that the precession of the electron orbit within its plane can vanish at some critical polar angle θ_c of the orbital plane. We calculated analytically the dependence of the critical angle on the angular momentum of the electron and on the parameters of the laser field.

For the particular situation, where the electron angular momentum is collinear with the linearly-polarized laser field, we demonstrated that the relativistic precession and the precession due to the laser field occur in the opposite directions. As a result, the combined effect of these two kinds of the precession is smaller than the absolute value of each of them. We showed that, by varying the ratio of the laser field strength F to the square of the laser field frequency ω, one can control the precession frequency of the electron orbit and even make the precession vanish, so that the elliptical orbit of the electron would become stationary. This is a counterintuitive result.

In general, any precession at some frequency Ω manifests in the radiation spectrum of a hydrogenic atom/ion as satellites separated from the unperturbed frequency of the spectral line by multiples of the precession frequency. So, the above situation where the elliptical orbit of the electron becomes stationary would manifest in the radiation spectrum as the disappearance of the satellites.

We hope that the fundamental nature of our analytical results for the hydrogenic atoms/ions, i.e., for atomic systems serving as the test bench of our understanding of atomic physics, makes the results significant.

Author Contributions: E.O.: conceptualization; investigation; writing—original draft preparation; writing—review and editing. N.K.: investigation; writing—original draft preparation; writing—review and editing. All authors have read and agreed to the published version of the manuscript.

Funding: This work received no funding.

Institutional Review Board Statement: Not applicable.

Informed Consent Statement: Not applicable.

Data Availability Statement: All data is included in the manuscript.

Conflicts of Interest: The authors declare no conflict of interest.

References

1. Oks, E. *Analytical Advances in Quantum and Celestial Mechanics: Separating Rapid and Slow Subsystems*; IOP Publishing: Bristol, UK, 2020.
2. Nadezhdin, B.B.; Oks, E. Highly excited atom in a high-frequency field of linearly-polarized electromagentic radiation. *Sov. Tech. Phys. Lett.* **1986**, *12*, 512–513.
3. Oks, E.; Davis, J.E.; Uzer, T. Rydberg electron dynamics in high-frequency, elliptically polarized microwave fields. *J. Phys. B At. Mol. Opt. Phys.* **2000**, *33*, 207–221. [CrossRef]
4. Beletsky, V.V. *Essays on the Motion of Celestial Bodies*; Section 1.7.; Birkhäuser/Springer: Basel, The Netherland, 2001.
5. Oks, E. Analytical solution for the three-dimensional motion of a circumbinary planet around a binary star. *New Astron.* **2020**, *74*, 101301. [CrossRef]
6. Roy, A.E. *Orbital Motion*; Section 10.4.; Adam Hilger: Bristol, UK, 1978.
7. Sommerfeld, A. *Atombau und Spektrallinien*; Section 5.1.; Friedr. Vieweg & Sohn: Braunschweig, Germany, 1951.

8. Nadezhdin, B.B. *Radiatsionnye i Relativistskie Effekty v Atomakh i Ionakh (Radiative and Relativistic Effects in Atoms and Ions)*; Scientific Council of the USSR Academy of Sciences on Spectroscopy: Moscow, Russia, 1986; p. 222. (In Russian)
9. Kapitza, P.L. Dynamic stability of the pendulum with vibrating suspension point. *Sov. Phys. JETP* **1951**, *21*, 588–597.
10. Kapitza, P.L. Pendulum with an oscillating pivot. *Usp. Fiz. Nauk* **1951**, *44*, 7–20. [CrossRef]
11. p. 95. Available online: https://pages.uoregon.edu/dbelitz/teaching/lecture_notes/Mechanics%20(DB)/611_2_Lecture_Notes.pdf (accessed on 13 October 2021).
12. Landau, L.D.; Lifshitz, E.M. *Mechanics*; Chapter 3, Problem 3; Pergamon: Oxford, UK, 1960.

foundations

MDPI

Article

The Frequency Fluctuation Model for the van der Waals Broadening

Andrei Letunov [1,2,*], Valery Lisitsa [1,2,3] and Valery Astapenko [3]

1 National Research Centre "Kurchatov Institute", 123182 Moscow, Russia; vlisitsa@yandex.ru
2 National Research Nuclear University MEPhI, 115409 Moscow, Russia
3 Moscow Institute of Physics and Technology, National Research University, Institutskij Per. 9, 141700 Dolgoprudnyj, Moscow Region, Russia; astval@mail.ru
* Correspondence: letunovandrey11@yandex.ru

Abstract: The effect of atomic and molecular microfield dynamics on spectral line shapes is under consideration. This problem is treated in the framework of the Frequency Fluctuation Model (FFM). For the first time, the FFM is tested for the broadening of a spectral line by neutral particles. The usage of the FFM allows one to derive simple analytical expressions and perform fast calculations of the intensity profile. The obtained results are compared with Chen and Takeo's theory (CT), which is in good agreement with experimental data. It is demonstrated that, for moderate values of temperature and density, the FFM successfully describes the effect of the microfield dynamics on a spectral line shape.

Keywords: spectroscopy; stark broadening; atomic physics

Citation: Letunov, A.; Lisitsa, V.; Astapenko, V. The Frequency Fluctuation Model for the van der Waals Broadening. *Foundations* **2021**, 2, 200–207. https://doi.org/10.3390/foundations1020015

Academic Editor: Eugene Oks

Received: 6 October 2021
Accepted: 25 October 2021
Published: 29 October 2021

1. Introduction

The problem of the microfield dynamics effect on a spectral line shape was recognized many years ago [1–3]. There are different methods which were developed to solve this problem [4–6]. However, the best agreement with the results of Molecular Dynamics (MD) simulations provides the Frequency Fluctuation Model (FFM) [7]. It is based on dividing a spectral line contour in a static field into separate regions, between which there is an exchange of intensities due to thermal motion. This model is widely used for spectral line shape calculations in plasmas (see, e.g., [8–13]). It was shown that the FFM is equivalent to the method of the quantum kinetic Equation [14]. This approach makes it possible to reformulate the FFM in terms of analytical expressions. Namely, the FFM spectral line shape can be considered as the functional of the static profile. This circumstance allows one to use simple analytical expressions and perform fast calculations of a spectral line shape for arbitrary values of temperatures and densities.

The FFM was formulated for spectral line shape calculations in plasmas. It was never used for the Stark broadening by neutral particles—to the best of our knowledge. Most of the material on the van der Waals broadening accumulated by the end of the 1960s is presented in the famous review of S. Chen and M. Takeo [15]. Moreover, in this paper, the authors present the line shape calculation method (based on the work of P.W. Anderson and J.D. Talman [16]) for arbitrary values of temperatures and densities. The Chen and Takeo results (CT) are very old. Nevertheless, they are in good agreement with experimental data and can be used for the examination of the FFM theory. Moreover, the analytical expressions for the line shape presented in [15] are very cumbersome. The direct use of the CT theory in numerical calculations requires significant computational resources. Therefore, the FFM may provide simple analytical expressions for the description of the van der Waals broadening.

The FFM has a drawback. It does not reproduce the impact width correctly. This circumstance was noted in the original paper [7]. Furthermore, it was directly shown in [12,17]. However, this problem was solved for the linear Stark effect. The resolution consists of the dependence of the jumping frequency, which characterizes the rate of change of the microfield on the energy shift. This dependence is obtained by comparing the analytical calculations with MD in the paper [12]. The authors of the work [17] overcame this problem in the alternative way. They used the asymptotic expression of the jumping frequency obtained by S. Chandrasekhar and J. von Neumann [18], who derived this for the description of the stellar dynamics.

The detailed analysis of the FFM results for different plasma parameters is presented in the paper [17]. It is demonstrated that the accounting for the dependence of the jumping frequency on the energy shift leads to the correct behaviour of the spectral line width in the impact limit for the linear Stark effect. The difference between the modified and the original versions of the FFM for moderate values of temperatures and densities is insignificant. Therefore, accounting for the jumping frequency dependence on the field strength yields the simple analytical theory which provides the correct results for a wide range of plasma parameters. The situation is more complicated for the quadratic Stark effect. In this case, the problem of reproducing the impact width is not solved in this way. However, for realistic plasma parameters, the FFM with the constant jumping frequency provides quite satisfactory results.

Usually, the van der Waals broadening can be described in terms of the impact approximation [19]. Furthermore, there are modern works related to the spectral line broadening by neutral atoms and molecules (see, e.g., [20,21]). Therefore, the main goal of the present paper is the examination of the FFM for the van der Waals interaction. The second aim of this work is to provide the simple analytical algorithm for the line shape calculations for low temperatures (less that 300 K) or high densities.

2. Description of the Method

In the present paper, we focused on the spectral broadening by neutral atoms and molecules. The detailed description of the analytical FFM formulation and the discussion of the dynamics effect on the spectral line formation are given in [17]. For convenience, an abridged version was presented below.

The effect of a multiparticle electric field on molecular and atomic spectra is characterized by the ratio of the jumping frequency and the Stark shift:

$$\nu = \frac{\Omega_J}{\Omega_S^n}, \tag{1}$$

where

$$\Omega_J = N^{1/3} v_T, \tag{2}$$

$$\Omega_S^n = C_n^S N^{\frac{n}{3}}. \tag{3}$$

Here, N and v_T are, respectively, the density and the thermal velocity of interacting particles; C_n^S is the constant of the Stark effect; Ω_J is called the jumping frequency. The potential of binary interacting particles has the following form:

$$V(r) = \frac{C_n^S}{R^n}. \tag{4}$$

In the present paper we considered the case of $n = 6$, which corresponds to a wide class of van der Waals interactions. The interaction between atoms and molecules often defers from (4). However, we focused on the examination of the FFM by the comparison with CT results, which were obtained for the potential (4).

The transition from binary to non-binary type of interactions is often characterized by the number of particles is the Weisskopf sphere h. For $n = 6$ in Expression (4), it equals to [19]

$$h = \left(\frac{3\pi}{8}\frac{C_6^S}{v_T}\right)^{3/5} N. \tag{5}$$

Parameter (5) also determines the transition from the static theory to the impact limit. The number of particles in the Weisskopf sphere is connected with ν by the simple relation

$$h \sim \nu^{-3/5}. \tag{6}$$

Using the results from [19], we could estimate the parameter ν. For $T \approx 300 - 5000$ K, ν approximately equals to:

$$\nu \approx (10^{35} - 10^{38}) N^{-5/3}, \tag{7}$$

where N is expressed in cm^{-3}.

All over the paper, atomic units were used ($e = m = \hbar = 1$, where: e—the elementary charge; m—the mass of an electron; $\hbar$ is the Planck constant). Furthermore, for the sake of simplicity, we used the reduced detuning: $z = \Delta\omega/\Omega_S^n$.

The FFM procedure is equivalent to the method of the kinetic equation with strong-collision integral, which describes an intensity exchange between different regions of a static profile [14]. The frequency of the exchange equals to ν. The solution of this equation led to the following expression for the resulting profile:

$$I(z) = \frac{1}{\pi} Re \frac{\int \frac{W(z')dz'}{\nu + i(z - z')}}{1 - \nu \int \frac{W(z')dz'}{\nu + i(z - z')}}, \tag{8}$$

where $W(z)$ is the normalized static profile. Note, that when $\nu \to 0$, $I(z)$ turns into $W(z)$.

The Formula (8) can be rewritten in more convenient form:

$$I(z) = \frac{\nu}{\pi} \frac{J_0(z)J_2(z) - J_1^2(z))}{J_2^2(z) + \nu^2 J_1^2(z)}, \tag{9}$$

where

$$J_k(z) = \int\limits_{-\infty}^{+\infty} \frac{W(z')(z - z')^k dz'}{\nu^2 + (z - z')^2}, \tag{10}$$

There is a useful relation between the functions $J_0(z)$ and $J_2(z)$:

$$J_2(z) - 1 - \nu^2 J_0(z). \tag{11}$$

S. Chen and M. Takeo presented the analytical expression for the intensity profile for an arbitrary value of ν for the van der Waals broadening [15]:

$$I^{(CT)}(z) = C \cdot Re\left\{ \int\limits_{-\infty}^{\infty} \exp\left[iz\tau - \frac{1}{\nu^{3/5}}\psi\left(\nu^{6/5}\tau\right)\right] d\tau \right\}, \tag{12}$$

where

$$C = \left[\int\limits_{-\infty}^{+\infty} Re\left\{ \int\limits_{-\infty}^{\infty} \exp\left[iz\tau - \frac{1}{\nu^{3/5}}\psi\left(\nu^{6/5}\tau\right)\right] d\tau \right\} dz \right]^{-1}.$$

The function $\psi(x)$ can be approximated by the simple functions:

$$\begin{cases} \psi(x \ll 1) = \dfrac{(2\pi)^{3/2}}{3}\sqrt{x}(1-i), \\ \psi(x \gg 1) = \dfrac{2\pi}{5}\left(\dfrac{3\pi}{8}\right)^{-3/5}\Gamma\left(\dfrac{2}{5}\right)\left(\cos\dfrac{\pi}{5} - i\sin\dfrac{\pi}{5}\right)x + 1.91 + 2.63i, \end{cases} \tag{13}$$

where $\Gamma(z)$ is the gamma function.

Expression (12) turned into the static profile $W(z)$ when ν approached zero. In order to derive $W(z)$, it was necessary to use the approximation for $\psi(x)$ for small values of x.

Using the first relation of (13) and the stationary phase approximation, one obtains the following formula for the static profile:

$$W(z) = \begin{cases} 0, z \geq 0 \\ \dfrac{2\pi}{3|z|^{3/2}}\exp\left[\dfrac{4\pi^3}{9z}\right], z < 0 \end{cases} \tag{14}$$

Result (14) is in agreement with the well-known formula derived by H. Margenau [22].

The examination of the FFM consisted of the comparison of the CT Formula (12) with Expression (9). In the integrand of (10), we substituted the static profile (14). In the impact limit ($\nu \gg 1$), we could perform a simple estimation of the impact width. Namely, for large values of ν, the spectral line shape turned into the Lorentzian:

$$L(z) = \frac{1}{\pi}\frac{\gamma}{\gamma^2 + (z - z_0)^2}, \tag{15}$$

where γ is the width of the profile and z_0 is the coordinate of the centre of the Lorentz profile.

Note, that the CT theory reproduced the impact limit. Indeed, the usage of the second relation from (13) led to Expression (15).

In order to estimate the impact width for the FFM, we used the following property of a wide class of normalized profiles:

$$F(0) \sim \frac{1}{\gamma}. \tag{16}$$

Relation (16) was valid here, because in the impact limit, γ and z_0 always have the same dependence on ν (see, e.g., [19]). Therefore, it was obvious that Property (16) worked for the Lorentz profile (15). Using the asymptotic behaviour of Formula (14), we could estimate $J_k(0)$:

$$\begin{gathered} J_0(0) \sim \frac{1}{\nu^2}, \\ J_1(0) \sim \int\limits_{\nu}^{+\infty} dz' \frac{1}{z'^{1/2}(z'^2 + \nu^2)} \sim \frac{1}{\nu^{3/2}}, \\ J_2(0) \sim \int\limits_{\nu}^{+\infty} dz' \frac{z'^{1/2}}{z'^2 + \nu^2} \sim \frac{1}{\nu^{1/2}}, \\ I(0) \sim \frac{1}{\nu^{1/2}}. \end{gathered} \tag{17}$$

From Relations (16) and (17), it was easy to see that $\gamma = \gamma_{FFM} \sim \nu^{1/2}$. Thus, the FFM did not reproduce the behaviour of the profile width in the impact limit. Indeed, it was the well-know result of the impact theory [19], which was $\gamma = 8.16\nu^{3/5}$. The difference in the dependence of the impact width on the power of velocity was very small $\gamma_{impact}/\gamma_{FFM} \sim \nu^{1/10}$.

3. The Results of Numerical Calculations

In order to test the FFM procedure, we compared it with the CT results. The numerical calculations showed that, for $\nu \sim 1$, the FFM was in good agreement with CT. The example of it is demonstrated in Figure 1, where one can see the comparison of two theories for $\nu = 1.5$. However, the FFM did not reproduce the impact width correctly (Figure 2) as it was shown in the previous section by the analytical estimations. The comparison of the CT theory (in the impact limit) with the FFM profile is presented in Figure 2. The full width at half maximum of the FFM profile was approximately 1.5 as large as the impact width for $\nu = 100$. With an increase in the parameter ν, the FFM profile slowly became wider than the impact line shape. This circumstance was connected with the slow growth of the ratio: $\gamma_{impact}/\gamma_{FFM} \sim \nu^{1/10}$. Moreover, we only knew how γ_{FFM} depended on ν. Uncertainty in the value of the prefactor played an important role in the impact width determination. However, the derivation of the certain value of the FFM impact width prefactor by analytical calculations is a complex problem. It might be determined numerically as in the work of [17], but it is not reasonable to do so until the dependence of the jumping frequency on the detuning is determined. The discrepancy between two graphs became insignificant for $\nu \sim 1000$ (Figure 3). According to Estimation (7), this value of ν for $T = 300$–5000 K corresponded to $N = 10^{19}$–10^{21} cm^{-3}.

The discrepancy in the impact limit might have most likely been eliminated by accounting for the dependence of the jumping frequency on the detuning. It was shown that, for the linear Stark effect, it led to the correct result for large values of ν [17]. However, in order to determine this dependence, it was necessary to describe complicated dynamics of the van der Waals forces.

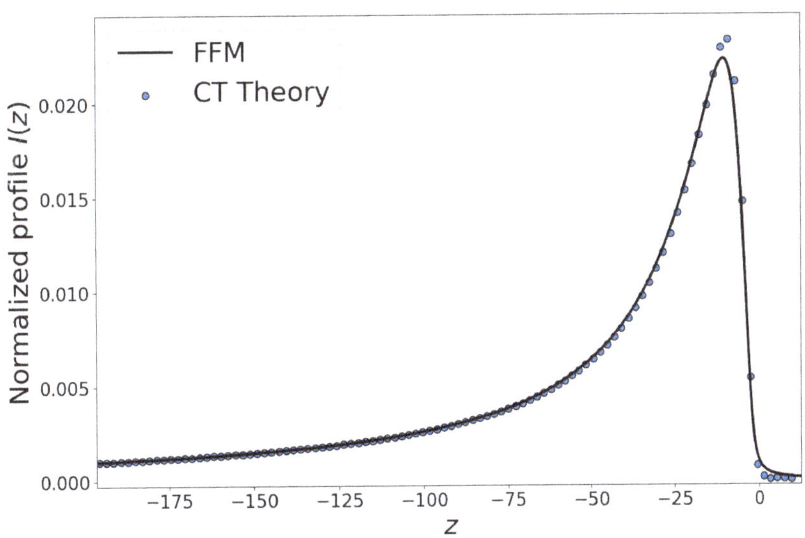

Figure 1. The normalized intensity profile of one spectral component as the function of the reduced energy shift. Comparison of the FFM profile with the Chen and Takeo theory; $\nu = 1.5$.

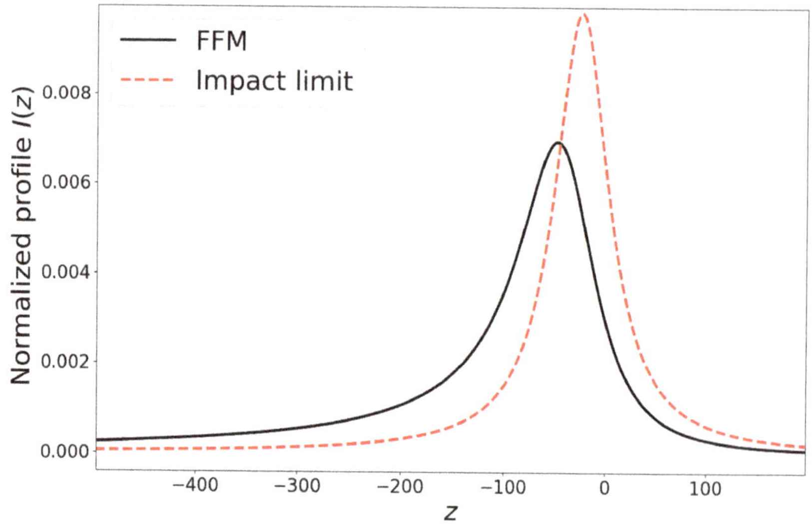

Figure 2. The normalized intensity profile of one spectral component as the function of the reduced energy shift. Comparison of the FFM profile with the impact theory; $\nu = 100$.

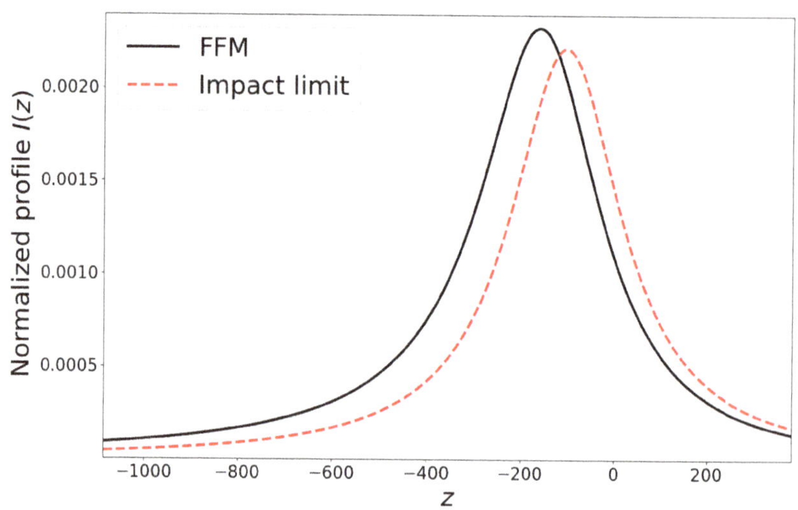

Figure 3. The normalized intensity profile of one spectral component as the function of the reduced energy shift. Comparison of the FFM profile with the impact theory; $\nu = 1200$.

4. Conclusions

The construction of the general analytical theory, which described the effect of the complicated microfield dynamics on the spectral line shape formation, faced the greatest difficulties. The Frequency Fluctuation Model is believed to be the most accurate and simplest method for account such effects. Therefore, it was necessary to examine this theory for different types of interactions. In the present paper, we tested the FFM for the van der Waals forces.

The FFM showed good results for moderate values of temperature and density. In the case of broadening by neutral atoms, it was no exception. The results of the FFM procedure were in agreement with the CT theory (Figure 1).

The FFM did not reproduce the impact width correctly (Figure 2). As it was shown in [17], for the linear Stark effect, this problem could be solved by accounting for the dependence of the jumping frequency on the field strength. Apparently, the resolution of the problem of the broadening by neutral particles consisted of the description of the complicated dynamics of the van der Waals field.

Author Contributions: Conceptualization, V.L., V.A. and A.L.; methodology, V.L.; software, A.L.; validation, V.L., A.L. and V.A.; formal analysis, V.L., V.A. and A.L.; investigation, A.L., V.A. and V.L.; resources, A.L.; data curation, V.L.; writing—original draft preparation, A.L.; writing—review and editing, V.L.; visualization, A.L.; supervision, V.L.; project administration, V.L. All authors have read and agreed to the published version of the manuscript.

Funding: This research received no external funding.

Institutional Review Board Statement: Not applicable.

Informed Consent Statement: Not applicable.

Conflicts of Interest: The authors declare no conflict of interest.

Abbreviations

The following abbreviations are used in this manuscript:

FFM	Frequency Fluctuation Model
MD	Molecular Dynamics
CT	Chen and Takeo

References

1. Griem, H.; Baranger, M.; Kolb, A.; Oertel, G. Stark broadening of neutral helium lines in a plasma. *Phys. Rev.* **1962**, *125*, 177. [CrossRef]
2. Kogan, V. Broadening of spectral lines in high-temperature plasma. *Plasma Phys. Probl. Control Fusion* **1958**, *4*, 305.
3. Stambulchik, E.; Maron, Y. Plasma line broadening and computer simulations: A mini-review. *High Energy Density Phys.* **2010**, *6*, 9–14. [CrossRef]
4. Brissaud, A.; Frisch, U. Theory of Stark broadening—II exact line profile with model microfield. *J. Quant. Spectrosc. Radiat. Transf.* **1971**, *11*, 1767–1783. [CrossRef]
5. Alexiou, S. Implementation of the Frequency Separation Technique in general lineshape codes. *High Energy Density Phys.* **2013**, *9*, 375–384. [CrossRef]
6. Ferri, S.; Calisti, A.; Mossé, C.; Rosato, J.; Talin, B.; Alexiou, S.; Gigosos, M.A.; González, M.A.; González-Herrero, D.; Lara, N.; et al. Ion Dynamics Effect on Stark-Broadened Line Shapes: A Cross-Comparison of Various Models. *Atoms* **2014**, *2*, 299–318. [CrossRef]
7. Talin, B.; Calisti, A.; Godbert, L.; Stamm, R.; Lee, R.; Klein, L. Frequency-fluctuation model for line-shape calculations in plasma spectroscopy. *Phys. Rev. A* **1995**, *51*, 1918. [CrossRef] [PubMed]
8. Mossé, C.; Calisti, A.; Stamm, R.; Talin, B.; Bureyeva, L.; Lisitsa, V. A universal approach to Rydberg spectral line shapes in plasmas. *J. Phys. At. Mol. Opt. Phys.* **2004**, *37*, 1343. [CrossRef]
9. Calisti, A.; Bureyeva, L.; Lisitsa, V.; Shuvaev, D.; Talin, B. Coupling and ionization effects on hydrogen spectral line shapes in dense plasmas. *Eur. Phys. J. D* **2007**, *42*, 387–392. [CrossRef]
10. Calisti, A.; Mossé, C.; Ferri, S.; Talin, B.; Rosmej, F.; Bureyeva, L.; Lisitsa, V. Dynamic Stark broadening as the Dicke narrowing effect. *Phys. Rev. E* **2010**, *81*, 016406. [CrossRef] [PubMed]
11. Ferri, S.; Calisti, A.; Mossé, C.; Mouret, L.; Talin, B.; Gigosos, M.A.; González, M.A.; Lisitsa, V. Frequency-fluctuation model applied to Stark-Zeeman spectral line shapes in plasmas. *Phys. Rev. E* **2011**, *84*, 026407. [CrossRef] [PubMed]
12. Stambulchik, E.; Maron, Y. Quasicontiguous frequency-fluctuation model for calculation of hydrogen and hydrogenlike Stark-broadened line shapes in plasmas. *Phys. Rev. E* **2013**, *87*, 053108. [CrossRef] [PubMed]
13. Letunov, A.; Lisitsa, V. The Coulomb Symmetry and a Universal Representation of Rydberg Spectral Line Shapes in Magnetized Plasmas. *Symmetry* **2020**, *12*, 1922. [CrossRef]
14. Bureeva, L.; Kadomtsev, M.; Levashova, M.; Lisitsa, V.; Calisti, A.; Talin, B.; Rosmej, F. Equivalence of the method of the kinetic equation and the fluctuating-frequency method in the theory of the broadening of spectral lines. *JETP Lett.* **2010**, *90*, 647–650. [CrossRef]
15. Takeo, M.; Chen, S. Broadening and shift of spectral lines due to the presence of foreign gases. *Rev. Mod. Phys.* **1957**, *29*, 20.
16. Anderson, P.; Talman, J. In Proceedings of the Conference on Broadening of Spectral Lines, Murray Hill, NJ, USA, 3 May 1956.

17. Astapenko, V.; Letunov, A.; Lisitsa, V. From the Vector to Scalar Perturbations Addition in the Stark Broadening Theory of Spectral Lines. *Universe* **2021**, *7*, 176. [CrossRef]
18. Chandrasekhar, S.; von Neumann, J. The Statistics of the Gravitational Field Arising from a Random Distribution of Stars. *Astrophys. J.* **1942**, *95*, 489–531. [CrossRef]
19. Sobelman, I.I. *Introduction to the Theory of Atomic Spectra: International Series of Monographs in Natural Philosophy*; Elsevier: Amsterdam, The Netherlands, 2016; Volume 40.
20. Demura, A.; Umanskii, S.Y.; Scherbinin, A.V.; Zaitsevskii, A.V.; Demichenko, G.V.; Astapenko, V.A.; Potapkin, B.V. Evaluation of Van der Waals Broadening Data. *Int. Rev. At. Mol. Phys.* **2011**, *2*, 109–150.
21. Omar, B.; González, M.Á.; Gigosos, M.A.; Ramazanov, T.S.; Jelbuldina, M.C.; Dzhumagulova, K.N.; Zammit, M.C.; Fursa, D.V.; Bray, I. Spectral line shapes of He I line 3889 Å. *Atoms* **2014**, *2*, 277–298. [CrossRef]
22. Margenau, H. Theory of pressure effects of foreign gases on spectral lines. *Phys. Rev.* **1935**, *48*, 755. [CrossRef]

foundations

MDPI

Communication

Possible Observational Evidence for the Existence of a Parallel Universe

Eugene Oks

Physics Department, Auburn University, 380 Duncan Drive, Auburn, AL 36849, USA; goks@physics.auburn.edu

Abstract: Many totally different kinds of astrophysical observations demonstrated that, in our universe, there exists a preferred direction. Specifically, from observations in a wide range of frequencies, the alignment of various preferred directions in different data sets was found. Moreover, the observed Cosmic Microwave Background (CMB) quadrupole, CMB octopole, radio and optical polarizations from distant sources also indicate the same preferred direction. While this hints at a gravitational pull from the "outside", the observational data from the Plank satellite showed that the bulk flow velocity was relatively small: much smaller than was initially thought. In the present paper we propose a configuration where two three-dimensional universes (one of which is ours) are embedded in a four-dimensional space and rotate about their barycenter in such a way that the centrifugal force nearly (but not exactly) compensates their mutual gravitational pull. This would explain not only the existence of a preferred direction for each of the three-dimensional universes (the direction to the other universe), but also the fact that the bulk flow velocity, observed in our universe, is relatively small. We point out that this configuration could also explain the perplexing features of the Unidentified Aerial Phenomena (UAP), previously called Unidentified Flying Objects (UFOs), recorded by various detection systems—the features presented in the latest official report by the US Office of the Director of National Intelligence. Thus, the proposed configuration of the two rotating, parallel three-dimensional universes seems to explain both the variety of astrophysical observations and (perhaps) the observed features of the UAP.

Keywords: parallel universes; multiverse; preferred direction in the universe; bulk flow; four spatial dimensions

Citation: Oks, E. Possible Observational Evidence for the Existence of a Parallel Universe. *Foundations* **2022**, *2*, 1–5. https://doi.org/10.3390/foundations2010001

Academic Editors: Kazuharu Bamba and Martin Bohner

Received: 19 November 2021
Accepted: 22 December 2021
Published: 23 December 2021

1. Introduction

The hypothesis for the existence of a parallel universe or universes (in the latter case called multiverses) has proponents and opponents among astrophysicists, e.g., see works [1–6] and references therein. The primary argument against this hypothesis was the lack of the observational evidence. In response, the proponents theorized (at different times) that the following two kinds of observations might constitute such evidence.

One theory was based on an early observation of "bulk flow" (i.e., a stream of galaxy clusters moving in the same direction), where the bulk flow velocity was found to be >4000 km/s [7]. These observations could be interpreted as evidence for the existence of a parallel universe. However, later, more precise observations (from the Plank satellite) revealed that the average clusters' velocities are "compatible with zero", being at the level of 120–160 km/s [8]. The authors of paper [8] wrote that this "constitutes an unprecedented and valuable confirmation of a prediction of the standard cosmological scenario"; therefore, this proposed evidence for the existence of a parallel universe should be discarded.

Another theory was that the observed cold spot in the Cosmic Microwave Background (CMB) radiation (nestled in the constellation Eridanus) is the remnant of a collision between our universe and another "bubble" universe during an early inflationary phase, e.g., see review [9] and references therein. Another hypothesis [10] was that the cold spot could be the imprint of another universe beyond our own, caused by quantum entanglement between universes before they are separated by cosmic inflation. However, a more thorough analysis of data from the WMAP (Wilkinson Microwave Anisotropy Probe) and from the Plank satellite, which has a resolution three times higher than WMAP, did not reveal any statistically significant evidence for such a bubble universe collision [11–13].

In the present paper we show that the existing observations, astrophysical or otherwise, might actually constitute proof for the existence of a parallel universe.

2. Possible Observational Evidence

Many totally different kinds of astrophysical observations demonstrated that, in our universe, there exists a preferred direction (called the "axis of evil" in paper [14]; see the review [15] and references therein). In particular, the author of the review [15] wrote:

"A very curious feature of SI [Statistical Isotropy] violations is the alignment of various preferred directions in different data sets. Several observations at wide range of frequencies suggest a preferred direction pointing roughly towards the Virgo supercluster, which is close to the direction of the observed CMB dipole. … Furthermore, the observed CMB quadrupole, CMB octopole, radio and optical polarizations from distant sources also indicate a preferred direction pointing roughly towards Virgo. … Statistical isotropy would imply that these are independent of one another as well of other multipoles, such as the dipole. However the preferred axis of both these multipoles points approximately in the direction of the CMB dipole. … This is rather surprising!"

It should be noted that the CMB dipole can be interpreted as the peculiar motion of the Earth toward the CMB. However, the CMB quadrupole and CMB octupole cannot be interpreted in this way.

Thus, the existence of the preferred direction (or axis) in our universe is undisputable. This hints at a gravitational pull from the "outside". However, the observational data from the Plank satellite showed (according to paper [8]) that the bulk flow velocity was no more than 160 km/s, i.e., much smaller than the previous observational result of >4000 km/s, and thus the actual gravitational pull from the outside was much smaller than was initially thought.

There is a possible way to reconcile the undisputable existence for the preferred direction of our universe with the relative smallness of the gravitational pull from the outside. Let us consider two three-dimensional universes (one of which is ours) embedded in a four-dimensional space. (By this, we mean only spatial dimensions.) The two universes rotate about their barycenter in such a way that the centrifugal force nearly (though not exactly) compensates their mutual gravitational pull. In this configuration, within each of the three-dimensional universes, there would be a preferred direction: the direction to the other universe. Additionally, in this configuration, the bulk flow velocity (in each of the universes) would be relatively small because the centrifugal force nearly compensates the gravitational force.

This scenario seems to offer a possible explanation for the above astrophysical observations; therefore, it seems to be self-sufficient. Nevertheless, it should be mentioned that there could also be non-astrophysical evidence for this scenario, as presented below.

Paper [16] focused on the following three perplexing features of Unidentified Aerial Phenomena (UAP), previously called Unidentified Flying Objects (UFOs), from the latest official report by the US Office of the Director of National Intelligence [17], where, out of 144 relatively resent observations of UAP by the US military recorded by various detection systems, 143 remained unexplained. First, some UAP demonstrated accelerations (measured by detection systems) of about 700 g. Humans, even those who are astronauts, can stand the acceleration of no more than about 10 g. Second, UAP can both appear suddenly and disappear suddenly (almost instantaneously), which is impossible for man-made objects. Third, these observed UAP were capable of traveling back and forth in air and water, without any significant change of the dynamics, which is also impossible for man-made objects.

For a more visual presentation of that author's main idea, one of his papers [16] first discussed the following. If an experimentalist shone a laser beam on a distant surface (e.g., on the surface of the Moon) and rotate the laser with some angular velocity, the bright spot can travel across the distant surface with a very large linear velocity, even exceeding the speed of light. (No physical law would be violated because it is the information that cannot be transmitted faster than the speed of light, while the bright spot cannot transmit any information from one place on the surface to another.)

If the experimentalist sharply changed the direction of the motion of the laser, the bright spot on a distant surface would exhibit an extremely sharp turn. If a hypothetical, two-dimensional observer residing on this surface calculated the "acceleration" of this "object" during the extremely sharp turn, the observer would achieve a very large value for this "acceleration", a number far exceeding the technological capabilities of the observer's community.

If the experimentalist (on the Earth) initially shone the laser beam parallel to a distant surface, and then abruptly changed the direction of the beam to hit the surface, the two-dimensional observer on this surface would register a sudden appearance of the bright spot. If later on, the experimentalist abruptly changed the direction of the laser beam to be parallel to the surface, that observer would register a sudden disappearance of the bright spot. In both cases, the observer would qualify this as being beyond the technological capabilities of the observer's community.

Further, let us picture that on that surface there are dry regions (the "air") and wet regions (the "water"). The bright spot can move through the "air", then through the "water", then again through the "air", without any change in its velocity (the velocity controlled by the motion of the laser in the third dimension). The two-dimensional observer on that surface would again qualify this as being beyond the technological capabilities of the observer's community.

At this point in paper [16], the following was written.

"Now let us add an extra spatial dimension both to the "surface" and to the space, from which the light is shined. Now the "surface" becomes our three-dimensional world, into which the light is incoming from the fourth spatial dimension. In our world we see a three-dimensional "bright spot". This "bright spot" is the projection of the light coming from the four-dimensional world on the three-dimensional "screen", the "screen" being our three-dimensional world."

Clearly, in this situation, all of the above three perplexing features of the three-dimensional "bright spot" would be observed and registered by detection systems, with features that are far beyond our technological capabilities, and we would consider such three-dimensional "bright spots" to be UAP. In other words, there is an explanation for all three perplexing features of the observed UAP; they are the three-dimensional projections of the light entering our world from the fourth dimension. In paper [16] the following was written:

"By varying the intensity distribution of the cross-section of the light beam at the source (for example, by using various filters), it would be possible to create any shape and form of the three-dimensional projection that we observe, including the shape of "flying saucers" and so on. By varying color filters or their combinations, it would be possible to make the three-dimensional projection of any color or their combinations."

Further in paper [16], it was explained that the detailed information on the properties of the electromagnetic radiation in four spatial dimensions was provided in paper [18]. According to paper [18], the only one difference of the electromagnetic wave in four spatial dimensions, compared to the electromagnetic wave in three spatial dimensions, is that, in the four-dimensional case, it is intertwined with a weak oscillatory gravitational field (the gravitational field oscillates in the direction of the propagation of the electromagnetic wave). As for the electric and magnetic components of the four-dimensional electromagnetic wave, they are the same as the three-dimensional world.

In paper [16], it was emphasized that, as of now, no new physical laws has been introduced: everything was based on the standard physics. Next in paper [16], the following was written:

"Next, it might seem that since the light is coming from the world of four spatial dimensions, then the source of light should be controlled by four-dimensional intelligent creatures (this would belong to realm of science fiction). However, this does not have to be the case ... the source of light could be located and controlled in a parallel three-dimensional world by three-dimensional relatively advanced civilization that developed the capability to manipulate the electromagnetic radiation in the way described above. By projecting the light into our three-dimensional world and detecting the reflected light, they monitor our technological capabilities."

Then, in paper [16], the above scenario was compared to the only existing alternative explanation for 143 unexplained UAP from the aforementioned official report—that the unexplained UAP could be drones. The perplexing features of these drones hint to their extra-terrestrial origins. There are three shortcomings of the drone hypothesis compared to the scenario where the UAP are the three-dimensional projections.

First, since our astrophysicists have not yet detected any extraterrestrial civilization located within hundreds of light years from the Earth, the advanced civilization controlling the extra-terrestrial drones would receive the many hundreds years or even in thousands years into the future. In contrast, in the scenario where the UAP are the three-dimensional projections, the information carried by the reflected light, could reach the origin of the light in just few years or less, since the parallel three-dimensional world could be just few light years (or less) away from our three-dimensional world across the four-dimensional space. Clearly, it would make much more sense to monitor our technological capabilities with just a few years of delay, as opposed to monitoring them with the delay of hundreds or thousand years.

Second, in the scenario where the observed UAP are extraterrestrial drones, the extraterrestrial civilization would be extremely advanced; otherwise, it would not be able to make spacecrafts that can withstand the acceleration of 700 g and can interchange the motion in the air and under water without any significant change of the velocity. In contrast, in the scenario of the UAP being the three-dimensional projections, it would be sufficient for the other civilization to be only slightly advanced—just capable of manipulating the electromagnetic radiation in the way described above.

Third, but most important: the scenario of extraterrestrial drones cannot explain the sudden, almost instantaneous appearance of the UAP and the subsequent sudden, almost instantaneous disappearance of the UAP. In contrast, in the scenario with the UAP being the three-dimensional projections, this perplexing feature is easy to explain.

Thus, the configuration of two parallel, three-dimensional universes embedded in a four-dimensional space (where they rotate about their barycenter) explains not only all relevant astrophysical observation, but perhaps also the mind-boggling features of the observed UAP (that have no consistent alternative explanation).

3. Conclusions

We started from the undisputable fact that, from various astrophysical observations of very different kinds, our universe has a preferred direction in space. We proposed a configuration where two three-dimensional universes (one of which is ours) are embedded in a four-dimensional space and rotate about their barycenter in such a way that the centrifugal force nearly (but not exactly) cancels out their mutual gravitational pull. This would explain not only the existence of a preferred direction within each of the three-dimensional universes (the direction to the other universe), but also the fact that the bulk flow velocity, observed in our universe, is relatively small.

We pointed out that this configuration could also explain the perplexing features of the UAP recorded by various detection systems—the features presented in the latest official report by the US Office of the Director of National Intelligence. Thus, the proposed configuration of the two rotating, parallel, three-dimensional universes seems to explain both the variety of the astrophysical observation and (perhaps) the observed features of the UAP.

We mention that one of the interpretations of quantum mechanics suggests the existence of parallel universes. It is the many-world interpretation, proposed as early as 1957 [19].

The totality of the astrophysical observations should be emphasized—those proving the existence of the preferred direction in our universe, as well as astrophysical observations where a weak but non-zero bulk flow (i.e., the gravitational pull from the "outside") was measured. This seems by itself to be sufficient for making the proposed configuration of the two rotating, parallel, three-dimensional universes viable. We hope that our work stimulates a further discussion of these issues.

Funding: This research received no external funding.

Institutional Review Board Statement: Not applicable.

Informed Consent Statement: Not applicable.

Data Availability Statement: All data is included in the paper.

Conflicts of Interest: The author declares no conflict of interest.

References

1. Kragh, H. Contemporary History of Cosmology and the Controversy over the Multiverse. *Ann. Sci.* **2009**, *66*, 529–551. [CrossRef]
2. Greene, B. *The Hidden Reality: Parallel Universes and the Deep Laws of the Cosmos*; A.A. Knopf: New York, NY, USA, 2011.
3. Nomura, Y. Physical theories, eternal inflation, and the quantum universe. *J. High Energ. Phys.* **2011**, *2011*, 63. [CrossRef]
4. Ellis, G.F.R. Does the Multiverse Really Exist? *Sci. Am.* **2011**, *305*, 38–43. [CrossRef] [PubMed]
5. Tegmark, M.; Vilenkin, A. The Case for Parallel Universes. *Sci. Am.* **2011**, *305*. Available online: https://www.scientificamerican.com/article/multiverse-the-case-for-parallel-universe/ (accessed on 16 November 2021).
6. Bousso, R.; Susskind, L. Multiverse Interpretation of Quantum Mechanics. *Phys. Rev. D* **2012**, *85*, 045007. [CrossRef]
7. Abate, A.; Feldman, H.A. Detected Fluctuations in Sloan Digital Sky Survey Luminous Red Galaxy Magnitudes: Bulk Flow Signature or Systematic? *Mon. Not. R. Astron. Soc.* **2010**, *419*, 3482–3490. [CrossRef]
8. Ade, P.A.R.; Aghanim, N.; Arnaud, M.; Ashdown, M.; Aumont, J.; Baccigalupi, C.; Balbi, A.; Banday, A.J.; Barreiro, R.B.; Battaner, E.; et al. Plank Intermediate Results XIII. Constraints on Peculiar Velocities. *Astron. Astrophys.* **2014**, *561*, A97.
9. Mackenzie, R.; Shanks, T.; Bremer, M.N.; Cai, Y.-C.; Gunawardhana, M.L.P.; Kovács, A.; Norberg, P.; Szapudi, I. Evidence Against a Supervoid Causing the CMB Cold Spot. *Mon. Not. R. Astronom. Soc.* **2017**, *470*, 2328–2338. [CrossRef]
10. Mersini-Houghton, L. Beyond the Horizon of the Universe. *Nautilus*, 2 January 2014.
11. Feeney, S.M.; Johnson, M.C.; Mortlock, D.J.; Peiris, H.V. First Observational Tests of Eternal Inflation. *Phys. Rev. Lett.* **2011**, *107*, 071301. [CrossRef] [PubMed]
12. Feeney, S.M.; Johnson, M.C.; Mortlock, D.J.; Peiris, H.V. First Observational Tests of Eternal Inflation: Analysis Methods and WMAP 7-Year Results. *Phys. Rev. D* **2011**, *84*, 43507. [CrossRef]
13. Bousso, R.; Harlow, D.; Senatore, L. Inflation after False Vacuum Decay: Observational Prospects after Planck. *Phys. Rev. D* **2015**, *91*, 083527. [CrossRef]
14. Land, K.; Maguijo, J. Examination of Evidence for a Preferred Axis in the Cosmic Radiation Anisotropy. *Phys. Rev. Lett.* **2005**, *95*, 071301. [CrossRef] [PubMed]
15. Ghosh, S.; Jain, P.; Kashyap, G.; Kothari, R.; Nadkarni-Ghosh, S.; Tiwari, P. Probing Statistical Isotropy of Cosmological Radio Sources using Square Kilometre Array. *J. Astrophys. Astron.* **2016**, *37*, 25. [CrossRef]
16. Oks, E. Explaining Mind-Boggling Features of UFOs from the Physical Point of View. *Intern. Rev. Atom. Mol. Phys.* **2021**, *11*, 115–117.
17. Office of the Director of National Intelligence. Preliminary Assessment: Unidentified Aerial Phenomena. 2021. Available online: https://www.dni.gov/files/ODNI/documents/assessments/Prelimary-Assessment-UAP-20210625.pdf (accessed on 16 November 2021).
18. Corben, H.C. A Classical Theory of Electromagnetism and Gravitation. *Phys. Rev.* **1946**, *69*, 225–234. [CrossRef]
19. Everett, H. Relative State Formulation of Quantum Mechanics. *Rev. Mod. Phys.* **1957**, *29*, 454–462. [CrossRef]

foundations

MDPI

Article

Greenhouse Effect in the Standard Atmosphere

Boris Michailovich Smirnov * and Dmitri Alexandrovich Zhilyaev *

Institute for High Temperatures of Russian Academy Sciences, Izhorskaya 13/19, 127412 Moscow, Russia
* Correspondence: bmsmirnov@gmail.com (B.M.S.); zhiliay@gmail.com (D.A.Z.)

Abstract: The "line-by-line" method is used for the evaluation of thermal emission of the standard atmosphere toward the Earth. Accounting for thermodynamic equilibrium of the radiation field with air molecules and considering the atmosphere as a weakly nonuniform layer, we reduce the emission at a given frequency for this layer containing molecules of various types to that of a uniform layer, which is characterized by a certain radiative temperature T_ω, an optical thickness u_ω and an opaque factor $g(u_\omega)$. Radiative parameters of molecules are taken from the HITRAN database, and an altitude of cloud location is taken from the energetic balance of the Earth. Within the framework of this model, we calculate the parameters of the greenhouse effect, including the partial radiative fluxes due to different greenhouse components in the frequency range up to 2600 cm^{-1}. In addition, the derivations are determined from the radiative flux from the atmosphere to the Earth over the concentration logarithm of greenhouse components. From this, it follows that the observed rate of growth of the amount of atmospheric carbon dioxide accounts for a contribution of approximately 30% to the observed increase in the global atmosphere during recent decades. If we assume that the basic part of the greenhouse effect is determined by an increase in the concentration $c(H_2O)$ of water atmospheric molecules, it is approximately $d\ln c(H_2O/dt) = 0.003\ \text{yr}^{-1}$. This corresponds to an increase in the average moisture of the atmosphere of 0.2%/yr.

Keywords: line-by-line; greenhouse effect; radiative fluxes; thermal emission

Citation: Smirnov, B.M.; Zhilyaev, D.A. Greenhouse Effect in the Standard Atmosphere. *Foundations* **2021**, *1*, 184–199. https://doi.org/10.3390/foundations1020014

Academic Editor: Eugene Oks

Received: 4 August 2021
Accepted: 22 September 2021
Published: 27 October 2021

1. Introduction

The goal of this paper is to evaluate the radiative fluxes for the greenhouse effect in the atmosphere. This evaluation is based on a developed algorithm formulated in [1]. The used model includes the following features.

1. The "line-by-line" model [2,3] is the basis of these evaluations, and integral radiative fluxes follow from these partial fluxes.

2. The model includes three basic greenhouse components, namely, H_2O molecules, CO_2 molecules and liquid water microdroplets, as the basic condensed phase in the atmosphere. In addition, trace components, such as CH_4 molecules and N_2O molecules, may be included in this scheme.

3. The model of standard atmosphere [4] is the basis of evaluations. In particular, the global temperature (the average temperature of the Earth's surface) is taken as $T_E = 288$ K, and its decrease with altitude h is $dT/dh = -6.5$ K/km. This model provides the altitude distribution for the number densities of atmospheric molecules. The model of standard atmosphere implies that atmospheric parameters depend only on the altitude.

4. Along with the local thermodynamic equilibrium for atmospheric components, this equilibrium takes place between the radiation field and atmospheric air.

5. Parameters of radiative transitions of greenhouse molecules are taken from the HITRAN data bank [5–7]; therefore, we use the formalism for the rates of molecular radiative processes of this data bank [8].

6. The energetic balance of the Earth and its atmosphere is taken into account. According to this balance, radiative fluxes toward the Earth and outside are determined by different atmospheric regions and are separated, i.e., the radiative fluxes to the Earth

that are connected with its temperatures do not depend on processes in high layers of the troposphere.

7. Basic greenhouse components are separated so that clouds are located starting from a certain altitude, and they are characterized by a sharp boundary. Radiation from greenhouse molecules is created in the gap between the Earth's surface and clouds.

Differences between these evaluations and those of [1] are twofold. First, the computer code presented in our paper does not separate frequencies in different ranges; this code relates to the total range of frequencies. Second, evaluations [1] are restricted by frequencies below 1200 cm^{-1}. Since the absorption in an additional frequency range is determined by water molecules, new evaluations may change the emission, due to water molecules. Of course, we account for this in our analysis for a contemporary understanding of atmospheric physics and processes in the atmosphere [2,3,9–22].

2. Model of Atmospheric Emission to the Earth

Thus, we evaluate the radiative fluxes on the basis of the algorithm, which is formulated in the introduction section. The character of radiative processes is given in Figure 1. Correspondingly, according to the model under consideration, we have for the radiative flux J_ω at a given frequency ω the following:

$$J_\omega = I_\omega(T_\omega) g(u_\omega) + I_\omega(T_{cl})[1 - g(u_\omega)] \tag{1}$$

where $I_\omega(T)$ is the radiative flux of a blackbody with a temperature T at this frequency that is given by the Planck formula as follows [23,24]:

$$I_\omega(T) = \frac{\hbar\omega^3}{4\pi^2 c^2 \left[\exp\left(\frac{\hbar\omega}{T}\right) - 1\right]}, \tag{2}$$

and the opaque factor $g(u_\omega)$ of a uniform gaseous layer is given by the following [25,26]:

$$g(u_\omega) = 2\int_0^1 \cos\theta d\cos\theta \left[1 - \exp\left(-\frac{u_\omega}{cos\theta}\right)\right], \tag{3}$$

Formula (3) takes into account the thermodynamic equilibrium of air molecules of the atmosphere with its optically active molecules, water microdroplets of clouds, and the radiation field.

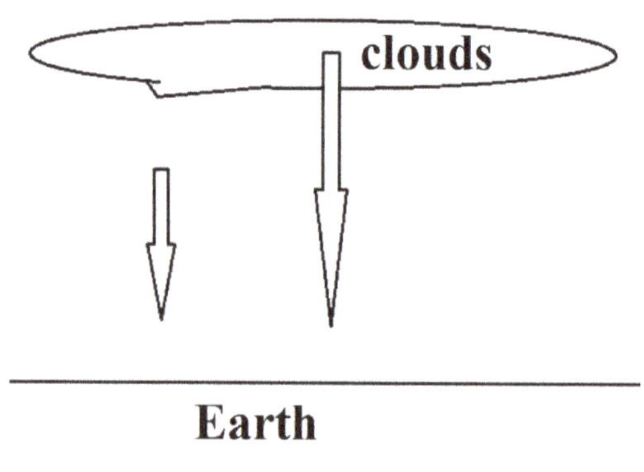

Figure 1. Character of emission of the atmosphere toward the Earth, according to which the radiative flux consists of those from clouds and atmospheric molecules located in the gap between the Earth and clouds.

We thus consider the radiating atmosphere as a shell of the Earth, whose thickness is small compared with the radius of the Earth. Hence, the radiating atmosphere is a weakly nonuniform gaseous layer [27]. One can reduce a radiating weakly nonuniform layer that contains optically active molecules to an effective uniform layer by introducing the effective temperature of radiation or the radiative temperature T_ω [28,29]. In addition, due to thermodynamic equilibrium in the atmosphere, the emission of clouds is characterized by the cloud + temperature T_{cl} or the temperature of the cloud boundary because of its sharp structure.

Formula (1) contains parameters that are expressed through the optical thickness u_ω of the radiated atmospheric layer, which in turn is connected with the optical parameters of optically active molecules, which are located in atmospheric air. There are three greenhouse parameters of the radiated atmospheric layer, namely, H_2O molecules, CO_2 molecules, and clouds consisting mostly of water microdroplets. Correspondingly, the optical thickness of the atmospheric gap located between the Earth and clouds is given by the following [1]:

$$u_\omega(h_\omega) = \sigma_\omega(H_2O)N(H_2O)\lambda\left[1-\exp\left(-\frac{h_\omega}{\lambda}\right)\right] + \sigma_\omega(CO_2)N(CO_2)\Lambda\left[1-\exp\left(-\frac{h_\omega}{\Lambda}\right)\right], \quad (4)$$

where the number densities of water molecules and carbon dioxide molecules are determined by the following formulas:

$$N(H_2O) = N_w \exp\left(-\frac{h}{\lambda}\right),\ N_w = 3.4\times10^{17}\,\mathrm{cm}^{-3},\ \lambda = 2.0\,\mathrm{km}$$
$$N(CO_2) = N_c \exp\left(-\frac{h}{\Lambda}\right),\ N_c = 1.1\times10^{16}\,\mathrm{cm}^{-3},\ \Lambda = 10\,\mathrm{km} \quad (5)$$

We use here the data of the model of standard atmosphere [4] and measured data on the basis of the NASA programs, which are analyzed in [30]. In addition, the equation for the effective altitude h_ω at a given frequency ω has the following form [1]:

$$u_\omega(h_\omega) = \frac{u_\omega(h_{cl})}{2\exp(-u_\omega(h_{cl}) + 1.5u_\omega(h_{cl}))} \quad (6)$$

Within the framework of the model of standard atmosphere, the radiative temperature T_ω for a given frequency follows from the given relation:

$$T_\omega = T_E - h_\omega\frac{dT}{dh}, \quad (7)$$

where the global temperature equals $T_E = 288$ K for the contemporary standard atmosphere, and its gradient is $dT/dh = 6.5$ K/km.

Formula (4) includes the absorption cross section σ_ω for molecules of a given sort, which is a sum of the cross sections for individual spectral lines due to this component, according to the following formula:

$$\sigma_\omega = \sum_j \frac{S_j\nu_j}{2\pi[(\omega-\omega_j)^2 + (\nu_j/2)^2]} \quad (8)$$

As is seen, for each radiative transition, this formula contains three parameters, namely, the transition intensity S_j, the frequency ω_j at the line center, and the width of this spectral line ν_j. We take these parameters from the HITRAN data bank [5–7], and these parameters allow one to determine the absorption cross section at a given frequency. In addition, for air pressures under consideration which are of the order of atmospheric one, the following criterion is fulfilled:

$$\Delta\omega \gg \nu_j, \quad (9)$$

where $\Delta\omega$ is a typical difference of frequencies for centers of neighboring spectral lines.

Note that, according to the Wien law [31], the maximum flux of photons for a blackbody corresponds to the wavelength λ as follows:

$$\lambda_{max}T = 0.3\,\mathrm{cm}\cdot\mathrm{K} \tag{10}$$

From this, it follows that a typical wavelength of atmospheric radiation is $\lambda \sim 10\,\mu\mathrm{m}$. Therefore, below, we are restricted by frequencies $\omega < 2600\,\mathrm{cm}^{-1}$.

In this evaluation, we are guided by strong spectral lines such that at the centers of these lines, the optical thickness satisfies the relation $u_\omega > 1$. In accordance with typical parameters of spectral lines due to water and carbon dioxide molecules, one can select from the HITRAN data bank radiative transitions for these molecules whose intensities satisfy the following relation:

$$S_j(H_2O) \gg 1\times 10^{-24}\,\mathrm{cm},\ S_j(CO_2) \gg 2\times 10^{-23}\,\mathrm{cm} \tag{11}$$

The problem in the evaluation of the radiative flux of the atmosphere is determining the parameters of clouds. In contrast to the number densities of optically active molecules, which are given by Formula (4), analogous information for clouds is absent. Indeed, clouds exist over a given surface point during a restricted time, and their distribution over altitudes has a random character. Within the framework of the model under consideration, we use one parameter of clouds, namely, the altitude h_{cl}, which determines the radiation of clouds, or the radiative temperature of clouds T_{cl}, which follows from the given equation below:

$$T_{cl} = T_E - \frac{dT}{dh}h_{cl}, \tag{12}$$

where the global temperature is $T_E = 288$ K, and the temperature gradient is $dT/dh = 6.5$ K/km.

One can determine the cloud parameters of this model by using the energetic balance of the Earth and atmosphere. The energetic balance includes the radiative flux $J_\downarrow$ from the atmosphere toward the Earth. The energetic balance and this radiative flux follows from different sources, which are presented in Table 1. The total radiative flux from the atmosphere to the Earth is given by the following:

$$J_\downarrow \equiv \int J_\omega d\omega \tag{13}$$

This analysis is used below for calculating the partial radiative fluxes from the atmosphere to the Earth.

Table 1. Values of the total radiative flux from the atmosphere to the Earth as it follows from different versions of the energetic balance for the Earth and its atmosphere with references that contain a certain version of the sources. The altitude h_{cl} of the cloud boundary and the temperature T_{cl} of the cloud emission relate to the corresponding version of Earth's energetic balance.

Number	1	2	3	4	5	Average
$J_\downarrow$, W/m^2	327	327	333	346	342	335 ± 7
Reference	[21,32–34]	[18,35–40]	[41–44]	[45]	[46]	-
h_{cl}, km	5.6	5.6	4.8	3.2	3.7	4.6 ± 0.7
T_{cl}, K	252	252	257	267	264	258 ± 6

3. Radiative Fluxes from the Standard Atmosphere

Evaluations of the radiative fluxes from the atmosphere to the Earth are based on the above algorithm [1]. In the previous analysis, the frequency range was separated over several ranges, and evaluations were fulfilled in each range independently. In this case, within the framework of a general computer code, one can make calculations in the total range of frequencies that determine the radiative flux. Let us choose the effective range of flux evaluations. If the Earth's surface with the temperature of the standard atmosphere $T_E = 288$ K emits as a blackbody, the radiative flux at frequencies above 1300 cm^{-1} is 10% of the total radiative flux, the radiative flux at frequencies above 2000 cm^{-1} is 1% of the total radiative flux, and the radiative flux at frequencies above 2600 cm^{-1} is 0.1% of the total radiative flux. In previous evaluations [1], calculations were made for frequencies below 1260 cm^{-1}, accounting for the radiation of molecules CO_2, CH_4 and N_2O at larger frequencies. We are now restricted by the frequency range below 2600 cm^{-1}.

By analogy with evaluations [1], in subsequent calculations, we use information about radiative transitions from the HITRAN data bank. Figure 2 contains the frequency dependence for the optical thickness of the atmospheric layer between the Earth's surface and clouds, which is determined by optically active atmospheric molecules in the infrared spectrum range. As it follows from this Figure, optical parameters of the atmosphere as a frequency function have a line character. One can determine from this the radiative temperature of molecules located in the atmospheric gap between the Earth and clouds, and its frequency dependence is presented in Figure 3.

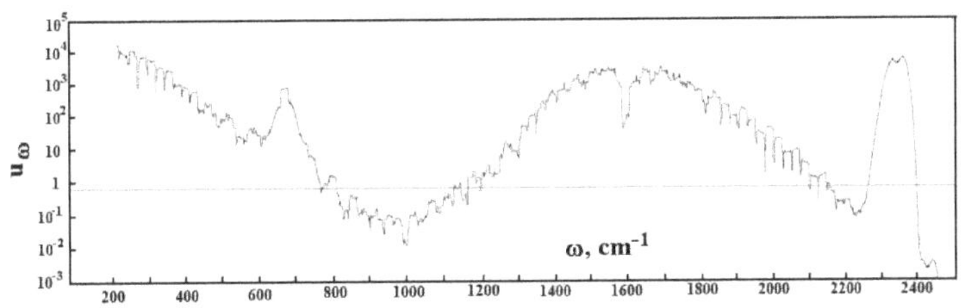

Figure 2. Optical thickness u_ω of the atmospheric gap between the Earth and clouds due to optically active atmospheric molecules. The solid curve corresponds to $u_\omega = 2/3$ and the cloud altitude is $h_{cl} = 4.6$ km.

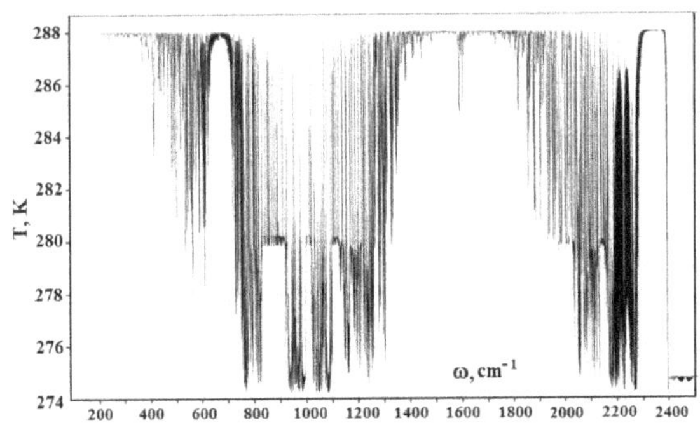

Figure 3. Radiative temperature T_ω due to atmospheric molecules located in the gap between the Earth's surface and clouds, whose boundary is located at the altitude $h_{cl} = 4.6$ km.

Figure 4 contains values of radiative temperatures, given in Figure 3, which are averaged over ranges of 20 cm^{-1} width. This averaging is made over the frequency range of the width of 10 cm^{-1} below and above the frequency under consideration.

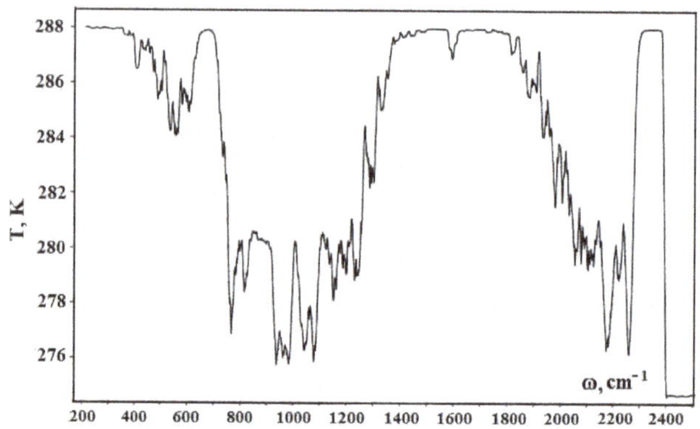

Figure 4. Averaged over a range of 20 cm^{-1}, radiative temperatures T_ω, which are determined by emission of atmospheric molecules, which are located in the gap between the Earth's surface and clouds. The cloud boundary is found at the altitude $h_{cl} = 4.6$ km.

The opaque factor $g(u_\omega)$ is determined by Formula (3), and its frequency dependence is given in Figure 5. The averaged value of the opaque factor is given in Figure 6. The averaging is made over the frequency range of 20 cm^{-1} by analogy with that for the radiative temperature.

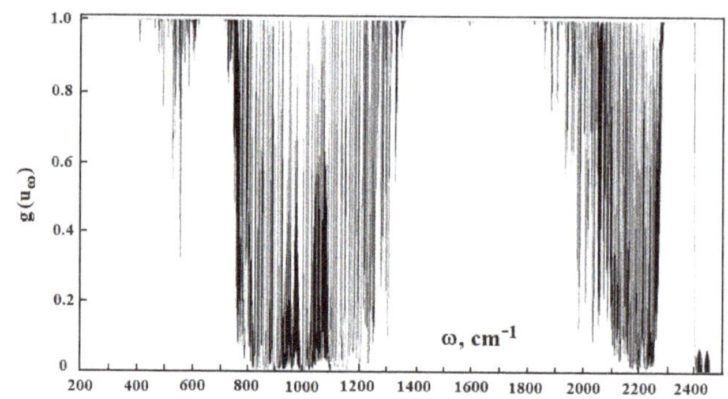

Figure 5. Opaque factor $g(u_\omega)$ due to absorption of atmospheric molecules in the gap between the Earth's surface and clouds if the cloud boundary is located at the altitude $h_{cl} = 4.6$ km. The model is shown, where the atmosphere is transparent in the frequency range between 780 cm^{-1} and 1250 cm^{-1}.

The opaque factor characterizes the part of the radiative flux that is emitted by the Earth's surface and attains the clouds. Figure 7 contains the radiative flux $I_\omega(T_E)$ that is emitted by the Earth as well as the radiative flux $\overline{g(u_\omega)}I_\omega(T_E)$ at a given frequency, which reaches the clouds for the model of standard atmosphere. We have also the average radiative flux, which is emitted by the Earth and reaches the cloud boundary as follows:

$$J_\uparrow = \int \overline{g(u_\omega)} I_\omega(T_E) d\omega \tag{14}$$

Taking $T_E = 288$ K on the basis of the model of standard atmosphere, we obtain the following:

$$J_E = \int I_\omega(T_E) d\omega = 390\,\mathrm{W/m^2}$$

in accordance with the energetic balance of the Earth, and $J_{\uparrow} = 120\ \mathrm{W/m^2}$. As is seen, a part of the thermal radiative flux that passes through the atmospheric layer below clouds and reaches the clouds is approximately 30% of the emitted radiative flux.

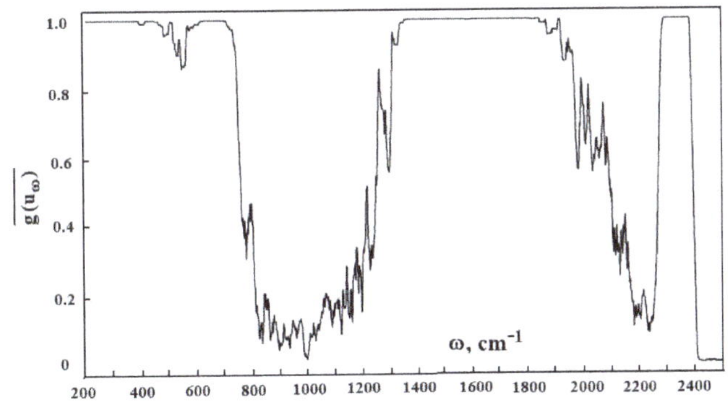

Figure 6. The averaged opaque factor $\overline{g(u_\omega)}$ over a frequency range of 20 cm^{-1} for the absorption of atmospheric molecules located in the gap between the Earth's surface and clouds. The cloud boundary is located at the altitude $h_{cl} = 4.6$ km.

From Formulas (1) and (13), one can determine the radiative flux from the atmosphere to the Earth's surface due to atmospheric molecules and clouds; this flux depends on the cloud temperature T_{cl} or the boundary altitude h_{cl} for clouds. This dependence is represented in Figure 8. In addition, these parameters are given in Table 1 for different versions of the energetic balance of the Earth and its atmosphere.

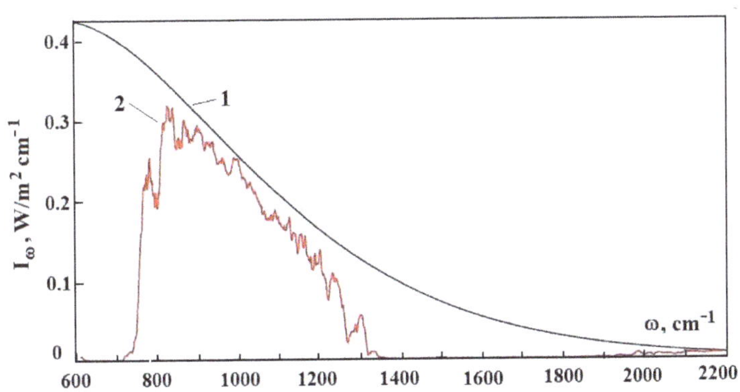

Figure 7. The frequency dependence for the radiative flux $I_\omega(T_E)$ from the Earth's surface (1) and its part $\overline{g(u_\omega)}I_\omega(T_E)$, which attains the boundary of clouds.

In previous evaluations [1], only data of the first source of Table 1 were used, whereas now we are based on a variety of these values. Because parameters of the Earth's energetic balance are different for these sources, this difference characterizes also the error in the final results. Note that the connection between the effective altitude h_{cl} of the cloud boundary and the cloud temperature T_{cl} is given by Formula (12).

A cursory glance at the emission parameters of the atmosphere according to Figures 2, 3 and 5 exhibits that the spectrum of atmospheric radiation consists of separate spectral lines, in contrast to climatological models with a smooth empiric frequency functions, which approximate the atmospheric spectrum. In addition, climatological models do not take into account the thermodynamics of the atmosphere and radiation field.

Therefore, computer codes of climatology are not reliable in the evaluation of the radiative parameters of the atmosphere.

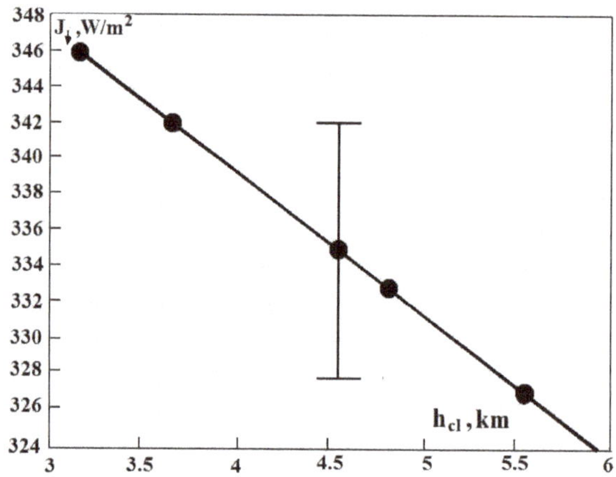

Figure 8. Dependence of the total radiative flux from the atmosphere to the Earth $J_\downarrow$ on the altitude of the cloud boundary h_{cl}. The average value of the radiative flux toward the Earth is marked.

One can divide the radiative flux J_ω from the atmosphere to the Earth in parts that are created by various components of the radiated atmosphere. For example, the partial radiative flux $J_\omega(CO_2)$, which is determined by CO_2 molecules, is given by the following formula:

$$J_\omega(CO_2) = J_\omega \frac{k_\omega(CO_2)}{k_\omega}, \tag{15}$$

where J_ω is the total radiative flux, $k_\omega(CO_2)$ is the absorption coefficient due to CO_2 molecules, and k_ω is the total absorption coefficient at a given frequency. These absorption coefficients are taken at the radiative temperature T_ω. Integrating the partial radiative flux as a frequency function over frequencies, one can determine the total radiative flux from the atmosphere to the Earth due to this component. Table 2 contains values of the total radiative flux from the atmosphere due to each greenhouse component.

Table 2. Radiative fluxes in W/m^2 for indicated frequency ranges and the total radiative flux due to a given component. Results of evaluations [1] are given in square parentheses.

Component	$\omega < 800$ cm^{-1}	1250 cm^{-1} > $\omega > 800$ cm^{-1}	$\omega > 1250$ cm^{-1}	Sum
H_2O-molecules	163	13	39	(215 ± 2) [166]
H_2O-droplets	8	52	1	(61 ± 7) [96]
water in total	171	65	40	(276 ± 10) [262]
CO_2-molecules	53	2	1	(56 ± 1) [60]
CH_4-molecules	0	1	1	2 [4]
N_2O-molecules	0	0	1	1 [3]

One can compare the results for radiative fluxes of these evaluations given in Table 2 and previous ones [1]. The basic difference is that the basic part of the emission, according to [1], is taken from the frequency range $\omega < 800$ cm^{-1}; in the other frequency range $\omega > 800$ cm^{-1}, which gives the contribution approximately 10% to the radiative flux, the emission of water molecules is ignored. In these evaluations, the HITRAN bank data are included up to $\omega = 2600$ cm^{-1}. As a result, a part of the emission transfers from clouds to water molecules. We also note that average radiative fluxes given in Table 2, according to evaluations [1], refer to the total radiative flux $J_{\downarrow} = 327\ \mathrm{W/m^2}$ from the atmosphere toward the Earth, whereas now, we are guided by this flux $J_{\downarrow} = (335 \pm 7)\ \mathrm{W/m^2}$; the results of Table 2 correspond to the total flux.

In addition, Figure 9 contains radiative fluxes due to individual greenhouse components. The frequency range is separated into two parts with the boundary of 800 cm^{-1}. One can evaluate separately the emission of trace greenhouse components, which are CH_4 and N_2O molecules. The number density of these molecules as an altitude function is analogous to formula (5) for CO_2 molecules and is given by the following formulas:

$$N(CH_4) = N_m \exp\left(-\frac{h}{\Lambda}\right),\ N_m = 3.5\times10^{13}\,\mathrm{cm}^{-3},\ N(N_2O) = N_n \exp\left(-\frac{h}{\Lambda}\right),\ N_n = 7.5\times10^{12}\,\mathrm{cm}^{-3}\ \Lambda = 10\,\mathrm{km} \tag{16}$$

The number densities of these molecules at the Earth's surface are taken from [47,48]. The absorption band for the CH_4 molecule is (1240–1360) cm^{-1}, the absorption bands for N_2O molecule are placed in the frequency ranges (1250–1310) cm^{-1} and (2180–2260) cm^{-1}. Figure 10 contains the partial optical thickness as a frequency function in the above absorption bands, due to individual greenhouse components, which include the above trace components and water molecules.

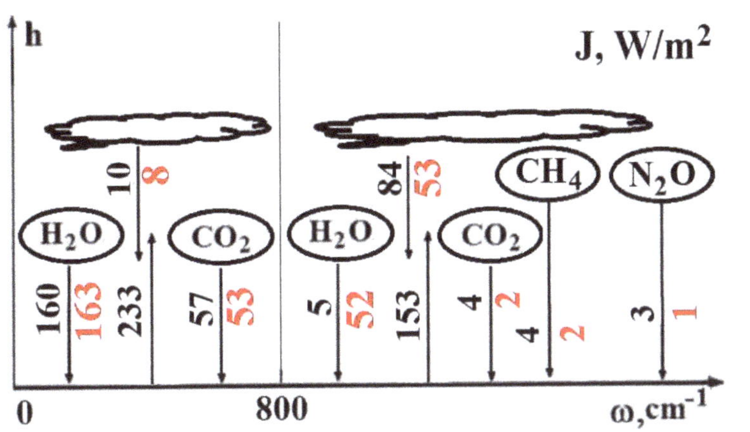

Figure 9. Radiative fluxes J_E from the Earth's surface for the model of standard atmosphere, and also fluxes from clouds and to indicated molecular components, where radiation attains the Earth's surface. A molecular component is indicated inside an ellipse; average values of the fluxes are expressed in W/m^2. Data of [1] are represented in black, and the results of these evaluations are given in red.

As it is seen, the optical thickness due to trace gases is not large and is screened by absorption of water molecules. Because we ignored in [1] the absorption of water molecules at the wing $\omega > 1200$ cm^{-1} of atmospheric thermal emission, evaluations [1] contain heightened values of radiative fluxes due to trace gases, as it follows from the data of Figure 9 and Table 2.

It should be noted that we exclude the ozone emission from the tropospheric radiation toward the Earth since the concentration of ozone molecules in the troposphere usually does not exceed 10^{12} cm^{-3}. Therefore, the optical thickness of the troposphere in the gap between the Earth's surface and clouds is small. The concentration of ozone molecules in the stratosphere is larger, but the stratospheric radiation does not reach the Earth because it is absorbed by clouds. Hence, we ignore the radiation of ozone molecules in the atmosphere.

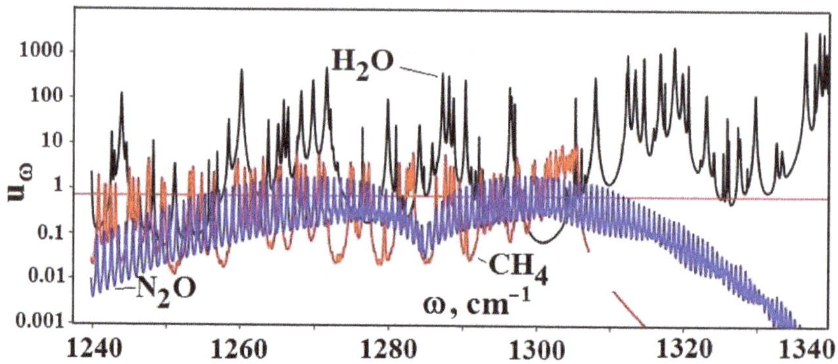

Figure 10. Optical thickness u_ω of the atmospheric gap between the Earth and clouds due to optically active atmospheric molecules CH_4, N_2O and H_2O in the absorption bands of trace gases.

4. Emission of Atmosphere with a Varied Composition

The code under consideration allows one to analyze variations in the radiative fluxes as a result of the change in the atmosphere composition. We start this analysis from the standard case, where the concentration of CO_2 molecules is doubled. This leads to the change in the global temperature that is named "Equilibrium Climate Sensitivity" [49]. Note that this operation implies that optically active gases have a uniform distribution over the globe that corresponds to the average altitude distribution. This criterion is fulfilled more or less for CO_2 molecules, and the uniform distribution is violated for H_2O molecules.

In this consideration, we take into account three basic greenhouse components, namely, CO_2 molecules, H_2O molecules, and water microdroplets of clouds. Table 3 contains the distribution over the spectrum for the change in radiative fluxes due to a given component as a result of doubling the concentration of CO_2 molecules. Quantities Δ_c, Δ_w, Δ_d of this table are the changes in the radiative fluxes to the Earth due to CO_2 molecules, H_2O molecules, and water microdroplets of clouds, correspondingly, according to the following relations:

$$\Delta_c = \int [J'_\omega(CO_2) - J_\omega(CO_2)]d\omega, \ \Delta_w = \int [J'_\omega(H_2O) - J_\omega(H_2O)]d\omega, \ \Delta_d = \int [J'_\omega(drop) - J_\omega(drop)]d\omega, \quad (17)$$

where $J_\omega(A)$ is the radiative flux due to an indicated component A at the current concentration of CO_2 molecules, and J'_ω is the radiative flux of this component at the doubling concentration of CO_2 molecules. The change in the total radiative flux Δ is introduced as follows:

$$\Delta = \Delta_c + \Delta_w + \Delta_d \quad (18)$$

It is evident that the change in radiative fluxes of components as well as the total change in the radiative flux from the atmosphere to the Earth takes place only in spectrum ranges where CO_2 molecules absorb. In addition, the basic change in the radiative flux as a result of the concentration change for some optically active components takes place in the frequency range where the optical thickness is of the order of one. The red line $u_\omega = 2/3$ in Figures 2, 10, and 11 separates the optical thicknesses, with large ones above this line and low ones below it. Comparing the data of Table 3 and Figure 11, one can conclude that the main contribution to the change in the radiative flux follows from a frequency range where $u_\omega \sim 1$. Next, changes in radiative fluxes resulted from the variation in the concentration of CO_2 molecules are nearby, according to these calculations and evaluations [1], because the absorption of CO_2 molecules is absent in the added frequency range 1200 cm^{-1} $< \omega <$ 2600 cm^{-1}. On the contrary, in the case of the change in the concentration of water molecules, the change in the total radiative flux differs by several times under identical conditions.

Table 3. Variations in radiative fluxes for the standard atmosphere as a result of doubling of the CO_2 concentration in an indicated spectrum range of CO_2 molecules. The frequency ranges $\Delta\omega$ are given in cm^{-1}, and radiative fluxes are expressed in W/m^2. Results of calculations [1] are given in parentheses.

Frequency Range, cm^{-1}	Δ_c	Δ_w	Δ_d	Δ
580–600	0.70	−0.60	−0.05	0.05 (0.03)
600–620	0.91	−0.75	−0.05	0.11 (0.04)
620–640	0.75	−0.71	0	0.04 (0.02)
640–660	0.23	−0.22	0	0.01 (0.01)
660–680	0.08	−0.08	0	0 (0)
680–700	0.27	−0.26	0	0.01 (0.01)
700–720	0.38	−0.32	0	0.06 (0.06)
720–740	0.56	−0.20	−0.12	0.24 (0.20)
740–760	1.03	−0.42	−0.37	0.24 (0.29)
760–780	0.65	−0.02	−0.45	0.18 (0.14)
780–800	0.25	0	−0.17	0.08 (0.06)
800–850	0.18	0.01	−0.15	0.04 (0.04)
900–950	0.20	0	−0.15	0.05 (0.04)
950–1000	0.35	0	−0.25	0.10 (0.22)
1000–1050	0.18	0	−0.12	0.06 (0.05)
1050–1100	0.45	0	−0.31	0.14
2250–2400	0.02	0	−0.01	0.01
total	7.19 (7.24)	−3.57 (−3.02)	−2.20 (−2.90)	1.42 (1.32)

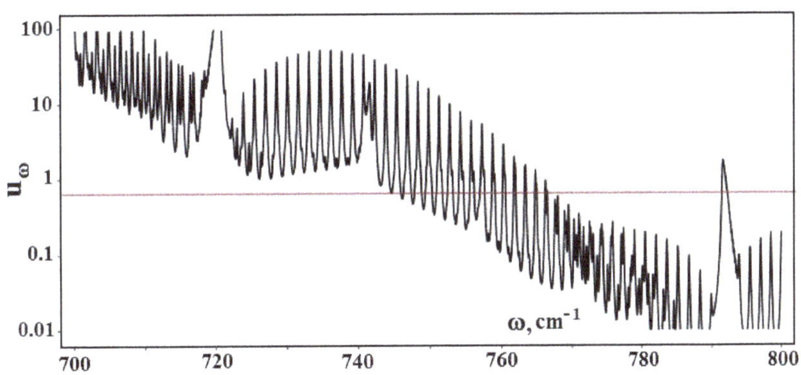

Figure 11. Optical thickness u_ω of the atmospheric gap between the Earth and clouds due to CO_2 and H_2O molecules in the frequency range where the optical thickness of the atmosphere is of the order of one.

One can see from the data of Table 3 that the change in the radiative flux Δ_c due to CO_2 molecules as a result of doubling the concentration of CO_2 molecules is five times larger than the change in the total radiative flux Δ. In climatological models, the Kirchhoff law [50], according to which radiators are simultaneously absorbers, is ignored. Therefore, the change in the radiative flux Δ_c due to CO_2 molecules is taken in climatological models instead of the change in the total radiative flux Δ, and the error in the change of the global temperature under these conditions is exceeded by five times. This results from neglecting the absorption of emitted radiative fluxes by water molecules and water microdroplets of clouds by added CO_2 molecules [51,52]). This large difference follows also from a general conclusion for a plane gaseous layer with a low temperature gradient [51,52]). Indeed, if we have a layer of a constant temperature and of a high optical thickness, where the emission is created by several components, this layer emits as a blackbody with the gaseous temperature. Therefore, a change in the concentration of some optically active component would not lead to the change in the outgoing radiative flux.

A comparison of the results of these evaluations with those of [1] allows one to estimate the accuracy of the change in radiative fluxes, which is estimated as $\sim$10%. The same accuracy corresponds to the derivative of the radiative flux over the concentration of a given component. From these calculations, in the case that the concentration of CO_2 molecules varies, while other atmospheric parameters are unvaried, we have the following:

$$\frac{dJ_\downarrow}{d\ln c(CO_2)} \approx 2\frac{W}{m^2}, \tag{19}$$

where $J_\downarrow$ is the radiative flux from the atmosphere to the Earth, and $c(CO_2)$ is the concentration of CO_2 molecules.

Along with variations in the concentration of CO_2 molecules, other changes take place in the real atmosphere, which influence the global temperature. Let us analyze this change in the real atmosphere on the basis of measurements within the framework of NASA programs. We use below the results of measurements for the concentration $c(CO_2)$ of carbon dioxide molecules [53,54]. Because the time of residence of a CO_2 molecule is large, we take the evolution of the atmospheric concentration of CO_2 molecules according to measurements in the Mauna Loa observatory as the global evolution of this quantity. Next, the evolution of the global temperature during the last 150 years is analyzed carefully by the Goddard Institute for Space Studies (GISS) (for example, [55]) within the framework of the NASA program. These results are collected in [1] and are as follows:

$$\frac{d\ln c(CO_2)}{dt} = 0.006\,\mathrm{yr}^{-1}, \quad \frac{d\Delta T}{dt} = 0.018\,\mathrm{K/yr} \tag{20}$$

One can connect the change in the radiative flux from the atmosphere to the Earth $J_\downarrow$ and the change in the global temperature ΔT through climate sensitivity S, according to relation [18,56]:

$$S = \frac{d\Delta T}{dJ_\downarrow} \tag{21}$$

We use the average value of these quantities, which follows from an average of various data and is $S \approx 0.5\,\mathrm{m}^2 \cdot \mathrm{K/W}$ [1]. One can estimate the accuracy of this value as $\approx$50%. From this, for the derivation of the radiative flux $J_\downarrow$ over the concentration of CO_2 molecules (cCO_2) in the real atmosphere, we have the following:

$$\frac{dJ_\downarrow}{d\ln c(CO_2)} \approx 6\frac{W}{m^2} \tag{22}$$

We note that in contrast to formula (19) where all atmospheric parameters, except for the concentration of CO_2 molecules, are unvaried, formula (22) takes into account other processes that lead to the change in the global temperature. As is seen, the contribution of carbon dioxide in the change in the global temperature for the evolution of the real temperature is approximately 30%.

By analogy with the CO_2 case, one can analyze the influence of the change in atmospheric water molecules on the change in the radiative flux from the atmosphere to the Earth. Some results that are analogous to those of Table 3 are given in Table 4. From the data of this table, we have the following:

$$\frac{dJ_\downarrow}{d\ln c(H_2O)} = 7\frac{W}{m^2}, \tag{23}$$

where $c(H_2O)$ is the concentration of atmospheric water molecules.

Table 4. Variations in radiative fluxes from the standard atmosphere to the Earth as a change in the concentration of H_2O molecules. Here, $c(H_2O)$ is the contemporary concentration of H_2O molecules, and $c'(H_2O)$ is that at a new concentration of water molecules. The radiative fluxes represented are expressed in W/m^2.

$c'(H_2O)/c(H_2O)$	Δ_c	Δ_w	Δ_d	Δ
0.5	3.84	−15.64	6.52	−4.98
0.7	2.05	−7.94	3.20	−2.53
0.9	0.62	−2.38	0.91	−0.74
1	0	0	0	0
1.1	−0.57	2.10	−0.81	0.66
1.5	−2.52	8.86	−3.36	2.78
2	−4.42	15.15	−5.65	4.71

Let us assume that along with an increase in the concentration of CO_2 molecules, the evolution of the global temperature of the real atmosphere is determined by the increase in the concentration of H_2O molecules. Then, water molecules account for the contribution of 2/3 of the change in the global temperature. On the basis of Formula (20), we have the following:

$$\frac{d\Delta T(H_2O)}{dt} = 12\frac{mK}{yr}, \tag{24}$$

where $\Delta T(H_2O)$ is the part of the global temperature change which follows from an increase in the concentration of H_2O molecules. Taking as early $S \approx 0.5\,\text{m}^2 \cdot \text{K/W}$, one can find the derivation part of the radiative flux from the atmosphere to the Earth due to H_2O molecules as follows:

$$\frac{dJ_\downarrow(H_2O)}{dt} = 24\frac{W}{\text{m}^2 \cdot \text{yr}}, \tag{25}$$

From this, it follows the rate of variation for the concentration of the water molecules $c(H_2O)$, which can provide this change:

$$\frac{d\ln c(H_2O)}{dt} = 0.003\ \text{yr}^{-1} \tag{26}$$

According to this result, the observed rate of an increase in the global temperature may be determined by an increase in the following average number density of water molecules:

$$\frac{dN(H_2O)}{dt} = 10^{15}\ \text{cm}^{-3}\ \text{yr}^{-1} \tag{27}$$

The annual change in the number density is small, compared to the contemporary number density of water molecules, which is equal to $3.4 \times 10^{17}\ \text{cm}^{-3}$. Therefore, a real determination of this derivation under contemporary conditions is problematic and requires additional efforts similar to those for the evolution of the global temperature.

It should be noted that the moisture η is the measured water parameter in the atmosphere that is given by the following:

$$\eta = \frac{N(H_2O)}{N_{sat}(H_2O)} = \frac{c(H_2O)}{c_{sat}(H_2O)}, \tag{28}$$

where $N_{sat}(H_2O)$ is the number density of water molecules at the saturated vapor pressure for a given temperature, $c(H_2O)$ is a current concentration of water molecules in atmospheric air, and $c_{sat}(H_2O)$ is its concentration for the saturated vapor pressure at the air temperature. The contemporary value of the atmosphere moisture is $\eta \approx 80\%$ [57].

Let us determine the rate of the moisture variation due to the temperature change. Because the amount of atmospheric water is small compared to that at the Earth's surface, we have, from an equilibrium between the atmospheric and surface water, the following:

$$c_{sat}(H_2O) \sim \exp\left(-\frac{E_a}{T}\right),$$

where $E_a = 0.43$ eV [58] is the binding energy for a water molecule with its bulk liquid. From this, for the rate of an increase in the atmosphere capacity with respect to water molecules due to a temperature increase, we have the following:

$$\frac{dc_{sat}(H_2O)}{dt} = \frac{E_a}{T^2} \cdot \frac{dT}{d\Delta t} \approx 0.001\,\mathrm{yr}^{-1} \quad (29)$$

The process of growth of the atmospheric moisture will be stopped through a time t_{ev}, according to the following equation:

$$t_{ev} = \frac{1 - c(H_2O)}{c(H_2O) - c_{sat}(H_2O)} \approx 100\,\mathrm{yr} \quad (30)$$

From this, it follows that $t_{ev} \approx 100\,\mathrm{yr}$, i.e., this time is sufficiently large to check this mechanism of the Earth's heating.

5. Conclusions

The goal of this paper is to apply the improved code for the evaluation of the radiative fluxes from the atmosphere to the Earth on the basis of the model of [1]. In these evaluations, we use the range of frequencies up to 2600 cm^{-1} in contrast to the previous calculations [1], where this range was restricted by 1200 cm^{-1}. Besides this difference, we use now various versions of the energetic balance of the Earth and its atmosphere (see Table 1), and the difference in radiative parameters due to various versions of the Earth's energetic balance may be considered an error of the results. Nevertheless, the cloud boundary altitude h_{cl}, as one of basic parameters of the model under consideration, according to [1], is $h_{cl} =$ 4.3 km, whereas from Table 1, the average altitude of the cloud boundary is $h_{cl} = 4.6$ km, i.e., these values coincide within the accuracy of these evaluations.

Because carbon dioxide molecules do not absorb in the additional spectrum range between 1200 cm^{-1} and 2600 cm^{-1}, the radiative parameters due to CO_2 molecules are close in these evaluations and in the previous one. In particular, the variation in radiative fluxes as a result of the change in the carbon dioxide amount in the atmosphere for these calculations are close. In this evaluation as well as previous evaluations, we have a contradiction with the results of climatological models in the analysis of the Earth's greenhouse effect, according to which the increase in the global temperature differs by five times. We show [51,52], so the large difference results from ignoring, in climatological models, the Kirchhoff law [50], according to which radiators are simultaneously the absorbers. In this case, we take the change in the radiative flux created by CO_2 molecules as the change of the total radiative flux.

Note the restrictions by the frequency range up to 1200 cm^{-1} in the previous calculations [1]; we thus assume that the atmosphere is transparent for larger frequencies, and the emission at larger frequencies is determined by clouds. However, according to the HITRAN data bank, water molecules absorb effectively in the enlarged frequency range. As a result, the derivative (22) is larger than that according to [1]. We proved early (for example, [1]) that atmospheric CO_2 molecules are not the main radiator of the atmosphere. From these evaluations, it follows that water molecules in the atmosphere may be responsible for the observed heating of the Earth.

Author Contributions: Writing—original draft, B.M.S. and D.A.Z. All authors have read and agreed to the published version of the manuscript.

Funding: This research received no external funding.

Conflicts of Interest: The authors declare no conflict of interest.

References

1. Smirnov, B.M. *Transport of Infrared Atmospheric Radiation*; De Gruyter: Berlin, Germany, 2020.
2. Goody, R.M. *Atmospheric Radiation: Theoretical Basis*; Oxford University Press: London, UK, 1964.
3. Goody, R.M.; Yung, Y.L. *Principles of Atmospheric Physics and Chemistry*; Oxford University Press: New York, NY, USA, 1995.
4. U.S. Government Printing Office. *U.S. Standard Atmosphere*; U.S. Government Printing Office: Washington, DC, USA, 1976.
5. Center for Astrophysics. Available online: https://www.cfa.harvard.edu/ (accessed on 4 August 2021).
6. HITRAN on the Web. Available online: http://www.hitran.iao.ru/home (accessed on 4 August 2021).
7. HITRANonline—Documentation: Units and Definitions. Available online: http://www.hitran.org/links/docs/definitions-and-units/ (accessed on 4 August 2021).
8. Simeckova, M.; Jacquemart, D.; Rothman, L.S.; Gamache, R.R.; Goldman, A. Einstein A-coefficients and statistical weights for molecular absorption transitions in the HITRAN database. *J. Quant. Spectrosc. Radiat. Transf.* **2006**, *98*, 130. [CrossRef]
9. Houghton, J.T. *The Physics of Atmospheres*; Cambridge University Press: Cambridge, UK, 1977.
10. Iribarne, J.V.; Cho, H.P. *Atmospheric Physics*; Reidel Publ.: Dordrecht, The Netherlands, 1980.
11. Fleagle, R.G.; Businger, J.A. *Introduction to Atmospheric Physics*; Academic Press: San Diego, CA, USA, 1980.
12. Salby, M.L. *Fundamentals of Atmospheric Physics*; Academic Press: San Diego, CA, USA, 1996.
13. Seinfeld, J.H.; Pandis, S.N. *Atmospheric Chemistry and Physics*; Wiley: New York, NY, USA, 1998.
14. Andrews, D.G. *An Introduction to Atmospheric Physics*; Cambridge University Press: Cambridge, UK, 2000.
15. Seinfeld, J.H.; Pandis, S.N. *Atmospheric Chemistry and Physics*; Wiley: Hoboken, NJ, USA, 2006.
16. Walace, J.M.; Hobbs, R. *Atmospheric Science. An Introductory Survey*; Elsevier: Amsterdam, The Netherlands, 2006.
17. Ambaum, M.H.P. *Thermal Physics of the Atmosphere*; Wiley-Blackwell: Oxford, UK, 2010.
18. Salby, M.L. *Physics of the Atmosphere and Climate*; Cambridge University Press: Cambridge, UK, 2012.
19. Lagzi, I.; Meszaros, R.; Gelybo, G.; Leelossy, A. *Atmospheric Chemistry*; Institute of Geography and Earth Science: Budapest, Germany, 2013.
20. Caballero, R. *Physics of the Atmosphere*; IOP Publishing: Bristol, UK, 2014.
21. Smirnov, B.M. *Microphysics of Atmospheric Phenomena*; Springer Atmospheric Series; Springer: Cham, Switzerland, 2017.
22. Visconti, G. *Fundamentals of Physics and Chemistry of the Atmosphere*; Springer Nature: Cham, Switzerland, 2017.
23. Reif, F. *Statistical and Thermal Physics*; McGrow Hill: Boston, MA, USA, 1965.
24. Landau, L.D.; Lifshitz, E.M. *Statistical Physics*; Pergamon Press: Oxford, UK, 1980; Volume 1.
25. Zel'dovich, B.Y.; Raizer, P.Y. *Physics of Shock Waves and High-Temperature Hydrodynamic Phenomena*; Acad. Press: New York, NY, USA, 1966.
26. Smirnov, B.M. *Physics of Weakly Ionized Gas*; Nauka: Moscow, Russia, 1972. (In Russian)
27. Smirnov, B.M. Radiation of Gas Layer over Hot Surface. *J. Exp. Theor. Phys.* **2018**, *126*, 446. [CrossRef]
28. Smirnov, B.M. *Physics of Weakly Ionized Gases*; Mir: Moscow, Russia, 1980.
29. Smirnov, B.M. *Physics of Ionized Gases*; Wiley: New York, NY, USA, 2001.
30. Smirnov, B.M. *Global Atmospheric Phenomena Involving Water*; Springer Atmospheric Series; Springer: Cham, Switzerland, 2020.
31. Wien, W. Ueber die Energievertheilung im Emissionsspectrum eines schwarzen Körpers. *Wied. Ann. Phys. Chem.* **1896**, *58*, 662. [CrossRef]
32. United States Committee for the Global Atmospheric Research Program. *Understanding Climate Change*; Nat. Acad. Science: Washington, DC, USA, 1975.
33. Smirnov, B.M. *Introduction to Plasma Physics*; Mir: Moscow, Russia, 1977.
34. Smirnov, B.M. *Energetics of Atmosphere*; Phys. Series N3; Znanie: Moscow, Russia, 1979. (In Russian)
35. Ramanathan, V. Atmospheric General Circulation and its Low Frequency Variance: Radiative Influences. *J. Meteorol. Soc. Jpn.* **1987**, *65*, 151. [CrossRef]
36. Ramanathan, V. The role of earth radiation budget studies in climate and general circulation research. *J. Geophys. Res.* **1987**, *92*, 4075. [CrossRef]
37. Ramanathan, V. The greenhouse theory of climate change: A test by an inadvertent global experiment. *Science* **1988**, *240*, 293. [CrossRef] [PubMed]
38. Raval, A.; Ramanathan, V. Observational determination of the greenhouse effect. *Nature* **1989**, *342*, 758. [CrossRef]
39. Ramanathan, V.; Cess, R.D.; Harrison, E.F.; Minnis, P.; Barkstrom, B.R.; Ahmad, E.; Hartmann, D. Cloud-radiative forcing and climate: Results from the Earth Radiation Budget Experiment. *Science* **1989**, *243*, 57. [CrossRef] [PubMed]
40. Ramanathan, V.; Harrison, E.F.; Barkstrom, B.R. Climate and the Earth's Radiation Budget. *Phys. Today* **1989**, *42*, 22. [CrossRef]
41. Kiehl, J.T.; Trenberth, K.E. Earth's Annual Global Mean Energy Budget. *Bull. Am. Meteorol. Soc.* **1997**, *78*, 197. [CrossRef]
42. Trenberth, K.E.; Fasullo, J.T.; Kiehl, J.T. Earth's Global Energy Budget. *Bull. Am. Meteorol. Soc.* **2009**, *90*, 311. [CrossRef]
43. Trenberth, K.E.; Fasullo, J.T. Tracking Earth's energy: From El Niño to global warming. *Surf. Geophys.* **2012**, **33**, 413. [CrossRef]
44. Fasullo, J.T.; Trenberth, K.E. A less cloudy future: The role of subtropical subsidence in climate sensitivity. *Science* **2012**, *338*, 792. [CrossRef] [PubMed]
45. Stephens, G.L.; Li, J.; Wild, M.; Clayson, C.A.; Loeb, N.; Kato, S.; L'ecuyer, T.; Stackhouse, P.W.; Lebsock, M.; Andrews, T. An update on Earth's energy balance in light of the latest global observations. *Nat. Geosci.* **2012**, *5*, 691. [CrossRef]

46. Wild, M.; Folini, D.; Schär, C.; Loeb, N.; Dutton, E.G.; König-Langlo, G. The global energy balance from a surface perspective. *Clim. Dyn.* **2013**, *40*, 3107. [CrossRef]
47. Trends in Atmospheric Methane. Available online: https://www.esrl.noaa.gov/gmd/ccgg/trends-ch4/ (accessed on 4 August 2021).
48. Machida, T.; Nakazawa, T.; Fujii, Y.; Aoki, S.; Watanabe, O. Increase in the atmospheric nitrous oxide concentration during the last 250 years. *Geophys. Res. Lett.* **1995**, *22*, 2921. [CrossRef]
49. Climate-Sensitivity. Available online: https://en.wikipedia.org/wiki/Climate-sensitivity (accessed on 4 August 2021).
50. Kirchhoff, G.; Bunsen, R. Zusammenhang von emission und absorption von Licht und Wärme. *Ann. Phys. Chem.* **1860**, *109*, 275. [CrossRef]
51. Smirnov, B.M. Interaction of radiative molecules in gas emission. *Int. Rev. At. Mol. Phys.* **2019**, *10*, 39.
52. Smirnov, B.M. Atmospheric carbon dioxide and climate. *J. Atmos. Sci. Res.* **2019**, *2*, N4. [CrossRef]
53. Mauna-Loa-Observatory. Available online: https://en.wikipedia.org/wiki/Mauna-Loa-Observatory (accessed on 4 August 2021).
54. Trends in Atmospheric Carbon Dioxide. Available online: https://www.ersl.noaa.gov/gmd/cggg/trends (accessed on 4 August 2021).
55. Hansen, J.; Sato, M.; Ruedy, R. Available online: http://www.columbia.edu/~jeh1/mailing/2016/20160120-Temperature2015 (accessed on 4 August 2021).
56. Rohling, E.J.; Sluijs, A.; Dijkstra, H.A.; van de Wal, R.S.W.; von der Heydt, A.S.; Bijl, P.K.; Zeebe, R. Making sense of palaeoclimate sensitivity. *Nature* **2012**, *491*, 683.
57. Moisture. Available online: https://en.wikipedia.org/wiki/Moisture (accessed on 4 August 2021).
58. Lide, D.R. (Ed.) *Handbook of Chemistry and Physics*, 84th ed.; CRC Press: London, UK, 2003.

foundations

MDPI

Communication

Analysis of Experimental Cross-Sections of Charge Exchange between Hydrogen Atoms and Protons Yields More Evidence of the Existence of the Second Flavor of Hydrogen Atoms

Eugene Oks

Physics Department, Auburn University, 380 Duncan Drive, Auburn, AL 36849, USA; goks@physics.auburn.edu

Abstract: Measurements of cross-sections of charge exchange between hydrogen atoms and low energy protons (down to the energy ~10 eV) revealed a noticeable discrepancy with previous theories. The experimental cross-sections were systematically slightly higher—beyond the error margins—than the theoretical predictions. In the present paper, we study whether this discrepancy can be eliminated or at least reduced by using the Second Flavor of Hydrogen Atoms (SFHA) in calculations. We show that for the SFHA, the corresponding cross-section is noticeably larger than for the usual hydrogen atoms. We demonstrate that the allowance for the SFHA does bring the theoretical cross-sections in a noticeably better agreement with the corresponding experiments within the experimental error margins. This seems to constitute yet another evidence from atomic experiments that the SFHA is present within the mixture of hydrogen atoms. In combination with the first corresponding piece of evidence from the analysis of atomic experiments (concerning the distribution of the linear momentum in the ground state of hydrogen atoms), as well as with the astrophysical evidence from two different kinds of observations (the anomalous absorption of the redshifted 21 cm radio line from the early universe and the smoother distribution of dark matter than that predicted by the standard cosmology), the results of the present paper reinforce the status of the SFHA as the candidate for dark matter, or at least for a part of it.

Keywords: charge exchange; second flavor of hydrogen atoms; dark matter; stark effect

Citation: Oks, E. Analysis of Experimental Cross-Sections of Charge Exchange between Hydrogen Atoms and Protons Yields More Evidence of the Existence of the Second Flavor of Hydrogen Atoms. *Foundations* **2021**, *1*, 265–270. https://doi.org/10.3390/foundations1020019

Academic Editors: Florentino Borondo and Ignazio Licata

Received: 25 October 2021
Accepted: 17 November 2021
Published: 19 November 2021

1. Introduction

Measurements of cross-sections of charge exchanges between hydrogen atoms and low energy protons (down to the energy ~10 eV), such as, e.g., experiments by Fite et al. [1], Fite et al. [2], and Belyaev et al. [3], revealed a noticeable discrepancy with previous theories. The experimental cross-sections were systematically slightly higher—beyond the error margins—than the theoretical predictions. The source of this discrepancy was stated as unknown [3]. Up to now, there has been no attention paid to this discrepancy, to the best of our knowledge.

In paper [4], we showed analytically that there is a slight difference in cross-sections of charge exchange between the usual hydrogen atoms and protons in comparison to charge exchange between the Second Flavor of Hydrogen Atoms (SFHA) and protons. The SFHA was discovered theoretically, as well as by the analysis of atomic experiments in paper [5], on which subsequent studies of the SFHA are based.

The basis of the theoretical discovery in paper [5] was a fresh analysis of the Dirac equation for hydrogen atoms. There are two solutions of the Dirac equation at a relatively small distance from the proton: the regular solution and the singular solution, with the latter usually being rejected. In paper [5], it was shown that at the proton boundary, the singular solution outside the proton can be tailored with the regular solution inside the proton for the ground state, so that the singular solution outside the proton is legitimate for the ground state. Using this fact, the author of paper [5] eliminated a huge discrepancy—by many orders of magnitude—between the experimental and theoretical distributions of the linear momentum in the ground state of hydrogen atoms. This constituted the first experimental evidence of the existence of the SFHA.

In a later paper [6], it was shown that the singular solution of the Dirac equation outside the proton is legitimate not only for the ground state, but for all discrete and continuous states of the zero angular momentum: for all S-states. This kind of hydrogen atom, having only S-states and described by the singular solution of the Dirac equation outside the proton, was called the "second flavor" in paper [7]—by an analogy with the flavors of quarks, so that there is an additional conserved physical quantity not commuting with some of the other ones and thus leading to an additional degeneracy.

By now, there are also two kinds of astrophysical evidence of the existence of the SFHA. First, the SFHA eliminated by a factor of two discrepancy between the absorption signal at the redshifted 21 cm radio line, recently observed by Bowman et al. [8] and the predictions of the standard cosmology—this was shown in paper [6]. Second, the SFHA explained the recent observations by Jeffrey et al. [9], where they found that the distribution of dark matter in the universe is noticeably smoother than predictions based on Einstein's relativity—this explanation was provided in paper [10].

The most striking feature of the SFHA is that since they have only S-states, then due to the well-known selection rules, the SFHA does not interact with the electromagnetic radiation (except for the radiative transition at a 21 cm wavelength between the two hyperfine structure substates of the ground state). Thus, the SFHA remains "dark". In combination with the above two kinds of astrophysical evidence, this makes the SFHA a good candidate for dark matter, or at least for a part of it.

The purpose of the present paper is to find out whether, by using the SFHA, one can eliminate, or at least reduce, the aforementioned discrepancy between the experimental and theoretical cross-sections of charge exchange involving hydrogen atoms and low energy protons. For this purpose, we extend the theory from paper [4] (where it was developed for excited states) specifically for the ground state since the corresponding experiments dealt with hydrogen atoms in the ground state. We demonstrate that for the SFHA, the corresponding cross-section is noticeably larger than for the usual hydrogen atoms. We show that the allowance for the SFHA does bring the theoretical cross-sections in a better agreement with the corresponding experiments within the experimental error margins.

2. Calculations and the Comparison with Experiments

According to paper [11], classically, the cross-section σ for the resonant transition of the electron from being associated with one ion to being associated with another ion is

$$\sigma = (8\pi/I^2)(1 - 0.8z^{2/5}),\ z = v/(2I)^{1/2} \tag{1}$$

Here, v is the relative velocity of the colliding nuclei, and I is the ionization potential from the particular atomic state. In the present paper we use atomic units unless noted otherwise. Formula (1) is valid for relatively small velocities:

$$v << v_{max} = (2I)^{1/2} \tag{2}$$

We study the resonant charge exchange between a hydrogen atom in the ground state and an incoming proton, so that from now on, I is the ionization potential of hydrogen atoms from the ground state. For calculating the quantity I, we take into account the Stark shift (if any) of the ground state due to the electric field the proton separated by the distance R from the atom.

It should be emphasized upfront that only the energy levels of the usual hydrogen atoms experience the Stark shift. In contrast, the energy levels of the SFHA have *no Stark shift* in the field of the proton *in any order of the multipole expansion*, as explained in paper [4]. This fact leads to the difference in the corresponding cross-sections of charge exchange for collisions of protons with the usual hydrogen atom compared to the collision of protons with the SFHA, as calculated below.

For a relatively large R, the ground state energy of *usual* hydrogen atoms can be represented in the form (according to Equation (4.59) from book [12]):

$$E_{large} = -1/2 - 9/(4R^4) - 15/(2R^6) \quad (3)$$

where we omitted higher order terms in this expansion; the subscript "large" signifies that this expansion is valid for relatively large R. So, the ionization potential for the ground state has the form:

$$I_{large} = 1/2 + 9/(4R^4) + 15/(2R^6) \quad (4)$$

In paper [11], it was noted that for the no-barrier transition of the electron from one nucleus to another nucleus, at the midpoint between the two nuclei, the interaction potential for the electron in the field of two nuclei should surpass the corresponding ionization potential. This condition translates in the following relation between the charge-exchange-effective distance R_0 of the proton from the atom and the ionization potential [11]:

$$I = 4/R_0 \quad (5)$$

Upon substituting Equation (5) in Equation (1), we obtain:

$$\sigma = (\pi R_0{}^2/2)\ [1 - 0.8v^{2/5}(R_0/8)^{1/5}] \quad (6)$$

Further, upon substituting Equation (5) into the left side of Equation (4) (and omitting the subscript "0"), we obtain:

$$4/R = 1/2 + 9/(4R^4) + 15/(2R^6) \quad (7)$$

The relevant root of this equation is

$$R_{large} = 7.991 \quad (8)$$

(We reiterate that the subscript "large" here and below simply refers to the fact that the results were obtained from the energy expansion for relatively large R).

In contrast, for the SFHA for a relatively large R, the energy of the ground state is

$$E_{large,2} = -1/2 \quad (9)$$

because the SFHA does not experience any Stark shift. (Here, the number 2 in the subscript signifies the "second flavor"). Therefore,

$$I_{large} = 1/2 \quad (10)$$

so that (according to Equation (5))

$$R_{large,2} = 8 \quad (11)$$

Now, let us consider the corresponding situation for relatively small R. In this case, the ground state energy of *usual* hydrogen atoms can be represented in the form (according to Equation (5.13) from book [12]):

$$E_{small} = -2 + 8R^2/3 - 16R^3/3 \quad (12)$$

where we omitted higher order terms in this expansion; the subscript "small" signifies that this expansion is valid for relatively small R. So, the ionization potential for the ground state has the form:

$$I_{small} = 2 - 8R^2/3 + 16R^3/3 \quad (13)$$

Further, on substituting Equation (5) in the left side of Equation (13) (and omitting the subscript "0"), we achieve:

$$4/R = 2 - 8R^2/3 + 16R^3/3 \quad (14)$$

The relevant root of this equation is

$$R_{small} = 0.953 \quad (15)$$

(we reiterate that the subscript "small" here and below simply refers to the fact that the results were obtained from the energy expansion for a relatively small R).

In contrast, for the SFHA for relatively small R, the energy of the ground state is

$$E_{small,2} = -2 \quad (16)$$

because the energy levels of the SFHA do not shift in the electric field (here, the number 2 in the subscript signifies the "second flavor"). Therefore,

$$I_{small,2} = 2 \quad (17)$$

so that (according to Equation (5))

$$R_{small,2} = 2 \quad (18)$$

We note that the ionization potentials, corresponding to the values of R from Equations (8), (11), (15) and (18), can be easily found from Equation (5).

For calculating the cross-sections of the charge exchange by using Equation (6), it is necessary to add the corresponding contributions from both channels, i.e., from both the "large R" case and the "small R" case. For obtaining the ratio of the resonant charge exchange cross-section from the ground state of the SFHA σ_{SFHA} to the corresponding result σ_{usual} for the usual hydrogen atoms in the simplest form—just to get the message across—we will consider the limit of v approaching zero. In this limit, the ratio $\sigma_{SFHA}/\sigma_{usual}$ simplifies to:

$$\sigma_{SFHA}/\sigma_{usual} = (R_{large,2}{}^2 + R_{small,2}{}^2)/(R_{large,}{}^2 + R_{small}{}^2) \quad (19)$$

On substituting into Equation (19) the data from Equations (8), (11), (15) and (18), we finally obtain:

$$\sigma_{SFHA}/\sigma_{usual} = 1.05 \quad (20)$$

so that σ_{SFHA} is by 5% greater than σ_{usual}.

Let us find out whether this result sufficiently improves the comparison with the corresponding experiments. The most precise experiment on the resonant charge exchange between hydrogen atoms in the ground state and relatively low energy protons was performed by Fite et al. [2] (the error margin was 9% or less, while for the later experiment by Belyaev et al. [3], the error margin reached 13%). Figure 1 presents:

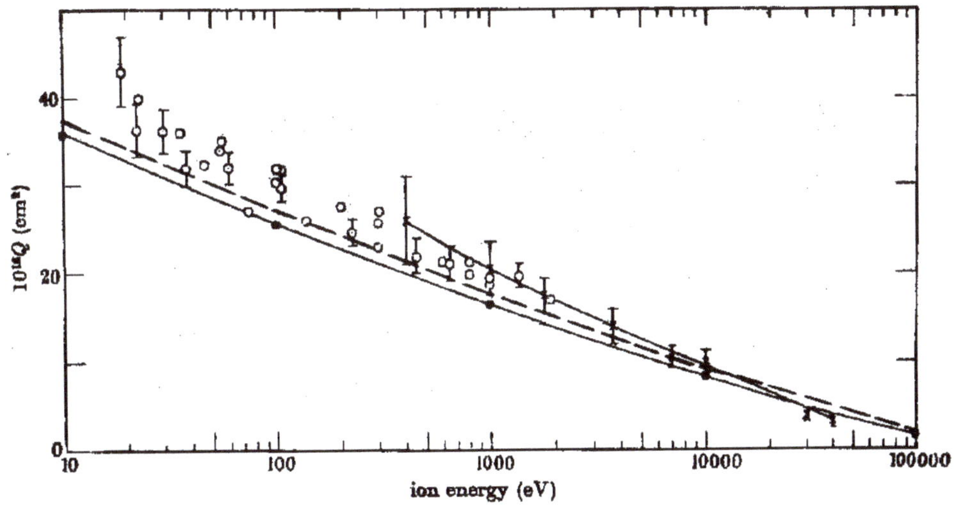

Figure 1. Comparison of the experimental cross-sections of the resonant charge exchange between hydrogen atoms and low energy protons with the corresponding theoretical cross-sections (the cross-sections are in units of 10–16 cm2). Circles (some with error margins)—experiment reported in paper [2]; crosses (with error margins) connected by a solid line—experiment reported in paper [1]; filled circles connected by a solid line—theory from paper [13], allowing only for the usual hydrogen atoms; the dashed line—theory from the present paper, allowing for the SFHA (adapted from [1,2,13]).

(1) the experimental cross-sections—circles, some with error margins—reproduced from Figure 2 of paper [2];
(2) the experimental cross-sections—crosses (with error margins) connected by a solid line—from the earlier measurements by Fite et al. [1], reproduced from Figure 2 of paper [2];
(3) the theoretical cross-sections—filled circles connected by a solid line—calculated by Dalgarno and Yadav [13], reproduced from Figure 2 of paper [2];
(4) the theoretical cross-sections for the case of the SFHA—the dashed line—from the present calculations.

It can be seen that the theoretical cross-sections for the case of the SFHA demonstrate a noticeably better agreement with the corresponding experimental cross-sections.

3. Conclusions

We studied whether the allowance for the SFHA can eliminate, or at least reduce, the noticeable discrepancy between the experimental and theoretical cross-sections of charge exchange involving hydrogen atoms and low energy protons: the discrepancy where the experimental cross-sections are systematically slightly higher than the corresponding theoretical cross-sections. We showed that, for the SFHA, the theoretical cross-sections are noticeably greater than for the usual hydrogen atoms. We demonstrated that the allowance for the SFHA leads to a noticeably better agreement with the experiments: the agreement with experiments within the experimental error margins.

This seems to constitute yet more evidence from atomic experiments that the SFHA is present within the mixture of hydrogen atoms. In combination with the first corresponding piece of evidence from the analysis of atomic experiments (presented in paper [6]), as well as with the astrophysical evidence from two different kinds of observations [9,11], the results of the present paper reinforce the status of the SFHA as the candidate for dark matter, or at least for a part of it.

Compared to other explanations of dark matter effects, the SFHA is favored by the Occam's razor principle. Indeed, it is based on the standard quantum mechanics (the Dirac equation), whereas other hypotheses either resort to mysterious, never-discovered particles beyond the standard model or require significant changes in the existing physical laws.

Funding: This research received no external funding.

Institutional Review Board Statement: Not Applicable.

Informed Consent Statement: Not applicable.

Data Availability Statement: All data is included in the paper.

Conflicts of Interest: The author declares no conflict of interest.

References

1. Fite, W.L.; Brackmann, R.T.; Snow, W.R. Charge Exchange in Proton-Hydrogen-Atom Collisions. *Phys. Rev.* **1958**, *112*, 1161–1169. [CrossRef]
2. Fite, W.L.; Smith, A.C.H.; Stebbings, R.F. Charge Transfer in Collisions Involving Symmetric and Asymmetric Resonance. *Proc. R. Soc.* **1962**, *A268*, 527–536.
3. Belyaev, V.A.; Brezhnev, B.G.; Erastov, E.M. Resonance Charge Exchange of Protons and Deuterons at Low Energies. *Sov. Phys. JETP* **1967**, *25*, 777–782.
4. Oks, E. Classical Description of Resonant Charge Exchange Involving the Second Flavor of Hydrogen Atoms. *Atoms* **2021**, *9*, 41. [CrossRef]
5. Oks, E. High-Energy Tail of the Linear Momentum Distribution in the Ground State of Hydrogen Atoms or Hydrogen-like Ions. *J. Phys. B At. Mol. Opt. Phys.* **2001**, *34*, 2235–2243. [CrossRef]
6. Oks, E. Alternative Kind of Hydrogen Atoms as a Possible Explanation of the Latest Puzzling Observation of the 21 cm Radio Line from the Early Universe. *Res. Astron. Astrophys.* **2020**, *20*, 109. [CrossRef]
7. Oks, E. Two Flavors of Hydrogen Atoms: A Possible Explanation of Dark Matter. *Atoms* **2020**, *8*, 33. [CrossRef]
8. Bowman, J.D.; Rogers, A.E.E.; Monsalve, R.A.; Mozdzen, T.J.; Mahesh, N. An Absorption Profile Centred at 78 Megahertz in the Sky-Averaged Spectrum. *Nature* **2018**, *555*, 67–70. [CrossRef] [PubMed]
9. Jeffrey, N.; Gatti, M.; Chang, C.; Whiteway, L.; Demirbozan, U.; Kovacs, A.; Pollina, G.; Bacon, D.; Hamaus, N.; Kacprzak, T.; et al. Dark Energy Survey Year 3 results: Curved-sky weak lensing mass map reconstruction. *Mon. Not. R. Astron. Soc.* **2021**, *505*, 4626–4645. [CrossRef]
10. Oks, E. DES Map Shows a Smoother Distribution of Matter than Expected: A Possible Explanation. *Res. Astron. Astrophys.* **2021**, *21*, 241–245.
11. Smirnov, B.M. The Classical Theory of Resonant Charge Exchange. *Sov. Phys. JETP* **1971**, *32*, 670–672.
12. Komarov, I.V.; Ponomarev, L.I.; Slavyanov, S.Y. *Spheroidal and Coulomb Spheroidal Functions*; Nauka: Moscow, Russia, 1976. (In Russian)
13. Dalgarno, A.; Yadav, H.N. Electron Capture II: Resonance Capture from Hydrogen Atoms by Slow Protons. *Proc. Phys. Soc.* **1953**, *A66*, 173. [CrossRef]

foundations

Article

Free Space Strange and Unipolar EM Pulses: Yes or No?

Nikolay L. Popov * and Alexander V. Vinogradov

P.N. Lebedev Physical Institute, Leninsky Prospekt 53, 119991 Moscow, Russia; vinograd@sci.lebedev.ru
* Correspondence: popovnl@sci.lebedev.ru

Abstract: E.G. Bessonov suggested the time integrated strength of an electric field $\int_{-\infty}^{\infty} \boldsymbol{E}(\boldsymbol{r},t)dt = \boldsymbol{S}_E(\boldsymbol{r})$ as a parameter to classify electromagnetic (EM) waves. Since then, this parameter has been studied and used in many works on microwave and laser physics, especially when it comes to unipolar, bipolar and few cycle EM pulses. In this paper, it is shown that $\boldsymbol{S}_E(\boldsymbol{r}) = 0$ is an identity for a wide class of free space pulses of finite total energy. This property can be useful in various applications of few cycle radiation and as a benchmark in EM and QED computations.

Keywords: space–time couplings; spatiotemporal; ultrafast optics; unipolar pulses; few cycle pulses

Citation: Popov, N.L.; Vinogradov, A.V. Free Space Strange and Unipolar EM Pulses: Yes or No? *Foundations* **2021**, *1*, 169–174. https://doi.org/10.3390/foundations1020012

Academic Editor: Eugene Oks

Received: 21 September 2021
Accepted: 11 October 2021
Published: 15 October 2021

1. Introduction

The development and widespread use of few cycle lasers [1] motivated the study of little-known and new properties of free electromagnetic (EM) waves, or more precisely, pulses [2–4]. By pulse, we mean the solution of Maxwell's equations in the absence of sources, corresponding to a finite energy. Obviously, such solutions should describe real pulses propagating in free space. The question posed in the title of the article also refers to real pulses. To answer it, it is necessary to turn to the full set of electromagnetic wave equations and add the only requirement: the finiteness of the energy of the electromagnetic field. Thus, monochromatic and plane waves are out of our consideration, since their energy is infinite, and also Gaussian beams, since they do not satisfy Maxwell's equations, as well as some other commonly used models.

Let's now go back to few cycle laser pulses of finite energy. Their distinctive property, which is clearly manifested and often discussed, is space–time couplings (STC). This implies that the spatial configuration of the field of the propagating pulse is constantly changing with time, while the time shape and the spectrum of the pulse change from point to point. For example, it is well known that when focusing, the shape and spectrum of the compressed pulses differ from the incident one, and it is not a matter of duration, but of the number of periods of the field. Any pulse of finite energy has the STC property. However, when the number of periods of the field N is large, these effects are weak and are not given importance. It is enough to characterize the structure of the pulse and its interaction with matter in terms of the average frequency ω and line width $\delta\omega$. If the pulse is short-period and N approaches 1, the values of ω and $\delta\omega$ become close. In this case, the result of the impact of a pulse on a substance essentially depends on its shape, and not on these average characteristics of the spectrum [5].

For a consistent description of STC effects, it is necessary to refer to models of finite energy pulses with nonseparable dependence on space and temporal coordinates. This can be achieved in exact solutions of Maxwell's equations, as well as in other rigorous methods. It should be noted here that, as a rule, finding exact solutions and studying their properties is also a difficult task [2,6–10]. On a qualitative level, we can say that STC is expressed in the extreme variability of any real electromagnetic pulse in time and space [11].

This work is devoted to another, in a sense, opposite property of electromagnetic pulses, which, like STC, is also important in ultrafast optics. This is due to the once rarely used characteristic of the EM field

$$S_E(\mathbf{r}) = \int_{-\infty}^{\infty} \mathbf{E}(\mathbf{r},t)dt, \tag{1}$$

which in the era of few cycle pulses, attracted increased attention. This value first appeared in the work [12] by E.G. Bessonov, who studied the radiation of particles in accelerators. To describe the radiation field $\mathbf{E}(\mathbf{r},t)$ of a system of charged particles, he introduced parameter (1) and proposed to use it to classify electromagnetic waves, calling waves with

$$S_E(\mathbf{r}) \neq 0 \tag{2}$$

strange. At the same time, as Bessonov showed, the radiation field of sources of a sufficiently wide class, namely, any system of charges performing a finite motion, satisfies the relation

$$S_E(\mathbf{r}) = 0. \tag{3}$$

He called EM waves, for which (3) is satisfied for all $\mathbf{r}$, usual. Thus, in the classification proposed by Bessonov, usual waves are necessarily bipolar. At the same time, strange waves, according to (2), can be both bipolar and unipolar (single sign). Their source can be charges that perform infinite motion, including bremsstrahlung, Compton scattering, the radiation of charged particles in bending magnets, the radiation of cosmic rays in the magnetic field of the Earth, radiation of electrons reflected from the surface of crystals, etc. [12]. The work of Bessonov received a noticeable response in accelerator and microwave communities. Since then, several theoretical and experimental papers on e-beam and other sources of bipolar and unipolar waves have been published (see [11] for references).

Interest in the topic increased sharply in the mid-1990s. The generation, application and study of unipolar pulses has become extremely relevant with the advent of the era of few cycle laser fields [13,14]. The main findings of [12] were again analyzed and confirmed [15,16]. The prospects and applications of unipolar pulses in microwave and optical ranges are presently being studied and reported in various new fields of science and technology. The parameter $S_E(\mathbf{r})$ is now also used in a broader sense than the criteria given by Bessonov Equations (1)–(3). For the characterization of bipolar strange waves, Arkhipov et al. [17] introduced the degree of unipolarity:

$$\xi(\mathbf{r}) = \frac{|S_E(\mathbf{r})|}{\int_{-\infty}^{+\infty} |\mathbf{E}(\mathbf{r},t)|dt} = \frac{\left|\int_{-\infty}^{+\infty} \mathbf{E}(\mathbf{r},t)dt\right|}{\int_{-\infty}^{+\infty} |\mathbf{E}(\mathbf{r},t)|dt}, \tag{4}$$

The method to measure the modulus of Bessonov vector (1) by observing quantum transitions induced by pulse interaction with a two-level system was proposed in [11].

In this work, it will be shown that in the absence of sources, any EM pulse is typically usual in the sense of Bessonov condition (3). In other words, condition (3) is the same inherent property of EM pulses in a vacuum as the invariants of energy, momentum, angular momentum, spin [2,18], the number of quanta [19–21] and STC, which was discussed above.

2. Materials and Methods: Some Relations for the Vector Potential in K-Space

First, let's repeat that we are talking about impulses of a general form in free space. The only requirement is the finiteness of total energy.

As in the method of expansion of an EM field in terms of field oscillators [22], it is convenient to use the Fourier transform. This allows, instead of the field strengths $\mathbf{E}(\mathbf{r},t)$ and $\mathbf{H}(\mathbf{r},t)$, satisfying the free space Maxwell equations:

$$\begin{aligned} \mathrm{rot}\mathbf{E}(\mathbf{r},t) &= -\tfrac{1}{c}\tfrac{\partial \mathbf{H}(\mathbf{r},t)}{\partial t}, \quad \mathrm{div}\mathbf{E}(\mathbf{r},t) = 0, \\ \mathrm{rot}\mathbf{H}(\mathbf{r},t) &= \tfrac{1}{c}\tfrac{\partial \mathbf{E}(\mathbf{r},t)}{\partial t}, \quad \mathrm{div}\mathbf{H}(\mathbf{r},t) = 0. \end{aligned} \tag{5}$$

consider only vector potential $A(r,t)$, which satisfies the equations:

$$\begin{cases} \Delta A(r,t) = \frac{1}{c^2}\ddot{A}(r,t) \\ \mathrm{div}A(r,t) = 0 \end{cases}, \tag{6}$$

where c is the speed of light in a vacuum. In this case, the vectors of the electric and magnetic fields are expressed in terms of A as follows:

$$E(r,t) = -\frac{1}{c}\dot{A}(r,t); \;\; H(r,t) = \mathrm{rot}\, A(r,t). \tag{7}$$

The transition to the Fourier space in Equation (6) after some transformations allows us to present the EM field as a set of independent harmonic oscillators. This is part of the standard procedure for quantizing the EM field. For the purposes of this work, it is sufficient to express the fields $E(r,t)$ and $H(r,t)$ in terms of the spatial harmonics of the vector potential $A(k,t)$; which, as is known, depend on time according to the harmonic law. As a result, the proof is reduced to the study of integrals of rapidly oscillating functions.

For the Fourier harmonics of the vector potential $A(k,\ t)$, from Maxwell's equations in the form of (6), it is easy to obtain the following equations:

$$\ddot{A}(k,t) + k^2c^2A(k,t) = 0, \tag{8}$$

$$A^*(k,t) = A(-k,t), \tag{9}$$

$$(k \cdot A(k,t)) = 0. \tag{10}$$

Expressions for $E(r,t)$ and $H(r,t)$ are easily found from the relations of (7):

$$E(r,t) = -\frac{1}{c}\int \dot{A}(k,\ t)e^{ikr}dk,\ H(r,t) = i\int (k \times A(k,t))e^{ikr}dk, A(r,\ t) = \int A(k,\ t)e^{ikr}dk. \tag{11}$$

The total energy ε does not depend on time:

$$\varepsilon = \int \epsilon(r,\ t)dr, \;\; \epsilon(r,\ t) = \frac{E^2\ (r,t) + H^2(r,t)}{8\pi}. \tag{12}$$

In k-space expression (12), taking into account (11) and (9), (10) can be written as

$$\varepsilon = \pi^2 \int \left[|E(k,t)|^2 + |H(k,t)|^2\right] d\vec{k} = \frac{\pi^2}{c^2} \int \left[\left|\dot{A}(k,t)\right|^2 + k^2c^2|A(k,t)|^2\right] dk\,. \tag{13}$$

We now use Equation (8), from which it obviously follows that

$$A(k,\ t) = A_-(k)e^{-ickt} + A_+(k)e^{+ickt}, \tag{14}$$

and, therefore, the total energy (13) is equal to

$$\varepsilon = 2\,\pi^2 \int k^2 \left[|A_-(k)|^2 + |A_+(k)|^2\right] dk. \tag{15}$$

The functions $A_-(k)$ and $A_+(k)$ are determined by the initial conditions for Equation (8). For each of them, the transversality condition similar to (10) is satisfied, and, in addition, they are linked by the relation

$$A_+(k) = A_-^*(-k) \tag{16}$$

Thus, formula (14) expresses the vector potential of an arbitrary EM pulse of finite energy in terms of the vector functions $A_-(k)$ and $A_+(k)$, satisfying the transversality condition and being square-integrable with weight k^2. In the next section, this will be used to prove relation (3).

3. Results: Free EM Pulses Are Not Strange

Along with the strangeness vector $S_E(r)$ (see (1)), consider the vector $S_E(r,T)$, so that

$$S_E(r) = \lim_{T\to\infty} S_E(r,T) = \lim_{T\to\infty} \int_{-T}^{T} E(r,t)dt = -\frac{1}{c}\lim_{T\to\infty}[A(r,\ T) - A(r,-T)]. \quad (17)$$

For the first term in (17), taking into account the last formula of (11), as well as formula (14), we obtain

$$A(r,\ T) = \int A(k,T)e^{ikr}dk = \int A_-(k)e^{ikr-ickT}dk + \int A_+(k)e^{ikr+ickT}dk. \quad (18)$$

According to the Riemann–Lebesgue theorem [23], both terms in (18) disappear, as $T \to \infty$ for any r. The condition for the applicability of this theorem is the convergence of the integrals

$$\int_0^\infty k^2 dk \left|\int d\hat{\Omega}\, A_-(k)\, e^{ikr}\right| \text{ and } \int_0^\infty k^2 dk \left|\int d\hat{\Omega}\, A_+(k)\, e^{ikr}\right|, \quad (19)$$

where $d\hat{\Omega}$ means the integral over the angles of the vector k. Taking into account the convergence of the energy integral (15), the requirement of convergence of the integrals in (19) does not seem excessively strict.

The second term on the right-hand side of (17) also disappears for $T \to \infty$. Thus, for a fairly large class of pulses in free space, we obtain

$$S_E(r) = \lim_{T\to\infty} \int_{-T}^{T} E(r,t)dt = 0, \quad (20)$$

and consequently, relation (3) is satisfied. Hence, we can summarize this section by saying that each of the field projections of the free EM pulse of a finite energy is a sign-variable function of time, the integral of which is zero at any point in space.

4. Conclusions and Discussion

So, the answer to the question in the title of the article is no. Free EM pulses are bipolar. Moreover, they satisfy Bessonov relation (3), i.e., their S-function (1) is equal to zero everywhere. According to the duality principle (see [24]), this relation is also valid for the magnetic field $H(r,t)$. The validity of what has been said is restricted by the conditions that determine the applicability of the Riemann–Lebesgue theorem for integrals of the vector potential (18).

Violations of relation (3) in the published exact solutions of the free Maxwell equations are not known to us (see [4,11,25]). However, we also failed to prove the convergence of integrals in (19) for an arbitrary electromagnetic pulse of finite energy. Therefore, the possibility of the existence of strange and unipolar pulses in a vacuum, although quite exotic, remains. This will happen if, despite the finiteness of the pulse energy (13), the conditions for the validity of the Riemann–Lebesgue theorem are broken; that is, the integral in (19) diverge. Note that the analysis of the radiation field of the laser medium in a cavity [14], as well as the radiation of charges performing a finite motion [12], showed the validity of relation (3).

Table 1 shows the classification of EM waves in terms of relations (2) and (3), as well as the terminology introduced in Section 1. Table 1 yields the conclusion that rigorous modeling of unipolar (single sign) EM pulses and their applications requires accurate consideration of both the nature of the source and the propagation of the wave, since both factors affect the time shape of the pulse at the point of interest.

Table 1. Bessonov characteristic $S_E(r)$ and classification of EM pulses from moving charges (2nd column) and EM pulses travelling in free space (3rd column).

	EM Pulses from Moving Charges [12]			Free Space EM Pulses (This Paper)
Charge Motion	Bounded	Unbounded		
$S_E(r)$	0	$\neq 0$	0	0
usual or strange	usual	strange	usual	usual
bipolar or unipolar	bipolar	both are possible	bipolar	bipolar

The 3rd column in Table 1, like the rest of the article, deals only with exact, finite-energy vector solutions to EM wave equations in free space. Important results beyond these restrictions (e.g., nondiffracting waves, scalar and 2D wave-packets, etc.) can be found in [26–29].

In conclusion, we note the coexistence of two properties inherent in EM pulses of finite energy. On the one hand, this is the extreme variability in space and time associated with STC, and on the other hand, the equality to zero of the S-function in the whole space is proved here. The latter can be used to control the accuracy of EM and QED computations with finite energy pulses (compare with [30,31]).

Author Contributions: Data curation, N.L.P. and A.V.V.; writing—review and editing, N.L.P. and A.V.V. All authors have read and agreed to the published version of the manuscript.

Funding: This research received no external funding.

Institutional Review Board Statement: Not applicable.

Informed Consent Statement: Not applicable.

Data Availability Statement: Not applicable.

Acknowledgments: The authors are indebted to I.A. Artyukov, S.G. Bochkarev, V.Yu. Bychenkov, N.V. Dyachkov, R.M. Feshchenko, A.P. Kiselev, A.B. Plachenov and I.V. Smetanin for fruitful discussions. We are also grateful to A.M. Fedotov who read the manuscript and made interesting comment.

Conflicts of Interest: The authors declare no conflict of interest.

References

1. Brabec, T.; Krausz, F. Intense few-cycle laser fields: Frontiers of nonlinear optics. *Rev. Mod. Phys.* **2000**, *72*, 545. [CrossRef]
2. Lekner, J. *Theory of Electromagnetic Pulses*; Morgan & Claypool Publishers: San Rafael, CA, USA, 2018.
3. Feng, S.; Winful, H.G.; Hellwarth, R.W. Spatiotemporal evolution of focused single-cycle electromagnetic pulses. *Phys. Rev. E* **1999**, *59*, 4630. [CrossRef]
4. April, A. Ultrashort Strongly Focused Laser Pulses in Free Space. In *Coherence and Ultrashort Pulse Laser Emission*; Duarte, F.J., Ed.; InTech: New York, NY, USA, 2010; pp. 355–382.
5. Keldysh, L.V. Multiphoton ionization by a very short pulse. *Phys. Usp.* **2017**, *60*, 1187–1193. [CrossRef]
6. Kiselev, A.P. Localized Light Waves: Paraxial and Exact Solutions of the Wave Equation (a Review). *Opt. Spectrosc.* **2007**, *102*, 603–622. [CrossRef]
7. Ziolkowski, R.W. LocalizedWaves: Historical and Personal Perspectives. In *Non-Diffracting Waves*; Hernández-Figueroa, H.E., Recami, E., Eds.; John Wiley & Sons: New York, NY, USA, 2013; Chapter 2; pp. 69–81.
8. Artyukov, I.A.; Dyachkov, N.V.; Feshchenko, R.M.; Vinogradov, A.V. Energy density and spectrum of single-cycle and sub-cycle electromagnetic pulses. *Quantum Electron.* **2020**, *2020*, 187–194. [CrossRef]
9. So, I.A.; Plachenov, A.B.; Kiselev, A.P. Simple unidirectional finite-energy pulses. *Phys. Rev. A* **2020**, *102*, 063529. [CrossRef]
10. Bialynicki-Birula, I. New solutions of the Dirac, Maxwell, and Weyl equations from the fractional Fourier transform. *Phys. Rev. D* **2021**, *103*, 085001. [CrossRef]
11. Popov, N.L.; Vinogradov, A.V. Space-Time Coupling: Current Concept and Two Example from Ultrafast Optics Studied Using Exact Solution of EM Equations. *Symmetry* **2021**, *13*, 529. [CrossRef]
12. Bessonov, E.G. On a class of electromagnetic waves. *Zh. Eksp. Teor. Fiz.* **1981**, *80*, 852–858.
13. Kaplan, A.E.; Shkolnikov, P.L. Electromagnetic "Bubbles" and Shock Waves: Unipolar, Nonoscillating EM Solitons. *Phys. Rev. Lett. D* **1995**, 2316–2319. [CrossRef]

14. Miloševi'c, D.B.; Paulus, G.G.; Bauer, D.; Becker, W.J. Above-threshold ionization by few-cycle pulses. *Phys. B At. Mol. Opt. Phys* **2006**, *39*, R203–R262. [CrossRef]
15. Kim, K.J.; McDonald, K.T.; Stupakov, G.V.; Zolotorev, M.S. A bounded source cannot emit a unipolar electromagnetic wave. *arXiv* **2000**, arXiv:physics/0003064.
16. Kim, K.J.; McDonald, K.T.; Stupakov, G.V.; Zolotorev, M.S. Comment on "Coherent Acceleration by Subcycle Laser Pulses". *Phys. Rev. Lett.* **2000**, *84*, 3210. [CrossRef] [PubMed]
17. Arkhipov, R.M.; Pakhomov, A.V.; Arkhipov, M.V.; Babushkin, I.; Tolmachev, Y.A.; Rosanov, N.N. Generation of unipolar pulses in nonlinear media. *JETP Lett.* **2017**, *105*, 408–418. [CrossRef]
18. Feshchenko, R.M.; Vinogradov, A.V. On the number and spin of photons in classical electromagnetic fields. *Phys. Scr.* **2019**, *94*, 065501. [CrossRef]
19. Zeldovich, Y.B. Number of quanta as an invariant of the classical electromagnetic field. *Sov. Phys.—Dokl.* **1966**, *10*, 771.
20. Bialynicki-Birula, I. Photon Wave Number. In *Progress in Optics*; Wolf, E., Ed.; Elsevier Science: New York, NY, USA, 1996; Volume 36, pp. 245–294.
21. Feshchenko, R.M.; Vinogradov, A.V. On the number of photons in a classical electromagnetic field. *J. Exp. Theor. Phys.* **2018**, *127*, 274–278. [CrossRef]
22. Landau, L.D.; Lifshitz, E.M. *The Classical Theory of Fields*, 2nd ed.; Addison-Wesley: Reading, MA, USA, 1962.
23. Olver, F. *Asymptotics and Special Functions*; Academic Press: New York, NY, USA, 1974.
24. Zangwill, A. *Modern Electrodynamics*; Cambridge University Press: Cambridge, UK, 2012.
25. Gonoskov, I.; Aiello, A.; Heugel, S.; Leuchs, G. Dipole pulse theory: Maximizing the field amplitude from 4π focused laser pulses. *Phys. Rev. A* **2014**, *86*, 053836. [CrossRef]
26. Kondakci, E.; Abouraddy, A.F. Diffraction-free space-time light sheets. *Nat. Photonics* **2017**, *11*, 733. [CrossRef]
27. Besieris, I.; Abdel-Rahman, M.; Shaarawi, A.; Chatzipetros, A. Two fundamental representations of localized pulse solutions to the scalar wave equation. *Prog. Electromagn. Res.* **1998**, *19*, 981–984. [CrossRef]
28. Fagerholm, S.J.; Friberg, A.T.; Salomaa, M.M. Unified description of nondiffracting X and Y waves. *Phys. Rev. E* **2000**, *62*, 4261.
29. Saari, P.; Reivelt, K. Generation and classification of localized waves by Lorentz transformations in Fourier space. *Phys. Rev. E* **2004**, *69*, 036612. [CrossRef] [PubMed]
30. Madsen, L.B. Gauge invariance in the interaction between atoms and few-cycle laser pulses. *Phys. Rev. A* **2002**, *65*, 053417. [CrossRef]
31. Fedotov, A.M.; Korolev, K.Y.; Legkov, M.V. Exact analytical expression for the electromagnetic field in a focused laser beam or pulse. *Proc. SPIE* **2007**, *6726*, 672613.

foundations

MDPI

Article

Geometric State Sum Models from Quasicrystals

Marcelo Amaral *, Fang Fang, Dugan Hammock and Klee Irwin

Quantum Gravity Research, Los Angeles, CA 90290, USA; Fang@QuantumGravityResearch.org (F.F.); Dugan@QuantumGravityResearch.org (D.H.); Klee@QuantumGravityResearch.org (K.I.)
* Correspondence: Marcelo@QuantumGravityResearch.org

Abstract: In light of the self-simulation hypothesis, a simple form of implementation of the principle of efficient language is discussed in a self-referential geometric quasicrystalline state sum model in three dimensions. Emergence is discussed in the context of geometric state sum models.

Keywords: self-simulation hypothesis; principle of efficient language; quasicrystals; empires; game of life; emergence; state sum models

Citation: Amaral, M.; Fang, F.; Hammock, D.; Irwin, K. Geometric State Sum Models from Quasicrystals. *Foundations* **2021**, *1*, 155–168. https://doi.org/10.3390/foundations1020011

Academic Editor: Eugene Oks

Received: 9 September 2021
Accepted: 9 October 2021
Published: 13 October 2021

1. Introduction

The self-simulation hypothesis (SSH) [1] posits that emergence is a core element in the engine of reality, down to the underlying code—considering spacetime and particles as secondary or emergent from this code. In 1938, Dirac [2] addressed the internal structure of the electron and how it affects the spacetime structure itself. Feynman later evolved this thinking to point out that a point of spacetime is like a computer, which lead Finkelstein to propose that reality is a code in action in his 1969 spacetime code paper [3]. Wheeler later wrote an interesting synthesis of this with ideas like the participatory universe as a self-excited circuit, law without law and it from bit [4–6]. Second test edit. This information theoretic line of thinking can lead to many ways of addressing the conundrum of quantum gravity and unification physics problems. We focus here on the specific perspective that a notion of pre-spacetime code or language in action leads to the physics and metaphysics idea of reality as a self-simulation [7–9] and requires a new principle to drive the evolution—the principle of efficient language (PEL) [1,7,9–11].

To address emergence of physical observables from an information theoretic framework governed by the PEL, we need to set up a concrete, constrained and rigorous mathematical substrate. We will consider the Three-Dimensional Penrose tiling quasicrystal (3DPT), also known as Amman or Ammann-Kramer-Neri tiling, projected from the Z_6 lattice [12–14]—a generalization of the two-dimensional Penrose tiling [15]. On this point set and associated tilings, we implement a state sum model [16,17], where the states are given by objects inherent in the quasicrystalline geometry. We consider geometric realism discussed in Section 2 as a new paradigm for state sum models following Einstein's program of geometrization of physics [18]. As we will see, quasicrystals are a natural substrate for geometric realism, where self-referential geometric symbols [1,19] are given from first principles for both kinematics and dynamics. Details on quasicrystals will be presented in Section 3. Essentially, quasicrystalline structures [20–24] are structures that exhibit a new kind of order—aperiodic order—which lies between disorder and periodicity. Statistical or quantum mechanical models defined on lattices can be generalized in a straightforward manner to quasicrystals [25]. The choice of Z_6 root lattice and its associated 3DPT quasicrystal provides a toy model for the conformal symmetry associated with the D_6 root system and the grand unified gauge theory associated with the exceptional Lie algebra E_8. Both D_6 and E_8 have similar quasicrystals associated with them [26–28]. The gauge symmetry represented in the root system is transformed to a 3D quasicrystal network, working as a toy model for quasicrystalline pre-spacetime code.

Quasicrystals come with a natural non-local structure called an empire [29–32], which can be considered as their defining property. The geometrical state sum model (GSS) proposed will make use of the concept of empire overlaps (called hits in this paper) built upon the rules driving the dynamics, which we will discuss in Section 4. One element that implements the PEL is that an empire position can save resources in the GSS evolution.

A simple implementation of GSS is given in Section 4.1 with a new kind of cellular automaton game of life, where different patterns emerge. Those patterns then act back on the more general GSS model in the form of observables discussed in Section 4.2. In Section 5, we conclude with some discussion about the notion of emergence within a GSS model that implements the PEL.

2. Geometric Realism

The construction of general relativity (GR) marked a breakthrough for the so-called program of geometrization of physics [18], which basically says that one should start with geometry to understand the physical world—there should be a one-to-one correspondence between physical quantities and geometric objects. Modern physics, starting with Einstein himself, follows a path to apply that idea to generalize GR by going beyond Riemannian manifolds (adding to the usual curvature variable, the torsion and nonmetricity variables). This is in essence Riemann's program as a unified view of geometry. Some variations of this approach include elements from Klein's program, which focus on symmetries and their associated groups. However, there is another path that we will consider in light of recent developments in theoretical physics, which brings the lesser known geometric program view of Fedorov/Delone that can address both local and global geometry when considering discrete systems [21]. For example, a problem that can be easily solved by this program is which shapes tile space and how. The idea is that regular point systems are determined by local settings. Global regularity results from the structure of local configurations.

Consider a clear route to the quantization of GR that is given by loop quantum gravity (LQG) [33], where the classical 4-dimensional manifold is foliated in 3-dimensional (3D) spacelike surfaces and the metric field $g_{\mu\nu}$ is decomposed in terms of connections and tetrads, which, in the 3D foliation, reduces to 3D-connection and the triad field. Then the connection and the triad are promoted to operators in a Hilbert space. This quantization procedure leads to the result that the main kinematic objects are spin network states spanning the Hilbert space. The dynamics can be achieved by the usual path integral procedure and leads to spin foam transition amplitudes. These objects can be described graph-theoretically with $SU(2)$-spin quantum numbers labeling the edges of the graphs and another group or algebraic $SU(2)$ data at the vertices. The spin foam path integral of quantum gravity can be understood as a sum over spin network states. In summary, the main object after the path integral quantization of GR is a sum over states of the quantum geometry. In fact, this formulation is a cornerstone of modern physics. Making use of the similarity of the path integral in quantum field theory and the partition function in statistical mechanics [34], many concrete computations of physical observable with different models, from Ising models to lattice gauge theory, condensed matter systems [35], and quantum gravity, is done with a state sum over discrete lattices or graphs. So, the other path to geometrization of physics is to address the geometry of state sum models at the quantum or statistical mechanics regime.

Geometric realism will dictate for us that the labels, for example a spin state ±1, that appear in those graphs, quasicrystals or lattices, must have a one-to-one correspondence with the underlying discrete geometry, in our case, the quasicrystal one. That is, the labels are directly related to the geometric building block of tilings. To be more concrete, let us consider the state sum object

$$W_{\triangle}(s_b) = \mathcal{N}_{\triangle} \sum_{s} \prod_{e} A_e(s) \prod_{v} A_v(s), \tag{1}$$

In spin foam models or some lattice gauge models, $W_\triangle(s_b)$ is called the quantum transition amplitude, or the partition function $Z_\triangle$ in some Ising like models where one sum also over the boundary states s_b. In Equation (1), $\mathcal{N}_\triangle$ is a normalization constant that depends on the discretization $\triangle$. There is weight or amplitude (Weight if $W_\triangle$ is considered as a partition function or quantum amplitudes in a path integral picture.) A_e for each edge of $\triangle$ and weight or amplitude A_v for each vertex. The sum goes over all the allowed configurations of states s. Usually A_e and A_v are built from the group and algebraic theoretic implementation of symmetries involved in the specific problem. Geometric realism requires that A_e and A_v are built from geometry of $\triangle$.

Let us consider some examples of standard state sum models:

- Ising models

The Ising model, constructed over a lattice $\triangle$ can be described using only weights A_e given by

$$A_e = e^{\beta g_{s(e)} g_{t(e)}^{-1}} \tag{2}$$

with the sum over spin states $s = \pm 1$, $g_{s(e)}$ representing the spin at the starting vertex and $g_{t(e)}$ representing the spin at the end vertex of every edge e. The β coupling constant is proportional to the inverse of temperature. General Ising models can be achieved by allowing the states, s, to take values in a large range of integers $0, 1, ..., n$. Further generalization can be considered by allowing the edge weights to locally vary as a general function of the states $g_{s(e)} g_{t(e)}^{-1}$, $A_e = f(g_{s(e)} g_{t(e)}^{-1})$.

- Lattice gauge theory (LGT):

LGT gives a non-perturbative formulation of the path integral quantization of gauge theories such as the standard model of particle physics. With LGT one gives up on Lorentz symmetry and works with gauge symmetries at the vertices of the lattice $\triangle$. Gauge invariant quantities are given by Wilson loops made up of edges around a two dimensional face of $\triangle$, $g_f = \prod_{e \in f} g_e$. For continuous groups of symmetry, the sum in Equation (1) is converted on an integral that goes over the infinite of group elements (gauge) symmetry g_e and the products of amplitudes on edges and vertices are converted to a product over faces

$$A_f = f(h_f) \tag{3}$$

where the amplitude functions A_f are class functions on the group of symmetry ($f(ghg^{-1}) = f(h)$). The explicit form of the function f depends on the specific gauge symmetry model. For example, for Yang-Mills theory, is given by

$$A_f = e^{\beta \sum_f \mathbb{R}(tr(U(h_f)))} \tag{4}$$

where U is a unitary finite dimensional matrix representation of the group, β here, is a coupling constant. Another example is given by topological models where A_f takes the simpler form

$$A_f = \delta(h_f) \tag{5}$$

where the delta function δ is taken with respect to the group measure.

- Spin foam

In spin foam models, $\triangle$ is a triangulation of spacetime manifold or its dual. Spin foam models also have amplitudes associated with faces (or edges), usually give by Equation (5), but also amplitudes A_v associated to vertices and constructed from more complicated group invariant objects. In LQG, the group of symmetry is given by the little group ($SU(2)$) of spacetime symmetries ($SL(2, C)$). A generalization of spin foam for quantum gravity is to include color charge symmetry ($SU(3)$), within the weights A_e and A_v in Equation (1). This can be done by defining the weights from invariants of $SU(2) \times SU(3)$ [17]. When $W_\triangle(s_b)$ depends only on the boundary states, the spin foam is topological and the amplitudes are constructed from topological invariants [16].

State sum models are built from algebraic and group-theoretic elements. Even the simple Ising model can be understood as a discrete Z_2 model. The underlying group and algebraic structures have a geometric correspondence. Geometric realism demands the values on the GSS to come from the geometry, more concretely from tiling of $\triangle$, which, for the specific model presented in next sections, will be a 3D quasicrystal. The states, coupling constants, potentials, weights and amplitudes themselves, will be determined by geometric objects as lengths, volumes and volumetric intersections.

3. Kinematics: The 3D Quasicrystal, Empire and Hits

The construction of the quasicrystal of interest here, the 3DPT, will make use of the canonical cut-and-project method [12,21]. To construct the 3DPT, $\triangle$, we consider the canonical hypercubic lattice $\mathbb{Z}^6$ in Euclidean space $\mathbb{R}^6$. Let ε be an irrational 3-dimensional subspace of $\mathbb{R}^6$ and $\varepsilon_\perp$ be its orthogonal complement. Let P be the orthogonal projector onto ε and $P_\perp$ onto $\varepsilon_\perp$. Now we fix a compact subset K of $\varepsilon_\perp$, called the cut-window. The canonical choice for the cut-window is the projection of the Voronoi cell of $\mathbb{Z}^6$ to $\varepsilon_\perp$. The Voronoi cell contains one lattice point, which lies at its center. Unlike the unit cells of lattices, these cells are unique and their symmetry groups are the stabilizer groups of the lattice points. The 3DPT quasicrystal $\triangle$ is constructed by projecting points $\lambda \in \mathbb{Z}^6$ to ε, $P(\lambda)$, such that $P_\perp(\lambda)$ lies inside K, the acceptance domain. $\mathbb{Z}^6$ lattice points are connected by the unit length edges. If two points λ_1 and λ_2 are connected in $\mathbb{Z}^6$ and $P_\perp(\lambda_1)$ and $P_\perp(\lambda_2)$ are accepted, then $P(\lambda_1)$ and $P(\lambda_2)$ are connected in $\triangle$. A vertex v_i can have different numbers of neighbors, here denoted v_{ij}, with j variating from 1 to the valence of v_i (The possible valences for any vertex at some 3DPT are: 4, 6, 8, 10, 12, 14, 16, 20.) The geometric lengths of the connections are labeled as l_{ij}. A tiling $\mathcal{T}$ of $\triangle$ is a set of possible points and connections given by this procedure. Different tiling configurations can be generated by doing a shift on $P_\perp(\lambda)$ in $\varepsilon_\perp$ before checking if $P_\perp(\lambda)$ lies inside K, so-called $\gamma_\perp$. The shift $\gamma_\perp$ can be used to generate different tilings and it is a continuous parameter that can be used to make the quasicrystal dynamic. A specific 3D tiling $\mathcal{T}$ of $\triangle$ has two different rhombohedral prototiles building blocks with 10 orientations each. Each vertex v_i of one $\mathcal{T}$ can be associated to different configurations of prototiles (up to 20 rhombohedral prototile around one vertex v_i). There are 24 possible different vertex types (VT), which appears with different frequencies and valence in a tiling. Most of the 24 VT have valence 20, but those with lower valence appear with more frequency in a tiling. See [12] for the explicit form of VT and its frequencies. A small tiling and 3 of the 3DPT VTs are shown in Figure 1.

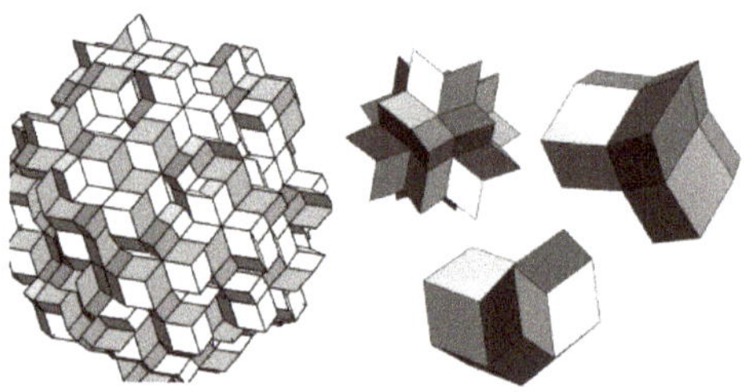

Figure 1. 3DPT tiling and some VTs.

The empire is an important property of quasicrystals that arises, in context, within the empire problem [29–32]. A quasicrystal itself arises in context of the the ancient problem of tiling space in an aperiodic way. To clarify the use of the 3DPT, we can think of $\triangle$ generated from K as a possibility point (tiling) space (PS) where points (or VT) can be turned ON or OFF. Or we can think of $\triangle$ as the selected points in $\mathbb{Z}^6$ but not projected yet. We are interested in project subsets of those points, the empires, which we define below. The empire problem asks what other vertices or VT of a quasicrystal tiling is forced to be ON if a specific vertex or VT is chosen to be ON—it is projected or actualized. By defining the empire problem and its solutions, one sets the kinematics of a given quasicrystal. For a specific vertex v_i in $\triangle$ there is associated a specific VT (from the 24 possible ones). We can define a window $E \subset K$ associated to any VT given at some vertex v_i, and so, the empire of a point $P(\lambda_i)$ of $\triangle$ will be the sub-set of $\triangle$ such that $P_\perp(\lambda)$ lies also inside E. The cut-window K can be volumetric partitioned in sub-windows E, which can overlap. The dynamic quasicrystal can be generated by projecting empires. Empires capture the non-local aspect of quasicrystals [36] in the sense that when a vertex v_i is ON, its empire—the whole set of points defined by window E_i—is also ON. An additional question can be asked now: If two vertices v_1 and v_2 and their empires are projected, what is the empire overlap between them? The answer is that a measure of the overlap of points sets of different empires is given geometrically by the overlap of empire windows E_1 and E_2 inside the cut-window K. We call this overlap the hit H_{12}. Consider the empire given by E_1 to be $\triangle_{E1} \subset \triangle$ and for E_2 to be $\triangle_{E2} \subset \triangle$, then $H_{12} = \triangle_{E1} \cap \triangle_{E2}$ or in terms of vertex window polytope intersection for any v_i and v_j we can compute a normalized measure of overlap by

$$H_{ij} = \frac{I_{Vol}(E_1, E_2)}{I_{Vol}(E_1)} \tag{6}$$

where I_{Vol} is a function that gives the volume of the intersection of different polytopes, computed here numerically, and if it has only one polytope as input, it returns the volume of that polytope. Computing hits H_{ij} between a VT at v_i and different v_j positions in one tiling and then changing v_i gives a hit map distribution for the specific tiling. One typical example is shown in Figure 2. In what follows we will consider only nearest neighbors H_{ij}.

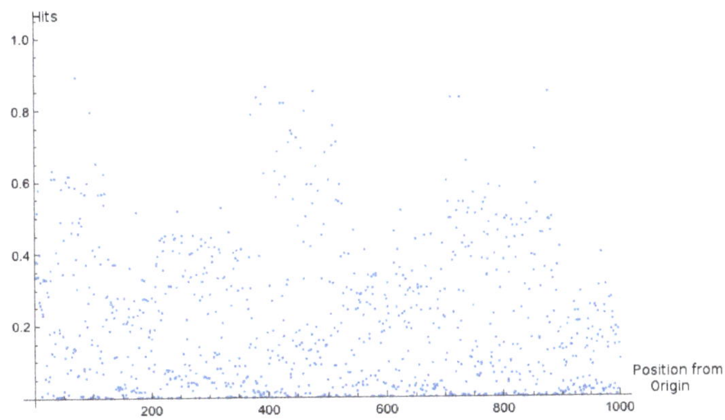

Figure 2. A typical hit map for the 3DPT quasicrystal. We consider a list with 1000 points of a 3DPT tiling and compute vertex window polytope overlap between them, Equation (6).

With the main construction of the 3DPT presented, we can establish that the main kinematics variables of interest are the hits H_{ij} between v_i and v_j, the length l_{ij} of the connection between v_i and v_j and the volume V_i of the VT polytope associated with v_i. We can turn now to the implementation of the dynamics aspects of the GSS $W_\triangle(s_b)$, Equation (1), to be done in the next section.

4. Dynamics: Geometric State Sum Model and the PEL

To implement dynamics of the GSS, we consider the state sum in Equation (1). Following the ideas from the previous sections, the weights A_e for an edge linking vertices v_i and v_j will be given by

$$A_e = l_{ij} H_{ij}, \tag{7}$$

where the states given at each edge $s_{ij} = H_{ij}$ are geometric quantities, specifically, volumetric polytope intersections. The length l_{ij} plays the role of the coupling constant or inverse of temperature. We can group these edge states by summing over nearest neighbors v_{ij}

$$A_{v_i} = \sum_j l_{ij} H_{ij}. \tag{8}$$

We will add an additional term to each vertex v_i, called the hit potential Y_i, which takes into account the PEL by implementing a look-ahead algorithm [37]. Consider a PS tiling $\mathcal{T}$ with only the central vertex v_c and its empire being ON, which means that the vertex window polytope E_{v_c} is being used to select the possible points of the $\mathbb{Z}^6$ that can be projected to the cut-window K. Now we will probe the quasicrystal possibility space $\triangle$ with random walks of the vertex type at v_c using rules based on H_{ij}. So, we start on step 1, with v_c and its empire being ON. In step 2, one of the neighbors v_{cj} will be ON according to a non-deterministic rule $R(H_{ij})$, which we will discuss in Section 4.2. Then we repeat these steps until step N. This defines one animation A_1. Next, we repeat this procedure, getting a new animation A_2, and so on until an animation A_M, so that we end with M animations, each with N steps. We call these animations possibility space random walks (PRW). The hits potential Y_i is defined at each vertex v_i as the number of PRWs that use that position. For consistency we will add a volume weight

$$\mathcal{Y}_i = V_i Y_i, \tag{9}$$

where V_i is the volume of the VT polytope at v_i. It encodes the coupling l_{ij} there. The hit potential comes from the idea of minimizing the cost of projection resources to turn ON points on $\triangle$. The vertices that have more PRWs going over them have more potential to save projection steps to generate animations for the look-ahead algorithm (or here: Look-savings-ahead algorithm). As there are more walks going over those positions they can be part of more possible emergent patterns. The hit potential can be defined from weights of entire animations. We can count hits and associate an integer to one animation A_m by counting how many empire vertices the PRW of that animation encounters. These vertices are already ON and don't need to be turned ON on the PRW. The hits H_{ij} and hit potential Y_i associated to a tiling $\mathcal{T}_k$ define one valid configuration of states given by

$$W_{\mathcal{T}_k} = \mathcal{N}_{\mathcal{T}_k} \prod_i A^k_{v_i} \mathcal{Y}^k_i, \tag{10}$$

where the index k means that the geometric quantities are computed on the specific tiling $\mathcal{T}_k$. The defining object of interest for the GSS is then, in a partition function form,

$$W_{\triangle} = \mathcal{N}_{\triangle} \sum_{\mathcal{T}_k} W_{\mathcal{T}_k}, \tag{11}$$

where the sum goes over the different allowed configurations, implemented by finding new allowed tilings. One way to find new tilings, which involves changing only the states H_{ij} and not the couplings l_{ij} or V_i, is by using the shift $\gamma_\perp$. To change the couplings l_{ij} or V_i from quasicrystalline first principles, one can use inflation [27], which requires changing the size of the windows. The state sum can also be defined as a transition function for state H_{ab} that fixes a subset of $\triangle$, where H_{ab} can be two disjointed subsets so that maintaining both fixed, and summing over the remaining part, would give the transition function between the states

$$W_{\triangle}(H_{ab}) = \mathcal{N}_{\triangle} \sum_{\mathcal{T}_k(ij/ab)} W_{\mathcal{T}_k}, \tag{12}$$

where the notation (ij/ab) indicates that all tilings and their accompanying H_{ij} are generated, but with H_{ab} remaining fixed. This implementation is given by finding values of the shift $\gamma_\perp$ that generate new tilings sharing a fixed configuration.

4.1. A New Kind of Game of Life in Quasicrystals

A concrete implementation of a dynamic GSS Equation (12) is given in the form of a cellular automaton game of life (GoL). In this case we let the quasicrystal tiling space $\triangle$ evolve according to local rules, which the same nature of $R(H_{ij})$ we use to generate the hit potential and which we discuss below.

Classical cellular automata are defined on regular lattices. The rules depend on the state of each site and its neighbors. The neighborhood structure looks the same across the lattice. For GoLs [38–42], which have outer totalistic cellular automaton rules, the next state of a site depends only on its current state, and the total number of neighbor sites in certain states. In 2-dimensional quasicrystalline GoLs [43–45] the neighborhoods are generalized and not the same for each site, but the dynamics are implemented with similar rules to the original GoL. GoL rules on Penrose tilings still have complex behavior.

The generalization to 3D that implements the GSS model is given by a 5-tuplet $\mathcal{G} = (\triangle, T, H, v, \mathcal{R})$ with elements as follows. $\triangle$ is a 3DPT; T includes the initial tiling condition (where at least one VT is ON) and the set of steps to update $\triangle$—we usually consider 1000 steps for 5000 3DPT point set; H is the set of states generalized to be a real number between 0 and 1 according to Equation (6); $v = v_{ij}$ are the neighbors of a vertex v_i, which vary between 4 and 20—we note that when a vertex is ON the whole VT associated to it is considered to be ON, and also that sometimes we can consider the point set for evolution to have only a specific VT instead of the full quasicrystal point set; $\mathcal{R}$ are the new local adapted rules for this kind of GoL. The rules measure empire overlap: If there is too much overlap with ON neighbor VTs then it turns or stays OFF (overpopulation condition); if there is too little overlap it also will be OFF (under-population condition); but if there is the right value of overlap with the mean value of all v_{ij}, then the current vertex will be ON. Good values for the normalized A_{v_i}, Equation (8), are found to be between 0.7 and 0.9. We also consider the information entropy associated to hits to drive the evolution as in Equation (14)—in this case the VT will be ON if its information entropy is close enough of the mean of its neighbors, which are ON (one additional option with this kind of GoL is that due to the non-local properties of empires we can allow the rules to be applied to the whole quasicrystal and not only with the connected vertices of a VT, which we will leave for future investigations).

These dynamics lead to different emergent patterns, most of which are oscillation patterns as in Figure 3. There are also different kinds of propagation as in Figure 4. However, varying the rules opens up possibilities of more complex dynamics, which are open for future systematic investigations, but which may also encounter problems of optimization of image processing.

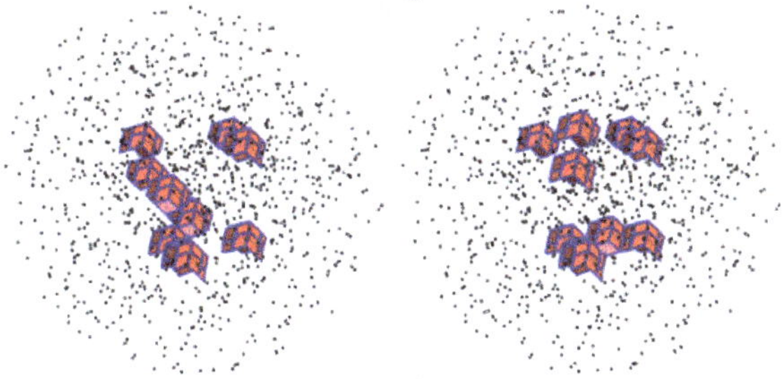

Figure 3. One 3DPT GoL oscillation pattern. The pattern is a oscillator period 2 and the two frames are shown.

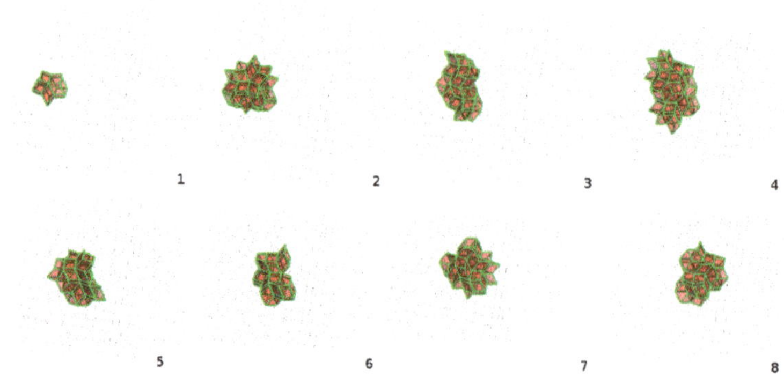

Figure 4. A recurrent pattern propagation for a 3DPT GoL made mainly from the VTs from Figure 3. (**1–8**) show eight frames of the dynamics where the pattern highlighted moves from left to right.

4.2. GSS Observables and Emergence

In light of more general dynamics, we can interpret Equation (12) as one observable on a GSS model. It is a fixed pattern over the space of geometric states. The specific emergent pattern and its properties can be addressed with Equation (12). The hit potential $\mathcal{Y}_i$ can, in fact, be considered as derived from one observable VT that is ON and that is following a family of PRWs. Let us consider the rules $R(H_{ij})$ in this context. The rules are used to guide the random walks to probe the possibility space and then define the hit potential. What is moving in the quasicrystal, or, what are those VT, which are being turned ON or OFF (being projected/actualized or not)? A GSS model aims to describe pre-spacetime physics—the Planck scale quantum gravity regime. This regime is considered to have the concept of holographic matter [46–49], which is proportional to information entropy

$$|I_{v_i} - I_{v_j}| = \triangle I_{ij} = \alpha m, \tag{13}$$

with α a constant of proportionality and m the mass crossing some horizon. In this Planck scale regime picture, each connection is considered to be crossing a holographic horizon and information is the stuff flowing between horizons and to be conserved. The local information entropy is given by

$$I_i = -\sum_j P_{ij} log P_{ij} \tag{14}$$

with $P_{ij} = \frac{H_{ij}}{N_i}$. Different rules can be implemented under I. The above motivation leads us to the notion of local conservation of I. We simply let the PRW, starting at the center of a tiling, be guided by a probability distribution constructed by choosing each successive position from a local subset whose values of I are within some selected range of the current value of I, which sets the non deterministic look-saving-ahead algorithm. The resultant hit potential $Y(r)$ as a function of the distance from the center is given in Figure 5 for 1000 animations over 1000 steps.

The specific form presented in Figure 5 has a mean value that drops with the inverse of distance away from the center but with a Gaussian contribution close to the center. In general its form depends on the distribution of VTs on a specific tiling, the local rules and random walk properties. We can also consider an additional second PRW starting at a different position and allow the $Y(r)$ to count overlap between the two walks—a synergistic effect [50] on the whole emergent hit potential not accounted for on the underlying GoL. See Figure 6.

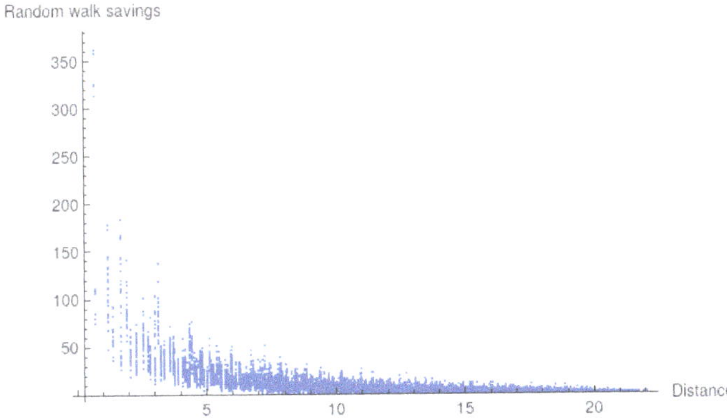

Figure 5. PRW hit potential.

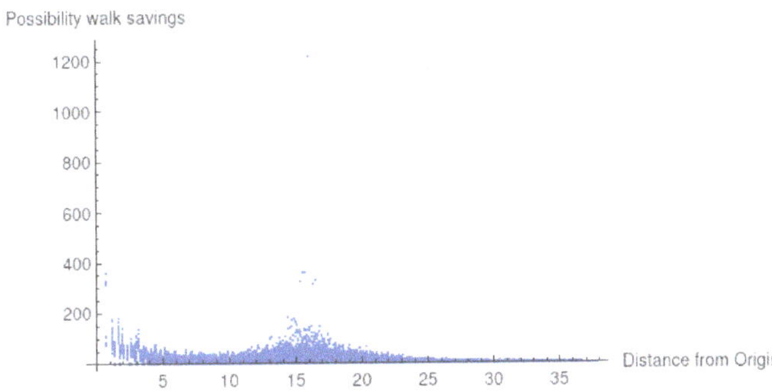

Figure 6. PRW hit potential from two patterns evolution starting at different positions with one at origin.

Note that we can use the PRWs to evolve $\triangle$ determining H, v and even $\mathcal{R}$ of the GoL $\mathcal{G}$. We can also allow local rules at site i to be determined from the full weights $A_{v_i}\mathcal{Y}_i$. This generates a more sophisticated stratified recursive game

$$\mathcal{G}_{n+1} = \mathcal{G}(\mathcal{G}_n). \tag{15}$$

As a result of the GSS evolution, we consider a candidate for an order parameter, which is analogous to magnetization in spin systems,

$$H_{\mathcal{T}}(l) = \frac{1}{\mathcal{N}_{\mathcal{T}}} \sum_i \frac{1}{\mathcal{N}_i} \sum_j H_{ij}(l_{ij}), \tag{16}$$

where l represents the length inflation scale that defines l_{ij}. As the cut-window K gets smaller, the points in the projected space $P(\lambda)$ get farther way from each other and in the perpendicular space the points $P_{\perp}(\lambda)$ get closer. l_{ij} and V_i get bigger but the intersection of vertex window polytopes in the perpendicular space also gets bigger, approaching 1. As a result, over many inflations, we see that there are two dominant regimes. One is "disordered", where there is not much overlap of vertex window polytopes and there are more local variations so that the information entropy rules depend on hits. The other regime is "ordered", where the overlaps approach 1 and therefore the information entropy rules should depend only on the vertex valences and not on hits. See Figure 7.

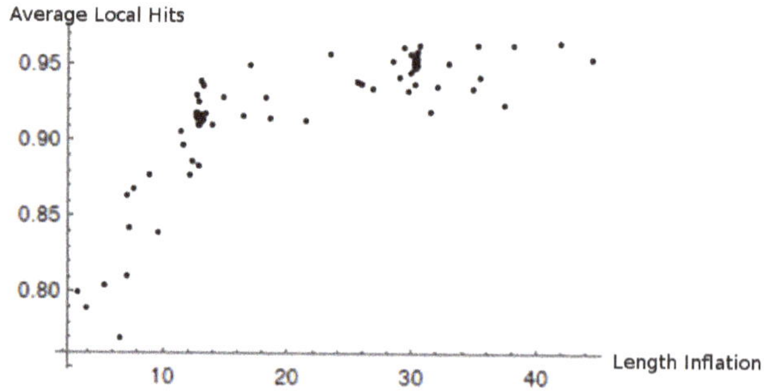

Figure 7. Average hits $H_{\mathcal{T}}(l)$ evolution under inflations.

Different emergent dynamics can be considered. For example, we can set a preferred direction to follow the rules and also consider the central hit potential, which leads to interesting curving path patterns as in Figure 8.

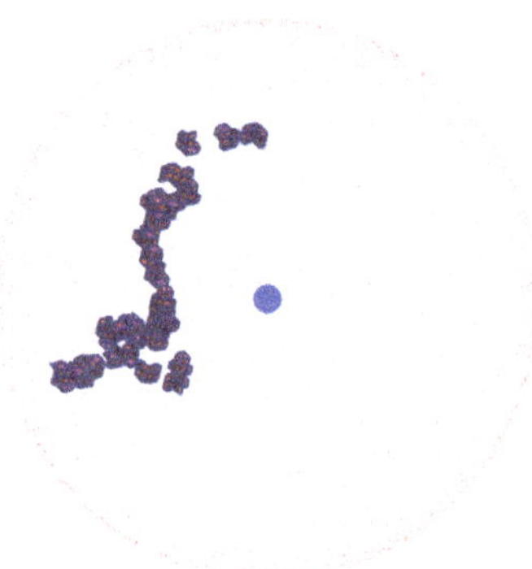

Figure 8. Evolution considering local information entropy conservation, initial preferred direction and the central hit potential.

As a further result, we consider an oscillation pattern around the center of a tiling with the large symmetric VT there. There are 12 of these same VTs around the center making the geometry of an icosahedron. In each frame step only one of the 12 VTs is ON. This gives a notion of an emergent quasiparticle (the whole icosahedron) with internal structure (the 12 symmetric VTs). See Figure 9.

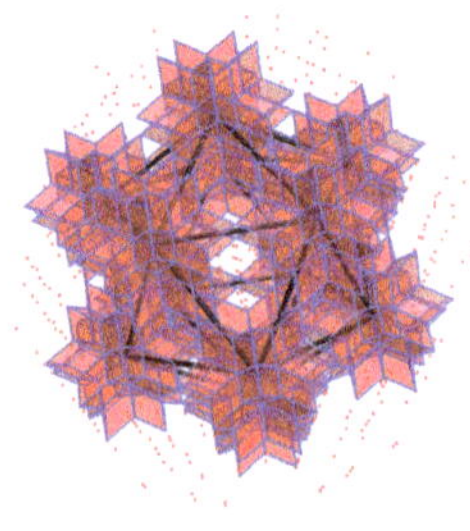

Figure 9. Central icosahedron oscillating pattern.

With this object we can consider a 4-dimensional dynamic by considering the influence of the empires of those 12 VTs at arbitrary 3D positions on $\triangle$. The oscillation on the 12 VTs plays the role of time. We consider a vertex $P(\lambda_1)$ around the icosahedron. The VT associated to that position depends on the $\gamma_\perp$ tiling $\mathcal{T}_k$, $VT_1^k = VT(P(\lambda_1))$. By shifting the projections by specific $\gamma_\perp$ values, we can change tiling k and VT_1^k, preserving the oscillating icosahedron pattern at the center. Figure 10 shows the cut-window K (the large window), the empire vertex window for the icosahedron (the middle window), and the small window limiting the available shifts.

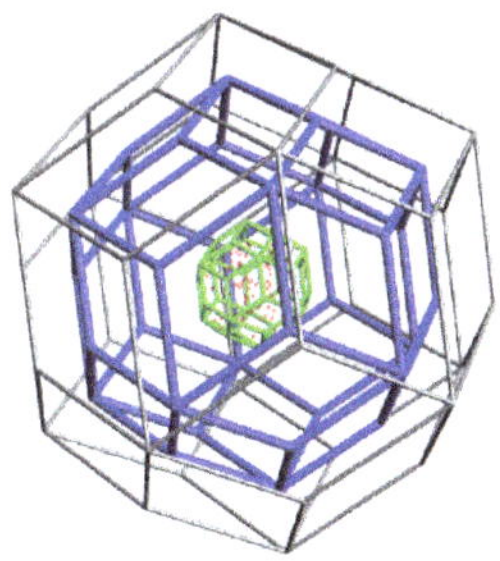

Figure 10. Available points for shifts $\gamma_\perp$ presented inside the small window.

So for each $P(\lambda_1)$ we compute H_{ij}, where i refers to one of the 12 central VTs and j to VT_1^k. We average this computation over a large number of tilings $\mathcal{T}_k$ and over 12 cycles of the central icosahedron, getting $\bar{H}_{ij}$, and then we go to a new position. We define the observable of interest to be

$$\mathcal{O}(l_{ij}) = \log(l_{ij}\bar{H}_{ij}), \tag{17}$$

which gives a different notion of emergent potential as shown in Figure 11. This one grows with distance while the hit potential drops.

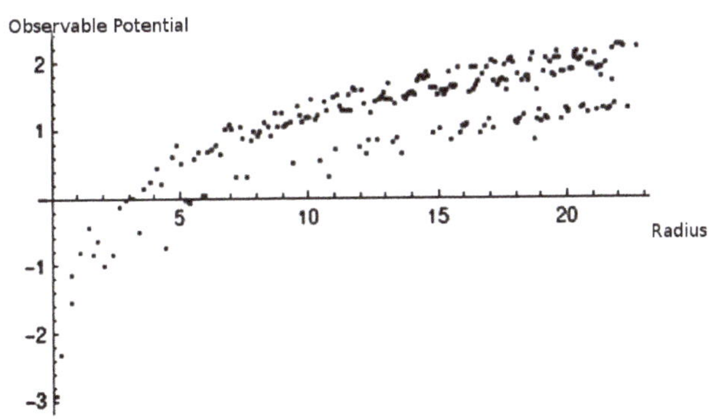

Figure 11. Potential due to composite quasiparticle oscillating at the center of the tiling.

As a last result, we consider how information entropy grows with the number of steps for patterns made of different VTs. We define the hit section (hs) to be a certain number of the same VTs within some distance r from the center of some tiling. We compute H_{ij}, where i refers to a certain VT at the center of this tiling and j refers to the same VT at different position p on that tiling, $p < r$, with $j \in hs$,

$$I_i(r) = -\sum_{j}^{hs(r)} P_{ij} log P_{ij}. \tag{18}$$

The result is that $I_i(r)$ distinguishes the patterns made of different VTs, which should be proportional to the frequency of appearance of the respective VTs, see Figure 12.

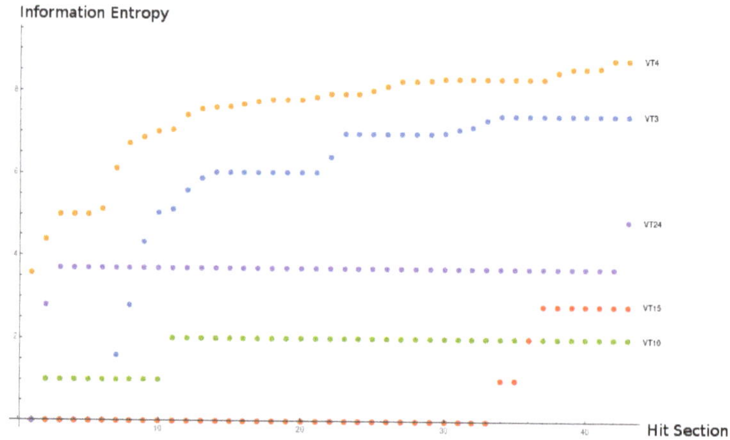

Figure 12. Information entropy order parameter for different VTn, where the integer n number the 3DPT VTs and we show only 5 of the 24. hs grows different with distance for the different VTs.

5. Discussions and Outlook

In this paper, we discussed state sum models under geometric realism. The SSH paradigm is looking to understand the emergence of spacetime and matter from a pre-spacetime code that takes into account stratification and recursion. GSS is discussed as a framework in this direction. State sum models implement the principles of locality and superposition. GSS adds geometric realism at the state level and the PEL for dynamics. With the GSS formulation, we made concrete the idea that emergence happens due to structure and not necessarily randomness. In the current implementation, the structure is given by a 3D quasicrystal with aperiodic order projected from a 6-dimensional lattice. By having dynamics where different points can be projected or actualized at different steps with their empires, quasiparticle patterns emerge in a simple GoL simulation as presented in Section 4.1. The patterns that emerge are built from known quasicrystal structures, the VTs. A stratified recursive feedback loop can then be established between, on the one hand, the quasicrystal level of projections, non-local empire overlaps and local information entropy rules and, on the other hand, the emergent level of quasiparticle patterns made of VTs. This is expressed as a potential derived from many evolutions of PRWs going over the underlying quasicrystal points, generating an emergent pattern. Overlap of PRW positions or PRWs and empire vertices is understood as potential optimization; it gives an opportunity to economize resources, which in this case are projections from 6 dimensions to 3 dimensions. This implements the PEL by allowing the simulation to express more emergent patterns over less projection step resources. The PEL is implemented on the structure of PRWs, which are considered here as virtual walks to probe the possibility space of walks. It aims to create "cognitive" structure, as in artificial neural networks, so that the simulation itself can decide the best sequence of movements for the emergent pattern. This kind of structure under the emergence paradigm within complex systems is what is observed in general: The emergent structure, with its new emergent properties, always occurs under the dynamics of a large underlying number of building blocks that have structure governing their dynamic interactions. This is seen from atoms to DNA to neurons to stars and galaxies.

In this work, we investigated some general code theoretic properties of general emergent patterns on the GSS. For future investigations, we aim to derive the physical emergent laws governing the dynamics of the emergent patterns, making more concrete the emergence of spacetime and matter. An important hint in this direction is that the cut-and-project method used to derive the quasicrystal makes clear the connection with the higher dimensional lattices. The main quasicrystals of interest are derived from the so-called root lattices of Lie algebras. Lie algebras and groups are the language of symmetry and so a core element in fundamental physics, as in quantum mechanics. The dynamics derived in this paper can be thought of as a dynamics that tails back to the root system of Lie algebras [17,27], suggesting a generalization from Z_6 to D_6 and E_8.

One different consideration to be addressed for GSS models is the computational efficiency on current classical computers. The values on usual state sum models come from algebraic or group-theoretic objects. There are more computational costs to obtain the values from the geometry.

Author Contributions: Conceptualization, M.A., F.F. and K.I.; methodology, M.A.; software, M.A. and D.H.; validation, F.F.; formal analysis, M.A.; investigation, M.A., F.F. and D.H.; writing—original draft preparation, M.A.; writing—review and editing, M.A.; visualization, F.F.; supervision, M.A. and K.I.; project administration, K.I.; funding acquisition, K.I. All authors have read and agreed to the published version of the manuscript.

Funding: This research received no external funding.

Acknowledgments: We acknowledge the many discussions had with David Chester, Raymond Aschheim and Richard Clawson and we thank them for their generous feedback in editing discussions.

Conflicts of Interest: The authors declare no conflict of interest.

Abbreviations

The following abbreviations are used in this manuscript:

SSH	Self-simulation hypothesis
PEL	Principle of Efficient Language
3DPT	3-Dimensional Penrose Tiling quasicrystal
PEL	Geometrical State Sum (GSS)
GR	General Relativity
LQG	Loop Quantum Gravity
3D	3-dimensional
LGT	Lattice Gauge Theory
VT	Vertex Type
PS	Possibility Space
PRW	Possibility Random Walk
GoL	Game of Life

References

1. Irwin, K.; Amaral, M.; Chester, D. The Self-Simulation Hypothesis Interpretation of Quantum Mechanics. *Entropy* **2020**, *22*, 247. [CrossRef]
2. Dirac, P.A.M. Classical theory of radiating electrons. *Proc. Roy. Soc. Lond. A* **1938**, *167*, 148–169.
3. Finkelstein, D. Space-time code. *Phys. Rev.* **1969**, *184*, 1261–1279. [CrossRef]
4. Wheeler, J.A. Beyond the Black Hole. In *Some Strangeness in the Proportion: A Centennial Symposium to Celebrate the Achievements of Albert Einstein*; Woolf, H., Ed.; Addison-Wesley: Reading, MA, USA, 1980.
5. Wheeler, J.A. Hermann Weyl and the Unity of Knowledge. *Am. Sci.* **1986**, *74*, 366–375.
6. Wheeler, J.A. Information, physics, quantum: The search for links. In *Complexity, Entropy, and the Physics of Information*; Addison-Wesley: Boston, MA, USA, 1990.
7. Langan, C.M. The Cognitive-Theoretic Model of the Universe: A New Kind of Reality Theory. *Prog. Complex. Inf. Des.* **2002**, *1*, 2–3.
8. Aschheim, R. Hacking Reality Code. FQXI Essay Contest 2011, Category: Is Reality Digital or Analog? Essay Contest (2010–2011), Number 929. Available online: https://fqxi.org/community/forum/category/31417 (accessed on 8 October 2021).
9. Irwin, K. A New Approach to the Hard Problem of Consciousness: A Quasicrystalline Language of Primitive Units of Consciousness in Quantized Spacetime. *J. Conscious Explor. Res.* **2014**, *5*, 483–497. [CrossRef]
10. Irwin, K. The Code-Theoretic Axiom: The Third Ontology. *Rep. Adv. Phys. Sci.* **2019**, *3*, 1950002. [CrossRef]
11. Irwin, K.; Amaral, M.M.; Aschleim, R.; Fang, F. Quantum walk on spin network and the golden ratio as the fundamental constant of nature. In Proceedings of the Fourth International Conference on the Nature and Ontology of Spacetime, Varna, Bulgaria, 30 May–2 June 2016; pp. 117–160.
12. Hammock, D.; Fang, F.; Irwin, K. Quasicrystal Tilings in Three Dimensions and Their Empires. *Crystals* **2018**, *8*, 370. [CrossRef]

13. Katz, A. Theory of Matching Rules for the 3-Dimensional Penrose Tilings. *Commun. Math. Phys.* **1988**, *118*, 263–288. [CrossRef]
14. Takakura, H.; Gomez, C.; Yamamoto, A.; De Boissieu, M.; Tsai, A.P. Atomic structure of the binary icosahedral Yb-Cd quasicrystal. *Nat. Mater* **2007**, *6*, 58–63. [CrossRef]
15. Penrose, R. The role of aesthetics in pure and applied mathematical research. *Bull. Inst. Math. Appl.* **1974**, *10*, 266–271.
16. Barrett, J.W. State sum models for quantum gravity. *arXiv* **2000**, arXiv:gr-qc/0010050.
17. Amaral, M.; Aschheim, R.; Irwin, K. Quantum Gravity at the Fifth Root of Unity. *arXiv* **2019**, arXiv:1903.10851.
18. Wanas, M.I. The accelerating expansion of the universe and torsion energy. *Int. J.Mod. Phys. A* **2007**, *31*, 5709–5716. [CrossRef]
19. Irwin, K. Toward the Unification of Physics and Number Theory. *Rep. Adv. Phys. Sci.* **2019**, *3*, 1950003. [CrossRef]
20. Baake, M.; Grimm, U. *Aperiodic Order*; Cambridge University Press: Cambridge, UK, 2013.
21. Senechal, M.J. *Quasicrystals and Geometry*; Cambrige University Press: Cambrige, UK, 1995.
22. Levine, D.; Steinhardt, P.J.; Quasicrystals, I. Definition and structure. *Phys. Rev. B* **1986**, *34*, 596. [CrossRef]
23. Gardner, M. Extraordinary nonperiodic tiling that enriches the theory of tiles. *Sci. Am.* **1977**, *236*, 110. [CrossRef]
24. de Bruijn, N.G. Algebraic theory of Penrose's non-periodic tilings. *Nederl Akad Wetensch Proc.* **1981**, *84*, 1–7. [CrossRef]
25. Grimm, U. Aperiodicity and Disorder-Do They Play a Role? In *Computational Statistical Physics: From Billiards to Monte Carlo*; Hoffmann, K.H., Schreiber, M., Eds.; Springer: Berlin/Heidelberg, Germany, 2002; pp. 191–210.
26. Elser, V.; Sloane, N.J.A. A highly symmetric four-dimensional quasicrystal. *J. Phys. A* **1987**, *20*, 6161–6168. [CrossRef]
27. Chen, L.; Moody, R.V.; Patera, J. Non-crystallographic root systems. In *Quasicrystals and Discrete Geometry*; Fields Institute Monographs; American Mathematical Society: Providence, RI, USA, 1998; Volume 10.
28. Fang, F.; Irwin, K. An Icosahedral Quasicrystal and E8 derived quasicrystals. *arXiv* **2016**, arXiv:1511.07786.
29. Conway, J.H. Triangle tessellations of the plane. *Amer. Math. Mon.* **1965**, *72*, 915.
30. Grunbaum, B.; Shephard, G.C. *Tilings and Patterns*; W. H. Freeman and Company: New York, NY, USA, 1987.
31. Effinger-Dean, L. The Empire Problem in Penrose Tilings. Bachelor's Thesis, Williams College, Williamstown, MA, USA, 2006.
32. Fang, F.; Hammock, D.; Irwin, K. Methods for Calculating Empires in Quasicrystals. *Crystals* **2017**, *7*, 304. [CrossRef]
33. Rovelli, C.; Vidotto, F. *Covariant Loop Quantum Gravity*, 1st ed.; Cambridge University Press: Cambridge, UK, 2014.
34. Ortiz, L.; Amaral, M.; Irwin, K. Aspects of aperiodicity and randomness in theoretical physics. *arXiv* **2020**, arXiv:2003.07282.
35. Bahr, B.; Dittrich, B.; Ryan, J.P. Spin foam models with finite groups. *J. Grav.* **2013**, *2013*, 549824. [CrossRef]
36. Fang, F.; Paduroiu, S.; Hammock, D.; Irwin, K. Empires: The Nonlocal Properties of Quasicrystals. In *Electron Crystallography, Devinder Singh and Simona Condurache-Bota*; IntechOpen: London, UK, 2019.
37. Alexander, S.; Cunningham, W.J.; Lanier, J.; Smolin, L.; Stanojevic, S.; Toomey, M.W.; Wecker, D. The Autodidactic Universe. *arXiv* **2021**, arXiv:2104.03902.
38. Gardner, M. Mathematical games—The fantastic combinations of John Conway's new solitaire game of 'life'. *Sci. Am.* **1970**, *223*, 120–123. [CrossRef]
39. Klein, J. Breve: A 3d environment for the simulation of decentralized systems and artificial life. In Proceedings of the 8th International Conference on Artificial Life, Dortmund, Germany, 14–17 September 2003.
40. Bak, P.; Chen, K.; Creutz, M. Self-organized criticality in the Game of Life. *Nature* **1989**, *342*, 780. [CrossRef]
41. Bays, C. Candidates for the game of life in three dimensions. *Complex Syst.* **1987**, *1*, 373–400.
42. Couclelis, H. Cellular worlds: A framework for modeling micro—macro dynamics. *Environ. Plan. A* **1985**, *17*, 585–596. [CrossRef]
43. Bailey, D.A.; Lindsey, K.A. Game of Life on Penrose Tilings. *arXiv* **2017**, arXiv:1708.09301.
44. Owens, N.; Stepney, S. Investigations of Game of Life Cellular Automata Rules on Penrose Tilings: Lifetime, Ash, and Oscillator Statistics. *J. Cell. Autom.* **2010**, *5*, 207–225.
45. Fang, F.; Paduroiu, S.; Hammock, D.; Irwin, K. Non-Local Game of Life in 2D Quasicrystals. *Crystals* **2018**, *8*, 416. [CrossRef]
46. Bekenstein, J.D. Black holes and entropy. *Phys. Rev. D* **1973**, *7*, 2333–2346. [CrossRef]
47. Verlinde, E. On the origin of gravity and the laws of Newton. *J. High Energ. Phys.* **2011**, *2011*, 29. [CrossRef]
48. García-Islas, J.M. Entropic Motion in Loop Quantum Gravity. *Can. J. Phys.* **2016**, *94*, 569–573. [CrossRef]
49. Amaral, M.M.; Aschheim, R.; Irwin, K. Quantum walk on a spin network. *arXiv* **2016**, arXiv:1602.07653.
50. Goertzel, B. Toward a Formal Model of Cognitive Synergy. *arXiv* **2018**, arXiv:1703.04361.

foundations

MDPI

Review

Foundations of Electromagnetism: A Review of Wilhelm Weber's Electrodynamic Force Law

Christof Baumgärtel and Simon Maher *

Department of Electrical Engineering and Electronics, University of Liverpool, Liverpool L69 3GJ, UK
* Correspondence: s.maher@liverpool.ac.uk

Abstract: This article reviews the electrodynamic force law of Wilhelm Weber and its importance in electromagnetic theory. An introduction is given to Weber's force and it is shown how it has been utilised in the literature to explain electromagnetism as well as phenomena in other disciplines of physics, where the force law has connections to the nuclear force, gravity, cosmology, inertia and quantum mechanics. Further, criticism of Weber's force is reviewed and common misconceptions addressed and rectified. It is found that, while the theory is not without criticism and has much room for improvement, within the limitations of its validity, it is equally as successful as Maxwell's theory in predicting certain phenomena. Moreover, it is discussed how Weber offers a valid alternative explanation of electromagnetic phenomena which can enrich and complement the field perspective of electromagnetism through a particle based approach.

Keywords: Weber's electrodynamics; Weber force; field theory; electromagnetism; electrodynamics; physics of elementary particles and fields; magnetic field; electric field; electrical engineering; fundamental physics

Citation: Baumgärtel, C.; Maher, S. Foundations of Electromagnetism: A Review of Wilhelm Weber's Electrodynamic Force Law. *Foundations* **2022**, *2*, 949–980. https://doi.org/10.3390/foundations2040065

Academic Editors: Eugene Oks and Martin Bohner

Received: 6 July 2022
Accepted: 12 October 2022
Published: 19 October 2022

1. Introduction

Wilhelm E. Weber formulated a generalised electrodynamic force law that was first published in 1846, only a few years prior to Maxwell's first works on electromagnetism. However, Weber's direct-action-at-a-distance theory is little known today and often dismissed *a priori* and without further thought in the scientific community, on the basis that it is old and superseded or taken as disproven. Whilst there are historic reasons that led to the dismissal of Weber's theory and the success of Maxwell-Lorentz field theory, there seems to be some common misconceptions surrounding the theories and the existing criticism.

This manuscript aims to give a balanced and comprehensive review about Weber's theory and show that it's dismissal is premature. Criticism will be reviewed and refuted where appropriate, and for the sake of context it will be shown that field theory is not without criticism either. After analysing the achievements of Weber's electrodynamics through its use in the literature and how it has been applied not only to electromagnetism, but several branches of physics, it will eventually be shown that both electromagnetic theories have great commonalities and they can be regarded as complementary rather than competing. Weber is shown to provide a viable alternative description of electrodynamics, and whilst it is not without limitations and has bounds of validity, as will be discussed, it is argued that neither theory is perfect and that their similarities far outweigh their differences.

Some have commented that the current state of physics is in a crisis, and that "new physics" is required to resolve current puzzles in particle physics and supersymmetry [1–4]. However, it may be possible that a re-examination of the foundations (i.e., electrodynamics in this case) can lead to new perspectives and insight which may guide and inform new solutions. It is argued that further pursuit and research of Weber electrodynamics can offer epistemological, physical and practical value.

2. Fundamentals of Weber's Theory

Weber's electrodynamic force is introduced to provide a mathematical overview and to familiarise the reader with Weber's direct-action approach as an explanation for electricity and magnetism. Weber's force describes the interaction of two point charges and was postulated before the electron was even discovered. Hence, it was originally based on Fechner's hypothesis that a current consists of equal amounts of positive and negative charges moving in opposite directions, which was the conventional wisdom at the time, as scientists imagined so called "electrical fluidae" moving through wires and circuits when subjected to electromotive forces. Fechner's hypothesis will be addressed separately (see Section 3.3) and it shall be noted that Weber's theory can still be used when we assume that only electrons are charge carriers in motion responsible for conduction currents in circuits. With this restriction lifted we can now explain the general workings of Weber's theory.

First, let us consider two charged particles, q_1 and q_2, in a Cartesian coordinate system (x, y, z), at their respective positions $\vec{r}_1$ and $\vec{r}_2$ (see Figure 1).

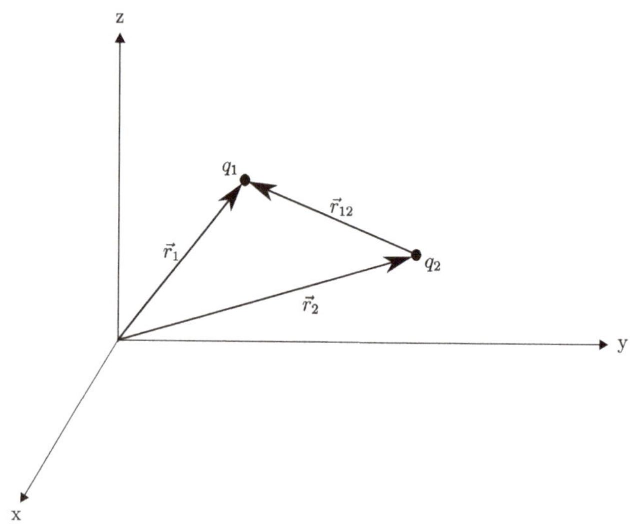

Figure 1. Two charged particles q_1 and q_2 in a Cartesian coordinate system (x, y, z) at positions $\vec{r}_1$ and $\vec{r}_2$. Their relative position $\vec{r}_{12}$ is a vector pointing from q_2 to q_1.

The position of these particles, which are time dependent spatial coordinates, can be expressed by

$$\vec{r}_1 = x_1(t)\hat{x} + y_1(t)\hat{y} + z_1(t)\hat{z}, \qquad \vec{r}_2 = x_2(t)\hat{x} + y_2(t)\hat{y} + z_2(t)\hat{z}, \tag{1}$$

where we have the unit vectors:

$$\hat{x} = \begin{pmatrix} 1 \\ 0 \\ 0 \end{pmatrix}, \qquad \hat{y} = \begin{pmatrix} 0 \\ 1 \\ 0 \end{pmatrix}, \qquad \hat{z} = \begin{pmatrix} 0 \\ 0 \\ 1 \end{pmatrix}. \tag{2}$$

Their relative position is the difference between $\vec{r}_1$ and $\vec{r}_2$:

$$\vec{r}_{12} = \vec{r}_1 - \vec{r}_2 \tag{3}$$

and a distance r_{12} given by the magnitude of $\vec{r}_{12}$

$$r_{12} = |\vec{r}_1 - \vec{r}_2| = \sqrt{(x_1(t) - x_2(t))^2 + (y_1(t) - y_2(t))^2 + (z_1(t) - z_2(t))^2}. \tag{4}$$

With this, the unit vector along $\vec{r}_{12}$ can be defined as

$$\hat{r}_{12} = \frac{\vec{r}_{12}}{r_{12}} \tag{5}$$

and both $\vec{r}_{12}$ as well as $\hat{r}_{12}$ are pointing from q_2 to q_1. The relative velocity $\vec{v}_{12}$ and relative acceleration $\vec{a}_{12}$ between the two charges can be calculated as first and second time derivatives of the relative position, Equations (6) and (7), respectively:

$$\frac{d\vec{r}_{12}}{dt} = \vec{v}_{12}, \tag{6}$$

$$\frac{d^2\vec{r}_{12}}{dt^2} = \frac{d\vec{v}_{12}}{dt} = \vec{a}_{12}. \tag{7}$$

To arrive at the time derivative $\dot{r}_{12}$ of r_{12} the chain rule is utilised:

$$\begin{aligned}
\dot{r}_{12} =& \frac{dr_{12}}{dt} = \frac{d}{dt}[(x_1(t) - x_2(t))^2 + (y_1(t) - y_2(t))^2 + (z_1(t) - z_2(t))^2]^{\frac{1}{2}} \\
=& [2x_1(t)\dot{x}_1 - 2(\dot{x}_1 x_2(t) + \dot{x}_2 x_1(t)) + 2x_2(t)\dot{x}_2 + 2y_1(t)\dot{y}_1 - 2(\dot{y}_1 y_2(t) + \dot{y}_2 y_1(t)) + 2y_2(t)\dot{y}_2 \\
& + 2z_1(t)\dot{z}_1 - 2(\dot{z}_1 z_2(t) + \dot{z}_2 z_1(t)) + 2z_2(t)\dot{z}_2] \\
& \cdot \frac{1}{2\sqrt{(x_1(t) - x_2(t))^2 + (y_1(t) - y_2(t))^2 + (z_1(t) - z_2(t))^2}} \\
=& \hat{r}_{12} \cdot \vec{v}_{12}.
\end{aligned} \tag{8}$$

By analogy, the same procedure can be applied to arrive at the second time derivative, or the quotient and product rule along with the substitutions $u = \vec{r}_{12}$, $v = \vec{v}_{12}$, $w = r_{12}$ can be used:

$$\left(\frac{u \cdot v}{w}\right)' = \frac{(u'v + v'u)w - w'uv}{w^2}, \tag{9}$$

$$\ddot{r}_{12} = \frac{d^2 r_{12}}{dt^2} = \frac{d\dot{r}_{12}}{dt} = \frac{[\vec{v}_{12} \cdot \vec{v}_{12} - (\hat{r}_{12} \cdot \vec{v}_{12})^2 + \vec{r}_{12} \cdot \vec{a}_{12}]}{r_{12}}. \tag{10}$$

With the help of these definitions, we can now examine Weber's potential between the charges in question which will eventually lead us to Weber's force. It was two years after Weber introduced his force law that he succeeded in showing that it could be derived from a potential, and this takes the form:

$$U = \frac{q_1 q_2}{4\pi\varepsilon_0} \frac{1}{r_{12}} \left(1 - \frac{\dot{r}_{12}^2}{2c^2}\right). \tag{11}$$

To arrive at the force, the principle of virtual work is invoked which states

$$\begin{aligned}
\vec{F}_{21} &= -\hat{r}_{12} \frac{dU}{dr_{12}} \\
\text{or} \quad \vec{v}_{12} \cdot \vec{F}_{21} &= -\frac{dU}{dt}.
\end{aligned} \tag{12}$$

(Note that the principle of virtual work by definition depends on a time dependent trajectory.) Now applying Equation (12) to (11) gives Weber's force law in the following way:

$$
\begin{aligned}
\vec{F}_{21} &= -\hat{r}_{12}\frac{dU}{dr_{12}} \\
&= \frac{q_1 q_2}{4\pi\varepsilon_0}\frac{\hat{r}_{12}}{r_{12}^2}\left(1-\frac{\dot{r}_{12}^2}{2c^2}+\frac{2r_{12}\ddot{r}_{12}}{2c^2}\right) \\
&= \frac{q_1 q_2}{4\pi\varepsilon_0}\frac{\hat{r}_{12}}{r_{12}^2}\left(1-\frac{\dot{r}_{12}^2}{2c^2}+\frac{r_{12}\ddot{r}_{12}}{c^2}\right) \\
&= \frac{q_1 q_2}{4\pi\varepsilon_0}\frac{\hat{r}_{12}}{r_{12}^2}\left[1+\frac{1}{c^2}\left(\vec{v}_{12}\cdot\vec{v}_{12}-\frac{3}{2}(\hat{r}_{12}\cdot\vec{v}_{12})^2+\vec{r}_{12}\cdot\vec{a}_{12}\right)\right],
\end{aligned}
\tag{13}
$$

where again the chain rule is necessary to derive the expression $\frac{d\dot{r}^2}{dr}$ correctly,

$$
\frac{d\dot{r}^2}{dr} = \frac{d(\dot{r}(t))^2}{dt}\frac{dt}{dr} = 2\dot{r}\frac{d\dot{r}}{dt}\frac{dt}{dr} = 2\ddot{r}. \tag{14}
$$

We can see in (13) different notations of Weber's force, for example, we can simply write the time derivatives indicated as dots, noting

$$
\vec{F}_{21} = \frac{q_1 q_2}{4\pi\varepsilon_0}\frac{\hat{r}_{12}}{r_{12}^2}\left(1-\frac{\dot{r}_{12}^2}{2c^2}+\frac{r_{12}\ddot{r}_{12}}{c^2}\right). \tag{15}
$$

Similarly, the following form can more readily be used to substitute quantities, which is especially handy when investigating the interaction between charge carriers for a given experiment or apparatus, for example:

$$
\vec{F}_{21} = \frac{q_1 q_2}{4\pi\varepsilon_0}\frac{\vec{r}_{12}}{r_{12}^3}\left(1-\frac{3}{2c^2}\left[\frac{\vec{r}_{12}\cdot\vec{v}_{12}}{r_{12}}\right]^2+\frac{1}{c^2}(\vec{v}_{12}\cdot\vec{v}_{12}+\vec{r}_{12}\cdot\vec{a}_{12})\right). \tag{16}
$$

We can see from these derivations that the force depends on the relative position, velocity and acceleration of the particles involved. The force is along the line joining them and follows Newton's third law in the strong form, that is every action has an equal and opposite reaction. Furthermore, conservation of energy as well as conservation of linear and angular momentum are followed by this law. Additionally the principle of superposition applies to this force law, similar to the superposition principle with electric and magnetic fields of Maxwellian field theory.

Other formulations of Weber's force formula exist in the literature, as will be discussed in Section 3.2.1, it has been shown that Weber can be formulated to incorporate electromagnetic fields [5–9] and especially in [8] the field-based Weber force is formulated with focus on the relation between source and test charges and how they define current elements and densities. Further, Hamiltonian and Lagrangian formulations of the force law have been obtained [6,10] and expressions like this have occasionally been used in the literature [11–13].

3. Literature Review

3.1. *Two Different Theories of Electrodynamics*

Historically, many scientists have worked on electrodynamics and electromagnetic phenomena, performing a wide range of experiments to investigate the nature of electricity and magnetism. From these experiments researchers have developed a multitude of hypotheses, laws and eventually attempted to merge them into cohesive theories, leading to several attempts to explain electrodynamics, some of them more successful than others. Many alternative theories have been proposed over the years with important contributions from several scientists. Some examples include Gauss, Neumann, Lorentz, Riemann, Weber, Helmholtz, Hertz, Ritz, Moon & Spencer, and Wheeler-Feynman direct-action theory, amongst others (see Appendix B of [6], also [14–21]).

In this section, some brief comments will be first given on the established theory of fields and ether by Maxwell, which forms the foundation of modern science and technological inventions, from particle accelerators to modern medical instrumentation. Following this, we will introduce Weber's force law in the context of its development before engaging with the wider literature relating to its development and application.

3.1.1. Maxwell's Equations and Field Theory

When James Clerk Maxwell presented his *magnum opus* on electromagnetic theory [22] in 1873, he formulated his ideas about the action of electric and magnetic fields partly in prose and partly as mathematical descriptions and equations he introduced. These can be summarised in the concise form of only four equations as widely disseminated in modern times [23]. The differential forms of Maxwell's equations in a vacuum and in SI-units are commonly given as:

$$\nabla \cdot \vec{E} = \frac{\rho}{\varepsilon_0}, \tag{17}$$

$$\nabla \cdot \vec{B} = 0, \tag{18}$$

$$\nabla \times \vec{E} = -\frac{\partial \vec{B}}{\partial t}, \tag{19}$$

$$\nabla \times \vec{B} = \mu_0 \left(\vec{J} + \varepsilon_0 \frac{\partial \vec{E}}{\partial t} \right). \tag{20}$$

Here, ρ is the local charge density, ε_0 and μ_0 are the vacuum permittivity and permeability and $\vec{J}$ is the current density. $\vec{E}$ and $\vec{B}$ are the electric and magnetic field, respectively, through which charged particles interact, meaning contact-action where a particle always interacts with the field as a medium and the fields themselves can interact, as for example in the transmission of electromagnetic waves.

Equation (17) is Gauss's law, which relates the electric field with the charge density, (18) is the law of non-existence of magnetic monopoles, (19) is the Maxwell-Faraday equation that expresses induction, and (20) is the Maxwell-Ampère equation that correlates currents and time-varying electric fields to magnetic fields, which is also a form of induction. Due to the time-dependent nature of (19) and (20) electromagnetic waves can be predicted, while Maxwell's approach is originally based on the ether through which electromotive forces and waves would propagate, from today's point of view the fields themselves have effectively replaced the ether as the dominant medium and are now considered to be responsible for interaction transmission. The ether as an original construct is largely and effectively ignored.

Further to the field equations, the Biot-Savart law is formulated to obtain the magnetic field for a current element integrated along a closed circuit path

$$\vec{B}_{BS} = \frac{\mu_0}{4\pi} \int \frac{I d\vec{l} \times \hat{r}}{r^2}, \tag{21}$$

where $Id\vec{l}$ is a current element and r is the distance from $d\vec{l}$ to the point where the field is evaluated and $\hat{r}$ is the corresponding unit vector. For forces between two current elements, Grassmann's force is normally utilised based on the Biot-Savart law,

$$\begin{aligned} d^2 F_{Grassmann} &= I_1 d\vec{l}_1 \times d\vec{B}_{BS} = I_1 d\vec{l}_1 \times \left(\frac{\mu_0}{4\pi} \frac{I_2 d\vec{l}_2 \times \hat{r}}{r^2} \right) \\ &= -\frac{\mu_0}{4\pi} \frac{I_1 I_2}{r^2} \left[(d\vec{l}_1 \cdot d\vec{l}_2)\hat{r} - (d\vec{l}_1 \cdot \hat{r}) d\vec{l}_2 \right]. \end{aligned} \tag{22}$$

For any charged particle q in more general situations moving in the presence of electric and magnetic fields the interaction is usually given by the Lorentz force:

$$\vec{F}_L = q\Big(\vec{E} + \vec{v} \times \vec{B}\Big). \tag{23}$$

In general, field theory and Maxwell's equations are a 'macroscopic' approach as they were developed from a continuous medium model (the ether). However, as we will see in the following section, Weber's force is 'microscopic' in that sense as it describes the interaction between two charged particles in its standard form. For a better comparison between Maxwell's and Weber's theory, Assis shows the derived force between two point charges from field theory [24] up to second order in v/c based on the work of Liénard, Wiechert and Schwarzschild, which was first obtained by O'Rahilly [25] as

$$\begin{aligned}\vec{F}_{21} =& q_1\vec{E}_2(\vec{r}_1) + q_1\vec{v}_1 \times \vec{B}_2(\vec{r}_1) = q_1\left\{\frac{q_2}{4\pi\varepsilon_0}\frac{1}{r^2}\left[\hat{r}\left(1 + \frac{\vec{v}_2\cdot\vec{v}_2}{2c^2}\right.\right.\right.\\ &\left.\left.\left. - \frac{3}{2}\frac{(\hat{r}\cdot\vec{v}_2)^2}{c^2} - \frac{\vec{r}\cdot\vec{a}_2}{2c^2}\right) - \frac{r\vec{a}_2}{2c^2}\right]\right\} + q_1\vec{v}_1 \times \left\{\frac{q_2}{4\pi\varepsilon_0}\frac{1}{r^2}\frac{\vec{v}_2\times\hat{r}}{c^2}\right\}.\end{aligned} \tag{24}$$

In this formula q_1 is the test charge and q_2 is the source charge generating the fields $\vec{E}_2$ and $\vec{B}_2$, where according to Assis time retardation, radiation and relativistic effects have been included. The constant c is the speed of light and $\vec{a}_2$ denotes the acceleration of the point charge. It is apparent that the expression depends on the square of the source charge velocity and on its acceleration, whereas Weber's force depends on the relative velocity and acceleration, as will be seen in Section 3.1.2. Assis also shows how this expression can be obtained from the Darwin Lagrangian ([6], Section 6.8), as the Darwin Lagrangian is more widely used in the literature to describe systems of point charges [26–28]. Both the Schwarzschild force (24) and Lorentz force (23) have been criticised as violating conservation of linear and angular momentum. To restore conservation it is usually argued that the energy is lost or gained by the electromagnetic field generated by the charges or that self force needs to be taken into account [29,30]. However, a system of two point charges seems extremely difficult to test. The general applicability (or non-applicability) of Newton's third law to the Lorentz force and generally in electrodynamics has been discussed in [31–33]. Cornille [32] also claims that if the electrodynamic force laws indeed violate Newton's third law, then it inevitably leads to the conclusion that energy can be extracted from the ether, as the ether exerts a force that is responsible for the violation.

It has further been criticised that the velocity $\vec{v}$ in the Lorentz force formula (23) is not clearly defined, that is what it is defined with respect to, was not even given by Lorentz himself [6,34]. It thus remains ambiguous if the definition is w.r.t a coordinate system or a source charge, which might itself be moving, although there seems to be support to the idea that Lorentz viewed the velocity as relative to the ether [6,34]. However, in relativistic treatments this is usually resolved by a chosen inertial frame of reference and regarding the velocity relative to the measuring device or observer.

In an interesting review about Maxwell's equations and the field approach Tran [35] concludes that there are only few experiments supporting the Maxwell-Ampère and Maxwell-Faraday equation, at least not to the same degree of accuracy that the continuity equation and the magnetic law are supported. There is further discussion about conceptual problems in classic electromagnetism and modern particle-field theories in the literature [36,37]. This mainly focuses on the problem of point charges and their diverging self-energy, as the calculated energy of an electron with its own field tends to infinity based on classic electromagnetism. One solution is the renormalisation approach in the quantum theory of Dirac where the point charge is treated as a singularity and the infinite energy is subtracted as a constant from the problem to renormalise the energy content. The other solution is the extended particle model, where elementary charges are not treated as point-like anymore and consequently the divergence in the singularity disappears. Pietsch [36] discusses both approaches and the associated cost of the proposed solutions, and for an interesting discussion of these approaches including a mathematical perspective, see [38], on which Pietsch bases their arguments. Pietsch then argues that both approaches are incompatible at a fundamental level and a better solution is needed, in which direct-action theories are proposed. Lazarovici [37] also discusses the self-energy problem and the Lorentz-Dirac as well as the extended particle solutions as unsatisfactory, but also involves free fields, among other philosophical, mathematical and physical arguments, and proposes the Wheeler-Feynman direct-action theory in particular as a solution to those problems. The renormalisation approach has also been criticised by other authors [39], including Feynman [40] and Dirac [41]. A similar argument has been made by Kastner about the Wheeler-Feynman direct-action theory, not only does it avoid self-energy problems, it is also not subject to Haag's theorem and the consequent problems of free and interacting fields in quantum field theory (QFT) [42].

This concludes the brief overview of field theory and Maxwell's equations and further reading where fields, waves, radiation and relativity are treated can be found in the works of other authors [26,43–46].

3.1.2. Weber's Theory of Electrodynamics

Wilhelm Eduard Weber first published his force law to describe the interaction of charged particles in 1846, some 15 years before Maxwell published his first work on electromagnetism, 'On physical lines of force' [47], a concept which would only tangentially relate to the field concept later introduced in Maxwell's *Treatise* [23]. Maxwell, as a contemporary of Weber, was well aware of his work and Maxwell positively mentions Weber in his *Treatise*, expressing admiration for Weber's work. As a 19th century scientist, Weber engaged with several physical disciplines, but a collection of his original work on electrodynamics can be found in [48,49] and English translations of his eight major memoirs on electromagnetism can be found in [50–57].

The main difference between Weber's theory and Maxwell's field equations is that Weber's is a direct-action-at-a-distance theory, such as Newton's law of gravity or Coulomb's force of electrostatic interaction, and the fields themselves are not conceptualised as a primary part of the mathematical description. Instead, Weber's force depends on the direct interaction and force transmission between charges themselves, as opposed to contact action in field theory, where the charges give rise to fields so a source charge and a test charge only interact with the field of the other. Some aspects of how fields can still be conceived with Weber will be discussed later in Section 3.2.

In modern vector notation and SI units, Weber's force can be expressed as

$$\vec{F}_{21} = \frac{q_1 q_2}{4\pi\varepsilon_0} \frac{\hat{r}_{12}}{r_{12}^2} \left(1 - \frac{\dot{r}_{12}^2}{2c^2} + \frac{r_{12}\ddot{r}_{12}}{c^2} \right). \qquad \text{((15) revisited)}$$

When Weber developed his force, the aim was to connect Coulomb's force and Ampère's force, arriving at a more general interaction law. Weber's force acts along the line joining two interacting point charges, following Newton's third law in the strong form with equal action and reaction, conserving linear and angular momentum. It depends only on the relative distance, relative velocity and relative acceleration of the interacting charges.

As this force is electrodynamic in nature, it contains electrostatic (i.e., Coulomb's force) and magnetic (i.e., Ampère's force) interactions, which is comparable with the Lorentz force (23) where a static (i.e., the electric field term) and a moving component (i.e., the magnetic field term) of the force are considered. The speed of light c in (15) was introduced as the ratio between electrostatic and electromagnetic units of charge, whose value was first determined experimentally in 1856 (10 years later) by Weber and Kohlrausch based on Weber's force. In 1848, two years after the presentation of the force law, Weber also showed that the force can be derived from a velocity dependent potential, as shown in Section 2.

A further analysis of the capabilities of Weber's theory will follow in Section 3.2 and the theory has undergone more development in recent decades. It is noteworthy that predominantly in the low velocity limit, Weber and Maxwell theory predict very similar results, if not the same results for a given phenomenon. Weber has also been shown to be consistent with field equations by a number of authors and the matter will be further addressed in Section 3.2.1.

However, it must be stressed, Weber's electrodynamic theory has not yet been developed to anywhere near the same degree that other theories have, which includes the high velocity regime near the speed of light where the theory has problems. When quantum interactions are considered, the Wheeler-Feynmann approach to direct-action has undergone development by Davies who introduced a quantum theory based on Wheeler-Feynman electrodynamics [58]. In the case of Weber, only some initial connections between Weber's theory and quantum mechanics have been made, as will be seen in Section 3.2.2, however, a rigorous treatment is yet to be developed. This is considered a work in progress and further research is needed before any conclusions can be drawn about Weber's theory in the quantum realm.

3.2. *Weber Electrodynamics in the Literature*

After Maxwell's success in the late 19th and early 20th century, Weber's electrodynamic force law has not received a lot of attention except from a few, with important contributions from O'Rahilly [25], Wesley [9,59,60], Assis [6] and others. Over the years, many connections have been made from Weber's theory of electrodynamics to different topics within physics, while Weber's force is electromagnetic in nature and has been used to describe phenomena in that field, it is also shown to interconnect with mechanics, the structure of the atom, gravity, quantum mechanics and even some effects of general relativity theory (GRT) and topics that are usually referred to as "breakthrough physics". A visual overview of the relations Weber has with electromagnetics and other disciplines can be seen in Figure 2.

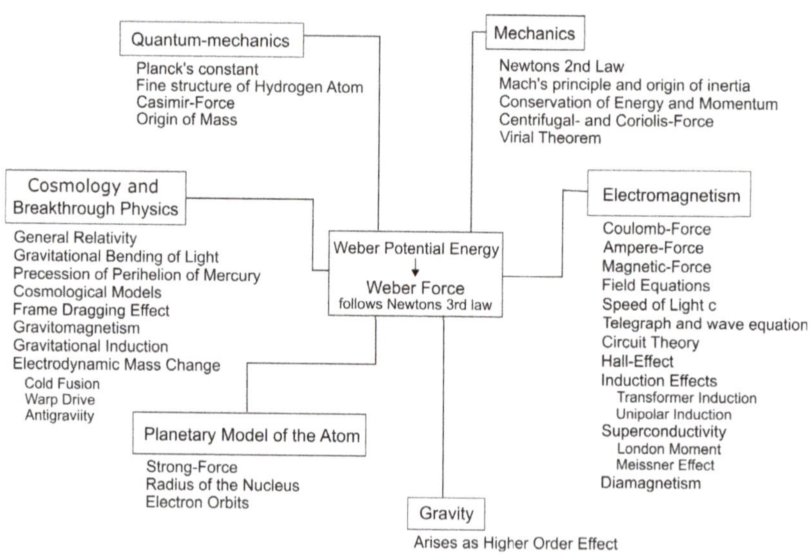

Figure 2. A visual map providing an overview of the different subsidiary fields and phenomena that Weber's theory has been shown to provide a basis for and connects with throughout the branches of physics and engineering.

3.2.1. Electromagnetic Phenomena

As Weber is essentially a theory of electrodynamics, it has been shown to explain many pure and applied phenomena in electricity and magnetism. From its basic form (15), it is easy to see that Weber's force reduces to the Coulomb force for stationary charges. That means for static charges where their velocities are zero, the formula can be simplified to

$$F_{Coulomb} = \frac{q_1 q_2}{4\pi\varepsilon_0} \frac{\hat{r}_{12}}{r_{12}^2}. \tag{25}$$

Thus, Weber can be readily seen to describe purely electrostatic interactions. When interacting charges start moving, the force then changes, and for two current elements in a circuit Ampère's force can be derived from Weber's force as shown by Assis [61]. For two steady current elements $I_1 d\vec{l}_1$ and $I_2 d\vec{l}_2$ this takes the form:

$$d^2 F_{Ampere} = -\frac{\mu_0}{4\pi} I_1 I_2 \frac{\hat{r}}{r^2}\left[2(d\vec{l}_1 \cdot d\vec{l}_2) - 3(\hat{r}d\vec{l}_1 \cdot \hat{r}d\vec{l}_2)\right] \tag{26}$$

Grassmann's force (22) from Section 3.1.1, for comparison, is slightly different in its interaction. Assis' extensive analysis indicates the similarities and differences between Ampère's and Grassmann's forces [6] and most notably shows, that Grassmann's force violates Newton's third law. However, it has been claimed that, when the respective force expressions are applied to any closed circuit, they are equivalent and lead to the same result [62]. There has been a discussion in the literature about which force is the correct one. For example, a paper by Cavalleri [63] claims that Grassmann's force gives the correct result for any given circuit and Ampère's does not; but as Assis commented in response [64], they did not consider all contributions of Ampère's force in their deductions and when carried out, both models predict the same force values [62,65]. This inevitably leads to the conclusion that it is impossible to distinguish between the two forces for any closed circuit, which has been verified for several configurations [66,67].

In this context, it seems adequate to briefly discuss the Ampère force and note its importance, as even Maxwell himself stated that it must remain the cornerstone of electrodynamics. The divide in the literature between Ampère's vs. Grassmann's force seems to stem from the nature of the Ampère force, which includes a longitudinal force component along the wire in the direction of movement of the current elements. This feature complies with Newton's third law but appears to be incompatible with the Lorentz force, whereas Grassmann's force for current elements does not include a longitudinal component and is in turn compatible with the Lorentz force, but violates Newton's third law. When Ampère conducted his original experiments, investigating the force between two wires [68–71], he found his force law as a result of these experiments and made sure to include the longitudinal component according to his observations. Further to the discussions about the general applicability of Newton's third law in electrodynamics [31,32], Chaib and Lima [72] have re-iterated that Ampère did not find any evidence in his experiments that would contradict Newton's third law and that it remains applicable in electrodynamics. They also clarify that Ampère regarded the third law as a consequence of his experiments, rather than an assumption he tried to conform to, and explain his philosophical reasoning in arriving at that conclusion. The authors of [72] also give a review of some of Ampère's main work in the manuscript and show that Ampère was the first to obtain an expression similar to the Biot-Savart law from his experiments.

This special quality of the Ampère force to incorporate longitudinal forces has sparked the interest of researchers and in modern times more experiments have been performed to investigate it, where the force has also been connected to a variety of effects and applications. In particular Graneau et al. have investigated Ampère's force [73–82], including water jet propulsion, exploding wires, fusion and railguns, as well as the electromagnetic impulse pendulum, liquid mercury experiments (e.g., Ampère's bridge) and homopolar motors [83–92]. Even though the exploding wire phenomenon has been investigated further [93,94] and the longitudinal Ampère force component does not seem responsible for the bursting of wires, it is still important in other situations. For example, the importance of Ampère's force to induction in general has been discussed along with the ability of the force to explain EM-waves in the near field [95]. A classical approach to derive longitudinal forces has been taken by Rambaut and Vigier [96], where they find longitudinal forces as an average effect of conduction electrons and lattice charges of a current element. In a follow-up paper [97] the authors then derive longitudinal forces with a different approach based on the Liénard-Wiechert potentials and discuss their influence on conductors in all phases, including plasmas and fusion applications. Recently Moyssides also succeeded in deriving longitudinal forces with the help of the Biot-Savart-Grassmann-Lorentz force law acting on a submarine projectile in mercury, showing that both forces are equivalent and both agree with experimental results presented in the paper [98]. Further experiments to detect longitudinal forces have been performed by Saumont [99], where difficulties in measurements of these forces due to thermal effects and rapid spurious forces are also addressed and the compatibility of longitudinal tensile forces with relativity theory and the Lorentz force is considered. An experiment by Graneau et al. [100] has investigated longitudinal forces utilising spark gaps in a circuit and found them consistent with Ampère's force, which has been interpreted as a direct proof of the existence of longitudinal forces. A new experiment has recently investigated how the charges in a neon glow lamp are influenced in the near field of a capacitor dependent on the signal frequency [101]. It was found that there is a longitudinal force component on the charges moving in the plasma that agrees with predictions of a Weber-Ampère model calculated in the paper. We can deduce from this brief analysis that longitudinal forces appear in both classic and Weber approaches which shows the similarity of the two theories.

Moving on from current elements, Weber's force in the general form (15) is a force between point charges and depends on the relative velocity between them, it is intrinsically electromagnetic in nature and so incorporates magnetic interaction by design. The magnetic force naturally arises from the movement of the charges, whereas the Lorentz force (23) is usually derived by considering special relativity theory (SRT) or Lorentz transformation of the Coulomb Force or electric field [46,102–104] and magnetism is considered to arise as a relativistic effect in this context. However, recently it has also been suggested that it may not be necessary to treat magnetism as a consequence of SRT and instead Maxwell's equations can be derived from Coulomb's force and time retardation without any further assumptions [105]. In Weber electrodynamics however, the intrinsic velocity dependence of the force can be used in combination with the concept of current elements to calculate magnetic interaction forces, such as has been applied to the fields of solenoids [106–109]. Specifically in the case of the magnetic field of a long straight wire [107] the Lorentz and Weber force on a charged particle have been found to be identical in the low velocity limit.

Further to the similarities between Ampère's and Grassmann's force as well as the Weber and the Lorentz force, Weber has been shown to be consistent with Maxwell's field equations by a number of authors [5,6,8,9,110,111], even though it does not conceptually depend on them. For example, Wesley [9] derives field equations from Weber by introducing charge densities and current densities into Weber's equation and integrates over a fixed volume. He arrives at the form

$$\begin{aligned}\frac{d^3F}{d^3r} &= -\rho\nabla\Phi + \vec{J}\times(\nabla\times\vec{A})\frac{1}{c} - \rho\frac{\partial\vec{A}}{\partial tc} - \vec{J}\nabla\cdot\vec{A}\frac{1}{c}\\ &+ \frac{\partial\vec{J}}{\partial t}\frac{\Phi}{c^2} + (\vec{J}\cdot\nabla)\nabla\Gamma\frac{1}{c} + \rho\nabla\frac{\partial\Gamma}{\partial tc} - (\frac{\partial\vec{J}}{\partial t}\cdot\nabla)\vec{G}\frac{1}{c^2},\end{aligned} \tag{27}$$

with

$$\Phi = \int d^3r'\rho'(\vec{r}',t)\frac{1}{R}, \tag{28}$$

$$\vec{A} = \int d^3r'\vec{J}'(\vec{r}',t)\frac{1}{cR}, \tag{29}$$

$$\Gamma = \int d^3r'\vec{R}\cdot\vec{J}'(\vec{r}',t)\frac{1}{cR}, \tag{30}$$

$$\vec{G} = \int d^3r'\vec{R}\rho'(\vec{r}',t)\frac{1}{R}. \tag{31}$$

The primed quantities in this formula are sources acting on the unprimed test currents and charges separated by distance R. It is especially interesting that in addition to the usual electric potential Φ and magnetic potential $\vec{A}$ two new potentials, Γ and $\vec{G}$ appear and Wesley points out that the Lorentz force and Maxwell's equations are a special case where only the first three terms on the left hand side of (27) appear. Wesley further argues that the representations through a force equation and field expression are mathematically isomorphic as long as the fields are intermediate without time retardation. However, when time retardation is introduced into (27), the field expressions then contain wave equations with velocity c.

While Assis [6] and Kinzer [5] have taken a similar approach to Wesley, starting from Weber's formula and deriving the field equations from it, the opposite approach, starting from Maxwell's equation and arriving at a Weber-type formulation, does also exist [8]. With extensive mathematical work the author of [8] shows that there are two implicit restrictions in Maxwell's field equations and without these restrictions a set of Weber equations can be obtained. One limitation is the condition that the charge density function ρ is a constant in time and the other is that the test charge velocity is required to be zero for mathematical consistency according to the author of [8]. The procedure of removing the restrictions from Maxwell's field theory is then to allow for moving test charges and time varying charge densities, which emphasises Wesley's argument that Maxwell's equations are a special case of Weber's law. Another important opposite approach has also been discussed where Weber's and Ampère's forces are obtained as a non-relativistic limit from the Liénard-Wichert potentials [97] and from a Fermi distribution of accelerated charges [112]. In the recent approach of Li [110] a field representation of Weber's force is developed with the help of Einstein notation, where the velocity and acceleration dependent terms in Weber's force can then each be identified with a respective Tensor field. Li states that this approach has the advantage of simplifying the necessary calculations in systems of many particles, reducing the number of required force calculations.

One of the strengths that is usually ascribed to Maxwell's field equations is that the velocity of light appears from the wave equations as does the relation between permittivity ε and μ. However, it was not Maxwell who first discovered the relation between the speed of light c and electromagnetic waves, in fact it was Weber and Kohlrausch who first predicted the value of c from Weber's equation and confirmed it experimentally [113,114]. Following this, Weber and Kirchhoff derived the telegraph equation for the propagation of electromagnetic signals through a wire independently of each other, and Assis has provided a modern derivation and analysis of the telegraph equation in this context [115–117]. This equation also reduces to a regular wave equation when the resistance of the wire goes to zero. Fukai [118] has further argued that modern views of the vacuum can be assumed, where the vacuum behaves as a medium with inductance and capacitance, similar to a coaxial cable or transmission line problem, Weber's theory predicts a signal propagation at light velocity in vacuum and thus should be able to predict radiation as well.

In relation to circuit theory, Assis [119] has analysed the self inductance of circuits with the help of Weber's force and Newton's second law and compared them to similar derivations based on the Lorentz force and the Liénard-Schwarzschild force for circuits. He manages to derive the self inductance from Weber's force by examining the acceleration that conductor charges will be subjected to. The result can be interpreted as an effective mass of the conduction electrons in the circuit, with the positive lattice ions opposing the motion of the accelerated electrons. In a further study [120] Assis and Bueno also show that the self inductance formulae for a single circuit of Weber and Maxwell are equivalent. This has added to the previously known fact that the mutual inductance formulae were the same, however when a single circuit is considered, the equivalence can be obtained by considering surface charge or volume charge elements instead of line charge elements. Weber himself also worked on circuit resistances [49], conducting many experiments to investigate their resistive behaviour and considered the derivation of an absolute unit of resistance based on his force law.

A possible derivation of the Hall effect from Weber's force has been investigated by McDonald [121] and it is deduced that, under the condition that only negative charges are assumed to be charge carriers in Weber's theory, and disregarding the original Fechner hypothesis, Weber does predict the Hall effect consistent with the Lorentz force derivation. Based on this assumption the Hall effect cannot distinguish between the two forces, as they are equivalent in this particular case. The Fechner hypothesis itself will be further discussed in Section 3.3.

In addition to self inductance and the generation of a Hall voltage, Weber's theory has also been applied to voltages arising through induction. Smith et al. [122,123] have developed a model for transformer induction with the assumption of conduction electrons following accelerated motion. They arrive at an expression identical to Faraday's law and predict voltages in receiver coils correctly. Maxwell even pointed out in his *Treatise* [22] that it is possible to derive Faraday's law from Weber's law, as Weber derived it himself from his force, and Wesley has also indicated the connection between Faraday's law and Weber's, besides consideration of induction in general [9,59].

Unipolar induction (also called homopolar induction), as another example, has been analysed by Assis [124] on the basis of Weber electrodynamics. He arrives at the conclusion that the phenomenon can only be predicted correctly if the closing wire is included in the analysis, thus considering the whole circuit that is influenced by the presence of a magnet in a Faraday generator setup. Recently, unipolar induction experiments have been found to be consistent with Weber electrodynamics [125,126] There are also recent claims in the literature [127] that not all observed induction effects can be explained by Faraday's law or the flux cutting rule, mentioning Weber as a possible fuller explanation. Assis has even presented an analysis [128] where he explores beyond the regular scope of unipolar induction and describes a situation where an additional voltage is induced due to the presence of an electrostatic potential, calling it Weberian induction. In this scenario a spinning disk is placed inside a charged spherical shell, with or without the magnet. From the perspective of field theory and the Lorentz force, the charge on the shell cannot induce a voltage in the disk due to the absence of a field inside the shell, but in Weber's electrodynamics the charge on the shell would still exert a force on the disk charges and an additional voltage should be induced. However, such an effect would be many orders of magnitude smaller than regular unipolar induction. This field-free electrostatic force is closely related to a criticism of Weber's theory and will be further discussed in Sections 3.2.3 and 3.3.

Related to unipolar induction is the so-called Feynman disk paradox [129], where metal spheres are mounted on a plastic disk free to rotate and a current carrying coil is situated on the same axis. Initially, the apparatus is at rest, but when the current is switched off, the paradox arises. Analysing the problem in terms of the change in magnetic flux one would expect the disk to rotate as an electric field is induced and seen by the spheres, however, analysing from the perspective of angular momentum would mean that the disk does not start to rotate, as the momentum is initially zero and since it must be conserved, the disk should remain stationary. The apparent contradiction in field theory is resolved by taking into account the electromagnetic angular momentum stored in the fields, so that initially the angular momentum was not zero [130]. Fukai argues that the problem is equally well explained by Weber's theory [118] and considers the force on the sphere exerted by the change of current in the solenoid, which turns out to be non-zero, so the disk will move accordingly.

It was also shown recently that superconductivity can be derived from Weber's theory according to two independent authors in two different ways. One approach by Prytz [131] considers a magnet or solenoid with a DC current at rest and the centripetal acceleration of the conductor charges. This acceleration causes the Meissner effect to appear in conduction materials according to their deductions. Assis and Tajmar [132] instead follow a more general treatment where AC currents are considered, and in turn the acceleration of the charges causes the Meissner effect and the London moment to appear from Weber's force.

The so called Aharonov-Bohm effect (AB-effect) that describes a phase shift experienced by electrons when they are scattered around a finite solenoid is usually explained through the quantum mechanical influence of the vector potential $\vec{A}$. Wesley [133] managed to establish a connection between Weber's force and the existence of the AB-effect. First, he takes a more classical approach and considers the force due to motional induction on the electron beam, which leads them to find a phase shift depending on the electron path due to this force component. Then, he shows that the same force is present when considering Weber's potential and force, causing the appearance of the phase shift and thus the AB-effect. Unfortunately in the digital version there seems to be a misprint and large parts of the paper are missing, but his essential deductions can still be found in the first and last sections of the manuscript [133].

An initial connection between Weber's force and antenna theory has been made by Prytz [134]. In this approach the concept of retarded time is considered to account for far distance effects and utilised with the acceleration dependent induction of Weber's force. This can describe loop antennae by itself and also contributes to dipole antennae, and simultaneously relates back to the transmission of radio waves. This is, of course, similar to the approach taken by Moon and Spencer [19] who had derived the behaviour of the loop and dipole antenna about 60 years prior, and also arrived at Neumann's inductance formula for circuits based on the acceleration dependent force in the process.

Weber also considered early forms of diamagnetism in his work, based on the molecular currents introduced by Ampère, where diamagnetism occurs when molecular currents are induced by an external field in a diamagnetic material that has otherwise no molecular currents. While the general idea of molecular currents remains to this day, present explanations necessitate quantum mechanical considerations to fully account for the existence of diamagnetism, due to the Bohr-Van Leeuwen theorem.

3.2.2. Relevance of Weber's Force Beyond Electromagnetism

There are several aspects of Weber's force that are not immediately obvious when just regarding it as an electrodynamic force law between charged particles. It holds wider consequences in its action and has been connected to a variety of topics in the literature. The first connection of note is the structure of the atom. Weber himself used his theory to devise a planetary model of the atom [135,136], where negative charges orbit around a positively charged centre, even before the electron was discovered. He realised a special feature of his force law by which an atomic structure emerges as a consequence, that is, a certain critical distance exists in the force law. It is most remarkable that below this critical distance like charges will start to attract instead of repel and opposite charges will still attract regardless. Furthermore, when two or more charges interact with only each other (no external forces are present in that scenario) they cannot transition to the inside or outside of that critical distance. If the interaction started below the critical radius, it would remain inside the radius, and if it began outside it would remain outside. Interestingly, the critical distance can be calculated to be of the order of the known diameter of the atomic nucleus when electron or positron interaction is considered. Thus, it seems possible to interpret Weber's force as relating to the strong nuclear force that is responsible for the attraction inside the nucleus, even though it is not yet known how it would work for protons. Additionally, for two charges of opposite sign orbiting each other in this model a confinement to elliptical orbits is found which would experience a precession of the elliptical axis [135–138] and remain within an upper and lower bound of the orbit radius [135,137,139–141].

Weber also speculated based on his model of the atom on the mechanism of heat conduction and a possible connection between light and electricity. He related the orbital frequency of charged particles in his model to the wavelength of excited heat- and light-waves. Zöllner [142] even considered that from Weber's model it would be possible to derive the spectral lines of chemical elements. Upon considering the possibility of multiple orbits in a molecule and that each orbit could be populated with a number of charges, Weber also deduced a classification system characterising the possible combinations of positive and negative charges in a table that has similarities to the periodic table of elements. He further concluded that these molecular configurations could attract other molecules and combine to form stable systems. Eventually he was led to the hypothesis that chemical bonds might have an electrical origin and arrived at a bond that is similar to a covalent bond between two atoms sharing an electron pair [135].

Unfortunately, Weber passed away before he could finish his planetary model of the atom, nevertheless it shows some very interesting properties, while Weber speculated about spectral lines emerging from their model, the fine structure of the atom was still in an early stage of discovery. There are relatively recent works in the literature [11,12] analysing the connections between Weber's theory and quantised energy levels, they generally tend to show slightly different splittings for the fine structure of the atom. In 2015, Torres-Silva et al. [11] used the Hamiltonian formalism of Weber's theory and also considered spin–orbit corrections, whereas Frauenfelder and J. Weber [12] used a Bohr-Sommerfeld quantisation on Weber's Hamiltonian for a hydrogen atom. Wesley obtained a similar result in [143] where he considered the perturbation energy on the electron orbit in the hydrogen atom by utilising the Schrödinger perturbation method, which leads to energy levels different from the experimentally known splittings. It is also reviewed by Post [144] that Phipps' modification to Weber's potential leads to twice the number of observed splittings. As the experimental determinations of quantum electrodynamics (QED), such as the fine structure, lamb shift, Rydberg constant and anomalous dipole moments are among the most well tested predictions derived from field theory, this will remain the greatest challenge for Weber-type theories as it seems that it cannot make the necessary predictions in its current form. However, it seems worth noting that Feynman has directly utilised ideas of direct-action and retarded time in his approach to develop QED [40].

The Weber force has also been shown to be similar to velocity dependent forces in nuclear physics, such as nucleon-nucleon forces [145]. These are central forces and obey Newton's third law, and in [145] the authors show that the momentum dependent inter-nucleon forces in the interaction potential and Hamiltonian are generalised Weber forces, pointing out that Weber's force can be considered a special case of the forces appearing in nuclear physics. They state further that a similarity to Ampères force exists in nuclear physics as well for scattering processes of spinless projectiles and that it cannot be understood on the basis of only the Lorentz force analogy, but is consistent with a Weber force analogy.

Another interesting similarity with the forces of nature exists between Weber's force and gravity, and Weber had already speculated about such a connection based on his atomic model. One of the underlying ideas is that the attractive force between charged particle assemblies is slightly larger than the repulsive force. Based on this idea, Assis [146,147] has developed a model to derive gravitational effects as a fourth order effect from the electromagnetic interaction through an extension of Weber's formula. He considers a series expansion of Weber's force law:

$$\vec{F}_{21} = \frac{q_1 q_2}{4\pi\varepsilon_0} \frac{\hat{r}_{12}}{r_{12}^2} \left(1 - \frac{\alpha}{c^2}(\dot{r}_{12}^2 - 2r_{12}\ddot{r}_{12}) - \frac{\beta}{c^4}(\dot{r}_{12}^4 - 4\dot{r}_{12}^2 r_{12}\ddot{r}_{12}) - \frac{\gamma}{c^6}(\dot{r}_{12}^6 - 6\dot{r}_{12}^4 r_{12}\ddot{r}_{12})\right), \tag{32}$$

with numerical constants α, β and γ and omitting any higher order terms of the series expansion. Equation (32) is then used to study the interaction of two neutral dipoles that consist of a positive charge in the centre and an oscillating negative charge. After an extensive analysis, Assis arrives at a non-zero force of attraction between the two dipoles which he proceeds to interpret as a gravitational interaction force. These derivations later led Tajmar [148] to analyse a minimum energy requirement between two neutral dipoles that is similar in nature to Planck's constant. The approach used by Assis is then further investigated by Baumgärtel and Tajmar [149] yielding a slightly different result to that originally obtained by Assis. The authors of [149] interpret the resulting force to contain the Casimir force and inertial effects as well. Further analysis of this result in regards to a minimum energy requirement is pursued and the study finds a connection to Planck's constant and the origin of mass establishing a general relation between Weber's theory and Planck's constant solely on electrodynamic properties of particles. One can generally see with the appearance of the Casimir force, Planck's constant and the fine structure of the atom that Weber's theory ties to certain quantum mechanical effects, however it still requires significant development on quantum theoretical aspects in the future.

There are other considerations about Weber's force and gravitational interaction. A common approach in the literature is to utilise a gravitational force of the Weber type, such as has been investigated by Tisserand [150], Gerber [151], Schrödinger [152] and others. Utilising such a gravitational form of Weber's law, Assis and Wesley [60,153] have shown that it is also possible to arrive at the origin of inertia in combination with Mach's principle. Their derivations generally follow the form

$$F = -Gm_1m_2\frac{\hat{r}_{12}}{r_{12}^2}\left[1+\frac{1}{c^2}\left(\vec{v}_{12}\cdot\vec{v}_{12}-\frac{3}{2}(\hat{r}_{12}\cdot\vec{v}_{12})^2+\vec{r}_{12}\cdot\vec{a}_{12}\right)\right], \tag{33}$$

where it is easy to see that the usual charges q_1 and q_2 have been replaced by masses m_1, m_2 and gravitational constant G. As a consequence of this approach however, Newtons second law, $F = ma$, follows naturally from this formulation, so that the interaction with the distant masses is responsible for the existence of inertia. In essence, Assis calculates the force on a body by investigating its interaction with what is more or less a celestial sphere and the cosmic microwave background, while Wesley uses a gravitational field in his derivations. Additionally, in Assis' model the resulting force contains terms that represent centrifugal and Coriolis forces.

Because there are many possible variations of Weber-type gravitational forces of similar nature, e.g., [60,150–152,154], their form has been further generalised by Bunchaft and Carneiro [155] as

$$F = k\frac{\hat{r}_{12}}{r_{12}^2}\left(1-\mu\dot{r}_{12}^2+\gamma r_{12}\ddot{r}_{12}\right). \tag{34}$$

Here, k is a parameter depending on the nature of the interacting bodies (like charges or masses) and the parameters μ and γ are positive constants of the velocity and acceleration terms usually of the order $1/c^2$. Further aspects of these approaches, with regards to cosmology and conservation of energy, will be discussed in the next section.

Weber's force law in its basic form (15), is a 'microscopic' force law which obeys certain general principles of mechanics. It intrinsically follows Newton's third law of action and reaction. Any action arising from Weber's force has an opposing force of equal magnitude and inverse sign, making it consistent with linear momentum conservation. It can also be shown that Weber's theory follows energy and angular momentum conservation [6,8], and Weber himself only succeeded in 1871 to show that conservation of energy is obeyed.

It is even possible to derive a virial theorem from Weber's force as has been shown by Mendes and Assis [156]. The virial theorem states that the time average of kinetic energy in a system of discrete particles is related to the potential energy of the particles. For Weber's formula this takes the form

$$2\langle T\rangle = -\langle U_w\rangle, \tag{35}$$

with the potential energy U_w being Weber's potential (see Section 2). As a consequence, this offers the possibility for statistical treatments of problems with Weber, such as the formations of galaxies on an astronomical scale or plasma physics, although limitations have been indicated regarding the applicability to cold plasmas [157].

Weber's force has also been used with a modified version of mechanics [158]. The idea is based on the relational nature of Weber's force, making mechanics relational as well. In essence, Newton's mechanical principles are extended with Mach's principle and a Weber-type force for interacting masses is implemented. This has consequences for cosmology, affecting the interpretation of Hubble's law and the cosmological redshift as light interacting on its journey. The cosmic microwave radiation background would then indicate an equilibrium state and the precession of mercury's perihelion is due to the interaction with distant fixed stars. Although the idea is intricate and intriguing, this would of course be less compatible with the standard cosmological interpretation of an expanding universe.

3.2.3. Cosmology and Breakthrough Physics

From the previous section it was shown that Weber has ties to other forces of nature, like nuclear forces and gravity, and it was suggested that Weber's force has significant links to cosmological theories and related phenomena. Based on the possibility of gravitational modifications to Weber's force, there have been many investigations about planetary motion and in particular the precession of mercury's perihelion [13,150–152,159–164]. In addition, some of these also investigate the bending of light in a gravitational field and the two effects are usually explained from GRT. From these investigations it can be seen that Weber generally offers a possible explanation for both observations, but the applicability has been shown to have limits.

The general form of Weber-type forces of the order $1/c^2$ (34), is further investigated by Bunchaft and Carneiro [155] with specific attention to energy conservation. The authors of [155] show that such a force is only conservative under a special condition when the parameters μ and γ of (34) obey the form $2\mu = \gamma$. They deduce from this analysis that a Weber-type gravitational force cannot predict the correct values for the precession of mercury's perihelion and the gravitational bending of light at the same time under the condition of being conservative. However, more generalised formulations of higher order might still be able to predict both effects whilst remaining conservative, and it is still evident from this that Weber's force offers an alternative approach to phenomena that are usually attributed to GRT. A somewhat more general discussion about how GRT and Weber-type theories are related on a fundamental level can be found by Giné [165], although that approach has been challenged [166] in not having a proper Einsteinian approach. A further analysis by Tiandho [167] came to the conclusion that Weber-type gravitational forces are a weak field approximation of GRT and thus a special case, nonetheless, it shows some similarity and connection between the two theories.

Recently, Weber-like gravitational interactions in combination with Mach's principle have been investigated with regards to their implications in cosmology [168]. The advantage of such an approach is that inertia arises naturally from the interaction with a celestial sphere and avoids the incompatibility of inertia in GRT with Mach's principle; and additionally maintains the equivalence of gravitational and inertial mass and the equivalence of kinematic and dynamic rotation rates of the Earth. On the other hand, the resulting universe is non-expanding and the redshift arises from energy loss of light on its journey. Another cosmology model based on Mach's principle has recently been developed and investigated by Das [169], where the universe is governed by Machian gravity. According to Das it is able to explain rotational curves and mass distribution of galaxies without dark matter or dark energy and when comparing the model to observed cosmological data it fits well with the measurements. Previous to these approaches, the general idea of a Machian cosmology had been suggested in the literature [158,170–173], where P. and N. Graneau [173] have even found a connection to the expanding universe, but Das seems to be the first to apply the concept to available data of galactic rotation curves.

Beyond the general connection of Weber's theory and GRT, some works have analysed more specific effects arising from gravitational perturbations, for example spacecraft flyby anomalies [154,174]. These approaches also utilise gravitational formulations of Weber's force, similar to the previously discussed modifications. The first of such gravitational-influenced effects is called frame dragging, also known as the Lense-Thirring-effect, which has its usual explanation in GRT. This effect adds a precession to a gyroscope or orbiting satellite in the presence of a large rotating mass. It is possible to arrive at a similar effect by utilising Weber's theory [174] for gravitational interactions. The authors of [174] present the formula for the action of a large spinning shell with mass M, radius R and angular velocity $\vec{\Omega}$ on a test body, mass m, at position $\vec{r}$, velocity $\vec{v}$ and acceleration $\vec{a}$ to be

$$\vec{F} = -\frac{2GmM}{Rc^2}\left[\vec{a} + \vec{\Omega} \times \left(\vec{\Omega} \times \vec{r}\right) + 2\vec{v} \times \vec{\Omega} + \vec{r} \times \frac{d\vec{\Omega}}{dt}\right], \tag{36}$$

where G and c are the gravitational constant and speed of light, respectively. As can be seen it is similar to the previously presented gravitational-type Weber forces, with terms resulting from the rotation and acceleration, for example the second and third term in square brackets of (36) represent centrifugal and Coriolis' force, which appear due to the implementation of Mach's principle. After analysing the problem with the help of Newton's second law, they arrive at a gravitationally induced azimuthal acceleration that will be experienced by the test body, similar to a frame dragging effect.

The gravitomagnetism arising from this principle has also led Tajmar and Assis to investigate flyby anomalies with a similar implementation of gravity-like Weber forces [154]. They found that flyby anomalies due to Weber-based interactions are several magnitudes smaller than currently measurable. They relate this to data of Rosetta flyby manoeuvres where expected anomalies have not been measured, even though the standard approach would have predicted them. However, if the effect is, according to the Weber-interaction, below the threshold of current detectability, this result would not be surprising.

Apart from the general and specific connections to GRT, an interesting possibility seems to exist in Weber's theory that pushes the boundaries of known physics. A situation can be created according to Assis, where the mass of charged particles can be manipulated under the influence of a field-free electrostatic force [6,175]. If it is indeed possible to change the inertial mass of a charged particle through electrodynamic means, it can possibly be applied to breakthrough propulsion physics technologies, such as the warp drive, anti-gravity or even cold fusion [176–178] which could revolutionise space travel, energy production and transportation in general.

The approach taken by Assis shows the influence of a charged spherical shell on a point charge according to Weber's force, whereby the force on the point charge can be interpreted as an inertial mass change of the particle. Assis considers a hollow spherical shell made of a dielectric with charge Q of radius R, with an angular velocity $\vec{\omega}$. This approach is similar in nature to the gravitational model of gravitomagnetism with Weber for a massive spherical shell producing a frame dragging effect. Here, the charged sphere acts on a point charge q at position $\vec{r}$, velocity $\vec{v}$ and acceleration $\vec{a}$. For the point charge inside of the spherical shell, he arrives at the expression

$$\vec{F} = \frac{\mu_0 qQ}{12\pi R}\left[\vec{a} + \vec{\omega} \times (\vec{\omega} \times \vec{r}) + 2\vec{v} \times \vec{\omega} + \vec{r} \times \frac{d\vec{\omega}}{dt}\right], \tag{37}$$

which can further be simplified with the restriction $\vec{\omega} = 0$, meaning the sphere being stationary, to:

$$\vec{F} = \frac{\mu_0 qQ}{12\pi R}\vec{a}. \tag{38}$$

With the help of Newton's second law this can then be interpreted as a change in effective inertial mass of the particle due to the potential on the surrounding shell.

This is an especially interesting prediction of Weber's theory, because in standard theory, the field inside a charged spherical shell is zero and such an effect is not intuitively expected. Assis also estimates an order of magnitude for the effect, which entails a sphere of radius 0.5 m and a potential on the shell of 1.5 MV to double the mass of an electron, which is generally in the realm of the possible to obtain in the laboratory. However, from these values compared with the size and charge of the electron it can be seen that the effect is still considerably small in nature. However, as a logical conclusion, in theory a particle can be made to have an effective negative mass through the influence of an electrostatic potential at the cost of sufficient energy expenditure. This behaviour might then be applied to future breakthrough propulsion applications [176–178].

As this is an interesting prediction and experiment to determine boundaries of the validity of Weber's force law, several experimental efforts have tried to investigate this predicted phenomenon, with recent evidence suggesting the non-existence of this effect. Mikhailov had first reported an experiment in 1999 [179] where he claims to have successfully observed the effect in question, with two follow-up experiments in 2001 and 2003 [180,181], which he reported to be equally successful. However, the attempts of independent researchers to repeat and, respectively, improve their experiments have not yielded positive results [182–184]. Since all of the re-evaluated experiments feature a refined methodology and uncover flaws in the settings of Mikhailov (e.g., coupling and detection methods), one inevitably concludes that the effect in question was not observed by Mikhailov. An important point is discussed by Lörincz and Tajmar [183] as to what degree a glow discharge is suited to produce and measure a possible mass change of charge carriers, because the discharge is always made up of a neutral plasma and hence probably not suited to show the effect, and Weikert and Tajmar also speculated that the oscillatory motion of electrons in a Barkhausen-Kurz configuration could mask the effect in question [184].

There is new evidence [185] that the sought after mass change effect does not seem to exist. Tajmar and Weikert tested electron beam deflections in a Perrin tube where under a certain magnetic field the beam would be deflected precisely into a Faraday cup measuring the beam current. The arrangement was located in an aluminium sphere that was charged up to ± 20 kV and they simultaneously observed the current feeding a set of Helmholtz coils generating the magnetic field and the measured beam current of the Faraday cup. By observing the necessary current to keep the beam consistently in the Faraday cup they concluded that the sought after effect can be ruled out by two orders of magnitude. This makes this topic a valid point of criticism against Weber electrodynamics which will be discussed further in the following Section 3.3.

These many investigations of Weber's force in the literature demonstrate the strength of the theory and how it links to many fields across physics, and therefore it is an interesting alternative model to standard electrodynamics. Weber's force law has even been suggested in the literature as a unified theory of nature [146,186], especially as it includes the electromagnetic force, a form of nuclear or strong force, a gravitational-type force and additionally an explanation of the origin of inertia. We can see from this that Weber offers an elegant path towards the unification of theories and, to quote O'Rahilly ([25], Volume II, p. 535): 'If any one man deserves credit for the synthetic idea which unifies the various branches of magnetic and electrical science, that man is Wilhelm Weber'. Weber's theory connecting so many disciplines is of course not an accident, since it is an electromagnetic theory. Field theory and relativity which the standard model of modern physics is based on also show these properties and strive for a unified theory of physics. It is well known that three out of the four forces of nature can be unified in the standard model (magnetic force, strong nuclear force and electroweak force) and attempts are made through quantum gravity to connect the remaining force to those three. So in general, any theory attempting to explain the natural phenomena on a larger scale is likely to show the characteristics of a unified theory. However, Weber's theory, despite the time that has elapsed since its inception, is still very much in an early stage of development and has not been researched to the same degree as conventional models. Nevertheless this should not be taken as discouraging, quite the opposite in fact, it gives motivation for further investigation of Weber's force to explore its capabilities to describe and predict the universe. Needless to say, Weber's theory is not without criticism.

3.3. Criticism of Weber's Theory

Weber's force is not pursued by mainstream physics as it was superseded by Maxwell's theory of fields and ether, and so we must question just why the theory was largely abandoned in favour of another. Historically there were three main points of criticism leading to the neglect of Weber's theory and a fourth factor of experimental nature. The points of criticism of Weber's force were: (i) It is based on the Fechner hypothesis, i.e., currents being comprised of equal amounts of moving positive and negative charges, (ii) Helmholtz first criticised Weber's force as violating energy conservation, and (iii) then criticised the theory for exhibiting unphysical behaviour in the form of negative mass and infinite acceleration. Finally, what was claimed as decisive evidence in favour of the field model were the successful experiments of Hertz, demonstrating electromagnetic waves and therefore taken as direct proof supporting Maxwell's theory as opposed to Weber's. We will now re-investigate these historical factors to see how they have aged and after that look at more modern criticisms or limitations of the theory. Another review of historic criticisms from a philosophical perspective of action-at-a-distance theories in general can be found in [187] with a strong focus on epistemologic aspects.

We will first consider the criticism that Weber's force is originally based on Fechner's hypothesis. In the middle of the 19th century it was assumed that a current in a circuit consisted of equal amounts of positive and negative charge (or electric "fluidae") moving in opposite directions. While it is true that Weber designed his force based on this assumption to derive Ampère's force, he himself moved away from the idea of a double current consisting of moving positive and negative charges towards a simple current where the positive charges remain fixed in the lattice and only the negative charges are considered to be moving in his later works. Despite this change in perspective, he did not alter the formulation of his force law because it would still remain valid in predicting observable effects. Assis showed [61] that it is possible to derive Ampère's force regardless of the Fechner hypothesis from Weber's force. The only assumptions made are the charge neutrality of current elements and the independence of velocities of positive and negative charge carriers therein. This means that it holds true for moving electrons and stationary lattice charges as well as oppositely moving positive and negative charges as, for example, in plasma states. Further the Hall effect can be explained with Weber's theory when the Fechner hypothesis is abandoned, as reviewed in the previous section.

When discussing this matter it should be mentioned that Clausius was the first to claim in 1877 that Weber's force would lead to unphysical situations when only one kind of charge is moving and the other kind is fixed in a conductor [188]. This criticism seems to have been taken as a decisive argument against the theory as reviewed by Woodruff [189] and persisted in the more recent literature when Pearson and Kilambi analysed the similarity with nuclear forces [145]. The argument is that in the situation where Weber's force is not balanced by oppositely moving charge carriers, so called electrostatic induction occurs outside of a conductor and this has been considered an exclusion criterion of Weber's theory as an explanation for electrodynamics in [145]. However, it must be said that an early extensive refutation of this argument has been given by Zöllner as early as 1877 [190] in response to Clausius, arguing that these electrostatic effects have been known since 1801 through experiments by Erman [191] and others, and Weber's theory is erroneously criticised. Assis and Hernandes [192–194] have given a modern review of electrostatic induction effects in theory and experiment, showing that these effects do indeed exist on different orders of magnitude and that it is too early to dismiss Weber's force law on such a basis. In fact, the existence of such effects could even provide experimental support for Weber's theory. In conclusion we can say that at present there is no requirement for Fechner's hypothesis in Weber's theory and the criticism is not valid.

Helmholtz originally criticised the theory by saying it did not obey conservation of energy. He considered the potential and kinetic energy, especially with particles in circular motion and came to the conclusion that there are possible situations in Weber's theory where energy can be lost or gained and thus objected to the theory. At the time, Maxwell was familiar with Weber's theory and knew of Helmholtz's argument, supporting the objection stated by him. It was only in 1869 and 1871 that Weber succeeded in showing that energy is conserved in his force law. After this proof Maxwell even acknowledged in his *Treatise* that Weber's theory was mistakenly criticised in that regard [22] and reconsidered it as a possible theory of electrodynamics. Helmholtz erroneously came to the conclusion that energy conservation is violated as he only included the velocities between interacting charges and did not consider the complete form of Weber's force which also depends on the acceleration of charges, leading him to an incomplete deduction [24].

Furthermore, Helmholtz issued a second major criticism of Weber's theory, where he describes a situation that leads a particle to exert behaviour of negative mass and upon movement it could accelerate infinitely in the presence of an external force, such as friction, and an infinite amount of work would be done. He describes a particle inside a charged spherical shell that experiences friction from a fluid and as it is infinitely accelerating, it would continue to heat the fluid due to friction and thus perform an infinite amount of work. An argument ensued between the two parties with Weber trying to defend against the criticism, but Helmholtz again countered his arguments and the criticism prevailed [189] and to this day remains an open question in Weber electrodynamics.

This is indeed a similar situation to the suggested inertial mass change of Assis due to a field-free electrostatic force and the criticism has been further analysed by Assis and Caluzi [195]. They have suggested three possible ways to resolve the problem: (1) If instead of using Weber's force and Newton's mechanics, a modification of mechanics to a relativistic type kinetic energy is assumed, then the particle will not accelerate *ad infinitum*. The modified expression for mechanics has also been obtained by Schrödinger [152] and Wesley [60] by considering Weber-type forces. (2) It is also possible to avoid the problem by a modification of Weber's potential energy, e.g., Phipps [196], which leads to a force expression where infinite acceleration does not occur. (3) It is still possible (though highly unlikely) that nature behaves this way, it just has not been sufficiently tested.

However, with the new experiment of Tajmar and Weikert [185] it seems very unlikely that nature behaves in a way where this change of mass or energy is involved. Especially when according to Helmholtz an infinite amount of work can be gained if a particle experiencing friction forces from a fluid would continue to increase the fluid's temperature due to the friction. By implication this would violate the first law of thermodynamics and allow for a perpetual motion machine under the condition that the energy can be extracted from the closed system formed by the charged shell, fluid and point charge. It could theoretically still be possible that in order for the effect to manifest, a friction force is needed, as this is what Helmholtz originally considered and was also included in the analysis of Assis and Caluzi [195]. However, this seems unlikely as Assis also made the prediction for a mass change without the consideration of friction forces at first [175] and further a situation like this seems very unlikely to manifest in practice. As far as we know, a similar system has never been observed in nature and the proposed scenario seems unlikely to occur naturally.

If we now consider resolutions (1) and (2) suggested by Assis and Caluzi, we can say for point (1) that the infinite acceleration of a particle can be avoided through the assumption of modified mechanics, but this would change the behaviour mainly for velocities near the speed of light, as pointed out by Weikert and Tajmar [185]. Additionally, in this case Assis and Caluzi still speak of an effective inertial mass influenced by the electrodynamics, so an apparent mass change would still occur in this instance. As for point (2), the modification of Weber's potential energy to Phipp's potential, this can generally present a solution to the problem, however Phipp's potential turns out to have other shortcomings [197]. Nevertheless, there might still be a more fundamental, more general Weber-type potential where the change in mass is avoided and the problem is resolved, such as suggested by Li [198]. So this criticism is a valid point, however, there might well be other ways not yet known to avoid the problem like a more general potential and here more research is needed. It must also be added to this discussion that due to the self-energy divergence in field theory, the Lorentz-Dirac renormalisation solution also leads to runaway behaviour with infinite self-acceleration of a particle [37]. Albeit a different situation where the particle interacts with its own field, as opposed to Weber's theory where the particle interacts with a field-free charge distribution, this shows that we can find the same unphysical behaviour through infinite acceleration exhibited in the framework of field theory as well. It would seem inadequate to use this singular behaviour as a decisive argument against Weber's theory when field theory is not rid of such a problem itself.

The fourth factor that was historically held against Weber's theory are Hertz's experiments that showed a finite propagation velocity of electromagnetic signals, and were taken as a direct verification of Maxwell's field theory. However, we have seen in the previous sections that Weber's theory has been shown to be consistent with fields and can indeed predict wave equations propagating with the speed of light and is related to radiation phenomena when the right physical methods and constraints are applied. So it remains questionable if Hertz's experiments should only be taken as a direct proof of Maxwell's equations. For example, the experiments are also consistent with Ritz's theory (as pointed out by O'Rahilly [25]), and it would be premature to exclude Weber's theory on that basis.

Additionally, it should be mentioned that Weber believed in a form of ether, i.e., the luminiferous ether, which nowadays has been effectively replaced by the electromagnetic field model. So one could argue that Weber's theory is based on the same ethereal premise as Maxwell's, as he tried to model the interactions of charges within the fluid of ether that was general scientific consensus at the time. Although it should be noted that Weber's force does not conceptually depend on the ether due to only involving relative velocities. Furthermore when invoking the principle of Ockham's razor, that is, entities should not be multiplied beyond necessity, it becomes apparent that Weber's theory is considered to be preferable over Maxwell's as it makes fewer assumptions. Weber being based only on the interaction of charges without invoking the concept of field entities conforms to minimal assumptions as only the charges and their motion is relevant. Additionally, there is an argument in the literature, when discussing contact-action through fields or emitted virtual particles, that entities like the ether or the field which cannot be observed directly should be avoided [199]. At its core this issue relates to a more general criticism of Weber's theory being a direct-action-at-a-distance theory with apparent instantaneous transmission unlike a field theory. The instantaneous nature of force laws like this is usually argued to be a problem since it violates causality and since the propagation of electromagnetic effects are clearly of finite velocity, hence there is an apparent problem with the theory. Take, for example, Newton's gravitational force law: if one body experiences a change in position, body two will immediately feel this change in force, no matter the distance between them. However, there exists a claim in the literature that Newtonian gravity actually propagates at the speed of light [200]. Although the analysis is based on dimensional, empirical and observational arguments, this is a remarkable postulate, and even though Newtonian gravity is formulated mathematically as instantaneous-action-at-a-distance, a finite propagation speed is implicitly contained within the formula, suggesting that the same could be true for other so called "instantaneous" force laws.

So the real question is how propagation velocities behave in Weber's theory or how they can be limited. There are multiple factors to consider here. First, Weber's force itself models a delay in propagation intrinsically with the constant c in the formula, and as argued by Brown [201], c should be viewed as a retardation constant at which cause and effect occur. Related to this Sokol'skii and Sadovnikov [13] have studied planetary orbits with a gravitational form of Weber and find that gravitational interaction propagates at the speed of light c in their model. In the previous Section 3.2.1 we have seen different considerations showing how Weber is able to predict wave propagation of electromagnetic signals at the speed of light, for example in combination with the principle of retarded time applied by Moon and Spencer, and Wesley who derived wave equations based on this premise. Additionally, Weber and Kohlrausch first obtained experimentally the value of c with Weber's force and Kirchhoff and Weber both arrived at the telegraph equation for propagation of signals in a circuit independently from each other based on Weber's force. From these considerations, we can see that Weber indeed expresses a delay in propagation and a finite propagation velocity despite it's action-at-a-distance origin, meaning causality is not violated.

The overall concept of action-at-a-distance theories on a more philosophical basis has been discussed by Pietsch [36]. The discussion is on a more general level about energy conservation, necessity of contact for transmission of energy, locality criteria and meta-physical considerations and does not favour Weber's theory over the Wheeler-Feynman type or other action-at-a-distance theories [21]. First the problem of divergent self energy of a point charge in classic electromagnetism is discussed, with the existing resolutions of renormalisation versus extension approaches. It is argued that both solutions come at a cost and that action-at-a-distance can offer an alternative to resolve either problem that arises in one approach or the other. Similarities of pure particle action and pure field theories are explored in the analysis [36] and then Pietsch continues to point out similarities of action-at-a-distance theories with modern particle-field theories of electrodynamics widely established today. Lentze [202] has argued that direct-action between charges as a general concept can provide deeper insight into explaining the origin of electromagnetic forces and the light speed principle, thus improving our understanding of nature. Further discussion of direct-action theories, also more focussed on the quantum version of the Wheeler-Feynman approach with respect to Haag's inconsistency theorem and how direct-action can overcome the mathematical inconsistencies and problems arising from QED and QFT can be found in [42], but are beyond the scope of this review.

From a modern perspective, Weber's force has been critiqued as it does not account for relativistic corrections of deflected electron beams, such as the Kaufmann-Bucherer or Bertozzi experiments [203,204] in its basic form. Additionally, Weber's force does not impose the same relativistic speed limit of c on a particle [205] as is expected from SRT, and it does not seem to be compatible with SRT and the Lorentz transformation based on these shortcomings. In its standard form Weber is only an approximation in v^2/c^2 for high speed particles compared with relativistic corrections, as shown by [107,203]. There are, however, some approaches in the literature to address and solve these inconsistencies of Weber's force. For example, this could suggest that the regular form of Weber's force (15) might still be an incomplete theory and only modified forms may be able to overcome this problem. Attempts on limiting the velocity have been made by Montes [206], Wesley [204] and Phipps [196] which manage to include relativistic corrections to varying degrees. Assis [207] investigated Phipp's potential and the inclusion of higher order terms similar to [146,147] in relation to their relativistic compatibility. He finds that these types of Weber theories either hold the speed limit c for an individual charge or their relative speed, but not both at the same time. In general these approaches consider some form of modification of Weber's theory, which might involve other costs as analysed for the gravitational type forces and their energy conservation [155], and even if a suitable modification is discovered, its properties should be studied further on this basis. In a recent study, Li [198] utilises an extension of Weber's potential and the additional assumption of modified mechanics and arrives at high velocity particle behaviour that is identical to the predictions made from SRT. This has certain similarities to Wesley's approach [143] of modified mechanics in conjunction with a Weber-type force and indicates that further research is necessary in relation to the high velocity regime.

Another interesting approach to potentially resolve the problem was suggested by Bush [137]. Bush investigated a direct-action force of the Weber-Ampère type and instead of a variation of mass with velocity assumes a variation of charge with velocity. He deduces from these investigations that the e/m ratio obtained with their approach is consistent with the Kaufmann-Bucherer experiments. Further, with this approach it is the force which tends to zero for a relative velocity between charges close or equal to the speed of light, rather than the mass tending to infinity. The reason why no superluminal particles are observed experimentally is then the lack of force transmission at the speed of light. From these different suggestions to reconcile Weber's force with relativistic corrections it can thus be concluded that the theory has not been fully developed for high speed interactions yet, and more research is needed if Weber can be correctly used in that sense.

Recently, a new study by Kühn found that it is also possible to resolve the problem with wave equations derived from Weber electrodynamics [111], where the wave solutions are obtained in the reference frame of the receiver. The article first shows how Maxwell's equations lead to Weber's force within the low velocity limit and then further investigates how the wave propagation in the far field can be limited to the velocity of light. Interestingly, the Lorentz transformation is not a necessity in that approach and rather than viewing the propagation velocity as constant w.r.t. the transmitter, it is perceived as constant by the receiver. The physical mechanism responsible for this behaviour is discussed with regard to a newly developed emission theory by Kühn [208] and its implications for electrodynamics and SRT. Concluding from these different approaches, while in its standard form Weber's force seems incompatible with SRT, we see that solutions have been suggested and further research is needed to investigate how Weber's theory and relativistic physics are related.

Other modern criticisms have been issued where Weber's force was shown to lead to unphysical results [157,209]. In the case of the investigation carried out by Clemente et al. Weber's force is applied to cold plasmas and analysed for the resulting propagation of waves in the plasma oscillations. The authors of [157] find that Weber predicts a type of longitudinal wave propagation that is unphysical and contradicts the well established experimental evidence [210] about dispersion relations in cold plasmas. On this basis it is also concluded that only the existence of electromagnetic fields can account for this effect and that Weber's force leads to erroneous results in this case because it strictly follows Newton's third law. This is, of course, an important point of criticism and it shows that Weber's theory has not yet been developed in relation to plasma physics at all. It might be interesting to see if Weber still leads to unphysical results if retarded action is taken into account (as it was discussed in Section 3.2.1 whereby Weber is consistent with fields in that case). It might be possible to arrive at the correct result by modification of Weber's equation or potentially Ritz's formula could account for the discrepancy, while this is a shortcoming of Weber's force it would be of interest to further develop a plasma physics approach based on Weber's force.

An analysis by Sherwin [209] compares the transmission and detection of an idealised radar system between Weber's force and the Liénard-Wiechert force in the standard theory. While the magnitude is found to be the same for both approaches, the angular dependence, direction of signal transmission and propagation delay is different between the two. Sherwin states that the standard result through the Liénard-Wiechert equation is well supported by experiments in radar technology and disagrees with the Weber prediction. On the other hand, the field-model result cannot easily account for longitudinal forces, which the Weber force can, and at least the propagation delay can be introduced through a retarded time approach, such as Wesley's [9,211]. From this, Sherwin concludes that neither theory agrees sufficiently well with observable phenomena yet. This seems to be a peculiar situation that neither theory fully agrees with expectations yet, as it is likely that the inclusion of longitudinal forces in the Liénard-Wiechert model would also change the angular dependence and transmission direction closer towards the Weber force result. Certainly, more research is needed to arrive at an answer to the problem and it would be advantageous to give Weber's theory a further rigorous development for signal transmission and radar applications, also on the basis of the approaches reviewed herein [9,19,134,211].

We have outlined how Weber's theory offers particular value in explaining phenomena arising in various areas of physics. Most of the original objections to the theory have been answered and the criticisms that remain are not yet completely resolved. Neither the

Maxwellian field approach nor Weber's electrodynamics are free of criticism and problems, each theory has its own advantages and disadvantages and unreconciled inconsistencies. In the same way that established theories are continually researched and developed further and eventually amended, we would make the argument that the same is necessary for direct-action theories, especially considering that there are significant parallels between both theories and many connections have been established between the two. Several authors have presented the view point that direct-action theories are a valid alternative to be considered in physics and can help overcome problems associated with field theory. None of the theories are perfect, so it is not wise to dismiss direct-action too early and it should be further developed to inform research and to investigate what explanations it can offer. Another useful approach might be to regard the two theories as complementary, and depending on the application to employ whichever theory is most suitable.

4. Perspective and Future Prospects

The present article has given a comprehensive review about an underrated contribution to the foundations of electrodynamics, namely the electrodynamic theory of Wilhelm Weber, and pointed out its many capabilities. However, even with the several strengths of Weber's formula presented, the current limitations of the theory have also been explored and criticism of the theory addressed. Overall, it is found that Weber's force in its standard form is only valid within certain limitations and further research is needed, but it can still complement Maxwell's field equations within those bounds.

In the literature analysis both Maxwell's theory of fields and Weber's electrodynamics have been introduced, and it was then reviewed how Weber is capable of explaining several electromagnetic phenomena, including Coulomb's force, Ampère's force, field equations, induction and the telegraph and wave equations. Moreover, the importance of longitudinal forces and their role in Weber electrodynamics was emphasised, as well as their analogy in field theory, as they have been shown to exist in both approaches.

Further to just explaining electricity and magnetism, Weber's theory provides a basis for other physics disciplines and has a unifying character connecting several branches of physics. It has been shown how Weber's force connects to gravity and the strong nuclear force, Newton's second law and Mach's principle and relativistic phenomena such as the bending of light or frame dragging, however it is not yet known how the weak force relates to Weber's force. For standard field theory it is well known that strong force and weak force can be related to the electromagnetic force but it is still unknown how gravity can be unified with the other forces in that approach, although it is being actively researched (e.g., quantum gravity). Lastly, field theory is compatible with SRT as well as quantum mechanics, but there is no extensive research yet how Weber's theory can be related to these topics, if at all. We can see from this juxtaposition that both theories are similar, yet different, and a table is given below summarising some of the similarities and differences of both theories (Table 1).

One can conclude that Weber's theory is not without limitations and is only valid within the low velocity regime, with standing problems in relativistic physics (special as well as general), radiation and plasma applications as well as quantum electrodynamics. However, if the force law is considered as an addition to Maxwellian field theory, it can enrich one's perspective on electromagnetic phenomena and beyond. It offers an explanation of observed phenomena from a particle perspective, as well as the prediction thereof. In this sense Weber's theory can be regarded as an important component of electrodynamic theory, even though it has not been developed anywhere near to the same level as field theory. While experimental evidence in relativistic electrodynamics and QED supports Maxwell's field theory, it has been found that Weber agrees with experiments in the near field and low velocity limit. This suggests, along with recent studies, that Weber's force is a low velocity approximation up to second order in v/c of a more fundamental underlying force, and needs further development. Several modifications are conceivable, with possible generalisations of the Ritz-type or corrections to the Weber potential as suggested by Li [198] and it will be subject to future work to investigate the full capabilities and boundaries of the theory and its modifications.

Table 1. A comparison between field and Weber theory in selected categories.

Category	Field Theory	Weber Electrodynamics
Point Particle Interaction	Liénard-Wiechert-Schwarzschild force	Weber force
Current Element Interaction	Grassmann's force	Ampère force
Longitudinal Forces	Can be obtained in various ways [96,97,112]	Intrinsic
Magnetic Force	Lorentz force	Weber-Ampère force
Conservation laws	Energy is stored or dissipated by the field to restore conservation laws	Follows linear momentum, angular momentum and energy conservation
Field Equations	Maxwell equations	Can be obtained in various ways [5,6,8,9,110]
Induction	Through changing electric and magnetic fields and flux	Through particle movement (velocity and acceleration)
Wave Equations	Arise from moving fields	Can be obtained through retarded time and also predicts telegraph equation
Compatibility with other forces of nature	Weak force, strong force, gravity not yet known	Gravity, strong force, weak force not yet known
Compatibility with SRT	Compatible through Lorentz transformation	Incompatible with Lorentz transformation
Extension to Quantum Mechanics	QED, QFT	Initial connections and some problems exist, but not yet fully known

A benefit of Weber's force is that it follows Newton's third law of motion, thus conserving linear momentum, and additionally conserving angular momentum as well as energy, so it does not violate conservation laws. In the field approach it is usually argued that conservation laws are not violated when the energy content of the field is taken into account, i.e., the field can obtain energy or momentum from a system and store it as well as release energy and momentum. Further, Weber's force accounts for longitudinal forces intrinsically, however, the absence of longitudinal forces seems to be desired in radar and plasma applications where the field approach benefits from this quality instead and Weber seems to fail to predict the expected results. However, Weber's force can be calculated directly from the movement of the charges involved in an interaction and does not necessitate the calculation of one or more fields of those charges from which the force is calculated, which offers clear force and particle-based explanations, which avoids problems such as the self-energy divergence. Another benefit in Weber's theory is that charge velocities are clearly defined, whereas in field theory there may be some ambiguity left as to what velocities are to be used in the Lorentz force equation. Lastly, Weber's force offers the possibility to unify gravitational forces with those of electromagnetism. It is thus considered constructive to use both theories in cooperation with each other, as each can compensate for the other's weakness and regarding a specific problem in question from both perspectives can potentially lead to new insight. Examples for this can be found not only in the flux cutting analogy of unipolar induction, but also transformer induction where the particle perspective considers the acceleration of the current electrons whereas the field perspective links the magnetic flux of either side of the transformer. Similarly, magnetic fields (e.g., of a solenoid) can be regarded from a particle perspective, where it is again the movement of the charges exerting an influence on the test body rather than the field mediating the force.

On the basis of the present review, one can identify certain aspects of Weber's electrodynamics that would be especially interesting to research further. Firstly, the importance of Ampère longitudinal forces to determine how they influence specific applications, for example the present limitations surrounding the radar equation and cold plasmas, and further development of Weber's force for signal transmission and radar applications would be of interest. Further, the general relevance of longitudinal forces to nuclear fusion applications has been suggested [82,97,212], and it remains to be determined how longitudinal forces could be applicable to plasma physics on a wider scale, especially as plasmas typically have a positive ion charge density and a negative electron charge density. There is not yet extensive experimental work giving a quantitative estimate of the influence of the longitudinal forces on plasmas of which the authors are aware. Some recent experiments [101] suggest that the influence is frequency dependent but relatively small. Further research into longitudinal forces would certainly be of general interest to clarify their overall role in electrodynamics.

Another aspect that should be further investigated is charged particle optics based on Weber's electrodynamics, in particular the deflection of high-speed electrons as in the Bertozzi experiment and further exploring how Weber relates to mass change with velocity and SRT. In order to advance the simulation in the high velocity regime, Weber's force will likely need modifications, such as suggested by Wesley, Assis, Montes and Li, along with other types of modification and generalisation still to be explored.

It could also be valuable to research further electrostatic induction and its connection to Weber's force, Assis notably mentions the experiments performed by Jefimenko and Edwards et al. [213–215]. Nowadays, there are more modern experiments that have been performed and extended the work of Edwards et al. with superconductors and electrostatic induction in general, but it seems like they have not yet been analysed from a Weber-perspective and similar experiments could help to further investigate the boundaries and validity of Weber's and field theory in these cases. Next to electrostatic induction, of course, investigating other induction experiments further and how Weber's force can predict them seems logical as Weber has been successfully applied to transformer and unipolar induction already.

There have been some initial connections between Weber's force and quantum mechanics (see Section 3.2.2), and it could be extremely valuable to extend Weber's force to the quantum realm, as this could resolve existing problems with Weber's force and might allow for new insight into particle physics from a Weber perspective relating to the nature and behaviour of the particles themselves. Similar to Weber's model of the atom, it might be conceivable that a Weber-type theory leads to a quantum wave equation which resembles Schrödinger's equation. Interestingly, a recent study by Zhao [216] has derived a wave equation similar to Schrödinger's, that shows interesting similarities to Weber, as both Weber's planetary model of the atom and Zhao's approach can account for precessing electron orbits. So it might be worth investigating what quantum mechanical capabilities Weber-type forces might have. It may well offer yet unknown solutions and insight regarding present problems surrounding supersymmetric expectations in particle physics.

In conclusion, Weber's force is an electrodynamic force law with limited validity that can complement Maxwell's field equations, and particularly in the low velocity and near field limit it is indistinguishable from field theory. It can offer explanations not only for electromagnetic phenomena but also general physics, with many connections yet to be explored, holding great potential for further development.

Author Contributions: Conceptualization, C.B. and S.M.; formal analysis, C.B.; writing—original draft preparation, C.B.; writing—review and editing, S.M.; supervision, S.M. All authors have read and agreed to the published version of the manuscript.

Funding: This research received no external funding.

Institutional Review Board Statement: Not applicable.

Informed Consent Statement: Not applicable.

Data Availability Statement: Not applicable.

Acknowledgments: One of the authors (CB) is funded by the School of Electrical Engineering, Electronics and Computer Science (EEE-CS), University of Liverpool. CB and SM are thankful to the EEE-CS School in this regard. The authors would also like to thank RT Smith for fruitful discussions.

Conflicts of Interest: The authors declare no conflict of interest.

References

1. Lykken, J.; Spiropulu, M. Supersymmetry and the Crisis in Physics. *Sci. Am.* **2014**, *310*, 34–39. [CrossRef]
2. Crowther, K. Defining a crisis: The roles of principles in the search for a theory of quantum gravity. *Synthese* **2021**, *198*, 3489–3516. [CrossRef]
3. Meucci, S. Physics has Evolved Beyond the Physical. *Cosm. Hist. J. Nat. Soc. Philos.* **2020**, *16*, 452–465.
4. Hossenfelder, S. *Lost in Math: How Beauty Leads Physics Astray*; Hachette: Paris, France, 2018.
5. Kinzer, E.; Fukai, J. Weber's force and Maxwell's equations. *Found. Phys. Lett.* **1996**, *9*, 457–461. [CrossRef]
6. Assis, A.K.T. *Weber's Electrodynamics*; Springer: Dordrecht, The Netherlands, 1994.
7. Gray, R.W. Candid Remarks On Introductory Classical Electrodynamics and Textbooks. *Preprint* **2020**, 1–27. [CrossRef]
8. Anonymous. *Advances in Weber and Maxwell Electrodynamics*; Amazon Fulfillment: Wroclaw, Poland, 2018.
9. Wesley, J.P. Weber electrodynamics, Part I—General theory, steady current effects. *Found. Phys. Lett.* **1990**, 3, 443–469. [CrossRef]
10. Holzmüller, G. *Über die Anwendung der Jacobi-Hamilton'schen Methode auf den Fall der Anziehung nach dem Electrodynamischen Gesetze von Weber*; Universität Halle-Wittenberg: Halle, Germany, 1870.
11. Torres-Silva, H.; López-Bonilla, J.; López-Vázquez, R.; Rivera-Rebolledo, J. Weber's electrodynamics for the hydrogen atom. *Indones. J. Appl. Phys.* **2015**, *5*, 39–46. [CrossRef]
12. Frauenfelder, U.; Weber, J. The fine structure of Weber's hydrogen atom: Bohr–Sommerfeld approach. *Z. Angew. Math. Phys.* **2019**, *70*, 105. [CrossRef]
13. Sokolskii, A.; Sadovnikov, A. Lagrangian Solutions for Weber's Law of Attraction. *Sov. Astron.* **1987**, *31*, 90.
14. Moon, P.; Spencer, D.E.; Mirchandaney, A.S.; Shama, U.Y.; Mann, P.J. The Electrodynamics of Gauss, Neumann, and Hertz. *Phys. Essays* **1994**, *7*, 28–33. [CrossRef]
15. Lorentz, H.A. *La Théorie Électromagnétique de Maxwell et son Application aux Corps Mouvants*; EJ Brill: Leiden, The Netherlands, 1892; Volume 25.
16. McCormmach, R. HA Lorentz and the electromagnetic view of nature. *Isis* **1970**, *61*, 459–497. [CrossRef]
17. Woodruff, A.E. The contributions of Hermann von Helmholtz to electrodynamics. *Isis* **1968**, *59*, 300–311. [CrossRef]
18. Martínez, A.A. Ritz, Einstein, and the emission hypothesis. *Phys. Perspect.* **2004**, *6*, 4–28. [CrossRef]
19. Moon, P.; Spencer, D.E. A new electrodynamics. *J. Frankl. Inst.* **1954**, *257*, 369–382. [CrossRef]
20. Wheeler, J.A.; Feynman, R.P. Interaction with the absorber as the mechanism of radiation. *Rev. Mod. Phys.* **1945**, *17*, 157. [CrossRef]
21. Wheeler, J.A.; Feynman, R.P. Classical electrodynamics in terms of direct interparticle action. *Rev. Mod. Phys.* **1949**, *21*, 425. [CrossRef]
22. Maxwell, J.C. *A Treatise on Electricity and Magnetism Unabridged*; Dover: New York, NY, USA, 1954.
23. Yaghjian, A.D. Reflections on Maxwell's treatise. *Prog. Electromagn. Res.* **2014**, *149*, 217–249. [CrossRef]
24. Assis, A.K.T.; Silva, H.T. Comparison between Weber's electrodynamics and classical electrodynamics. *Pramana* **2000**, *55*, 393–404. [CrossRef]
25. O'Rahilly, A. *Electromagnetic Theory: A Critical Examination of Fundamentals*; Dover Publications: New York, NY, USA, 1965; Volumes I and II.
26. Jackson, J.D. *Classical Electrodynamics*; Springer Science & Business Media: New York, NY, USA, 1999.
27. Darwin, C.G. LI. The dynamical motions of charged particles. *Lond. Edinb. Dublin Philos. Mag. J. Sci.* **1920**, *39*, 537–551. [CrossRef]
28. Landau, L.; Lifshitz, E. *Electrodynamics of Continuous Media*, 2nd ed.; Pergamon: Oxford, UK, 1985.
29. Page, L.; Adams, N.I., Jr. Action and reaction between moving charges. *Am. J. Phys.* **1945**, *13*, 141–147. [CrossRef]
30. Steane, A.M. Nonexistence of the self-accelerating dipole and related questions. *Phys. Rev. D* **2014**, *89*, 125006. [CrossRef]
31. Cornille, P. The Lorentz force and Newton's third principle. *Can. J. Phys.* **1995**, *73*, 619–625. [CrossRef]
32. Cornille, P. Review of the application of Newton's third law in physics. *Prog. Energy Combust. Sci.* **1999**, *25*, 161–210. [CrossRef]
33. Sebens, C.T. Forces on fields. *Stud. Hist. Philos. Sci. Part B Stud. Hist. Philos. Mod. Phys.* **2018**, *63*, 1–11. [CrossRef]
34. Assis, A.; Peixoto, F. On the velocity in the Lorentz force law. *Phys. Teach.* **1992**, *30*, 480–483. [CrossRef]
35. Tran, M. Evidence for Maxwell's equations, fields, force laws and alternative theories of classical electrodynamics. *Eur. J. Phys.* **2018**, *39*, 063001. [CrossRef]
36. Pietsch, W. On conceptual issues in classical electrodynamics: Prospects and problems of an action-at-a-distance interpretation. *Stud. Hist. Philos. Sci. Part B Stud. Hist. Philos. Mod. Phys.* **2010**, *41*, 67–77. [CrossRef]
37. Lazarovici, D. Against fields. *Eur. J. Philos. Sci.* **2018**, *8*, 145–170. [CrossRef]
38. Muller, F.A. Inconsistency in classical electrodynamics? *Philos. Sci.* **2007**, *74*, 253–277. [CrossRef]
39. Frisch, M. *Inconsistency, Asymmetry, and Non-Locality: A Philosophical Investigation of Classical Electrodynamics*; Oxford University Press: New York, NY, USA, 2005.
40. Feynman, R.P. Space-time approach to quantum electrodynamics. In *Selected Papers of Richard Feynman*; Brown, L.M., Ed.; World Scientific: Singapore, 2000; pp. 99–119.
41. Dirac, P.A.M. The evolution of the physicist's picture of nature. *Sci. Am.* **1963**, *208*, 45–53. [CrossRef]
42. Kastner, R.E. Haag's Theorem as a Reason to Reconsider Direct-Action Theories. *arXiv* **2015**, arXiv:1502.03814.

43. Griffiths, D.J. *Introduction to Electrodynamics*; Pearson Prentice Hall: Hoboken, NJ, USA, 2005.
44. Cullwick, E.G. *Electromagnetism and Relativity: With Particular Reference to Moving Media and Electromagnetic Induction*; Longmans: London, UK, 1959.
45. Moon, P.H.; Spencer, D.E. *Foundations of Electrodynamics*; Boston Technical Publishers Inc: Cambridge, MA, USA, 1965.
46. Lorrain, P.; Corson, D.R. *Electromagnetic Fields and Waves*; Courier Corporation: North Chelmsford, MA, USA, 1970.
47. Maxwell, J.C. XXV. On physical lines of force: Part I—the theory of molecular vortices applied to magnetic phenomena. *Lond. Edinb. Dublin Philos. Mag. J. Sci.* **1861**, *21*, 161–175. [CrossRef]
48. Weber, W.E. *Wilhelm Weber's Werke*; First Part; Julius Springer: Berlin, Germany, 1893; Volume 3.
49. Weber, W.E. *Wilhelm Weber's Werke*; Second Part; Julius Springer: Berlin, Germany, 1893; Volume 4.
50. Weber, W. *Determinations of Electrodynamic Measure: Concerning a Universal Law of Electrical Action*; 21st Century Science and Technology: Leesburg, VA, USA, 2007.
51. Weber, W. Electrodynamic Measurements, Especially Measures of Resistance. 2021. Available online: https://www.ifi.unicamp.br/~assis/Weber-EM2.pdf (accessed on 5 July 2022).
52. Weber, W. Electrodynamic Measurements, Especially on Diamagnetism. 2021. Available online: https://www.ifi.unicamp.br/~assis/Weber-EM3.pdf (accessed on 5 July 2022).
53. Kohlrausch, R.; Weber, W. Electrodynamic Measurements, Especially Attributing Mechanical Units to Measures of Current Intensity. 2021. Available online: http://www.ifi.unicamp.br/~assis/Weber-EM4.pdf (accessed on 5 July 2022).
54. Weber, W. *Electrodynamic Measurements, Especially on Electric Oscillations*. 2021. Available online: http://www.ifi.unicamp.br/~assis/Weber-EM5.pdf (accessed on 5 July 2022).
55. Weber, W. Electrodynamic measurements—Sixth memoir, relating specially to the principle of the conservation of energy. *Lond. Edinb. Dublin Philos. Mag. J. Sci. Fourth Ser.* **1872**, *43*, 1–20.
56. Weber, W. Electrodynamic Measurements, Especially on the Energy of Interaction. 2021. Available online: http://www.ifi.unicamp.br/~assis/Weber-EM7.pdf (accessed on 5 July 2022).
57. Weber, W. *Determinations of Electrodynamic Measure: Particularly in Respect to the Connection of the Fundamental Laws of Electricity with the Law of Gravitation*; 21st Century Science and Technology: Leesburg, VA, USA, 2021. Available online: https://www.ifi.unicamp.br/~assis/Weber_EM8.pdf (accessed on 5 July 2022).
58. Davies, P. A quantum theory of Wheeler–Feynman electrodynamics. In *Proceedings of the Mathematical Proceedings of the Cambridge Philosophical Society*; Cambridge University Press: Cambridge, UK, 1970; Volume 68, pp. 751–764. [CrossRef]
59. Wesley, J. Weber electrodynamics, part II unipolar induction, Z-antenna. *Found. Phys. Lett.* **1990**, *3*, 471–490. [CrossRef]
60. Wesley, J.P. Weber electrodynamics: Part III. mechanics, gravitation. *Found. Phys. Lett.* **1990**, *3*, 581–605. [CrossRef]
61. Assis, A.K.T. Deriving Ampere's law from Weber's law. *Hadron. J.* **1990**, *13*, 441–451.
62. Moyssides, P.G. Calculation of the sixfold integrals of the Ampere force law in a closed circuit. *IEEE Trans. Magn.* **1989**, *25*, 4307–4312. [CrossRef]
63. Cavalleri, G.; Bettoni, G.; Tonni, E.; Spavieri, G. Experimental proof of standard electrodynamics by measuring the self-force on a part of a current loop. *Phys. Rev. E* **1998**, *58*, 2505. [CrossRef]
64. Assis, A.K.T. Comment on "Experimental proof of standard electrodynamics by measuring the self-force on a part of a current loop". *Phys. Rev. E* **2000**, *62*, 7544. [CrossRef]
65. Assis, A.K.T.; Bueno, M.A. Equivalence between Ampere and Grassmann's forces. *IEEE Trans. Magn.* **1996**, *32*, 431–436. [CrossRef]
66. Bueno, M.; Assis, A.K.T. Proof of the identity between Ampère and Grassmann's forces. *Phys. Scr.* **1997**, *56*, 554. [CrossRef]
67. Assis, A.K.; Bueno, M. Bootstrap effect in classical electrodynamics. *Rev. Fac. Ing.* **2000**, *7*, 49–55.
68. Hofmann, J.R. *André-Marie Ampère: Enlightenment and Electrodynamics*; Cambridge University Press: Cambridge, UK, 1996; Volume 7.
69. Ampère, A.M. Mémoire présenté à l'Académie royale des Sciences, le 2 octobre 1820, où se trouve compris le résumé de ce qui avait été lu à la même Académie les 18 et 25 septembre 1820, sur les effets des courans électriques. In Proceedings of the Annales de Chimie et de Physique, Paris, France, 2 October 1820; Volume 15, pp. 59–76.
70. Ampère, A. The effects of electric currents. In *Early Electrodynamics—The First Law of Circulation*; Tricker, R.A.R., Ed.; Pergamon: Oxford, UK, 1965; pp. 140–154.
71. Assis, A.; Souza Filho, M.; Caluzi, J.; Chaib, J. From Electromagnetism to Electrodynamics: Ampère's Demonstration of the Interaction between Current Carrying Wires. In Proceedings of the 4th International Conference on Hands-on Science, University of Azores, Ponta Delgada, Portugal, 23 July 2007; pp. 9–16.
72. Chaib, J.; Lima, F. Resuming Ampère's experimental investigation of the validity of Newton's third law in electrodynamics. *Ann. Fond. Louis Broglie* **2020**, *45*, 19.
73. Graneau, P. Electromagnetic jet-propulsion in the direction of current flow. *Nature* **1982**, *295*, 311. [CrossRef]
74. Graneau, P. Application of Ampere's force law to railgun accelerators. *J. Appl. Phys.* **1982**, *53*, 6648–6654. [CrossRef]
75. Graneau, P. First indication of Ampere tension in solid electric conductors. *Phys. Lett. A* **1983**, *97*, 253–255. [CrossRef]
76. Graneau, P. Ampere tension in electric conductors. *IEEE Trans. Magn.* **1984**, *20*, 444–455. [CrossRef]
77. Graneau, P. *Ampère-Neumann Electrodynamics of Metals*; Hadronic Press, Inc.: Nonantum, MA, USA, 1985.
78. Graneau, P. The Ampere-Neumann Electrodynamics of Metallic Conductors. *Fortschritte Der Phys.* **1986**, *34*, 457–501.

79. Graneau, P.; Graneau, N. The electromagnetic impulse pendulum and momentum conservation. *Il Nuovo Cimento D* **1986**, *7*, 31–45. [CrossRef]
80. Graneau, P. Amperian recoil and the efficiency of railguns. *J. Appl. Phys.* **1987**, *62*, 3006–3009. [CrossRef]
81. Graneau, P.; Thompson, D.S.; Morrill, S.L. The motionally induced back-emf in railguns. *Phys. Lett. A* **1990**, *145*, 396–400. [CrossRef]
82. Graneau, P.; Graneau, N. The role of Ampere forces in nuclear fusion. *Phys. Lett. A* **1992**, *165*, 1–13. [CrossRef]
83. Assis, A.K.T. On the mechanism of railguns. *Galilean Electrodyn.* **1992**, *3*, 93–95.
84. Nasiłowski, J. A note on longitudinal Ampere forces in gaseous conductors. *Phys. Lett. A* **1985**, *111*, 315–316. [CrossRef]
85. Pappas, P. The original Ampere force and Biot-Savart and Lorentz forces. *Il Nuovo Cimento B (1971–1996)* **1983**, *76*, 189–197. [CrossRef]
86. Phipps, T.; Phipps, T., Jr. Observation of Ampere forces in mercury. *Phys. Lett. A* **1990**, *146*, 6–14. [CrossRef]
87. Phipps, T.E. New evidence for Ampère longitudinal forces. *Phys. Essays* **1990**, *3*, 198. [CrossRef]
88. Wesley, J. Pinch effect and Ampere tension to drive Hering's pump. *Found. Phys. Lett.* **1994**, *7*, 95–104. [CrossRef]
89. Peoglos, V. Measurement of the magnetostatic force of a current circuit on a part of itself. *J. Phys. D Appl. Phys.* **1988**, *21*, 1055. [CrossRef]
90. Achilles, R. Again on the Guala-Valverde homopolar induction experiments. *Spacetime Subst.* **2002**, *3*, 235.
91. Achilles, R.; Guala-Valverde, J. Action at a Distance: A Key to Homopolar Induction. *Apeiron Stud. Infin. Nat.* **2007**, *14*, 169–183.
92. Guala-Valverde, J.; Achilles, R. A Manifest Failure of Grassmann's Force. *Apeiron Stud. Infin. Nat.* **2008**, *15*, 202–205.
93. Papageorgiou, C.; Raptis, T. Fragmentation of Thin Wires Under High Power Pulses and Bipolar Fusion. In *AIP Conference Proceedings*; AIP Publishing LLC: Melville, NY, USA, 2010; Volume 1203, pp. 955–960. [CrossRef]
94. Alfaro, A.D. *Ampre's Longitudinal Forces Revisited*; Technical Report; Naval Postgraduate School, United States: Monterey, CA, USA, 2018.
95. Pappas, P.; Pappas, L.; Pappas, T. Ampère cardinal forces-electrodynamics—Proof and prediction of empirical Faraday induction. *Phys. Essays* **2014**, *27*, 570–579. [CrossRef]
96. Rambaut, M.; Vigier, J. The simultaneous existence of EM Grassmann-Lorentz forces (acting on charged particles) and Ampère forces (acting on charged conducting elements) does not contradict relativity theory. *Phys. Lett. A* **1989**, *142*, 447–452. [CrossRef]
97. Rambaut, M.; Vigier, J. Ampere forces considered as collective non-relativistic limit of the sum of all Lorentz interactions acting on individual current elements: Possible consequences for electromagnetic discharge stability and tokamak behaviour. *Phys. Lett. A* **1990**, *148*, 229–238. [CrossRef]
98. Moyssides, P.G. The anticipated longitudinal forces by the Biot-Savart-Grassmann-Lorentz force law are in complete agreement with the longitudinal Ampère forces. *Eur. Phys. J. Plus* **2014**, *129*, 1–17. [CrossRef]
99. Saumont, R. Ampère Force: Experimental Tests. In *Advanced Electromagnetism: Foundations, Theory and Applications*; World Scientific: Singapore, 1995; pp. 620–635.
100. Graneau, N.; Phipps, T., Jr.; Roscoe, D. An experimental confirmation of longitudinal electrodynamic forces. *Eur. Phys. J. D At. Mol. Opt. Plasma Phys.* **2001**, *15*, 87–97. [CrossRef]
101. Kühn, S. Experimental investigation of an unusual induction effect and its interpretation as a necessary consequence of Weber electrodynamics. *J. Electr. Eng.* **2021**, *72*, 366–373. [CrossRef]
102. Einstein, A. Zur elektrodynamik bewegter körper. *Ann. Der Phys.* **1905**, *354*, 769–822. [CrossRef]
103. Frisch, D.H.; Wilets, L. Development of the Maxwell-Lorentz equations from special relativity and Gauss's law. *Am. J. Phys.* **1956**, *24*, 574–579. [CrossRef]
104. Field, J. Derivation of the Lorentz force law, the magnetic field concept and the Faraday–Lenz and magnetic Gauss laws using an invariant formulation of the Lorentz transformation. *Phys. Scr.* **2006**, *73*, 639. [CrossRef]
105. Dodig, H. Direct derivation of Lienard–Wiechert potentials, Maxwell's equations and Lorentz force from Coulomb's law. *Mathematics* **2021**, *9*, 237. [CrossRef]
106. Smith, R.T.; Jjunju, F.P.; Maher, S. Evaluation of Electron Beam Deflections across a Solenoid Using Weber-Ritz and Maxwell-Lorentz Electrodynamics. *Prog. Electromagn. Res.* **2015**, *151*, 83–93. [CrossRef]
107. Smith, R.T.; Maher, S. Investigating Electron Beam Deflections by a Long Straight Wire Carrying a Constant Current Using Direct Action, Emission-Based and Field Theory Approaches of Electrodynamics. *Prog. Electromagn. Res.* **2017**, *75*, 79–89. [CrossRef]
108. Baumgärtel, C.; Smith, R.T.; Maher, S. Accurately predicting electron beam deflections in fringing fields of a solenoid. *Sci. Rep.* **2020**, *10*, 10903. [CrossRef] [PubMed]
109. Baumgärtel, C.; Maher, S. A Charged Particle Model Based on Weber Electrodynamics for Electron Beam Trajectories in Coil and Solenoid Elements. *Prog. Electromagn. Res. C* **2022**, *123*, 151–166. [CrossRef]
110. Li, Q. Electric Field Theory Based on Weber's Electrodynamics. *Int. J. Magn. Electromagn.* **2021**, *7*, 1–6.
111. Kühn, S. The Inhomogeneous Wave Equation, Liénard-Wiechert Potentials, and Hertzian Dipole in Weber Electrodynamics. 2022. Available online: https://www.researchgate.net/publication/358747323_The_inhomogeneous_wave_equation_Lienard-Wiechert_potentials_and_Hertzian_dipole_in_Weber_electrodynamics (accessed on 5 July 2022). [CrossRef]
112. Rambaut, M. Macroscopic non-relativistic Ampère EM interactions between current elements reflect the conducting electron accelerations by the ion's electric fields. *Phys. Lett. A* **1991**, *154*, 210–214. [CrossRef]

113. Weber, W.; Kohlrausch, R. Ueber die Elektricitätsmenge, welche bei galvanischen Strömen durch den Querschnitt der Kette fliesst. *Ann. Phys.* **1856**, *175*, 10–25. [CrossRef]
114. Assis, A.K. On the first electromagnetic measurement of the velocity of light by Wilhelm Weber and Rudolf Kohlrausch. In *Volta and the History of Electricity*; Università degli Studi di Pavia and Editore Ulrico Hoepli: Milano, Italy, 2003; pp. 267–286.
115. Kirchhoff, G. Ueber die Bewegung der Elektricität in Leitern. *Ann. Phys.* **1857**, *178*, 529–544. [CrossRef]
116. Weber, W. *Elektrodynamische Maassbestimmungen Insbesondere über Elektrische Schwingungen*; Teubner: Leipzig, Germany, 1864; Volume 6, pp. 105–241.
117. Assis, A.K.T.; Hernandes, J. Telegraphy equation from Weber's electrodynamics. *IEEE Trans. Circuits Syst. II Express Briefs* **2005**, *52*, 289–292. [CrossRef]
118. Fukai, J. *A Promenade Along Electrodynamics*; Vales Lake Publishing: Pueblo West, CO, USA, 2003.
119. Assis, A.K.T. Circuit theory in Weber electrodynamics. *Eur. J. Phys.* **1997**, *18*, 241. [CrossRef]
120. Bueno, M.; Assis, A.K.T. Equivalence between the formulas for inductance calculation. *Can. J. Phys.* **1997**, *75*, 357–362. [CrossRef]
121. McDonald, K.T. Weber's Electrodynamics and the Hall Effect. 2020; pp. 1–15. Available online: http://kirkmcd.princeton.edu/examples/weber.pdf (accessed on 5 July 2022).
122. Smith, R.T.; Taylor, S.; Maher, S. Modelling electromagnetic induction via accelerated electron motion. *Can. J. Phys.* **2014**, *93*, 802–806. [CrossRef]
123. Smith, R.T.; Jjunju, F.P.; Young, I.S.; Taylor, S.; Maher, S. A physical model for low-frequency electromagnetic induction in the near field based on direct interaction between transmitter and receiver electrons. *Proc. R. Soc. A Math. Phys. Eng. Sci.* **2016**, *472*, 20160338. [CrossRef]
124. Assis, A.K.; Thober, D.S. Unipolar induction and Weber's electrodynamics. In *Frontiers of Fundamental Physics*; Springer: Berlin, Germany, 1994; pp. 409–414.
125. Baumgärtel, C.; Maher, S. A Novel Model of Unipolar Induction Phenomena Based on Direct Interaction Between Conductor Charges. *Prog. Electromagn. Res.* **2021**, *171*, 123–135. [CrossRef]
126. Baumgärtel, C.; Maher, S. Resolving the Paradox of Unipolar Induction—New Experimental Evidence on the Influence of the Test Circuit. *Sci. Rep.* **2022**, *12*, 16791. [CrossRef]
127. Härtel, H. Unipolar Induction-a Messy Corner of Electromagnetism. *Eur. J. Phys. Educ.* **2020**, *11*, 47–59.
128. Assis, A.K.T.; Fukai, J.; Carvalho, H. Weberian induction. *Phys. Lett. A* **2000**, *268*, 274–278. [CrossRef]
129. Feynman, R. *Lectures in Physics*; Chapter 17 The Laws of Induction; California Institute of Technology: Pasadena, CA, USA, 1963; Volume II.
130. Ma, T.C.E. Field angular momentum in Feynman's disk paradox. *Am. J. Phys.* **1986**, *54*, 949–950. [CrossRef]
131. Prytz, K.A. Meissner effect in classical physics. *Prog. Electromagn. Res.* **2018**, *64*, 1–7. [CrossRef]
132. Assis, A.; Tajmar, M. Superconductivity with Weber's electrodynamics: The London moment and the Meissner effect. *Ann. Fond. Louis Broglie* **2017**, *42*, 307.
133. Wesley, J. Induction Produces Aharonov-Bohm Effect. *Apeiron* **1998**, *5*, 73–82.
134. Prytz, K. Antenna Theory—The Loop and the Dipole. In *Electrodynamics: The Field-Free Approach*; Springer: Berlin, Germany, 2015; pp. 219–238.
135. Assis, A.K.T.; Wiederkehr, K.H.; Wolfschmidt, G. *Weber's Planetary Model of the Atom*; Tredition Science: Hamburg, Germany, 2011.
136. Weber, W.E. *Electrodynamische Maassbestimmungen*; S. Hirzel: Leipzig, Germany, 1871; Volume 2.
137. Bush, V. The force between moving charges. *J. Math. Phys.* **1926**, *5*, 129–157. [CrossRef]
138. Clemente, R.; Assis, A.K.T. Two-body problem for Weber-like interactions. *Int. J. Theor. Phys.* **1991**, *30*, 537–545. [CrossRef]
139. Riecke, E. Ueber Molecularbewegung zweier Theilchen, deren Wechselwirkung durch das webersche Gesetz der elektrischen Kraft bestimmt wird. In *Nachrichten von der Königl. Gesellschaft der Wissenschaften und der Georg-Augusts-Universität zu Göttingen*; Dieterich: Göttingen, Germany, 1874; p. 665
140. Lolling, G. Ueber Bewegungen elektrischer Theilchen nach dem Weber'schen Grundgesetz der Elektrodynamik. In *Verhandlungen der Kaiserlichen Leopoldinisch-Carolinischen Deutschen Akademie der Naturforscher*; Kaiserlich-Leopoldinisch-Carolinische Deutsche Akademie der Naturforscher: Halle, Germany, 1882; Volume 44, pp. 273–336.
141. Ritter, E., II. Bewegung eines materiellen mit Eletricität geladenen Theilchens unter der Einwirkung eines ruhenden Centrum bei Giltigkeit des Weber'schen Gesetzes. *Z. Math. Und Phys. Organ Angew. Math.* **1892**, *37*, 8–24.
142. Zöllner, J.F. *Principien einer Elektrodynamischen Theorie der Materie*; Engelmann: Leipzig, Germany, 1876; Volume 1.
143. Wesley, J.P. Evidence for Weber–Wesley electrodynamics. In Proceedings of the Conference on Foundations of Mathematics and Physics, Perugia, Italy, 27–29 September 1989; pp. 27–29.
144. Post, E.J. A Tale of Fine Structure Coincidences. In *Quantum Reprogramming*; Springer: Berlin, Germany, 1995; pp. 151–168.
145. Pearson, J.; Kilambi, A. Velocity-dependent nuclear forces and Weber's electrodynamics. *Am. J. Phys.* **1974**, *42*, 971–975. [CrossRef]
146. Assis, A.K.T. Deriving gravitation from electromagnetism. *Can. J. Phys.* **1992**, *70*, 330–340. [CrossRef]
147. Assis, A.K.T. Gravitation as a fourth order electromagnetic effect. In *Advanced Electromagnetism: Foundations, Theory and Applications*; Barret, T.W., Gales, D.M., Eds.; World Scientific Publishers: Singapore, 1995; pp. 314–331.
148. Tajmar, M. Derivation of the Planck and Fine-Structure Constant from Assis's Gravity Model. *J. Adv. Phys.* **2015**, *4*, 219–221. [CrossRef]

149. Baumgärtel, C.; Tajmar, M. The Planck Constant and the Origin of Mass Due to a Higher Order Casimir Effect. *J. Adv. Phys.* **2018**, *7*, 135–140. [CrossRef]
150. Tisserand, F. *Sur le Mouvement des Planètes Autour du Soleil, D'après la loi Électrodynamique de Weber*; Gauthier-Villars: Paris, France, 1872.
151. Gerber, P. Die räumliche und zeitliche Ausbreitung der Gravitation. *Z. Math. Phys.* **1898**, *43*, 415–444.
152. Schrödinger, E. Die Erfüllbarkeit der Relativitätsforderung in der klassischen Mechanik. *Ann. Phys.* **1925**, *382*, 325–336. [CrossRef]
153. Assis, A.K.T. On Mach's principle. *Found. Phys. Lett.* **1989**, *2*, 301–318. [CrossRef]
154. Tajmar, M.; Assis, A.K.T. Influence of Rotation on the Weight of Gyroscopes as an Explanation for Flyby Anomalies. *J. Adv. Phys.* **2016**, *5*, 176–179. [CrossRef]
155. Bunchaft, F.; Carneiro, S. Weber-like interactions and energy conservation. *Found. Phys. Lett.* **1997**, *10*, 393–401. [CrossRef]
156. Mendes, R.; Malacarne, L.; Assis, A. Virial Theorem For Weber's Law. In *Has the Last Word Been Said on Classical Electrodynamics? New Horizons*; Chubykalo, A., Espinoza, A., Onoochin, V., Smirnov-Rueda, R., Eds.; Rinton Press: Paramus, NJ, USA, 2004; pp. 67–70.
157. Clemente, R.A.; Cesar, R.G. Cold plasma oscillations according to Weber's law. An unphysical result. *Int. J. Theor. Phys.* **1993**, *32*, 1257–1260. [CrossRef]
158. Assis, A.K.T. *Relational Mechanics and Implementation of Mach's Principle with Weber's Gravitational Force*; C. Roy Keys Inc.: Montreal, QC, Canada, 2014.
159. Seegers, C. *Über die Bewegung und die Störungen der Planeten, Wenn Dieselben Sich Nach dem Weberschen Elektrodynamischen Gesetz um die Sonne Bewegen*; Springer: Berlin, Germany, 2013.
160. Servus, H. Untersuchungen über die Bahn und die Störungen der Himmelskörper mit Zugrundelegung des Weber'schen Electrodynamischen Gesetzes. Ph.D. Thesis, Universität Halle-Wittenberg, Halle, Germany, 1885.
161. Zöllner, J.K.F. *Über die Natur der Cometen: Beiträge zur Geschichte und Theorie der Erkenntniss*; Engelmann: Leipzig, Germany, 1872.
162. Eby, P. On the perihelion precession as a Machian effect. *Nuovo C. Lett.* **1977**, *18*, 93–96. [CrossRef]
163. Ragusa, S. Gravitation with a modified Weber force. *Found. Phys. Lett.* **1992**, *5*, 585–589. [CrossRef]
164. Lévy, M. Sur l'application des lois électrodynamiques au mouvement des planétes. *Comput. Rend. Acad. Sci.* **1890**, *110*, 545.
165. Giné, J. Is gravitational quantization another consequence of General Relativity? *Chaos Solitons Fractals* **2009**, *42*, 1893–1899. [CrossRef]
166. Kholodenko, A. Newtonian Limit of Einsteinian Gravity and Dynamics of Solar System; In *Relativity, Gravitation, Cosmology: Foundations*; Nova Science Publishers: Hauppauge, NY, USA, 2015; pp. 101–152.
167. Tiandho, Y. Weber's gravitational force as static weak field approximation. In *AIP Conference Proceedings*; AIP Publishing LLC: Melville, NY, USA, 2016; Volume 1708, p. 070012.
168. Lima, F. Nonzero Gravitational Force Exerted by a Spherical Shell on a Body Moving Inside It, and Cosmological Implications. *Gravit. Cosmol.* **2020**, *26*, 387–398. [CrossRef]
169. Das, S. Mach Principle and a new theory of gravitation. *arXiv* **2012**, arXiv:1206.6755.
170. Roscoe, D. Mach'S Principle and Instantaneous Action at a Distance. In *Instantaneous Action at a Distance in Modern Physics: "Pro" and "Contra"*; Andrew E. Chubykalo, V.P., Smirnov-Rueda, R., Eds.; Nova Science Publishers, Inc.: Commack, NY, USA, 1999; pp. 175–188.
171. Assis, A. A steady-state cosmology. In *Progress in New Cosmologies*; Springer: Berlin, Germany, 1993; pp. 153–167.
172. Assis, A.K.; Graneau, P. Nonlocal forces of inertia in cosmology. *Found. Phys.* **1996**, *26*, 271–283. [CrossRef]
173. Graneau, P.; Graneau, N. Machian inertia and the isotropic universe. *Gen. Relativ. Gravit.* **2003**, *35*, 751–770. [CrossRef]
174. Tajmar, M.; Assis, A. Gravitational induction with Weber's force. *Can. J. Phys.* **2015**, *93*, 1571–1573. [CrossRef]
175. Assis, A.K.T. Changing the inertial mass of a charged particle. *J. Phys. Soc. Jpn.* **1993**, *62*, 1418–1422. [CrossRef]
176. Tajmar, M. Propellantless propulsion with negative matter generated by electric charges. In Proceedings of the AIAA Joint Propulsion Conference, AIAA-2013, San Jose, CA, USA, 14–17 July 2013; Volume 3913.
177. Tajmar, M.; Assis, A.K.T. Particles with negative mass: Production, properties and applications for nuclear fusion and self-acceleration. *J. Adv. Phys.* **2015**, *4*, 77–82. [CrossRef]
178. Tajmar, M. Revolutionary Propulsion Research at TU Dresden. In Proceedings of the Advanced Propulsion Workshop, Estes Park, CO, USA, 19–22 September 2016; pp. 20–22.
179. Mikhailov, V. The Action of an Electrostatic Potential on the Electron Mass. *Ann. Fond. Louis Broglie* **1999**, *24*, 161.
180. Mikhailov, V.F. Influence of an electrostatic potential on the inertial electron mass. *Ann. Fond. Louis Broglie* **2001**, *26*, 633–638.
181. Mikhailov, V. Influence of a field-less electrostatic potential on the inertial electron mass. *Ann. Fond. Louis Broglie* **2003**, *28*, 231.
182. Junginger, J.E.; Popovic, Z.D. An experimental investigation of the influence of an electrostatic potential on electron mass as predicted by Weber's force law. *Can. J. Phys.* **2004**, *82*, 731–735. [CrossRef]
183. Lőrincz, I.; Tajmar, M. Experimental Investigation of the Influence of Spatially Distributed Charges on the Inertial Mass of Moving Electrons as Predicted by Weber's Electrodynamics. *Can. J. Phys.* **2017**, *95*, 1023. [CrossRef]
184. Weikert, M.; Tajmar, M. Investigation of the Influence of a field-free electrostatic Potential on the Electron Mass with Barkhausen-Kurz Oscillation. *arXiv* **2019**, arXiv:1902.05419.
185. Tajmar, M.; Weikert, M. Evaluation of the Influence of a Field-Less Electrostatic Potential on Electron Beam Deflection as Predicted by Weber Electrodynamics. *Prog. Electromagn. Res. M* **2021**, *105*, 1–8. [CrossRef]

186. Assis, A. On the unification of forces of nature. *Ann. Fond. Louis Broglie* **2002**, *27*, 149–161.
187. Fricke, H. Two Rival Programmes in 19th. Century Classical Electrodynamics Action-at-a-Distance versus field Theories. Ph.D. Thesis, The London School of Economics and Political Science (LSE), London, UK, 1982.
188. Clausius, R. Ueber die Ableitung eines neuen elektrodynamischen Grundgesetzes. *J. Reine Angew. Math.* **1877**, *82*, 85.
189. Woodruff, A.E. Action at a distance in nineteenth century electrodynamics. *Isis* **1962**, *53*, 439–459. [CrossRef]
190. Zöllner, F. Ueber die Einwendungen von Clausius gegen das Weber'sche Gesetz. *Ann. Phys.* **1877**, *236*, 514–537. [CrossRef]
191. Erman, P. Über die electroskopischen Phänomene des Gasapparats an der Voltaischen Säule. *Ann. Phys.* **1802**, *10*, 1–23. [CrossRef]
192. Assis, A.K.T.; Hernandes, J.A. *The Electric Force of a Current: Weber and the Surface Charges of Resistive Conductors Carrying Steady Currents*; Apeiron, C. Roy Keys Inc.: Montreal, QC, Canada, 2007.
193. Hernandes, J.; Assis, A.K.T. Electric potential due to an infinite conducting cylinder with internal or external point charge. *J. Electrost.* **2005**, *63*, 1115–1131. [CrossRef]
194. Assis, A.K.T.; Rodrigues, W.; Mania, A. The electric field outside a stationary resistive wire carrying a constant current. *Found. Phys.* **1999**, *29*, 729–753. [CrossRef]
195. Caluzi, J.; Assis, A.K.T. A critical analysis of Helmholtz's argument against Weber's electrodynamics. *Found. Phys.* **1997**, *27*, 1445–1452. [CrossRef]
196. Phipps, T.E. Toward modernization of Weber's force law. *Phys. Essays* **1990**, *3*, 414. [CrossRef]
197. Caluzi, J.; Assis, A.K.T. An analysis of Phipps's potential energy. *J. Frankl. Inst.* **1995**, *332*, 747–753. [CrossRef]
198. Li, Q. Extending Weber's Electrodynamics to High Velocity Particles. *Int. J. Magn. Electromagn.* **2022**, *8*, 1–9.
199. Assis, A.K.T. Arguments in favour of action at a distance. In *Instantaneous Action at a Distance in Modern Physics: "Pro" and "Contra"*; Chubykalo, A.E., Pope, V., Smimov-Rueda, R., Eds.; Nova Science Publishers: Commack, NY, USA, 1999; pp. 45–56.
200. Haug, E.G. Demonstration that Newtonian gravity moves at the speed of light and not instantaneously (infinite speed) as thought! *J. Phys. Commun.* **2021**, *5*, 025005. [CrossRef]
201. Brown, G.B. *Retarded Action-at-a-Distance: The Change of Force with Motion*; Cortney Publications: Luton, Bedfordshire, UK, 1982.
202. Lentze, G. Dialogue concerning magnetic forces. *Stud. Hist. Philos. Sci. Part B Stud. Hist. Philos. Mod. Phys.* **2019**, *68*, 158–162. [CrossRef]
203. Assis, A.K.T. Weber's law and mass variation. *Phys. Lett. A* **1989**, *136*, 277–280. [CrossRef]
204. Wesley, J.P. Empirically correct electrodynamics. *Found. Phys. Lett.* **1997**, *10*, 189–204. [CrossRef]
205. Assis, A.K.T.; Caluzi, J. A limitation of Weber's law. *Phys. Lett. A* **1991**, *160*, 25–30. [CrossRef]
206. Montes, J. On Limiting Velocity with Weber-like Potentials. *Can. J. Phys.* **2017**, *95*, 770–776. [CrossRef]
207. Assis, A.K.T.; Clemente, R. The ultimate speed implied by theories of Weber's type. *Int. J. Theor. Phys.* **1992**, *31*, 1063–1073. [CrossRef]
208. Kühn, S. Analysis of a stochastic emission theory regarding its ability to explain the effects of special relativity. *J. Electromagn. Anal. Appl.* **2020**, *12*, 169–187. [CrossRef]
209. Sherwin, C.W. The Weber Radar. *Phys. Essays* **1991**, *4*, 417–419. [CrossRef]
210. Barrett, P.; Jones, H.; Franklin, R. Dispersion of electron plasma waves. *Plasma Phys.* **1968**, *10*, 911. [CrossRef]
211. Wesley, J. Weber electrodynamics extended to include radiation. *Specul. Sci. Technol.* **1987**, *10*, 50–53.
212. Phipps, T. Electromagnetic Force Laws and Hot Fusion. *Galilean Electrodyn.* **2016**, *27*, 79–80.
213. Edwards, W.F.; Kenyon, C.; Lemon, D. Continuing investigation into possible electric fields arising from steady conduction currents. *Phys. Rev. D* **1976**, *14*, 922. [CrossRef]
214. Jefimenko, O. Demonstration of the electric fields of current-carrying conductors. *Am. J. Phys.* **1962**, *30*, 19–21. [CrossRef]
215. Jefimenko, O.D. *Electricity and Magnetism: An Introduction to the Theory of Electric and Magnetic Fields*, 2nd ed.; Electret Scientific Company: Star City, WV, USA, 1989.
216. Zhao, Y. A Novel Two-Particle Steady-State Wave Equation. *IRanian J. Sci. Technol. Trans. A Sci.* **2022**, **46**, 1019–1026. [CrossRef]

foundations

Review

The Role of Axions in the Formation of the Photoluminescence Spectrum in Dispersive Media

Valeriy Evgenjevich Ogluzdin

Federal Research Center A. M. Prokhorov Institute of General Physics of the Russian Academy of Science, 119991 Moscow, Russia; ogluzdin@kapella.gpi.ru

Abstract: In the review, based on the analysis of the results published in the works of domestic and foreign researchers, a variant of an unconventional interpretation of the photoluminescence of dispersive media in the energy range of 0.5–3 eV is proposed. The interpretation meets the requirements of the energy conservation law for photons and axions participating in the photoluminescence process. The participation of axions in the process is consistent with Primakov's hypothesis. The role of nonradiative relaxation at the stage of axion decay is noted. The axion lifetimes are estimated for a number of dispersive media.

Keywords: photoluminescence; dispersive media; axion

Citation: Ogluzdin, V.E. The Role of Axions in the Formation of the Photoluminescence Spectrum in Dispersive Media. *Foundations* **2022**, 2, 184–198. https://doi.org/10.3390/foundations2010011

Academic Editor: Eugene Oks

Received: 14 November 2021
Accepted: 4 January 2022
Published: 29 January 2022

1. Introduction

The review is devoted to photoluminescence (PL) and the axion problem [1–5]. The processes of interaction of light, usually laser monochromatic radiation with various media in the visible and near infrared regions of the spectrum (0.5–3 eV), are considered. The appearance as a result of the light exposure of a broadband PL at the output of the dispersive medium under study, in the author's opinion, can serve as evidence of the presence of axions in such processes. According to the existing theoretical concepts, the circumstance confirming the possibility of the presence of axions in the specified spectral region can be their decay, leading to the appearance of radiation at the output of the environment under study at frequencies that are absent at the entrance to the medium.

In this paper, the definition of an axion is based on Primakov's hypothesis [6,7], according to which the fusion of two primary photons in a strong electromagnetic field of an atomic nucleus can lead to the birth of a pseudoscalar particle (axion) and vice versa, a pseudoscalar particle (axion, neutral pion) can decay into two secondary photons—annihilation (forward and reverse) processes. The light beam, falling into the dispersion medium, loses laminarity due to the presence of atoms of the alloying element. When photons penetrate into the atom into the region of the nucleus, a meeting, a collision of photons becomes inevitable. The strength of the axion–photon interaction is determined by the energy characteristics of the outer and inner electron shells of an atom. If for the outer electron shells, the value of the binding energy of electrons is from several eV to several tens of eV, then for the inner shells of an atom, the order of this value is 102–104 eV. For the photon energy range that I am considering (0.5–3 eV), the interaction force turns out to be sufficient for the generation of axions.

For reference, we point out that in theoretical physics that there is no consensus on the nature of the axion. At this stage, all that matters to us is that the mass of the axion is small. This fact corresponds to the statement: "In the theory modified within the framework of the Grand Unification axion must be a particle of small mass" [4], apparently comparable in size to the small mass of a moving photon. Therefore, it becomes possible to extract information about the presence of axions in certain optical processes initiated by photons, which we will be guided by in the next part of the review.

This work was written by an experimenter. In our work, we do not limit ourselves to analyzing our own results. The availability of published experimental results obtained by many research groups in the visible and near infrared regions of the spectrum (0.5–3 eV) allows them to be compared and analyzed. In this paper, I discuss physical processes, mechanisms, and models that illustrate and confirm the position put forward by the author. After considering the materials included in the review, the readers will be offered a scheme according to which the axion lifetime estimates for a number of objects are estimated from published sources [1,7].

The traditional approach to the description of optical phenomena using wave representations [8] does not allow for the full interpretation of the results observed at the junction of optics and elementary particle physics. The involvement of methods used in the physics of the quantum world can help solve the problem. If we turn to photons (quanta), which successfully moved from particle physics to optics with the advent of lasers, then the connection between optics and the physics of the quantum world turns out to be real. This fact will help us get closer to finding out the nature of the axiom.

Here, it is appropriate to recall the generation of harmonics of optical radiation, parametric light generation. The authors of [9], the subtitle of which is "electromagnetic waves in nonlinear dispersive media (DM), argued that when using the quantum interpretation, the birth of harmonics and the appearance of combinational frequencies indicates the essential role of the processes of fusion or division of photons (light quanta), noting the need to take into account the laws of conservation of energy and momentum in nonlinear optical transformations. When the second harmonic is generated, two pump radiation photons merge in an elementary act. In a parametric light generator [10], the decay of a pump photon in an elementary act ensures the appearance of two photons at the output of a nonlinear medium: a signal and an idle one.

The purpose of this review is as follows: based on the analysis of experimental data published by foreign and Russian researchers, to present a scheme, a model reflecting the contribution of axions to the PL process. The analysis of the results obtained in the study of PL, according to the author [1,7,8], admits the possibility of the presence of axions at the stage of interaction of pumping radiation with atoms used for doping the medium. As a rule, the medium under study is a dispersive medium (glass or crystal) in which the pump radiation is transformed, leading to the appearance of PL quanta. The process of conversion of pumping radiation is associated with the birth of axions and their decay. The energy of the PL quanta, as a rule, is different from the energy of the pump radiation quanta.

In our review, we will deal with a DM. The continuous phase DM consists of glass, crystal structure, and liquid. The second component of DM distributed (suspended) in the volume of the continuous phase consists of atoms or nanoparticles of alloying material: atoms of holmium, bismuth or silicon. That is to say, this phase consists of electrons and nucleis of atoms of the alloying material.

From chemistry, it is necessary to also recall the complex environment [11]. The complex environment is a substance composed of complex particles (atoms or nanoparticles) capable of independent or connected existence in a liquid, amorphous (glass), or crystalline environment, acting as a matrix. The complex particle itself, in turn, can be heformed from other, simpler particles (nuclei, surrounded by electrons). In optics, it is natural to talk about a dispersive medium—DM, a medium that is associated with the phenomenon of light dispersion.

Returning to the problem we are interested in, we inform you that the review will consider PL in DM, that is, in glasses doped with bismuth and holmium atoms, in crystal samples doped with holmium, and in PL suspended in ethanol silicon nanoparticles.

2. The Bohr Frequency and Its Relation to the Lorentz Harmonic Oscillator Model

In this section, we will consider the conditions for the propagation of a light beam in DM. In particular, these are the reasons that change the laminar propagation of a light beam to a turbulent one. This circumstance can change the phase velocity of photon propagation in the DM, which leads to the interaction of primary photons with each other. Multiphoton nonlinear optical processes are considered as an example of such interaction.

2.1. Niels Bohr's Postulates

When considering the atom model, Niels Bohr proposed the postulates of quantization. Following [12,13], we recall their contents. The first postulate is that there are a certain number of electron orbits in an atom, which they call stationary states (levels). Moreover, in each of these states, an electron can stay indefinitely without radiating at all. This position was postulated arbitrarily and obviously contradicted the classical electromagnetic theory. Second, he postulated that an electron could be knocked out by a blow from the lower stationary, normal state E_1 to an energetically higher E_2, which Bohr called excited. The transition between these states is characterized by the Bohr frequency—ν_{21} [14], the value of which is determined from Equation (1)

$$\nu_{21} = (E_2 - E_1)/h, \quad (1)$$

h is Planck's constant, and $(E_2 - E_1) = \Delta E$ is the energy of the electron transition between levels. Such relations can link many levels together in pairs, and the role of the lower state can be assigned not only to the lower level but also to the higher ones. The Equation (1) in this case should have the following form:

$$\nu_{mp} = (E_m - E_p)/h \quad (2)$$

i corresponds to the level located below, and j corresponds to the level, located above. Thus, $m = 1,2,3 \ldots$, and $p = i + 1,2,3 \ldots$

Bohr's ideas proved useful for explaining the structure of optical spectra and found application in practical spectroscopy.

The next part of the review is devoted to the consideration of the processes of interaction of a monochromatic radiation beam with resonant and almost resonant media, which link the frequency of exciting radiation at the entrance to the medium to the frequencies of the medium levels responsible for scattered radiation at the exit of the medium under study with Bohr's frequencies. This circumstance occurs, as a rule, in nonresonant conditions, when the detuning of the pump radiation frequency relative to the frequency of the electronic transition in the ν_{mn} atom does not exceed 10–20 cm^{-1}. When considering the PL process, in various environments with volumetric placement of multidimensional oscillators, and, consequently, having a large set of Bohr frequencies, the proposed approach turns out to be useful for practical use.

As a rule, on spectrograms, the line corresponding to the Bohr frequency ν_{ij} is not observed at the output of DM.

2.2. The Bohr Frequency and Its Relation to the Lorentz Harmonic Oscillator Model

The Bohr frequency is a characteristic of the medium that appeared with the birth of quantum mechanics. Note that an expression similar to Equations (1) and (2 bnvvm) has been used in classical physics for a long time to describe the characteristics of a harmonic oscillator. We have to figure out how the refractive index n of a medium consisting of identical and different-sized oscillators behaves depending on the frequency. To do this, we should refer to the works that used a model in which the oscillations of electrons in an atom near the equilibrium position are considered [15–19] and compare with the work in which the author deals with the Bohr frequency [14]. In the first case, the book "Fundamentals of Optics" by M. Born and E. Wolf can help us [20]. Following to her we note that "the phase velocity V and, consequently, the refractive index $n(\nu)$ cannot be constant in the volume of the entire medium under study." These values depend on the detuning value of the pump frequency n relative to the Bohr frequency ν_{ij}. E. Fermi pointed out that the refractive index $n(\nu)$ and the phase velocity V are not quantities that have a constant value in the medium under study [14]. For the classical harmonic Lorentz oscillator, according to [16] we have the Equation (2):

$$\nu_{mp} = (E_m - E_p)/h,$$

ν_{mp} is the natural frequency of the electron oscillations near the stable equilibrium position, E_m is the value of the electron energy at the "m" level, and E_p is the value of the electron energy at the "p" level [16]. Planck's constant $h = 6.6252 \times 10^{-27}$ erg × s.

According to Fermi [14], a similar ratio has the form:

$$\nu_{mp} = (E^{(m)} - E^{(p)})/\hbar \tag{3}$$

ν_{mp} is the Bohr transition frequency, $E^{(m)}$ is the energy of an electron occupying level "m", and $E^{(p)}$ is the energy of an electron occupying level "p". In this case, $\hbar = h/2\pi = 1.0544 \times 10^{-27}$ erg × s.

It is essential for us that, regardless of the type of model, the refractive index $n(\nu)$ depends on the pumping frequency.

Note that when photons pass through the electron shell surrounding the nucleus, both the magnitude and the sign of the pump frequency detuning relative to the frequency of the transition under study have an effect on the magnitude of their velocity.

If, for a harmonic oscillator, the refractive index $n(\nu)$ is greater than 1 (the pumping frequency ν is less than the Bohr frequency ν_{mp}, then the photon velocity slows down. If the value of the refractive index $n(\nu)$ is greater than the Bohr frequency ν_{mp}, then photons are reflected toward the pump light beam, which assists in the addition of photons, leading to the birth of an axion.

The results of consideration of the question of the relationship between the frequencies of the medium, its refractive index, and the characteristics of the radiation scattered by the medium are considered in detail in [21–26] on the example of atomic potassium vapors. Below, we will see that the considered dependencies will bring us closer to solving the problem: finding out the nature of the PL, the axion, and its place in the physical picture of the world. The Bohr frequencies are a real characteristic of the medium under study. The numerical values of spectral lines included in reference books and atlases are simple coincide with the Bohr frequencies of interlevel transitions in atoms. This position is valid for linear to laser optics. For high-power laser radiation used in experiments of this kind, shifts of electronic levels in atoms may occur [24], which naturally affects the spectral characteristics of scattered radiation and PL radiation at the outlet of the medium. As a rule, the level shift is associated with multiphoton processes. For example, in atomic pairs of potassium, these are three-photon electron Raman scattering [22] and six-photon parametric scattering (SPR) [25,26].

The frequency of the three-photon electron Raman ν_3 in a two-level medium can be calculated in accordance with the law of conservation of energy:

$$\nu_3 = 2\nu - \nu_{21} \tag{4}$$

where the multiplier h is omitted, and the frequency ν_{21} corresponds to the tabular value of the frequency of the interlevel transition.

Equation (4) does not take the shift of the levels of the medium under study in the pumping field into account; however, it allows comparing experimental data and tabular values of wavelengths or frequencies of interlevel transitions.

The results of experiments related to the three-photon electron Raman process can be found in [22–26].The similar results were obtain also in work [27]. Theoretical estimates of the three-photon process in two-level media are made in the book by Pantel R. and Puthof G. [18]. For further consideration, it should be noted that in the elementary act of three-photon Raman scattering Equation (4), two pump radiation photons participate. This fact will help us in solving the problem of PL and axion.

2.3. Starting the Analysis of the PL Spectra in the DM

In this section, we need to find out what happens to the radiation at the output of the DM under study, in which the atoms of the element alloying the medium are "weighted", while not forgetting that the atoms are nuclei surrounded by electron shells. Namely, the electrons of the shells determine the behavior of photons in the medium, and the behavior of the refractive index $n(\nu)$ DM near the resonance transition region is determined by electrons. Resonant transition and harmonic oscillator? According to Niels Bohr, there is a correlation of complementarity between "the unambiguous application of stationary states and the mechanical analysis of intra-atomic motions … " [12,13].

Taking this into account, we combine the ideas about the interlevel transitions of electrons of doping DM atoms with the oscillations of different-frequency harmonic oscillators. Namely, the electrons of the atoms doping the medium, responsible for the optical properties of the medium, are an ensemble of identical or different-frequency classical oscillators. Above, we have partially discussed the consequences of this circumstance.

For the future, we must consider three cases concerning the processes of interaction of the pump radiation ν with DM, namely:

$$(a)\ \nu = \nu_{21}, n(\nu) \cong 1;$$
$$(b)\ \nu > \nu_{21}, n(\nu) > 1;$$
$$(c)\ \nu < \nu_{21}, n(\nu) > 1,$$

where $n(\nu)$ is the refractive index of the medium at the frequency of the pump radiation ν; ν_{21}—the Bohr frequency resonant transition.

According to the theory of dispersion for the case:

(a) when $\nu = \nu_{21}$, the refractive index $n(\nu)$ of the medium is close to unity [15–18]. The reflection coefficient of such a medium increases as the excitation frequency ν approaches the Bohr frequency ν_{21}, reaching a maximum value at $\nu = \nu_{21}$ [16]. A qualitative illustration of this circumstance is Wood's experience in observing resonant radiation in the case of atomic sodium vapors placed in a cuvette illuminated by a sodium lamp. This experience practically illustrates the case of exposure to a two-level medium of photons whose frequency ν is equal to the Bohr frequency ($\nu = \nu_{21}$).

In fact, the scheme of forced (induced) photon emission proposed by A. Einstein in 1916 was implemented. Indeed, in a resonant two-level medium, in the elementary act of photon absorption, an electron leaves the ground level to the excited level.

By the next photon of the beam, electron can be knocked out from the excited level, i.e., we have:

$$h\nu_{21} + h\nu = 2h\nu \tag{5}$$

Here, $\nu = \nu_{21}$, where $\nu_{21} = (E_2 - E_1)/h$ is the transition frequency of the investigated duplex environment; the value of $h\nu_{21}$—energy of an electron in the excited level ($h\nu_{12} = h\nu_{21}$).

The left part of the Equation (5) open mechanism of enhancement of light emission in the inverted environment by doubling the number of photons, right side of Equation (5) in each elementary act of forced (induced) radiation. This process is implemented in tasks related to the amplification and generation of monochromatic radiation [20,28]. It should be noted that even at the dawn of the development of laser physics, the model of the classical harmonic Lorentz oscillator that we have attracted has already been used to consider the laser generation process [29].

In case (b), when $\nu > \nu_{21}$, the feature limiting the propagation of photons is due to the fact that, according to the theory of dispersion, the refractive index $n(\nu) < 1$. The propagation of photons of monochromatic radiation in this region of the spectrum, generally speaking, is impossible, because otherwise their velocity $V = c/n(\nu)$ would exceed the speed of light c, which contradicts existing concepts. If a high-power laser is used in the experiment, then due to three-photon electron Raman scattering [22] after leveling the populations of the levels of the transition under study, we will get $n(\nu) \cong 1$, which will allow part of the pumping to pass through such a medium. If the intensity distribution across the cross-section of a powerful beam is Gaussian (single-mode laser), then we have self-focusing at the output of such a medium [21].

Finally, consider case (c) $\nu < \nu 21$, where $\nu 21$ is the resonant transition frequency. In this region of the spectrum, the refractive index is n(ν) > 1. According to [15–19], the refractive index n(ν) increases with decreasing detuning in the low-frequency spectral region with respect to the resonance frequency. This case was considered in [22]. Below is a summary of it.

If the condition $\nu \neq \nu 21$ is fulfilled in an elementary act, the energy of two pump photons ν should add up, which forms a kind of "virtual" level in the medium $(\nu + \nu) = 2\,\nu$, the breakdown of which is accompanied by the emission of one light quantum at a new frequency. PL is associated with this process. The fate of the second quantum is nonradiative relaxation (electron leaves "virtual" level for transition on one of real levels of doping atom—heating of the medium) [19].

When using high-power laser radiation with a frequency close to the transition frequency, the populations of the ground and excited levels of the atomic medium can be equalized due to three-photon electron Raman. In this case, according to Equation (4) radiation at the frequency ν_3 is recorded on the spectrograms [22,27].

When summing up the results under point (c) we note, that the process of three-photon electron Raman process is accompanied by a decrease of the refractive index of the medium $n(\nu) \to 1$ at the frequency of exciting radiation, ν.

If the pumping of the medium under study is carried out by a single-mode beam, in the cross-section of which the energy distribution has a Gaussian structure, then the consequence of this is a sagging refractive index on the beam axis. The consequence of this is a deviation from the axis of the particle beam of the pump radiation, which leads to the appearance of a conical structure of the beam at the outlet of the medium [21,22].

With an increase in the power of the exciting radiation, the process of three-photon electron Raman process is supplemented by a six-photon parametric process [25,26] in this case, and a second cone structure can be registered at the outlet of the cuvette with potassium vapor. The conditions of their appearance are discussed in detail. The first reason is the propagation of "superluminal" photons [23]. Naturally, the speed of propagation of these photons does not exceed the speed of light. The second reason is the propagation in a medium with superluminal velocity along the beam axis of nonlinear polarization associated with the six-photon parametric scattering [26].

We have shown that this situation can be explained by the dependence of the refractive index of the medium on the intensity of the pump radiation and the Gaussian intensity distribution over the cross-section of a single-mode beam. The propagation of photons in a medium with a phase velocity $V = c/n(\nu)$ makes them "superluminal" [30]. The results of experiments in atomic vapors of alkali metals in the frequency range of the main doublet can serve as a visual proof of the processes considered.

The propagation of radiation in a two-level medium occurs according to the law of conservation of energy. This applies equally to Raman and parametric multiphoton processes. The same correspondence takes place if we have an environment with a set of different frequency oscillators, which corresponds to DS. In comparison with the considered model of a two-level medium, in this case, due to a significant increase in the number of Bohr frequencies, the spectra of the radiation scattered by the medium become more complicated (broadening, overlapping). Nevertheless, the decoding of spectrograms turns out to be feasible within the framework of the proposed model.

The above information is necessary so that the reader can analyze the nonstandard.

3. Photoluminescence (PL)

In order to expand our understanding of PL, let us first dwell on the definition of this process. PL is luminescence excited by optical radiation [31]. Remaining within the framework of this definition, it is necessary to consider three options: (1) resonant radiation, (2) PL, corresponding to the Stokes rule, and (3) anti-Stokes PL.

The energy conservation law for PL [31,32] can be written in a form convenient for further work:

$$h\nu_{pl} = 2h\nu - h\nu_{ij}, \tag{6}$$

where $h\nu_{pl}$—the energy quanta (photons) PL, $h\nu$—the energy of the pumping quanta (photons)—the light radiation used to excite the PL, $2h\nu$—the energy of "virtual" level, ν—the frequency of pumping radiation, and $h\nu_{ij}$—the energy of electron, expended for nonradiative relaxation.

The index i, j correspond to a set of levels involved in the relaxation process; index i—real level, index j—"virtual" level.

According to (6), each of these transitions corresponds to a component in the PL spectrum at the outlet of the medium. The value of ν_{ij} is a combination of a large number of transition frequencies associated with nonradiative relaxation and transfer of thermal energy to the medium. Since the PL spectrum in DM is usually broadened, it is natural to assume that the PL process is accompanied by a multitude of electronic interlevel transitions responsible for nonradiative relaxation in the atoms of the element used for doping DM. Naturally, the law of conservation of energy is fulfilled for each frequency component of the broadened PL spectrum.

$h\nu_{pl}$—the quantum energy of the PL radiation,

ν_{pl}—the frequencies filling the broadened spectrum of the PL.

The index "*pl*" corresponds to the whole set of frequencies involved in PL. The same number of frequencies of nonradiative transitions correspond to the index "*ij*".

From the Equation (6), it follows that the value of the pump radiation frequency is the arithmetic mean between each pair of frequencies ν_{pl} and ν_{ij}.

Resonant radiation is the simplest case of PL. In this case we have an implementation of the Einstein scheme: (1) absorption of a light quantum, (2) spontaneous emission of a quantum, and (3) stimulated radiation [17,18]. The experiments of the American optician R. Wood, who studied the resonant interaction of light radiation in a cuvette with atomic sodium vapor in the frequency range of the main doublet, were mentioned above. The process of absorption by sodium atoms of quanta falling on the window of a cuvette with atomic vapors is accompanied by the processes of spontaneous and forced resonant radiation.

Note that, in contrast to the resonant case, the PL spectrum at the DM output is, as a rule, broadened. Most often, it is broadened relative to the pumping frequency in the Stokes region of the spectrum. This fact is considered fundamental for the definition of PL. We will return to this issue below, since the processes of PL have not yet, in fact, been fully investigated. This was noted by Paul R.V. [17], referring to the experiments of Vavilov S.I. [33]. Finally, the well-known rule of mirror symmetry between the PL spectrum and the absorption spectrum is uniquely associated with the law of conservation of energy [28]. Indeed, in compliance with Equation (6) from the law of conservation of energy frequencies involved in the PL process, for the Stokes region of the PL spectrum, we have mirror symmetry:

$$\nu_{ij} - \nu = \nu - \nu_{pl} \tag{7}$$

here: $\nu > \nu_{pl}$.

A similar ratio can be written for the anti-Stokes region of the PL spectrum:

$$\nu_{pl} - \nu = \nu - \nu_{ij} \tag{8}$$

here $\nu < \nu_{pl}$.

The Equations (7) and (8) helped to author to understand the PL spectra, obtained by him [34–36], or borrowed from the materials of various authors, in which the results on PL are presented, and to which we will return below. Unfortunately, the accuracy of the frequency estimation at the stage of analysis of the PL spectrograms, published in the journals in the DM could not be high enough. However, this did not prevent the author from establishing coincidences between the tabular values of wavelengths or frequencies responsible for nonradiative relaxation and the values of frequencies or wavelengths calculated from the experimental results.

If a frequency-nontunable laser is used as a pump, then the coincidence of the "virtual" level, whose energy is $2h\nu$ and the real level in the alloying atom is unlikely. That is why we do not have an exact match in the case of PL. In the next section of the paper, the case of an exact match is considered. The section is devoted to studies of PL and laser generation (LG) in holmium-doped media.

The next step is to discuss, within the framework of the above model, a number of experimental results published by domestic and foreign researchers and the consequences arising from the proposed model of PL.

4. On the Results of Studies of PL and LG in Holmium-Doped Media

We will focus on the studies of PL in DM performed in different laboratories. The purpose is to test the model, proposed by the author on the material of publications that contain information about the spectra of PL or laser generation (LG) in media activated (doped) with holmium. The author is aware that a small part of the publications on this topic has been considered. Nevertheless, the information extracted from the considered works confirms the position of the considered nontraditional model of PL.

The cases of using quartz glass fiber optical fibers [37,38], crystal samples, and chips [34,39–41] as DM are considered. Information about the preliminary review of the results of these works can be found in [42].

Both PL and LG processes starting at frequencies corresponding to the maximum peaks and humps on graphs reflecting the spectral dependence of intensity (power) will be considered PL at the DM output depends on the wavelength of the pump radiation. The transition of the overhead line to the LG mode in fiber light guides is carried out due to their length by retaining pumping and fluorescence radiation in the light guide. For chips, the LG mode is carried out using powerful pumping [34]. The use of holmium as an alloying additive is due to the search for media whose radiation is safe enough for vision, since the wavelengths of PL and LG radiation are in the IR region of the spectrum (~2 μm). In this area of the spectrum, it is convenient to conduct remote sensing, the development of differential absorption radars.

The reader should pay attention to the fact that to obtain PL in the infrared region of the spectrum (~2 μm), different sources whose wavelengths and radiation power differ significantly from different authors can be used for pumping.

The task of the analysis will be to determine the specific interlevel electronic transitions responsible for PL and LG in the specified spectral region (~2 μm). The use of holmium as an alloying additive is due to the search for media whose radiation is safe enough for vision, since the wavelengths of PL and LG radiation are in the IR region of the spectrum (~2 μm). Let us focus on the experiments performed using fiber light guides. Let us consider the results presented in the works of A.S. Kurkov and his collaborators [37,38]. According to these results, PL was obtained in the infrared region of the spectrum (~2 μm), and at wavelengths of 2.02 μm, 2.05 μm, 2.07 μm, 2.1 μm, 2.13 μm, 2.15 μm—LG. The experiments used a pump laser with a wavelength of 1125 nm, as well as 1147.35 nm. It is interesting to determine the wavelength of the electronic transition, which is associated with the heating of the glass fiber due to nonradiative relaxation. In particular, the transition responsible for LG in this region of the spectrum (~2 μm) corresponds, according to tabular data, to the wavelength value 755.09 nm [43]. This is the transition $^5\,\mathrm{I}\,(9 - 5^5/_2) - {}^4\,\mathrm{I}^0\,(^{15}/_2 - {}^9/_2)$ [44].

This result is obtained as follows. Previously, it was possible to obtain the value of the frequency ν_{pl}.

To calculate the frequency of the transition responsible for this process, it is necessary to use the following equation from the law of conservation of energy:

$$\nu_{pl} - \nu = \nu - \nu_{ij},$$

where:

ν is the frequency source of pump;

ν_{pl} is the frequency of the maximum peak on the spectrogam PL.

ν_{ij} is the frequency electronic transition between electronic levels (in case LG, for the holmium atom). Recall that the frequency ν_{ij} corresponds to transitions between any pair of electronic levels of opposite parity.

A slight (0.028 eV) discrepancy between the tabular data and the calculated value allows us to conclude that the results of the work [37,38] correspond to the PL model proposed by the author. Let us consider examples of obtaining radiation in the region (~2 μm) in crystal structures doped with holmium. In [41], when using a crystal made of yttrium–aluminum oxide ($YAlO_3$, YAP), radiation in the region (~2 μm) was obtained by pumping, the wavelength of which is 791 nm.

In [34], LG was obtained at a wavelength of 2050.5 nm (~2 μm) in a laser on a *Tm,Ho:YLF* microchip, when using a diode laser with a wavelength of 785 nm as a pump. In this case, the difference between the energy of the interlevel transition calculated according to Equation (7) corresponding to the frequency ν_{lm}, and, in fact, responsible for the PL, and the energy for the table value of the wavelength 4939.01 Å [43] does not exceed 0.043 eV. Thus, in the case of glass fiber for the 755.09 nm pumping wavelength used in the experiment, the transition, $^5I\,(9 - 5^5/2) - {}^4I\,^0({}^{15}/_2 - {}^9/_2)$, is responsible for the LG mode, corresponding to the frequency of 13243.35 cm^{-1}.

In the crystal structures, using the pumping 493,901 nm LG is responsible for transition $^5I\,(9 - 5^5/2) - {}^4I\,^0({}^{15}/_2 - {}^9/_2)$; its frequency—20246.97 cm^{-1}.

In [41], generation at a wavelength of 2044 nm was obtained using a holmium-doped *Tm, Ho:*$YALO_3$ (*YAP*) crystal, where a laser diode was used for pumping, the radiation wavelength of which is 794.8 nm. Numerical calculation shows in this case that the difference between the calculated value of the energy of the interlevel transition responsible for the PL and the tabular value does not exceed 0.027 eV.

In [34], the case of obtaining LG at a wavelength of 2050.5 nm in a laser on a *Tm,Ho:YLF* microchip when using a diode laser with a wavelength of 785 nm as a pumping source is considered. The calculation shows that in this case, the difference between the calculated value of the energy of the interlevel transition corresponding to the frequency ν_{pl} and, in fact, responsible for photoluminescence, and the energy of the table value of the wavelength 4939.01 Å (see [43]) does not exceed 0.043 eV.

Thus, in the case of glass fiber [37,38], for the 755.09 nm pumping used in the experiment, the transition $^5I\,(9 - 5^5/2) - {}^4I\,^0({}^{15}/_2 - {}^9/_2)$, corresponds to the LG mode, corresponding to the frequency of 13243.35 cm $^{-1}$. In crystal structures, when pumping 493.901 nm, the transition $^5I\,(9 - 5^5/2) - {}^4I\,^0({}^{15}/_2 - {}^9/_2)$ is responsible for LG, its frequency is 20,246.97 cm^{-1}. Some results of the section are presented in [42].

5. Analysis of the Results, Observed in Bismuth-Doped Media

Are the results and conclusions of the previous section random? To make sure that the PL model we are considering is viable, let us consider the papers that present the results of the study of PL and LG in media doped with atomic bismuth. It is either fiberglass or monolithic glass. In fiberglass, as a rule, we are talking about PL and LG; in glass samples, we are talking about PL.

Table 1 presents information about PL and LG in the visible and IR spectral ranges. The criterion confirming the validity of our assumptions can be proximity (or even coincidence, which is not necessary) calculated transition frequencies associated with nonradiative relaxation. Naturally, the results of calculations of the frequencies of inter-level transitions are compared with the values of the frequencies presented in the reference literature, which has generally recognized fame and reliability [43,44].

Table 1. Juxtaposition tabular value of the transition wavelength with the result of calculating the wavelength of the transition.

1	2	3	4	5
Link number in the list literatures	Wavelength corresponding to the maximal valued ordinal curve of PL (LG): λ_{pf} (nm)	Wavelength of excitation source PL: $-\lambda$ (nm)	The result of calculating the wavelength of the transition: $-\lambda_{lm}$ (nm)	Tabular value of the transition wavelength: $-\lambda_t$ (nm)
			which is associated with non-radiative relaxation.	
[45]	~1130 nm	1058 nm	994.6 nm	982.8 nm
[46,47]	~720 nm	514 nm	399.6 nm	388.6 nm
[48,49]	~750 nm	500 nm	374.99 nm	359.6 nm
[48,49]	~1140 nm	500 nm	320.2 nm	323.9 nm
[48,50]	~1300 nm	800 nm	577.7 nm	574.2 nm
[48,51]	~1315 nm	808 nm	583.1 nm	574.2 nm
[48,52,53]	~1310 nm	808 nm	584.1 nm	574.2 nm
[48,54]	~1150 nm	980 nm	853.8 nm	854.4 nm
[48,55]	~1210 nm	405 nm	243.2 nm	243.3 nm
[56]	~1260 nm	798 nm	584 nm	574 nm
[56]	~1153.5 nm	502 nm	314.9 nm	306 nm
[56]	~1153,5 nm	525 nm	339.8 nm	339.7 nm
[56]	~1085.4 nm	680 nm	472 nm	472.2 nm
[56]	~1171.6 nm	738 nm	540 nm	555.2 nm
[56]	~1260 nm	798 nm	584 nm	527.4 nm

The works in which pumping was used in the 405–1230 nm spectral regions and the PL spectra were recorded in the 720–1650 nm range are considered. Bismuth atoms are embedded in a homogeneous amorphous isotropic medium (example: quartz, aluminosilicate glass), the temperature of which depends on the environment. For reference: bismuth has an ionization potential of 7.3 eV, which corresponds to 58,765 cm^{-1}.

The pumping radiation that initiates PL in a medium, containing bismuth atoms, due to the addition of two quanta of light in the elementary act, transfers electrons to "virtual" levels that occur near the levels of $^2D_{3/2}$, $^2D_{5/2}$, $^4P_{5/2}$, etc. These levels of bismuth correspond to the energy region of 0.5–3 eV.

The addition of two pump radiation quanta in the atomic nucleus field, according to Primakov [6], may indicate the birth of an axion. The condition for its appearance is the high intensity of intra-atomic fields. The "virtual" levels cannot be long-lived. The born axion decays into two new quanta, one of which leaves the medium—a contribution to the PL, and the second gives its energy to heating the medium due to nonradiative relaxation of electrons. Here, we should recall the direct and reverse effects of Primakov. Theorists are familiar with Feynman diagrams that illustrate the possibility of the existence of such processes.

In general, in bismuth-doped glass fiber, the structure of the PL spectrum can be quite complex, as evidenced by the published results [45–55]. Nevertheless, it is possible to distinguish frequencies that correspond to the maximum values of the spectral curve of the PL. The exact numerical value of the frequency corresponding to the top of the hump of the spectral curve PL is quite difficult to obtain from illustrations in journal publications. For this reason, in Table 1, depending on the pumping frequency, the discrepancy between the exact tabular value (column No. 5) of the transition frequency responsible for nonradiative relaxation and the one calculated on the basis of publications (column No. 4) cannot be ideal. However, the results of the calculation and the tabular values for the viewed publications are quite close to each other.

Column 1 of Table 1 contains a reference to the source. Column 2 indicates either the wavelength of the line on which the LG is obtained or the wavelength corresponding to the hump on the PL spectrogram. The wavelength of the excitation source PL is given in column 3. The calculated wavelength corresponding to the position of the hump having the maximum power value on the PL spectrogram (column 4). Tabular values of wavelengths of atomic bismuth lines with which radiative relaxation is associated are presented in column 5. The results from [45–55] were obtained in fiberglass, and only in the work [56] PL was investigated in a monoblock.

Before proceeding to the axion problem, let us briefly consider two more elements of the periodic system of Mendeleev, used for doping media used in optics and quantum electronics and related to PL.

First of all, I will focus on neodymium [57]. A new generation has come to replace the neodymium laser with lamp-pumped rods cooled by running water, in which laser LEDs are used to pump rods doped with neodymium. According to the author, the choice of the operating frequency of a narrow-band neodymium laser pump source corresponds to the PL model considered in the review.

A few words about silicon, the main material of electronics: The review [31,34,35] discussed the problem of mirror symmetry of PL and absorption of multi-dimensional silicon nanoparticles suspended in ethanol. The contribution of the anti-Stokes component in the PL spectrum is noted. The PL excitation of silicon nanoparticles suspended in ethanol was carried out using an argon laser λ = 488 nm.

The author [59], using lasers with different wavelengths (365 nm, 456 nm, 532 nm, 660 nm), showed that the peaks of the PL spectra (660 nm, 690 nm, 720 nm, 820 nm) of silicon nanoparticles are shifted, new sections of the spectrum are filled.

The relation (8) is valid in the case of LG. In the case of broadband PL, indices "i,j" correspond to electron transitions between the "virtual" level, whose energy is $2h\nu$, and the set of electronic levels of the doping atom. The frequencies of these transitions are not reflected in the tables and depend on the radiation frequency of the pump source. If the value of the pump radiation frequency ν is greater than the frequencies "ν_{ij}" of these transitions, then the conditions for anti-Stokes PL are met.

6. Axions in the Optical Range of the Spectrum and Their Lifetime

We have constantly noted above that when using monochromatic radiation to pump DM, the PL spectrum is, as a rule, broadened. The width of the PL spectrum is undoubtedly related to transitions in the atom of the alloying element. Among the reasons for the broadening of the PL spectrum are: the finite width of the electronic levels involved in the elementary act of excitation of the atoms of the element used for doping the medium, shifts of these levels in the pumping field [24], temperature conditions of the experiment.

The virtual level, whose energy is determined by the sum of the energies of the two pumping quanta, as a rule, does not coincide with the levels of the atom. According to Primakov [6], a meeting in an elementary act of two photons in the field of an atomic nucleus can lead to the birth of an axion. The time of its life is extremely limited. The instability of the virtual level due to the interaction with the levels of the atom of the alloying element in the field of the atomic nucleus leads to the decay of the axion into two quanta.

The energy of one of them is spent on heating the medium due to non-radiative relaxation. Unfortunately, the issue of heating, heat transfer of the DM is little discussed. But when dye lasers were being developed, the issue of heat dissipation was the main one. The same can be said about neodymium and other solid-state lasers [57].

As follows from the experiment, the main contribution to the width of the PL spectrum is due to the fact that the number of allowed transitions of electrons from the virtual level corresponds to the set of levels in the atomic shell. A lot of non-radiative electron transitions are the reason for heating the luminescent medium.

The second quantum of the decayed axion, according to the ratio (6), leaves the DM at a slow speed, which depends on the structure of the energy spectrum of the atoms of the alloying element, its energy, i.e., frequency. In addition, it is necessary to remember about the multitude of different-frequency harmonic oscillators (electrons) encountered on his way, and the law of dispersion for each of them. At the output of the DM, we get, as a rule, an expanded spectrum of PL. For example, when silicon nanoparticles placed in ethanol [34,58,59] are excited by argon laser radiation, a PL spectrum with a width from 464 nm to 670 nm is obtained.

What can the width of the PL spectrum give us and what information can we extract from it?

In fact, due to the quantum-mechanical uncertainty of the position of the energy levels of the atom, including the virtual one, we have the right to believe that the width of the PL spectrum recorded by the photodetector corresponds to the natural width [60]. If τ is the average lifetime of the axion at the virtual level, ΔW is the energy of this level, then for this situation there is a known uncertainty relation.

$$\Delta W\tau \approx h/2\pi \tag{9}$$

According to [61], the uncertainty ratio due to the smallness of h is significant only for microsystems, which we are dealing with. Let us use this relation to determine the lifetime of the axion.

The contour of a wide spectrum of PL $I(v)$, reflecting the dependence of the power of the PL on the frequency, usually has a maximum (hump) and falling wings. For the width of the spectral line ΔW, a difference of frequencies is taken, which corresponds to a halving of its peak power $I(\nu)_{\mathrm{max}}/2$. Table 2 shows several examples illustrating the relationship between the width of the spectral band of the PL radiation and the lifetime of the axion. To estimate the lifetime of the axion, the ratio was used: 1 $\mathrm{cm}^{-1} \approx 2.99793 \times 10^{10}\ \mathrm{s}^{-1}$. The lifetime of the axion is significantly shorter than the lifetime of the excited energy levels, which, according to reference data [62], corresponds to a value of 10^{-8} s.

Table 2. Axion lifetime for Alloying Materials the Test Sample.

References	Alloying Material; in Parentheses—Environment of the Test Sample	The Wavelength of the Pump Radiation, nm	Frequency Corresponding to the Maximum Value of Intensity PL, cm^{-1}	Width of the PL Spectrum at Half-Length, ΔW, cm^{-1}	Axion Lifetime τ, s
[35,36]	Silicon (ethanol)	488	~5480	~1160	~3.48 × 10−13
[63]	Bismuth-Bi (glasscorderite))	514	~8547	~420 (~700) ~1120	2.93 × 10−13
[64]	Bismuth-Bi (aluminosil-rolled glass T = 77 K	1075	~8880	~1213	3.64 × 10−13
[65]	Bismuth-Bi (phosphorus-silicate glass)	1240	~7463	~340	~1.02 × 10−13
[66]	Bismuth-Bi T = 1.4 K	375	~6803	~340	1.02 × 10−13

7. Summing Up

Analysis of experimental results shows that the definition of PL in DM needs to be supplemented and clarified. We have previously settled on the definition according to which PL is a glow generated by optical radiation. By definition, Vavilov S.I. luminescence is an excess of radiation over temperature, provided that the excess radiation has a finite duration exceeding the period of light oscillations (10^{-10} s). Stokes's law states that PL light has a longer wavelength compared to the light used for excitation. According to Lommel, the PL spectrum as a whole and its maximum are always shifted in comparison with the excitation spectrum and its maximum towards long waves.

For a long time, it was believed that the substance has a completely defined luminescence spectrum, which is not sensitive to changes in the wavelength of exciting light. For DM, this rule is violated. Experiments in silicon, holmium, and bismuth doped media have not confirmed this position. On the other hand, the validity of the rule of mirror symmetry of absorption and PL spectra established by Levshin V.L. is confirmed, which follows from the unconventional definition of PL, considered in the work.

Finally, the main thing: the laminar luminous flux of the pump radiation, when propagated in a dispersing medium, becomes turbulent, which allows photons to collide with each other, forming a virtual energy level in the atom shell. This circumstance ensures the appearance of axions with a short lifetime at this level. As a result of exciton annihilation, new pairs of photons appear. One of the pair of photons leaves the medium—a contribution to PL. The second throws an electron to one of the higher levels of the alloying atom. The non-radiative relaxation of such an electron to the lower levels is the reason for the heating of the medium.

Funding: This review received no external funding.

Data Availability Statement: No new data were created or analyzed in this study. Data sharing is not applicable to this review.

Conflicts of Interest: The author declares no conflict of interest.

References

1. Ogluzdin, V.E. Axion N and Photoluminescence. *Acta Sci. Med. Sci.* **2021**, *5*, 38–42.
2. Peccei, R.D. The Strong CP Problem and Axions. *Phys. Rev. Lett.* **1977**, *40*, 279.
3. Wilczek, F.A. Problem of Strong P and T Invariance in the Presence of Instantons. *Phys. Rev. Lett.* **1978**, *40*, 279–282. [CrossRef]
4. Wilczek, F. Nobel Lecture: Asymptotic freedom: From paradox to paradigm. *Rev. Mod. Phys.* **2005**, *77*, 857–870. [CrossRef]
5. Sikivie, P. Experimental Tests of the "Invisible" Axion. *Phys. Rev. Lett.* **1983**, *51*, 1415–1417. [CrossRef]
6. Primakoff, H. Photo-Production of Neutral Mesons in Nuclear Electric Fields and the Mean Life of the Neutral Meson. *Phys. Rev.* **1951**, *81*, 86–89. [CrossRef]
7. Ogluzdin, V.E. Axions in optical experiments. *Inzhenernaya phisika (Eng. Phys.)* **2015**, *9*, 16–22. (In Russian)
8. Feynman, R.P.; Leighton, R.B.; Sands, M. *The Feinman Lectures on Physics*; Mir: Moscow, Russia, 1965; Volume 1.
9. Akhmanov, S.A.; Khokhlov, R.V. *Problems of Nonlinear Optics 1962–1963*; Institute of Scientific Information: Moscow, Russia, 1964.
10. Akhmanov, S.A.; Khokhlov, R.V. Parametric amplifiers and generators of light. *Sov. Phys. Usp* **1966**, *9*, 210–222. [CrossRef]
11. Glinca, N.L. *Obzhaya Himiya (General Chemistry)*; Vyscshaya Shkola: Moscow, Russia, 2003.
12. Bor, N. *Atomnaya Fisika I Chelovecheskoe Posnanie (Atomic Physics and Human Cognition)*; Isd-vo inostrannoi literatury: Moscow, Russia, 1961.
13. Tolansky, S. *Revolution in Optics*; Penguin: Moscow, Russia, 1968.
14. Fermi, E. *Notes on Quantum Mechanics. A Course Given by Enrico Fermi at the University of Chicago*; The University of Chicago Press: Chicago, IL, USA, 1964.
15. Ditchburn, R.W. *Light*; Blackie & son LIMITED: London, UK, 1963.
16. Korolev, F.A. *Teoreticheskaya Optika (Teoretical Optics)*; Vyscshaya shkola: Moscow, Russia, 1965.
17. Pol, R.V. *Vvedenie v Optiku*; State Technical and Theoretical Press: Moscow, Russia, 1947.
18. Pantell, R.H.; Puthoff, H.E. *Fundamentals of Quantum Electronics*; Wiley: New-York, NY, USA, 1969.
19. Karlov, N.V. *Lekzii po Kvantovoi Elektronike (Notes on Quantum Electronics)*; Fizmatlit: Moscow, Russia, 1988.
20. Born, M.; Wolf, E. *Principles of Optics*; Nauka: Moscow, Russia, 1970.
21. Akhmanov, S.A.; Kovrigin, A.I.; Maksimov, C.A.; Ogluzdin, V.E. Dispersion of resonant nonlinear susceptibility in potassium vapors. *Pis'ma Zh. Eksp. Teor. Fiz.* **1972**, *15*, 180–185.
22. Anikin, V.I.; Kryuchkov, S.V.; Ogluzdin, V.E. Resonant electron forced raman scattering in potassium vapor. Dispersion near the main doublet and the influence of four-photon processes. *Kvantovaya Electron.* **1974**, *1*, 1991–1998.
23. Ogluzdin, V.E. The role of Bohr frequencies in the scattering, luminescence, and generation of radiation in different media. *Physics-Uspekhi* **2006**, *49*, 401–405. [CrossRef]
24. Ogluzdin, V.E. Direct observation of the shift of the 4P(1/2) electronic level in atomic potassium vapor upon the saturation (redistribution of the population of levels) of the 4S(1/2)-4P(3/2) transition under the conditions for nearly resonant laser pumping. *Laser Phys.* **2006**, *16*, 1178–1183. [CrossRef]
25. Ogluzdin, V.E. On the angular structure of quasi-monochromatic radiation scattered in a resonant medium. *Pis'ma Zh. Tech.Fiz.* **1975**, *1*, 563–566.

26. Ogluzdin, V.E. Vavilov-Cherenkov effect under conditions of near resonant interaction of intense light beams with atomic potassium vapor. *Zh. Eksp. Teor. Phyz.* **1985**, *79*, 361–367.
27. Badalyan, A.M.; Dabagyan, A.A.; Movsesyan, M.E. Investigation of the dynamics of the development of nonlinear resonance processes in potassium vapor. *Zh. Eksp. Teor. Phyz.* **1976**, *70*, 1178–1184.
28. Levshin, V.L. *Photoluminescence zhidkich i tverdych veczhestv (Photoluminescence of Liguid and Solid Substances)*; GITTL Publication: Moscow, Russia, 1951.
29. Garbuny, M. *Optical Physics*; Academic Press: London, UK, 1965.
30. Bennett, U.R. *Gasovye Lasery (Gaseous Lasers)*; Mir: Moscow, Russia, 1964.
31. Ogluzdin, V.E. Interpretation of PL in the visible region of the spectrum of differently sized silicon nanoparticles suspended in ethanol. using the model of the classical harmonic Lorentz oscillator. *Bull. Lebedev Phys. Inst.* **2003**, *59*, 3–17.
32. Ogluzdin, V.E. Photons traveling at the speed of light in a two-level atomic medium as a source of Cherenkov radiation cones. *Physics-Uspekhi* **2004**, *47*, 829–832. [CrossRef]
33. Vavilov, S.I. *Mikrostructurasveta (Microstructure of Light)*; Publishing House of the USSR Academy of Sciences: Moscow, Russia, 1950.
34. Ogluzdin, V.E. Interpretation of the visible photoluminescence of inequisized silicon nanoparticles suspended in ethanol. *Semiconductors* **2005**, *39*, 884–890. [CrossRef]
35. Ogluzdin, V.E. Interpretation of the visible photoluminescence of inequisized d in ethanol excited by argon laser radiation. In *Book of Abstract 4th International Conference "Amorphous and Microcrystalline Semiconductors"-Section C*; Springer International Publishing AG: St.-Petersburg, Russia, 2004.
36. Meng, P.B. Diode-pumped room temperature single longitudinal mode lasing of Tm,Ho:YLF microchip laser at 2050.5 μm. *Laser Phys.* **2011**, *21*, 643. [CrossRef]
37. Kurkov, A.S.; Sholokhov, E.M.; Medvedkov, O.I.I. All-fiber Yb-Ho pulsed laser. *Laser Phys. Lett.* **2009**, *6*, 135–138. [CrossRef]
38. Kurkov, A.S.; Sholokhov, E.M.; Tsvetkov, V.B.; Marakulin, A.B.; Minashina, L.A.; Medvedkov, O.I.; Kosolapov, A.F. Holmium fiber laser with record quantum efficiency. *Quantum Electron.* **2011**, *41*, 492–494. [CrossRef]
39. Ju, Y.L.; Zhang, C.H.; Chen, F.; Li, G.; Yao, B.Q.; Tian, L.X.; Wang, Y.Z. Room temperature single longitudinal mode Tm,Ho:Yap microchip laser at 2102.6 nm. *Laser Phys.* **2011**, *21*, 97–100. [CrossRef]
40. Li, L.J.; Yao, B.Q.; Wu, D.Y.; Wang, J.; Gang, L.; Wang, Y.Z.; Zhang, Z.G. High efficient double end-pumped b-cut Tm,Ho:YAlO3 laser. *Laser Phys.* **2011**, *21*, 446–449. [CrossRef]
41. Li, L.J.; Yao, B.Q.; Qin, J.P.; Wu, D.Y.; Wang, Y.M.; Wang, J.; He, Z.L.; Liu, W.Y.; Chen, J.J.; Wang, Y.Z.; et al. High power and efficiency of a 2044-nm c-cut Tm, Ho:YAlO3 laser. *Laser Phys.* **2011**, *21*, 489–492. [CrossRef]
42. Ogluzdin, V.E. Photoluminecsence and Generation of Golmium (Ho) Atoms in the Glassfiber and in the Cristall Media. *arXiv Prepr.* **2011**, arXiv:1109.1673.
43. Seidel, A.N. *Tablizy spektral'nych linii (Tables of Spectral Lines)*; Nauka: Moscow, Russia, 1977.
44. Yazenko, A.C. *Diagrammy Grotriana neutral'nych atomov (Grotrian Diagrams of Neutral Atoms)*; Nauka: Novosibirsk, Russia, 1993.
45. Suzuki, N.; Anan, T.; Hatakeyama, H.; Tsuji, M. Low resistance tunnel junctions with type-II heterostructures. *Appl. Phys. Lett.* **2006**, *88*, 231103. [CrossRef]
46. Dianov, E.M.; Firstov, S.V.; Khopin, V.F.; Medvedkov, O.I.; Gur'yanov, A.N.; Bufetov, I.A. Bi-doped fibre lasers operating in the range 1470–1550 nm. *Quantum Electron.* **2009**, *39*, 299. [CrossRef]
47. Dianov, E.M.; Firstov, S.V.; Medvedkov, O.I.; Bufetov, I.A.; Khopin, V.F.; Guryanov, V.N. Luminescence and laser generation in Bi-doped fibers in a spectral region of 1300-1520 nm. In Proceedings of the Name of the Conference (Optical Fiber Communication Conference), San Diego, CA, USA, 22–26 March 2009.
48. Bulatov, L.I. *(Abstract of dissertation of the Faculty of Physical and Mathematical Sciences.) Absorption and Luminescent Properties of Bismuth Centers in Aluminum and Phosphorosilicate Light Guides*; MSU: Moscow, Russia, 2009.
49. Bufetov, I.A.; Dianov, E.M. Bi-doped fiber lasers. *Laser Phys. Lett.* **2009**, *6*, 487. [CrossRef]
50. Fujimoto, Y.; Nakatsuka, M. Infrared Luminescence from Bismuth-Doped Silica Glass. *Jpn. J. Appl. Phys.* **2001**, *40*, L279. [CrossRef]
51. Peng, M.; Qiu, J.; Chen, D.; Meng, X.; Yang, I.; Jiang, X.; Zhu, C. Bismuth- and aluminum-codoped germanium oxide glasses for super-broadband optical amplification. *Opt. Lett.* **2004**, *29*, 1998–2000. [CrossRef]
52. Peng, M.; Meng, X.; Qiu, J.; Zhao, Q.; Zhu, C. GeO_2: Bi, M (M = Ga, B) glasses with super-wide infrared luminescence. *Chem. Phys. Lett.* **2005**, *403*, 410–414. [CrossRef]
53. Peng, M.; Qiu, J.; Chen, D.; Meng, X.; Zhu, C. Superbroadband 1310 nm emission from bismuth and tantalum codoped germanium oxide glasses. *Opt. Lett.* **2005**, *30*, 2433–2435. [CrossRef] [PubMed]
54. Ren, J.; Qiu, J.; Wu, B.; Chen, D. Ultrabroad infrared luminescences from Bi-doped alkaline earth metal germanate glasses. *J. Mater. Res.* **2007**, *22*, 1574–1578. [CrossRef]
55. Meng, X.W.; Qiu, J.-R.; Peng, M.-Y.; Chen, D.-P.; Zhao, Q.-Z.; Jiang, X.-W.; Zhu, C.-S. Near infrared broadband emission of bismuth-doped aluminophosphate glass. *Opt. Express* **2005**, *13*, 1628–1634. [CrossRef] [PubMed]
56. Denker, B.; Galagan, B.; Osiko, V.; Shulman, I.; Sverchkov, S.; Dianov, E. Absorption and emission properties of Bi-doped Mg–Al–Si oxide glass system. *Appl. Phys. B* **2009**, *95*, 801–805. [CrossRef]
57. Krylov, K.I. *Osnovy lasernoi tehniki (Fundamentals of Laser Technology)*; Mashinostroenie: Leningpad, Russia, 1990.

58. Ogluzdin, V.E. Change energy photons of radiation, stimulating a photoluminescence in glasses and optical fiber, activated by bismuth. *arXiv Prepr.* **2011**, arXiv:1109.1373.
59. Vladimirov, A.G.; Korovin, S.; Surkov, A.; Kelm, E.; Pustovoy, V.; Borsella, E. Tunable luminecsence of silicon nanoparticles. In Proceedings of the Breakthroughs in Nanoparticles for Bio-Imaging, Frascati, Italy, 8–9 April 2010.
60. Seidel, A.N. *Tehnikaipraktikaspectroskopii (Spectroscopy Technique and Practice)*; Nauka: Moscow, Russia, 1972.
61. Davydov, A.S. *KvantovayaMekhanika (Quantummechanics)*; Fizmatgiz: Moscow, Russia, 1965.
62. *Physical Encyclopedic Dictionary*; Soviet; Encyclopedia: Moscow, Russia, 1984.
63. Sokolov, V.O.; Plotnichenko, V.G.; Koltashev, V.V.; Dianov, E.M. Centres of broadband near-IR luminescence in bismuth-doped glasses. *J. Phys. D Appl. Phys.* **2009**, *42*, 095410. [CrossRef]
64. Krylov, A.A.; Kryukov, P.G.; Dianov, E.M.; Okhotnikov, O.G.; Guina, M. Pulsed bismuth fibre laser with the intracavity-compensated group velocity dispersion. *Quantum Electron.* **2009**, *39*, 21. [CrossRef]
65. Bufetov, I.A.; Melkumov, S.V.; Firstov, K.E.; Ryumkin, K.E.; Shubin, A.V.; Khopin, V.F.; Gur'yanov, A.N.; Dianov, E.M. Bi-doped optical fibers and fiber lasers. *Quantum Electron.* **2014**, *44*, 700. [CrossRef]
66. Laguta, O.V.; Hamzaoui, H.E.; Bouazaoui, M.; Arion, V.B.; Razdobreev, I.M. On the nature of photoluminescence in Bismuth-doped silica glass. *Sci. Rep.* **2017**, *7*, 3178. [CrossRef]

foundations

MDPI

Review

Neutron Stars and Gravitational Waves: The Key Role of Nuclear Equation of State

Polychronis S. Koliogiannis *,†,‡, Alkiviadis Kanakis-Pegios †,‡ and Charalampos C. Moustakidis †,‡

Department of Theoretical Physics, Aristotle University of Thessaloniki, 54124 Thessaloniki, Greece; alkanaki@auth.gr (A.K.-P.); moustaki@auth.gr (C.C.M.)

* Correspondence: pkoliogi@physics.auth.gr

† Current address: Aristotle University of Thessaloniki, 54124 Thessaloniki, Greece.

‡ These authors contributed equally to this work.

Abstract: Neutron stars are the densest known objects in the universe and an ideal laboratory for the strange physics of super-condensed matter. Theoretical studies in connection with recent observational data of isolated neutron stars, as well as binary neutron stars systems, offer an excellent opportunity to provide robust solutions on the dense nuclear problem. In the present work, we review recent studies concerning the applications of various theoretical nuclear models on a few recent observations of binary neutron stars or neutron-star–black-hole systems. In particular, using a simple and well-established model, we parametrize the stiffness of the equation of state with the help of the speed of sound. Moreover, in comparison to the recent observations of two events by LIGO/VIRGO collaboration, GW170817 and GW190425, we suggest possible robust constraints. We also concentrate our theoretical study on the resent observation of a compact object with mass $\sim 2.59^{+0.08}_{-0.09}\ M_{\odot}$ (GW190814 event), as a component of a system where the main companion was a black hole with mass $\sim 23\ M_{\odot}$. There is scientific debate concerning the identification of the low mass component, as it falls into the neutron-star–black-hole mass gap. This is an important issue since understanding the nature of GW190814 event will offer rich information concerning the upper limit of the speed of sound in dense matter and the possible phase transition into other degrees of freedom. We systematically study the tidal deformability of a possible high-mass candidate existing as an individual star or as a component in a binary neutron star system. Finally, we provide some applications of equations of state of hot, dense nuclear matter in hot neutron stars (nonrotating and rapidly rotating with the Kepler frequency neutron stars), protoneutron stars, and binary neutron star merger remnants.

Keywords: neutron stars; nuclear equation of state; gravitational waves; speed of sound; tidal polarizability

PACS: 26.60.-c; 21.30.Fe; 21.65.Cd; 26.60.Kp

Citation: Koliogiannis, P.; Kanakis-Pegios, A.; Moustakidis, C.C. Neutron Stars and Gravitational Waves: The Key Role of Nuclear Equation of State. *Foundations* **2021**, *1*, 217–255. https://doi.org/10.3390/foundations1020017

Academic Editor: Eugene Oks

Received: 24 September 2021
Accepted: 25 October 2021
Published: 5 November 2021

1. Introduction

One of the currently unsolved open problems in nuclear physics is the properties of dense nuclear matter. In particular, compact objects, such as white dwarfs and especially neutron stars, offer the opportunity to study the behavior of nuclear matter at high densities [1–4]. Neutron stars are a very promising tool for studying the properties of dense nuclear matter, such as the speed of sound and its possible upper bound.

The main assumption for the speed of sound is that it cannot exceed the speed of light because of the causality. However, this is not determinant, as Zel'dovich [5,6] showed the importance of defining a rigorous limit of speed of sound upon the equation of state (EoS). To be more specific, in the electromagnetic interaction, the main assumption is that $v_s \leq c/\sqrt{3}$ is generally low in nature. Moreover, by considering the interaction of baryons through a vector field, he noticed that the upper limit of the speed of sound is the causality, $v_s = c$. Therefore, the only restriction imposed by general principles is that $v_s \leq c$ [5,6].

On the other hand, Hartle noticed that the causality is not enough to constrain the high-density part of the EoS [7], while Weinberg pointed out that the speed of sound is much less than the speed of light for a cold nonrelativistic fluid [4]. In addition, for nonrelativistic and/or weakly coupled theories, the bound $v_s = c/\sqrt{3}$, according to Bedaque and Steiner seems to be valid, while in conformal theories, the upper bound is saturated [8]. According to these authors, the existence of a 2 $M_\odot$ neutron star, in combination with the knowledge of the EoS of hadronic matter at the low density region, is not consistent with the limit $c/\sqrt{3}$. We notice that various recent studies have been conducted regarding the speed of sound and the tidal deformability of neutron stars [9–12].

One of the goals of this study is to apply a method that directly relates the observed tidal deformability, derived from binary neutron star mergers, to the maximum mass of neutron stars, aiming to obtain constraints on the upper bound of the speed of sound. The main idea is the fact that while the measured upper limit of the effective tidal deformability favors softer EoSs, recent measurements of high neutron star masses favor stiffer EoSs. As a basis in our study, we used a model in which we parametrized the EoS through the various bounds of the speed of sound (stiffness). Hence, the EoS is a functional of the transition density and the speed of sound bound. In our approach, we used the observation of two recent events, GW170817 [13] and GW190425 [14], as well as the current observed maximum neutron star masses ($1.908 \pm 0.016 M_\odot$ [15], $2.01 \pm 0.04\ M_\odot$ [16], $2.14^{+0.10}_{-0.09}\ M_\odot$ [17], and $2.27^{+0.17}_{-0.15}\ M_\odot$ [18]). The need for (a) a soft EoS for the low density region (to be in accordance with the observed upper limit of the effective tidal deformability) and (b) a stiff EoS for the high density region (to provide the high neutron star masses) leads to robust constraints on the EoS. In addition, this method allows making postulations about the kind of future measurements that would be more informative and help to improve our knowledge.

Furthermore, we highlight the very recent observation of the GW190814 event, where a gravitational wave has been detected from the merger of a 22.2–24.3 $M_\odot$ black hole with a non-identified compact object with mass 2.5–2.67 $M_\odot$ [19,20]. Although the authors of the mentioned references suggest that is unlikely for the second component's mass to belong to a neutron star, they do leave *open the window* that the improved knowledge of the neutron star EoS and further observations of the astrophysical population of compact objects could alter this assessment.

It is worth pointing out that the observation of the GW190814 event has some additional general benefits, apart from the measurement of 2.6 $M_\odot$ of the second partner [19]. Firstly, this binary system has the most unequal mass ratio yet measured with gravitational waves close to the value of 0.112. Secondly, the dimensionless spin of the primary black hole is constrained to ≤ 0.07, where various tests of general relativity confirm this value, as well as its predictions of higher multiple emissions at high confidence intervals. Moreover, the GW190814 event poses a challenge for the understanding of the population of merging compact binaries. It was found, after systematic analysis, that the merger rate density of the GW190814-like binary system was $7^{+16}_{-6}\ \mathrm{Gpc}^{-3}\mathrm{yr}^{-1}$ [19]. More relevant to the present study, the observation of the GW190814 event led to the following conclusion: due to the source's asymmetric masses, the lack of detection of an electromagnetic counterpart and of clear signature of tides or the spin-induced quadrupole effect in the waveform of the gravitational waves, we are not able to distinguish between a black-hole–black-hole and black-hole–neutron-star system [19]. In this case, one must count only the comparison between the mass of the second partner with the estimation of maximum neutron star mass [21]. This is one of the main subjects which has been revised in the present work. It should be emphasized that the measurements of neutron star mass can also inform us about a bound on the maximum gravitational mass independently of the assumptions of the specific EoS. For example, Alsing et al. [22], fitting the known population of neutron stars in binaries to double-Gaussian mass distribution, obtained the empirical constraint that $M_{\rm max} \leq 2.6\ M_\odot$ (with 90% confidence interval).

Farr and Chatziionannou updated knowledge from previous studies, including recent measurements [23]. Their study constrains the maximum mass $M_{\max} = 2.25^{+0.81}_{-0.26}\, M_{\odot}$, leading to the conclusion that the posterior probability (for the mass of the second partner $m_2 \leq M_{\max}$) is around only 29%. However, the prediction of $M_{\max}$ is sensitive to the selection mass rules of neutron stars (not only on binary systems but also isolated) as well as to the discovery of new events, and this consequently remains an open problem. Finally, the conclusion of the recent GW190814 event in comparison with previous ones (for example the GW170817 event [24]) may shed light on the problem of the $M_{\max}$. For example, the spectral EoSs, which are conditioned by the GW170817 event, are once more elaborated to include the possibility that the prediction of $M_{\max}$ is at least equal to m_2. This approach leads to significant constraints on the radius and tidal deformability of a neutron star with mass of 1.4 $M_{\odot}$ ($R_{1.4} = 12.9^{+0.8}_{-0.7}$ km and $\Lambda_{1.4} = 616^{+273}_{-158}$, respectively, [19]).

The matter of the finite temperature and its effect on the nuclear EoS, as well as astrophysical applications, has been extensively studied by Bethe et al. [25], Brown et al. [26], Lamb et al. [27], Lattimer and Ravenhall [28], and Lattimer [29]. Lattimer and Swesty [30], as well as Shen et al. [31], constructed the most used hot neutron star EoSs. The first one is based on the liquid drop-type model, and the second one is based on the relativistic mean field model. Afterwards, Shen et al. [31] broadened their study in order to study supernovae, binary neutron star mergers, and black hole formations by developing EoSs for various temperatures and proton fractions [32]. In addition, the density and temperature dependence of the nuclear symmetry free energy using microscopic two- and three-body nuclear potentials constructed from Chiral effective field theory have been studied in a series of works, including the one of Wellenhofer et al. [33]. Furthermore, hot neutron star and supernova properties have been studied by Constantinou et al. [34,35], where a suitable hot EoS is produced. Finally, the interplay between the temperature and the neutron star matter was probed by Sammarruca et al. [36], by considering the framework of chiral effective field theory.

EoSs at finite temperature constructed within the Brueckner–Hartree–Fock approach and the properties of hot β-stable nuclear matter were studied in a series of papers [37–45]. In addition, a model for cold nucleonic EoSs, extended to include temperature and proton fractions for simulations of astrophysical phenomena, was constructed by Raithel et al. [46]. Pons et al. [47], as well as Prakash et al. [48], focused on describing the thermal and chemical evolution of protoneutron stars by considering neutrino opacities consistently calculated with the EoS. Finally, neutron stars, along with the hot EoS of dense matter, are presented in a recent review of Lattimer and Prakash [49].

The processes that occupy the stages of the merger and postmerger phases of a binary neutron star system have been extensively studied in recent years. However, matters that concern the remnant evolution are still under consideration or even unsolved. In particular, the remnant evolution contains (a) the collapse time, (b) the threshold mass, (c) the possible phase transition in the interior of the star, and (d) the disk ejecta and neutrino emission (for an extended discussion and applications, see Perego et al. [50]). It has to be noted here that the possibility of a phase transition will affect the signal of the emitted gravitational wave. Relevant previous work is also be available in Bauswein et al. [51], Kaplan et al. [52], Tsokaros et al. [53], Yasin et al. [54], Radice et al. [55], Sarin et al. [56], Soma and Bandyopadhyay [57], and Sen [58].

In the present work, we review some applications of the thermal effects on neutron star properties. In particular, we apply a momentum-dependent effective interaction (MDI) model, where thermal effects can be studied simultaneously on the kinetic part of the energy and also on the interaction one. In addition, the extension of the proposed model can lead to EoSs with varying stiffness with respect to the parameterized symmetry energy. In fact, Gale et al. [59] presented a model aimed at the influence of MDI on the momentum flow of heavy-ion collisions. However, the model has been successfully applied in studying the properties of cold and hot nuclear and neutron star matter (for an extensive review of the model, see References [60–62]).

Moreover, we review specific properties of neutron stars (including mainly the mass and radius, moment of inertia, Kerr parameter, etc.) with respect to the EoS, both at nonrotating and rapidly rotating (considering the mass-shedding limit) configurations. The above properties are well applied in studying hot neutron stars, protoneutron stars, and the remnants of binary neutron star systems.

To summarize, in the present work, we review the applications of various theoretical nuclear models on a few recent observations of binary neutron stars or neutron-star–black-hole systems, including mainly the GW170817, GW190425, and GW190814 events. Our results have been published recently in sections, in the following journals [63–66].

It has to be noted that the present study is mainly dedicated to the case of the emission of the gravitation waves due to the merger of a binary neutron star system. However, there are other mechanisms by which gravitational waves are emitted by a neutron star and thus, have additional ways of studying its internal structure (for a recent review, see Reference [67]). These mechanisms, which are considered as continuous gravitational waves sources, include, for example, (a) the case of radiation of gravitational waves by a rigidly rotating aligned triaxial ellipsoid (radiation of purely quadrupolar waves), (b) the emission of gravitational waves due to asymmetry in the magnetic field distribution in the interior of the neutron star, and (c) the radiation of gravitational waves from the rapidly rotating neutron stars. In this case, neutron stars may suffer a number of different instabilities with a general feature in common: they can be directly associated with unstable modes of oscillation (for exable g-modes, f-modes, w-modes, r-modes; for a review see Reference [68]). The more notable mechanism is the r-mode oscillations. In these oscillations, the restoring force is the Coriolis force. The r-mode mechanics have been proposed as an explanation for the observed relatively low spin frequencies of young neutron stars, as well as of accreting neutron stars in low-mass X-ray binaries. This instability only occurs when the gravitational radiation driving timescale is shorter compared to the ones of the various dissipation mechanisms, which occur in the neutron star matter [69]. The free procession may cause deformation of the neutron star, leading to a better understanding of some neutron star matter properties, including breaking strain, viscosity, rigidity, and elasticity [67]. We expect that in the future, the development of the sensitivity of LIGO and Virgo detectors in cooperation with and new instruments will help to significantly improve our knowledge of neutron star interiors with the detection of the emitted gravitational waves [67].

Another way to study the properties of dense nuclear matter that exists inside neutron stars can be done with the help of the statistical study of the observed properties of neutron stars. To be more specific, we refer to the observational data concerning both isolated neutron stars and those that exist in binary systems. In the first case, there are extensive studies where observational data are applied to estimations with the help of statistical studies, for example, the maximum possible mass of a neutron star, but mainly the radius of neutron stars with masses of about 1.4 $M_\odot$. In each case, valuable information and ideas can be extracted and utilized in terms of knowledge of the EoS of neutron star matter, in order to evaluate the reliability of the existing EoS (see the review article [70] and references therein). In the case of binary systems, the analysis of the emitted gravitational waves from the fusion of a binary system of neutron stars, where a large amount of information is received by studying the amplitude and phase of gravitational waves, is utilized. These studies mainly focus on the measurement (and utilization of the measurement) of tidal deformability. In any case, extensive and systematic statistical estimation of data can lead to valuable knowledge of the structure and composition of neutron stars (for a recent review see [71] and references therein).

The article is organized as follows: in Section 2, we present the theory concerning the EoS and the structure of cold neutron stars. In Section 3, we present the construction of the hot EoSs (both isothermal and isoentropic) and briefly discuss the stability equations of hot rapidly rotating neutron stars. The results and the discussion are provided in Section 4, while Section 5 includes the most noteworthy concluding remarks of the present review.

2. Cold Neutron Stars

2.1. The Momentum-Dependent Interaction Nuclear Model

The description of the interior structure of neutron stars demands the use of a nuclear model suitable to describe the properties of dense nuclear matter. In the present work, the EoS of nuclear matter is studied using the MDI model. In this model, the energy per baryon is given by the formulae [60,72]

$$\begin{aligned} E(n,I) &= \frac{3}{10}E_F^0 u^{2/3}\left[(1+I)^{5/3}+(1-I)^{5/3}\right]+\frac{1}{3}AQ_0 u+\frac{\frac{2}{3}BQ_3 u^{\sigma}}{1+\frac{2}{3}B'Q_3 u^{\sigma-1}} \\ &+ \frac{3}{2}\sum_{i=1,2}\left[C_i+\frac{C_i-8Z_i}{5}I\right]\left(\frac{\Lambda_i}{k_F^0}\right)^3\left(\frac{((1+I)u)^{1/3}}{\frac{\Lambda_i}{k_F^0}}-\tan^{-1}\frac{((1+I)u)^{1/3}}{\frac{\Lambda_i}{k_F^0}}\right) \\ &+ \frac{3}{2}\sum_{i=1,2}\left[C_i-\frac{C_i-8Z_i}{5}I\right]\left(\frac{\Lambda_i}{k_F^0}\right)^3\left(\frac{((1-I)u)^{1/3}}{\frac{\Lambda_i}{k_F^0}}-\tan^{-1}\frac{((1-I)u)^{1/3}}{\frac{\Lambda_i}{k_F^0}}\right), \end{aligned} \quad (1)$$

where u is the baryon density normalized with respect to the saturation density ($n_s = 0.16$ fm^{-3}), $I = (n_n - n_p)/n$ is the asymmetry parameter, $X_0 = x_0 + 1/2$, $X_3 = x_3 + 1/2$, $Q_0 = \frac{3}{2} - X_0 I^2$, and $Q_3 = \frac{3}{2} - X_3 I^2$. The parameters A, B, σ, C_i, and B' appear in the description of symmetric nuclear matter (SNM) and are determined so that the relation $E(n_s, 0) = -16$ MeV holds. Λ_1 and Λ_2 are finite range parameters equal to $1.5k_F^0$ and $3k_F^0$, respectively, where k_F^0 is the Fermi momentum at the saturation density. The remaining parameters, x_0, x_3, Z_i, appear in the description of asymmetric nuclear matter (ANM) and, with a suitable parametrization, are used in order to obtain different forms for the density dependence of symmetry energy, as well as the value of the slope parameter L and the value of the symmetry energy $E_{\rm sym}$ at the saturation density, defined as [63]

$$L = 3n_s \frac{dE_{\rm sym}(n)}{dn}\bigg|_{n_s} \quad \text{and} \quad E_{\rm sym}(n) = \frac{1}{2}\frac{\partial^2 E(n,I)}{\partial I^2}\bigg|_{I=0}, \quad (2)$$

and consequently different parametrizations of the EoS stiffness.

The specific choice of the MDI model is based on the following: (a) it combines both density and momentum dependent interaction among the nucleons, (b) it is suitable for studying neutron star matter at zero and finite temperature (due to the momentum term), (c) it reproduces with high accuracy the properties of SNM at the saturation density, including isovector quantities, (d) it reproduces the microscopic properties of the Chiral model for pure neutron matter (PNM) and the results of *state-of-the-art* calculations of Akmal et al. [73] with suitable parametrizations, (e) it predicts higher maximum neutron star mass than the observed ones [16–18], and (f) it maintains the causal behavior of the EoS even at densities higher than the ones that correspond to the maximum mass configuration.

2.2. Speed of Sound Formalism

An EoS can be parametrized in order to reproduce specific values of the speed of sound in the interior of the neutron star. This parametrization is possible following the formula available from References [22,74–80]:

$$P(\mathcal{E}) = \begin{cases} P_{\rm crust}(\mathcal{E}), & \mathcal{E} \le \mathcal{E}_{\rm c-edge} \\ P_{\rm NM}(\mathcal{E}), & \mathcal{E}_{\rm c-edge} \le \mathcal{E} \le \mathcal{E}_{\rm tr} \\ \left(\frac{v_s}{c}\right)^2(\mathcal{E}-\mathcal{E}_{\rm tr})+P_{\rm NM}(\mathcal{E}_{\rm tr}), & \mathcal{E}_{\rm tr} \le \mathcal{E}, \end{cases} \quad (3)$$

where P and $\mathcal{E}$ denote the pressure and the energy density, respectively, and the corresponding subscript "tr" in energy density denotes the energy density at the transition density.

However, following the approach in Equation (3), only the continuity in the EoS is ensured. The artificial character of Equation (3) does not take into account the continuity in the speed of sound. Therefore, in order to ensure the continuity and a smooth transition, we followed the method presented in Reference [81]. We proceeded with the matching of the EoSs with the transition density by considering that, above this value, the speed of sound is parametrized as follows (for more details, see Reference [81]):

$$\frac{v_s}{c} = \left(a - c_1 \exp\left[-\frac{(n-c_2)^2}{w^2}\right]\right)^{1/2}, \quad a \in [1/3, 1] \tag{4}$$

where the parameters c_1 and c_2 are fit to the speed of sound and its derivative at $n_{\rm tr}$ and also to the demands $v_s(n_{\rm tr}) = [c/\sqrt{3}, c]$ [74] according to the value of α. The remaining parameter w controls the width of the curve, which in our case is equal to 10^{-3} fm^{-3}, in order to preserve the neutron star properties. Using Equation (4), the EoS for $n \geq n_{\rm tr}$ can be constructed with the help of the following [81]:

$$\mathcal{E}_{i+1} = \mathcal{E}_i + \Delta\mathcal{E}, \quad P_{i+1} = P_i + \left(\frac{v_s}{c}(n_i)\right)^2 \Delta\mathcal{E}, \tag{5}$$

$$\Delta\mathcal{E} = \Delta n \left(\frac{\mathcal{E}_i + P_i}{n_i}\right), \tag{6}$$

$$\Delta n = n_{i+1} - n_i. \tag{7}$$

2.3. Construction of the EoS

The construction of the EoS for the interior of neutron stars is based on the MDI model and data provided by Akmal et al. [73]. More specifically, we utilized the data for the A18+UIX (hereafter APR-1) EoS from Akmal et al. [73] for the energy per particle of SNM and PNM in the density range [0.04, 0.96] fm^{-3}. Due to the complexity of the microscopic data, we divided the density region into three sections—(a) low-density region [0.04, 0.2] fm^{-3}, (b) medium-density region [0.2, 0.56] fm^{-3}, and (c) high-density region [0.56, 0.96] fm^{-3}—in order to acquire the best fitting using Equation (1). From this process emerged the coefficients for the SNM and ANM, and eventually the EoS, hereafter MDI-APR.

In the case of the speed-of-sound-parametrized EoSs, the construction of the EoSs follows the procedure: (a) in region $\mathcal{E} \leq \mathcal{E}_{\rm c-edge}$, we used the equation of Feynman et al. [82] and also of Baym et al. [83] for the crust and low densities of neutron star; (b) in the intermediate region, $\mathcal{E}_{\rm c-edge} \leq \mathcal{E} \leq \mathcal{E}_{\rm tr}$, we employed the MDI-APR EoS; and (c) for $\mathcal{E}_{\rm tr} \geq \mathcal{E}$ region, the EoS is maximally stiff with the speed of sound, defined as $v_s = c\sqrt{(\partial P/\partial \mathcal{E})_S}$ (where S is the entropy) fixed in the present work in the range $[c/\sqrt{3}, c]$. The lowest allowed value of the speed of sound, that is $(v_s/c)^2 = 1/3$, is introduced in order to be consistent with the possibility of a phase transition in quark matter. In this case, the theoretical predictions lead to this value as an upper bound of the speed of sound. The implementation of speed of sound values between the limited ones will lead to results well constrained by the two mentioned limits. Although the energy densities below the $\mathcal{E}_{\rm c-edge}$ have negligible effects on the maximum mass configuration, we used them in calculations for the accurate estimation of the tidal deformability.

In this study, two cases, based on the transition density, $n_{\rm tr} = p n_s$ and the speed of sound, are employed, in particular, (a) the ones where p takes the values [1.5, 2, 3, 4, 5] while the speed of sound is parametrized in the two limiting cases, $(v_s/c)^2 = 1/3$ and $(v_s/c)^2 = 1$, and (b) the ones where p takes the values [1.5, 2] while the speed of sound is parametrized in the range $(v_s/c)^2 = [1/3, 1]$.

For reasons of completeness, the treatment with both discontinuity and continuity in the speed of sound is presented in Table V of Reference [74]. The main point was that the two approaches converge and consequently the effects of the discontinuity are negligible.

In Figure 1, we present the pressure as a function of the rest mass density ($\rho_{\text{rest}} = n_b m_n$) and the square speed of sound in units of speed of light as a function of the transition density for the EoSs constructed in cases (a) and (b). In addition, we display the credibility intervals proposed by Reference [24] from LIGO/Virgo collaboration for the GW170817 event. It is clear from these figures that the pure MDI-APR EoS is well-defined in the proposed limits of LIGO/Virgo collaboration and also fulfills the speed of light limit at high densities.

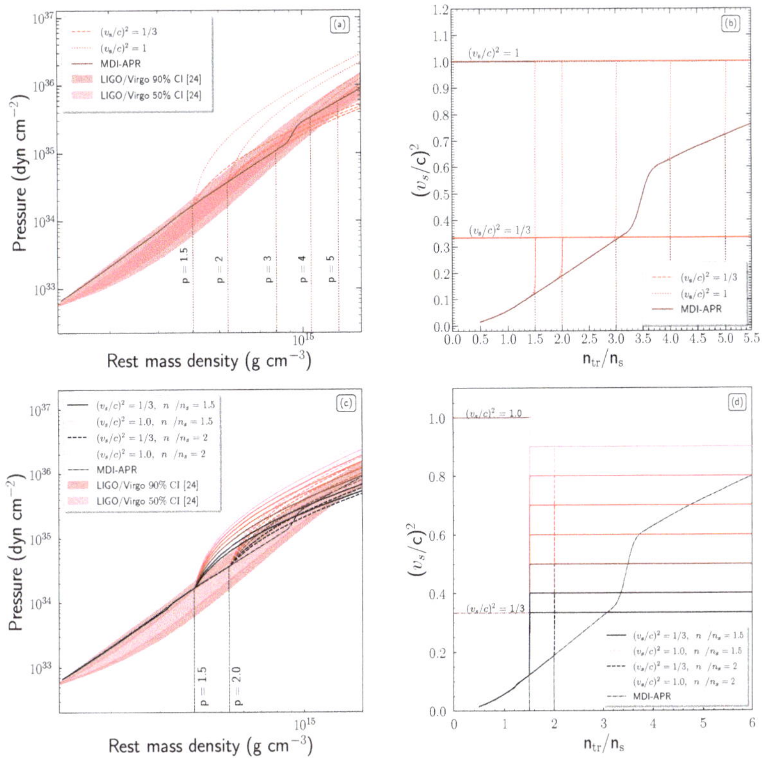

Figure 1. (**a**,**c**) Dependence of the pressure on the rest mass density and (**b**,**d**) dependence of the square sound speed in units of light speed on the transition density. (**a**,**b**) The speed of sound is fixed at the two boundary cases, $(v_s/c)^2 = 1/3$ and $(v_s/c)^2 = 1$, and the value p takes the arguments $[1.5, 2, 3, 4, 5]$. (**c**,**d**) The value p takes the arguments $[1.5, 2]$, and the speed of sound is parametrized in the range $(v_s/c)^2 = [1/3, 1]$ (the lower values of the speed of sound correspond to the darker colored curves). In all figures, the vertical dotted lines indicate the transition cases, while the shaded regions note the credibility interval derived from Reference [24].

2.4. *Structure Equations*

2.4.1. Nonrotating Neutron Stars

For a static spherical symmetric system, which is the case of a nonrotating neutron star, the metric can be written as follows [1,2]:

$$ds^2 = e^{\nu(r)} dt^2 - e^{\lambda(r)} dr^2 - r^2 \left(d\theta^2 + \sin^2\theta d\phi^2 \right). \tag{8}$$

The density distribution and the local pressure is related to the metric functions $\lambda(r)$ and $\nu(r)$ according to the relations [1,2]

$$\frac{8\pi G}{c^2}\rho(r) = \frac{1}{r^2}\left(1 - e^{-\lambda(r)}\right) + e^{-\lambda(r)}\frac{\lambda'(r)}{r}, \quad (9)$$

$$\frac{8\pi G}{c^4}P(r) = -\frac{1}{r^2}\left(1 - e^{-\lambda(r)}\right) + e^{-\lambda(r)}\frac{\nu'(r)}{r}, \quad (10)$$

where derivatives with respect to the radius are denoted by $'$. The combination of Equations (9) and (10) leads to the well known Tolman–Oppenheimer–Volkoff (TOV) equations [1,2]:

$$\frac{dP(r)}{dr} = -\frac{G\rho(r)M(r)}{r^2}\left(1+\frac{P(r)}{\rho(r)c^2}\right)\left(1+\frac{4\pi P(r)r^3}{M(r)c^2}\right)\left(1-\frac{2GM(r)}{c^2 r}\right)^{-1}, \quad (11)$$

$$\frac{dM(r)}{dr} = 4\pi r^2\rho(r). \quad (12)$$

It is difficult to obtain exact solutions to TOV equations in closed analytical form, and they are solved numerically with an equation of state specified [84]. Actually, there are hundreds of analytical solutions of TOV equations, but there are three that satisfy the criteria that the pressure and energy density vanish on the surface of the star, and they also both decrease monotonically with increasing radius. These three solutions are the Tolman VII, the Buchdahl, and the Nariai IV. Actually, the Tolman VII and the Buchdahl have already been analyzed and employed in Reference [85]. It is worth pointing out that in the present work we use the Tolman VII solution, which contains two parameters, that is, the central density ρ_c and the compactness parameter $\beta = GM/Rc^2$. All the mentioned solutions have been presented and analyzed in detail in References [85–88].

One of the most significant sources for the detectors of terrestrial gravitational waves is the gravitational waves from inspiraling binary neutron star systems before their merger [86,89–96]. The component masses of these binary systems can be measured. Additionally, during the last orbits before the merger, the tidal effects that are present can also be measured [90].

The dimensionless parameter that describes the response of a neutron star to the induced tidal field is called tidal Love number k_2. This parameter depends on the neutron star structure (i.e., the mass of the neutron star, and the EoS). Specifically, the tidal Love number k_2 is a proportional parameter between the induced quadrupole moment Q_{ij} and the applied tidal field E_{ij} [90,97] given below:

$$Q_{ij} = -\frac{2}{3}k_2\frac{R^5}{G}E_{ij} \equiv -\lambda E_{ij}, \quad (13)$$

where R is the neutron star's radius and $\lambda = 2R^5k_2/3G$ is a key-role quantity, which is called tidal deformability. The tidal Love number k_2 is given by [90,91]

$$\begin{aligned} k_2 &= \frac{8\beta^5}{5}(1-2\beta)^2[2-y_R+(y_R-1)2\beta] \times \Big[2\beta(6-3y_R+3\beta(5y_R-8)) \\ &+ 4\beta^3\left(13-11y_R+\beta(3y_R-2)+2\beta^2(1+y_R)\right) \\ &+ 3(1-2\beta)^2[2-y_R+2\beta(y_R-1)]\ln(1-2\beta)\Big]^{-1}. \end{aligned} \quad (14)$$

The quantity y_R is determined by solving the following differential equation

$$r\frac{dy(r)}{dr} + y^2(r) + y(r)F(r) + r^2Q(r) = 0, \quad (15)$$

with the initial condition $y(0) = 2$ [93]. $F(r)$ and $Q(r)$ are functionals of $\mathcal{E}(r)$, $P(r)$, and $M(r)$, defined as [86,93]

$$F(r) = \left[1 - \frac{4\pi r^2 G}{c^4}(\mathcal{E}(r) - P(r))\right]\left(1 - \frac{2M(r)G}{rc^2}\right)^{-1}, \tag{16}$$

and

$$\begin{aligned} r^2 Q(r) &= \frac{4\pi r^2 G}{c^4}\left[5\mathcal{E}(r) + 9P(r) + \frac{\mathcal{E}(r) + P(r)}{\partial P(r)/\partial \mathcal{E}(r)}\right] \times \left(1 - \frac{2M(r)G}{rc^2}\right)^{-1} \\ &- 6\left(1 - \frac{2M(r)G}{rc^2}\right)^{-1} - \frac{4M^2(r)G^2}{r^2c^4}\left(1 + \frac{4\pi r^3 P(r)}{M(r)c^2}\right)^2\left(1 - \frac{2M(r)G}{rc^2}\right)^{-2}. \end{aligned} \tag{17}$$

The Equation (15) must be integrated self-consistently with the TOV equations using the boundary conditions $y(0) = 2$, $P(0) = P_c$ and $M(0) = 0$ [86,91]. The numerical solution of these equations provides the mass M, the radius R of the neutron star, and the value of $y_R = y(R)$. The latter parameter along with the quantity β are the basic ingredients of the tidal Love number k_2.

Moving on to the parameters of a binary neutron star system, a well-measured quantity by the gravitational waves detectors is the chirp mass $\mathcal{M}_c$ of the system [13,98]:

$$\mathcal{M}_c = \frac{(m_1 m_2)^{3/5}}{(m_1 + m_2)^{1/5}} = m_1 \frac{q^{3/5}}{(1+q)^{1/5}}, \tag{18}$$

where m_1 and m_2 are the masses of the heavier and lighter components. Therefore, the binary mass ratio $q = m_2/m_1$ is within the range $0 \leq q \leq 1$.

Another quantity that can be constrained from the analysis of the gravitational wave signal and is of great interest, is the effective tidal deformability [13,98]

$$\tilde{\Lambda} = \frac{16}{13}\frac{(12q+1)\Lambda_1 + (12+q)q^4\Lambda_2}{(1+q)^5}, \tag{19}$$

where Λ_i is the dimensionless tidal deformability, defined as [13,98]

$$\Lambda_i = \frac{2}{3}k_2\left(\frac{R_i c^2}{M_i G}\right)^5 \equiv \frac{2}{3}k_2\beta_i^{-5}, \quad i = 1, 2. \tag{20}$$

By observing Equations (14) and (20), one can find that Λ_i depends both on the star's compactness and the value of $y(R)$. More specifically, Λ_i depends directly on the stiffness of the EoS through the compactness β and indirectly through the speed of sound which appears in Equation (17). In addition, the applied EoS also affects the behavior of Λ regarding the neutron star's mass M and radius R.

2.4.2. Rotating Neutron Stars

In a fully general relativistic framework, the rotating neutron stars are studied with the use of the stationary axisymmetric spacetime metric, which is given by [99]

$$ds^2 = -e^{2\nu}dt^2 + e^{2\psi}(d\phi - \omega dt)^2 + e^{2\mu}\left(dr^2 + r^2 d\theta^2\right) \tag{21}$$

where the metric functions ν, ψ, ω, and μ depend only on the coordinates r and θ. In order to describe a rapidly rotating neutron star, in addition to the above metric, we need the matter inside the neutron star described as a perfect fluid. By neglecting sources of non-isotropic stresses, as well as viscous ones and heat transport, then the matter inside the neutron star can be fully described by the stress-energy tensor [99],

$$T^{\alpha\beta} = (\mathcal{E} + P)u^{\alpha}u^{\beta} + Pg^{\alpha\beta} \tag{22}$$

where u^{α} is the fluid's 4-velocity and $\mathcal{E}$ and P are the energy density and pressure.

For the stability of cold rotating neutron stars, the turning-point criterion is being used. It has to be noted that this is only a sufficient and not a necessary condition. Actually, the neutral stability line is positioned to the left of the turning-point line in (M, ρ_c) space, which implies that the star will collapse before reaching the turning-point line [100,101].

The numerical integration of the equilibrium equations was conducted under the RNS code [102] by Stergioulas and Friedman [103], which is actually based on the method developed by Komatsu, Eriguchi, and Hachisu [104] and modifications introduced by Cook, Shapiro, and Teukolsky [105].

3. Hot Neutron Stars

3.1. Thermodynamical Description of Hot Neutron Star Matter

The description of nuclear matter at finite temperature includes the Helmholtz free energy, where the differentials of the total free energy, as well as the total internal energy (assuming that the baryons are contained in volume V) are given by [106,107]

$$dF_{\text{tot}} = -S_{\text{tot}}dT - PdV + \sum_i \mu_i dN_i, \tag{23}$$

$$dE_{\text{tot}} = TdS_{\text{tot}} - PdV + \sum_i \mu_i dN_i, \tag{24}$$

where S_{tot} is the total entropy of baryons, μ_i is the chemical potential of each species, and N_i is the number of particles of each species, respectively. Furthermore, the free energy per particle can be expressed as

$$F(n,T,I) = E(n,T,I) - TS(n,T,I) = \frac{1}{n}[\mathcal{E}(n,T,I) - Ts(n,T,I)], \tag{25}$$

where $E = \mathcal{E}/n$ and $S = s/n$ are the internal energy and entropy per particle, respectively. It has to be noted here that for $T = 0$ MeV, Equation (25) leads to the equality between free and internal energy.

In addition, the entropy density s has the same functional form as a noninteracting gas system, calculated through the form

$$s_{\tau}(n,T,I) = -g\int \frac{d^3k}{(2\pi)^3}[f_{\tau}\ln f_{\tau} + (1-f_{\tau})\ln(1-f_{\tau})], \tag{26}$$

with spin degeneracy $g = [1,2]$. The first case corresponds to protons, neutrons, electrons, and muons, and the second case corresponds to neutrinos. Finally, the pressure and chemical potentials, which depend on the above quantities, are described as:

$$P = -\left.\frac{\partial E_{\text{tot}}}{\partial V}\right|_{S,N_i} = n^2\left.\frac{\partial(\mathcal{E}/n)}{\partial n}\right|_{S,N_i}, \tag{27}$$

$$\mu_i = \left.\frac{\partial E_{\text{tot}}}{\partial N_i}\right|_{S,V,N_{j\neq i}} = \left.\frac{\partial \mathcal{E}}{\partial n_i}\right|_{S,V,n_{j\neq i}}. \tag{28}$$

3.2. Bulk Thermodynamic Quantities

The pressure and chemical potentials, which are essential for the thermodynamical description of nuclear matter, can also be connected to the key quantity, that is, the free energy, as

$$P = -\frac{\partial F_{\rm tot}}{\partial V}\bigg|_{T,N_i} = n^2 \frac{\partial (f/n)}{\partial n}\bigg|_{T,N_i}, \tag{29}$$

$$\mu_i = \frac{\partial F_{\rm tot}}{\partial N_i}\bigg|_{T,V,N_{j\neq i}} = \frac{\partial f}{\partial n_i}\bigg|_{T,V,n_{j\neq i}}, \tag{30}$$

where f denotes the free energy density. It is worth mentioning that the pressure P can also be determined by [106,107],

$$P = Ts - \mathcal{E} + \sum_i \mu_i n_i. \tag{31}$$

In this case, the calculation of the entropy per particle is possible by differentiating the free energy density f with respect to the temperature,

$$S(n,T) = -\frac{\partial (f/n)}{\partial T}\bigg|_{V,N_i} = -\frac{\partial F}{\partial T}\bigg|_{n}. \tag{32}$$

By applying Equation (30), the chemical potentials take the form [37,108,109]

$$\mu_n = F + u\frac{\partial F}{\partial u}\bigg|_{Y_p,T} - Y_p \frac{\partial F}{\partial Y_p}\bigg|_{n,T}, \tag{33}$$

$$\mu_p = \mu_n + \frac{\partial F}{\partial Y_p}\bigg|_{n,T}, \tag{34}$$

$$\hat{\mu} = \mu_n - \mu_p = -\frac{\partial F}{\partial Y_p}\bigg|_{n,T}. \tag{35}$$

The free energy $F(n,T,I)$ and the internal energy $E(n,T,I)$, as well as the entropy, which depends on the latter quantities, can have the following quadratic dependence from the asymmetry parameter [37,109–112]:

$$F(n,T,I) = F(n,T,I=0) + I^2 F_{\rm sym}(n,T), \tag{36}$$

$$E(n,T,I) = E(n,T,I=0) + I^2 E_{\rm sym}(n,T), \tag{37}$$

$$S(n,T,I) = S(n,T,I=0) + I^2 S_{\rm sym}(n,T), \tag{38}$$

as the parabolic approximation (PA) suggests, where

$$F_{\rm sym}(n,T) = F(n,T,I=1) - F(n,T,I=0), \tag{39}$$

$$E_{\rm sym}(n,T) = E(n,T,I=1) - E(n,T,I=0), \tag{40}$$

$$S_{\rm sym}(n,T) = S(n,T,I=1) - S(n,T,I=0) = \frac{1}{T}(E_{\rm sym}(n,T) - F_{\rm sym}(n,T)). \tag{41}$$

While the above approximation is valid for the specific model, in general, it is mandatory to check it. For completeness, the PA is satisfied in both the internal energy and the free energy, as References [37,109–112] state. However, there are studies [113] in which the PA contains uncertainties. In sum, the PA and its validity strongly depend on the specific character of the selected nuclear model.

Equation (35), which is mandatory for this approximation, can be acquired by substituting Equation (36) as

$$\hat{\mu} = \mu_n - \mu_p = 4(1 - 2Y_p) F_{\text{sym}}(n, T). \tag{42}$$

This equation is similar to that obtained for cold catalyzed nuclear matter by replacing $E_{\text{sym}}(n)$ with $F_{\text{sym}}(n, T)$.

3.3. Lepton's Contribution

Stable nuclear matter and a chemical equilibrium state are explicitly connected at high densities for all reactions. Specifically, electron capture and β decay take place simultaneously

$$p + e^- \longrightarrow n + \nu_e \quad \text{and} \quad n \longrightarrow p + e^- + \bar{\nu}_e. \tag{43}$$

In consequence, a change in the electron fraction Y_e is in order. Considering that the generated neutrinos have left the system, a significant effect on the EoS is presented by changing the values of proton fraction Y_p [114,115]. In particular, the absence of neutrinos implies that

$$\hat{\mu} = \mu_n - \mu_p = \mu_e. \tag{44}$$

It is has to be mentioned that, in principle, the matter contains neutrons, protons, electrons, muons, photos, and antiparticles, which are in a thermal equilibrium state. However, in the present study, we consider only the contribution of neutron, protons, and electrons, as the contribution of the remaining particles is negligible [114]. As a consequence, the following relation holds:

$$\mu_n = \mu_p + \mu_e. \tag{45}$$

The energy density and pressure of leptons are calculated through the following formulae:

$$\mathcal{E}_l(n_l, T) = \frac{g}{(2\pi)^3} \int \frac{d^3k \sqrt{\hbar^2 k^2 c^2 + m_l^2 c^4}}{1 + \exp\left[\frac{\sqrt{\hbar^2 k^2 c^2 + m_l^2 c^4} - \mu_l}{T}\right]}, \tag{46}$$

$$P_l(n_l, T) = \frac{1}{3} \frac{g(\hbar c)^2}{(2\pi)^3} \int \frac{1}{\sqrt{\hbar^2 k^2 c^2 + m_l^2 c^4}} \times \frac{d^3k\, k^2}{1 + \exp\left[\frac{\sqrt{\hbar^2 k^2 c^2 + m_l^2 c^4} - \mu_l}{T}\right]}. \tag{47}$$

The chemical potential of electrons is available through the Equations (42) and (45), as

$$\mu_e = \mu_n - \mu_p = 4I(n, T) F_{\text{sym}}(n, T), \tag{48}$$

which is crucial for the calculation of the proton fraction as a function of both the baryon density and the temperature. The construction of the EoS of hot nuclear matter in the β-equilibrium state is provided through the calculation of the total energy density $\mathcal{E}_t$, as well as the total pressure P_t. The total energy density (with all terms) is given by

$$\mathcal{E}_t(n, T, I) = \mathcal{E}_b(n, T, I) + \sum_l \mathcal{E}_l(n, T, I) + \sum_{\bar{l}} \mathcal{E}_{\bar{l}}(n, T, I) + \mathcal{E}_\gamma(n, T), \tag{49}$$

where $\mathcal{E}_b(n, T, I)$ is the contribution of baryons, $\mathcal{E}_l(n, T, I)$, $\mathcal{E}_{\bar{l}}(n, T, I)$ are the contributions of particles and antiparticles of leptons, and $\mathcal{E}_\gamma(n, T)$ is the contribution of photons. The total pressure (with all terms) is

$$P_t(n, T, I) = P_b(n, T, I) + \sum_l P_l(n, T, I) + \sum_{\bar{l}} P_{\bar{l}}(n, T, I) + P_\gamma(T), \tag{50}$$

where $P_b(n,T,I)$ is the contribution of baryons,

$$P_b(n,T,I) = T \sum_{\tau=p,n} s_\tau(n,T,I) + \sum_{\tau=n,p} n_\tau \mu_\tau(n,T,I) - \mathcal{E}_b(n,T,I), \tag{51}$$

while $P_l(n,T,I)$, $P_{\bar{l}}(n,T,I)$ are the contributions of particles and antiparticles of leptons, and $P_\gamma(T)$ is the contribution of photons.

3.4. Isothermal Configuration

In the isothermal temperature profile, by considering that for each value of temperature, the value of the proton fraction is a well-known function of the baryon density, the total energy density is evaluated as

$$\mathcal{E}_t(n,T,Y_p) = \mathcal{E}_b(n,T,Y_p) + \mathcal{E}_e(n,T,Y_p), \tag{52}$$

where

$$\mathcal{E}_b(n,T,Y_p) = nF_{PA} + nTS_{PA}, \tag{53}$$

$\mathcal{E}_e(n,T,Y_p)$ is given by Equation (46), replacing the leptons with electrons and μ_e from Equation (48), and F_{PA} and S_{PA} are given by Equations (36) and (38), respectively. Moreover, the total pressure is evaluated as

$$P_t(n,T,Y_p) = P_b(n,T,Y_p) + P_e(n,T,Y_p), \tag{54}$$

where

$$P_b(n,T,Y_p) = n^2 \frac{\partial F_{PA}(n,T,Y_p)}{\partial n}\bigg|_{T,n_i}, \tag{55}$$

and $P_e(n,T,Y_p)$ is given by Equation (47), replacing the leptons with the electrons and μ_e from Equation (48).

Thus, the Equations (52) and (54) for the energy density and the pressure, respectively, correspond to the ingredients for the construction of isothermal EoSs of hot nuclear matter in β equilibrium state.

3.5. Isentropic Configuration and Neutrino Trapping

In an isentropic configuration, we assume that the entropy per baryon and lepton fraction are constant in the interior of the neutron star (protoneutron star). Specifically, according to Equation (43), the neutrinos are trapped in the interior of the star and the proton fraction is significantly increased. The relevant chemical equilibrium can be expressed in terms of the chemical potentials for the four species,

$$\mu_n + \mu_{\nu_e} = \mu_p + \mu_e. \tag{56}$$

In addition, the charge neutrality demands the equality between proton and electron fraction, while the total fraction of leptons is equal to $Y_l = Y_e + Y_{\nu_e}$. Henceforth, the chemical equilibrium is expressed as

$$\mu_e - \mu_{\nu_e} = \mu_n - \mu_p = 4(1-2Y_p)F_{\rm sym}(n,T). \tag{57}$$

In this case too, the relevant system of equations can provide us with the density and temperature dependence of proton and neutrino fractions, and their chemical potentials, by assuming a constant entropy per baryon. Nonetheless, we applied the approximation of $Y_p \simeq 2/3Y_l + 0.05$ (3% accuracy) introduced by Takatsuka et al. [114], in order to avoid computational complications. The ingredients for the construction of isentropic EoSs are given by the Equations (49) and (50).

3.6. Construction of the Hot EoSs

The construction of the EoS for the interior of neutron stars at finite temperature and entropy per baryon is based on the MDI model and data provided by Akmal et al. [73]. In particular, we utilized the APR-1 EoS data from Akmal et al. [73] for the energy per particle of SNM and PNM in the density range $[0.04, 0.96]$ fm^{-3}. The process leads to the evaluation of the coefficients for the symmetric and asymmetric nuclear matter, and finally, to the construction of the EoS, hereafter MDI+APR1.

In the case of the isothermal temperature profile, we have constructed 10 EoSs in the temperature range $[1, 60]$ MeV, while in the isentropic case, we have constructed nine EoSs in entropy per baryon and lepton fraction ranges $[1, 3]$ k_B and $[0.2, 0.4]$, respectively. For the crust region of the finite temperature cases and the low-density region ($n_b \leq 0.08$ fm^{-3}), as well as the finite entropies per baryon and lepton fractions, the EoSs of Lattimer and Swesty [30] and the specific model corresponding to the incomprehensibility modulus at the saturation density of SNM $K_s = 220$ MeV are used (https://www.stellarcollapse.org, accessed on 4 March 2020).

3.7. Rapidly Rotating Hot Neutron Stars

The stability of hot neutron stars is acquired via a specific version of the secular instability criterion of Friedman et al. [116], which follows Theorem I of Sorkin [117]. In a continuous sequence of equilibria at a fixed baryon number $N_{\rm bar}$ and total entropy of the neutron star $S_{\rm t}^{\rm ns}$, the extremal point of the stability loss is found when [118]

$$\left.\frac{\partial J}{\partial n_b^c}\right|_{N_{\rm bar}, S_{\rm t}^{\rm ns}} = 0, \tag{58}$$

where J and n_b^c are the angular momentum and central baryon density of the star, respectively.

Furthermore, in a sequence, the turning-point appears in the case where three out of four following derivatives vanish,

$$\frac{\partial M_{\rm gr}}{\partial n_b^c}, \quad \frac{\partial M_b}{\partial n_b^c}, \quad \frac{\partial J}{\partial n_b^c}, \quad \text{and} \quad \frac{\partial S_{\rm t}^{\rm ns}}{\partial n_b^c}, \tag{59}$$

with $M_{\rm gr}$ and $M_{\rm b}$ denoting the gravitational and baryon mass [52,119]. In addition, the turning-point theorem shows that at this point, the fourth derivative also vanishes, and the sequence has transitioned from stable to unstable.

It has to be mentioned that the criterion for secularly stable/unstable configurations is essential only for constant entropy per baryon or temperature [119]. In this review, the entropy per baryon and the temperature in each case are constant throughout the neutron star. Therefore, the remaining criteria vanish at the maximum mass configuration (the last stable point). Furthermore, we considered that in the rotating configuration, the maximum mass and the maximum angular velocity coincide, which generally is not the case [99]. However, the existing difference is very small, and it could not be detected within the precision of our calculations [118].

The numerical integration of the equilibrium equations was conducted under the publicly available numerical code *nrotstar* from the C++ Lorene/Nrotstar library [120].

4. Results and Discussion

4.1. Speed of Sound and Tidal Deformability

In our study, we used two cases for the value of speed of sound, the lower bound of $(v_s/c)^2 = 1/3$ and the upper one of $(v_s/c)^2 = 1$, and four transition densities $n_{\rm tr} = \{1, 1.5, 2, 3\} n_s$ [65].

In Figure 2, we display the corresponding mass-radius (M-R) diagram, which we obtained from the numerical solution to the TOV system of equations. The green colored lines correspond to the $(v_s/c)^2 = 1/3$ limit, while the blue ones correspond to the

$(v_s/c)^2 = 1$ limit. One can observe that each transition density leads to bifurcations in the M-R diagram. Between the same kind of linestyle, the lower and upper bounds, $(v_s/c)^2 = 1/3$ and $(v_s/c)^2 = 1$ of speed of sound correspond to lower and higher masses, respectively. In general, the higher the transition density, the softer the EoS, with the lower bound of $(v_s/c)^2 = 1/3$ leading to a more soft EoS compared to the $(v_s/c)^2 = 1$ case. In addition, the estimation of the GW170817 event and the NICER's data are also displayed [24,121]. We notice that while there is an accordance between the two observations, the GW170817 event (from the gravitational-waves perspective) is more informative for our study than the NICER's detection, as it restricts the cases leading to the exclusion of EoS at least with transition density $n_{tr} = n_s$, for both bounds of speed of sound.

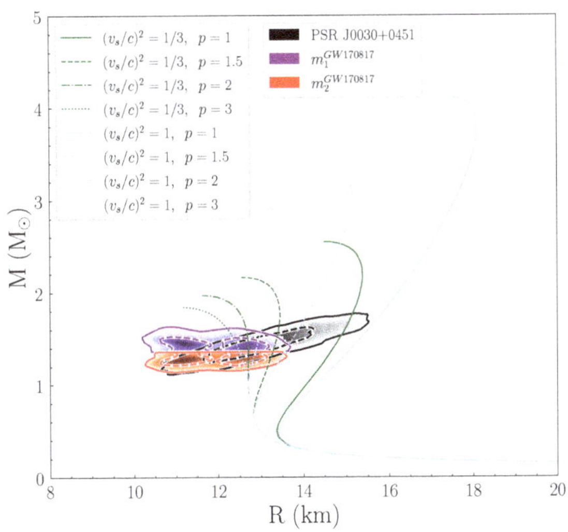

Figure 2. Mass vs. radius for an isolated neutron star and for the two cases of speed of sound bounds. The blue (green) lines correspond to the upper (lower) bound. The black diagonal shaded region corresponds to NICER's observation (data taken from Reference [121]), while the purple upper (orange lower) shaded region corresponds to the higher (smaller) component of GW170817 event (data taken from Reference [24]). The solid (dashed) contour lines describe the 90% (50%) confidence interval.

Our study takes into consideration the observation of binary neutron stars mergers from the gravitational waves detectors. Therefore, we used the measured upper limit of the effective tidal deformability $\tilde{\Lambda}$ provided by the events GW170817 and GW190425 [14,24,98]. The chirp masses for the two events are $\mathcal{M}_c = 1.186\ M_\odot$ [13] and $\mathcal{M}_c = 1.44\ M_\odot$ [14], respectively. The component masses vary in the ranges $m_1 \in (1.36, 1.60)\ M_\odot$ and $m_2 \in (1.16, 1.36)\ M_\odot$ [98] (GW170817) and $m_1 \in (1.654, 1.894)\ M_\odot$ and $m_2 \in (1.45, 1.654)\ M_\odot$ (GW190425). We notice that we modified the range of the component masses (especially in the second event) to have an equal mass boundary, i.e., $q \leq 1$.

In Figure 3, we display the effective tidal deformability $\tilde{\Lambda}$ as a function of q for both events. In Figure 3a, we observe that the upper limit on $\tilde{\Lambda}$ (derived from the GW170817 event) leads to the exclusion of both cases of speed of sound with transition density $n_{tr} = 1, 1.5\ n_s$. By comparing to Figure 2, the constraints on the upper limit of $\tilde{\Lambda}$ in Figure 3 make more clear which cases must be excluded. For the second event in Figure 3b, we observe that all the EoSs are shifted to lower values of $\tilde{\Lambda}$, compared to the GW170817 event. This is because of the higher value of chirp mass in the second event (GW190425). Contrary to the GW170817 event, the upper limit on $\tilde{\Lambda}$, provided by GW190425 event, excludes only these EoSs with transition density $n_{tr} = n_s$, for both bounds of the speed of sound.

In general, for both events, the EoSs corresponding to higher values of transition density $n_{\rm tr}$ lead to smaller values of $\tilde{\Lambda}$. Therefore, the measured upper limits of $\tilde{\Lambda}$ favor softer EoSs. We have to notice that for the GW190425 event, we did not take into consideration the cases with transition density $n_{\rm tr} = 3\ n_s$ because the EoS with $(v_s/c)^2 = 1/3$ and $n_{\rm tr} = 3n_s$ cannot reproduce the masses of this event.

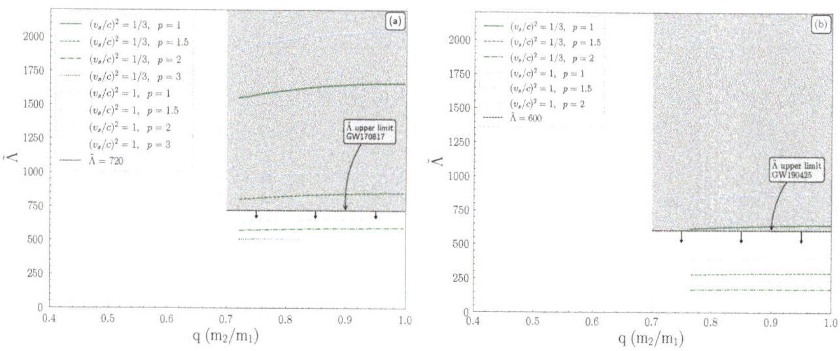

Figure 3. The effective tidal deformability $\tilde{\Lambda}$ as a function of the binary mass ratio q for the event (**a**) GW170817 and (**b**) GW190425. The measured upper limits for $\tilde{\Lambda}$ are also indicated, with the grey shaded region corresponding to the excluded area. The green (blue) curves correspond to the $(v_s/c)^2 = 1/3$ ($(v_s/c)^2 = 1$) case.

Beyond the useful constraints that we obtained so far by the study of the EoSs through the observed upper limit of $\tilde{\Lambda}$, a more direct connection between this quantity and the speed of sound bounds is still needed. This idea, which lies at the very heart of our study, was the main motive. Such a direct relation between the referred quantities can be accomplished if we treat the variation of $\tilde{\Lambda}$ in Figure 3 as a function of the transition density $n_{\rm tr}$, i.e., the $\tilde{\Lambda}^{(1/3,1)}_{\rm min} - n_{\rm tr}$ and $\tilde{\Lambda}^{(1/3,1)}_{\rm max} - n_{\rm tr}$ relations.

In Figure 4, we display the relation between the effective tidal deformability $\tilde{\Lambda}$ and the transition density $n_{\rm tr}$ at the maximum mass configuration for the two bounds of the speed of sound, $v_s = c/\sqrt{3}$ and $v_s = c$, and the two events GW170817 (Figure 4a) and GW190425 (Figure 4b). The corresponding upper measured limits for $\tilde{\Lambda}$, as well as the compatible lower transition density values, are also indicated. The predictions on the bound of the speed of sound which are considered between the two referred limits correspond to the middle region.

The main remarks from the observation of Figure 4 are the following

1. The overall thickness decreases as the transition density $n_{\rm tr}$ reaches higher values. This behavior can be explained by the variation of the radius $M(R)$ presented in the M-R diagram (see Figure 2).
2. The thickness of each shaded region decreases as the $n_{\rm tr}$ reaches higher values.
3. The shaded areas are shifted downwards in the GW190425 event, compared with the GW170817 event. This behavior is due to the increase in the component masses. A similar behavior was observed in Figure 3b, compared to Figure 3a.

According to our findings, for the GW170817 event, the lower limit for the transition density is 1.626 n_s for $v_s = c/\sqrt{3}$ and 1.805 n_s for $v_s = c$. In the case of the second event, GW190425, the corresponding limits are 1.015 n_s for $v_s = c/\sqrt{3}$ and 1.216 n_s for $v_s = c$. Therefore, the first event imposes more stringent constraints on the EoS. In particular, the value of the speed of sound must be lower than $v_s = c/\sqrt{3}$, at least up to density 1.626 n_s (so the EoS is soft enough to predict the tidal deformability). Furthermore, the EoS must remain casual at least up to density 1.805 n_s. In addition, according to the Fermi liquid theory (FLT), the speed of sound must be $v^2_{s,FLT} \leq 0.163c^2$ for $n = 1.5n_s$ [122], meaning that the EoS cannot exceed this value for $n \leq 1.5n_s$, which is in agreement with our finding of the lower limit $n_{tr} = 1.626n_s$ for the case of $v_s = c/\sqrt{3}$.

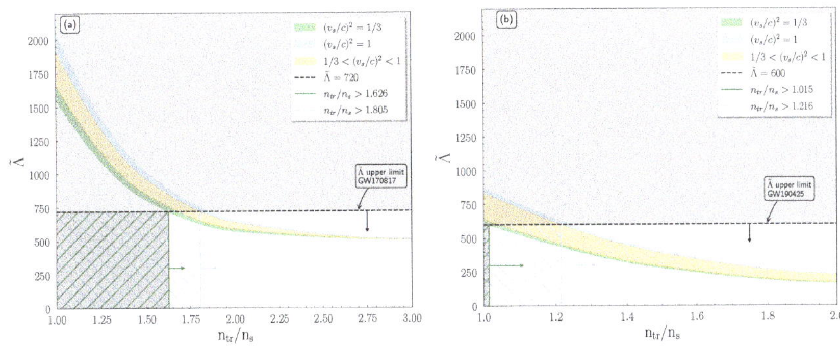

Figure 4. $\tilde{\Lambda}$ as a function of the transition density n_{tr} (in units of saturation density n_s) at the maximum mass configuration for the two speed of sound bounds $v_s = c/\sqrt{3}$ and $v_s = c$ and for the events (**a**) GW170817 and (**b**) GW190425. The measured upper limits for $\tilde{\Lambda}$ [14,98] as well as the corresponding lower values of transition density are also indicated for both events. The green (blue) arrow marks the accepted region of transition density for the $v_s = c/\sqrt{3}$ ($v_s = c$) case. The green lower (blue upper) curved shaded region corresponds to the $v_s = c/\sqrt{3}$ ($v_s = c$) limit. The yellow shaded area indicates the region between the two cases of bounds of the speed of sound.

We notice that so far we used the upper limit on $\tilde{\Lambda}$ to impose stringent constraints on the n_{tr}. The existence of a lower limit on $\tilde{\Lambda}$ could provide further information. Indeed, for the GW170817 event, such a lower limit is provided both by the gravitational wave data [24,98] and the electromagnetic (EM) counterpart of the merger [123–128]. Most et al. [129] used the bound of Reference [123] and demonstrated its significance in order to further constrain the tidal deformability $\tilde{\Lambda}_{1.4}$ and the radius $R_{1.4}$ of a $M = 1.4\, M_{\odot}$ neutron star. For our case of interest and especially for the GW170817 event, a lower limit on $\tilde{\Lambda}$ similar to the proposed values, could not provide any further constraint, even if we consider the more optimistic boundary of $\tilde{\Lambda} \geq 400$.

On the contrary, for the second event GW190425, its higher component masses lead to smaller values of $\tilde{\Lambda}$. Hence, there is an inability for the upper limit of $\tilde{\Lambda}$ to provide further constraints. We speculate that the existence of a lower limit on $\tilde{\Lambda}$ would be able to provide constraints, especially leading to an upper limit for n_{tr}. Hence, binary neutron stars coalescences with heavy masses would be helpful to constrain the upper limit of n_{tr} via the lower limit of $\tilde{\Lambda}$ as provided by the EM counterpart. Unfortunately, an EM counterpart for the GW190425 event was not detected [14,130].

Furthermore, we provide in Figure 4 an expression for the $\tilde{\Lambda}^{(1/3)}_{\rm min}$ and $\tilde{\Lambda}^{(1)}_{\rm min}$ boundary curves of the green (lower) and blue (upper) shaded regions, respectively. This expression gives the exact value of the lower limit on $n^{(1/3)}_{\rm tr}$ and $n^{(1)}_{\rm tr}$, respectively. The expression is given by the following equation, and the coefficients on Table 1,

$$\tilde{\Lambda} = c_1 \coth\left[c_2 \left(\frac{n_{tr}}{n_s}\right)^2\right]. \tag{60}$$

As one can observe from Figure 2, the highest mass is provided by the stiffest EoSs, i.e., the higher value of speed of sound. Therefore, the behavior of the maximum mass M_{max} and the speed of sound v_s^2 has to be studied further.

In Figure 5, the behavior of the effective tidal deformability $\tilde{\Lambda}$ as a function of the maximum mass for the two speed of sound bounds and for both events is displayed. The corresponding upper observational limit for $\tilde{\Lambda}$ (black dashed horizontal line), the compatible maximum mass in each case (horizontal arrows), and the current observed maximum neutron star mass $M = 2.14^{+0.10}_{-0.09}\, M_\odot$ (vertical purple shaded region) are also indicated.

Table 1. Parameters of the Equations (60) and (61) for both events and all bounds of the speed of sound.

Bounds	GW170817				GW190425			
	c_1	c_2	c_3	c_4	c_1	c_2	c_3	c_4
c	500.835	0.258	53.457	0.873	47.821	0.055	10.651	1.068
$c/\sqrt{3}$	503.115	0.325	38.991	1.493	43.195	0.069	5.024	1.950

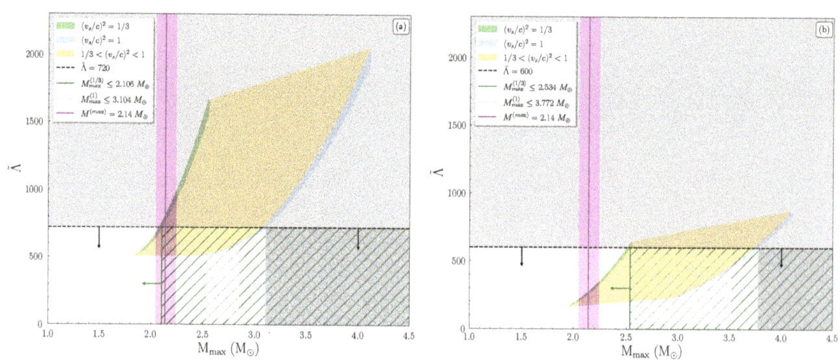

Figure 5. The effective tidal deformability $\tilde{\Lambda}$ as a function of the maximum mass for the two speed of sound bounds $v_s = c/\sqrt{3}$ and $v_s = c$ and for the events (**a**) GW170817 and (**b**) GW190425. The measured upper limits for $\tilde{\Lambda}$ (black dashed lines with arrows; see References [14,98]); the corresponding maximum mass shaded regions, for the $v_s = c/\sqrt{3}$ (left green) case, for the $v_s = c$ case (right blue), and for the middle cases (yellow); and the current observed maximum neutron star mass $M = 2.14^{+0.10}_{-0.09}\, M_\odot$ (purple shaded vertical area; see Reference [17]) are also displayed. The green left (blue right) arrow marks the accepted region of maximum mass $M_{\rm max}$ for $v_s = c/\sqrt{3}$ ($v_s = c$) case.

At first glance in Figure 5, there is a contradiction between the maximum mass and the upper limit of the observed $\tilde{\Lambda}$. For the first event shown in Figure 5a, the upper limit of $\tilde{\Lambda}$ is compatible with a maximum mass value $2.106 M_\odot$ for $v_s = c/\sqrt{3}$ and $3.104 M_\odot$ for $v_s = c$. Nonetheless, this bound corresponds to transition density in approximation 1.5 n_s. Experimental evidence disfavors this value. Therefore, the simultaneous derivation of the maximum mass combined with the experimental knowledge that the EoS cannot take this bound of sound speed for $n_{\rm tr} = 1.5 n_s$ are in contradiction. Furthermore, the upper limit on $M_{\rm max}$ for the case of $(v_s/c)^2 = 1/3$ lies roughly inside the estimation of the measured maximum mass. In the general perspective, we notice that two different points of view antagonize each other. The constraints derived by the upper limit on $\tilde{\Lambda}$ lead to softer EoSs, contrary to the observational estimations of the maximum mass of neutron stars, which lead to stiffer EoSs. As we move to higher values of the speed of sound, this contrast decreases, with the causal scenario of $v_s = c$ leading to a very wide area for the maximum mass.

In the case of the second event GW190425 displayed in Figure 5b, the constraints provided by the measured $\tilde{\Lambda}$ are less stringent than the GW170817 event, with a maximum mass value of $M_{\rm max} \leq 2.534\ M_\odot$ for $v_s = c/\sqrt{3}$ and $M_{\rm max} \leq 3.772\ M_\odot$ for $v_s = c$. HOwever, the presence of a lower limit on $\tilde{\Lambda}$, especially in the case of events with heavy components (such as GW190425), could constrain the lower maximum mass.

In addition, we provide an expression that describes the $\tilde{\Lambda}$ as a function of the maximum mass $M_{\rm max}$. The expression is given by the following equation and the coefficients in Table 1,

$$\tilde{\Lambda} = c_3 \left(e^{M_{\rm max}} - 1 \right)^{c_4}. \tag{61}$$

The expression in this form means that $M_{\text{max}} \to 0 \Rightarrow \tilde{\Lambda} \to 0$. Moreover, the adoption of an upper limit on the maximum mass M_{max} in Figure 5 could provide an additional constraint on the behavior of the speed of sound. Specifically, by applying the estimated upper limit $M_{\text{max}} \leq 2.33\ M_{\odot}$ [131], the case of $(v_s/c)^2 = 1$ in Figure 5a for the GW170817 event should be excluded. On the contrary, the estimated upper limit $M_{\text{max}} \leq 2.106\ M_{\odot}$ for the $(v_s/c)^2 = 1/3$ bound, is a more tight constraint. Additionally, an upper limit such as $M_{\text{max}} \leq 2.33\ M_{\odot}$ imposes a general upper bound on the possible intermediate values of speed of sound (intermediate shaded area in the figure). Concerning the second event in Figure 5b, a strict upper limit on M_{max} could constrain even the $(v_s/c)^2 = 1/3$ case.

Another interesting relation is the $\tilde{\Lambda}$ as a function of the radius of a 1.4 $M_{\odot}$ neutron star, for both events, which is displayed in Figure 6. First of all, the upper limit on $\tilde{\Lambda}$ leads to a limitation on the maximum values of the radius, especially in the case of $v_s = c/\sqrt{3}$. Furthermore, there is a trend between $\tilde{\Lambda}$ and $R_{1.4}$, which was also remarked on by Raithel et al. [132], mentioning that the effective tidal deformability depends strongly on the radii of the stars rather on the component masses. This strong dependence can be observed in Figure 6.

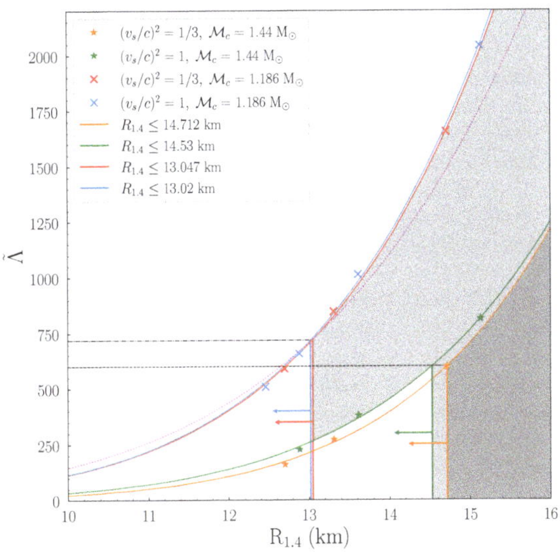

Figure 6. The effective tidal deformability $\tilde{\Lambda}$ as a function of $R_{1.4}$ for both events and all the bounds of the speed of sound. The dashed and dash-dotted horizontal black lines correspond to the upper limit on $\tilde{\Lambda}$ for the GW190425 and GW170817 events, respectively, taken from References [14,98]. The grey shaded regions correspond to the excluded areas. The horizontal arrows indicate the allowed area for $R_{1.4}$ in each case. The purple dotted curve demonstrates the proposed expression by Reference [133,134].

In particular, for the GW170817 event, the curves of the two limited cases, red and blue for the $(v_s/c)^2 = 1/3$ and $(v_s/c)^2 = 1$ bounds, respectively, of the speed of sound are almost identical. The cross marks correspond to the specific values for each case. In our study, we considered four cases of transition density n_{tr}; therefore, eight marks are expected in total, but in the diagram, only seven can be seen. This is because of the identical values for the $n_{\text{tr}} = 3n_s$ case that the two bounds provide. This is clear from their behavior in Figure 2 in which, for the mass range of GW170817 event, their M-R curves are identical.

Moreover, as the effective tidal deformability $\tilde{\Lambda}$ shows higher values, the distance between them also increases. The same behavior is present in the curves of Figure 3a, in which their in-between distance increases for higher values of $\tilde{\Lambda}$. This increment is related to the

n_{tr}, meaning that the differentiation for small values of n_{tr} is more obvious. Hence, in these cases, the effect of each bound of the speed of sound is easier to be manifested.

The dotted purple line corresponds the approximate relation of References [133,134]. We notice that this approximate relation is valid only for the first event and for specific assumptions on the components' radii. In particular, the main assumption of this approximation consists on the $R_1 \approx R_2$ relation. From the comparison of Figure 6 with the M-R diagram of Figure 2, it follows that for smaller values of n_{tr} (i.e., more stiff EoSs), (a) the inclination of the curves increases and (b) the difference between the M-R curves of boundary cases also increases. Therefore, these remarks, combined with the strong dependence of Λ_i to R (see Equation (20)), show the inability of the proposed expression to accurately reproduce the values of $\tilde{\Lambda}$ in the high-values region can be explained.

The grey shaded area indicates the excluded area due to the upper limit on $\tilde{\Lambda}$, provided by Reference [98]. The upper limit of $\tilde{\Lambda}$ leads to constraints on the radius $R_{1.4}$, especially $R_{1.4} \leq 13.047$ km for the $(v_s/c)^2 = 1/3$ bound and $R_{1.4} \leq 13.02$ km for the $(v_s/c)^2 = 1$ bound. These upper limits are consistent with other analyses [24,124,128,129,132,134–137].

For the second event (GW190425), we notice that the exact range of the component masses is not determinant [130]. The orange and green lines and marks correspond to the $(v_s/c)^2 = 1/3$ and $(v_s/c)^2 = 1$ bounds, respectively, of the speed of sound. The shaded grey region indicates the excluded region by the upper limit of $\tilde{\Lambda}$ [14]. The orange and green arrows indicate the allowed region for each case. For $(v_s/c)^2 = 1/3$, the constraint on the radius is $R_{1.4} \leq 14.712$ km, while for $(v_s/c)^2 = 1$ is $R_{1.4} \leq 14.53$ km. These are more stringent constraints compared to the 15 km and 16 km of Reference [14]. We notice that recently it was found in Reference [138] that the joint contribution of gravitational waves and NICER data favors the violation of the conformal limit $(v_s/c)^2 < 1/3$. In particular, this analysis suggests the violation of the conformal limit around $4\rho_{nuc}$ density, where $\rho_{nuc} = 2.8 \times 10^{14} \mathrm{g/cm^3}$ is the nuclear saturation density.

In addition, one can observe the similarity of the curves' behavior between the two events. For higher values of n_{tr}, the distance between the points grows. One of the main differences is that for the second event, the curves and the points are shifted to smaller values of $\tilde{\Lambda}$ because of the higher chirp mass $\mathcal{M}_c$ of the system. Another observation is that the fitting lines are more distinct from each other, contrary to the GW170817 event, in which they were almost identical; nevertheless, there is a common trend (see also Reference [132]). For this reason, we applied the following expression

$$\tilde{\Lambda} = c_1 R_{1.4}^{c_2}, \tag{62}$$

where $R_{1.4}$ is in km, similar to the proposed relations of References [133,134]. The coefficients for each case are given in Table 2.

Table 2. Coefficients of Equation (62) for the two bounds of the speed of sound.

Event	Bounds	c_1	c_2
GW170817	c	0.12357×10^{-4}	6.967
	$c/\sqrt{3}$	0.12179×10^{-4}	6.967
GW190425	c	0.870×10^{-6}	7.605
	$c/\sqrt{3}$	0.088×10^{-6}	8.422

4.2. *GW190814: A Postulation of the Most Massive Neutron Star*

The GW190814 event that arose from the merger of a $\sim$23 $M_\odot$ black hole with a $\sim$2.6 $M_\odot$ compact object has provided various scenarios for the nature of the second component. In particular, the possibilities for the second merger component are that of (a) the lightest black hole, (b) the most compact neutron star, (c) a rapidly rotating neutron star, and (d) an exotic compact object. We note that in the present review, we consider only the scenarios where the compact object is a nonrotating (most compact) neutron star or is a rapidly rotating one [66].

In Figure 7, we display the gravitational mass as a function of the Kerr parameter for the pure MDI-APR EoS. In addition, we note the universal relation

$$M_{\rm rot} = M_{\rm TOV}\left[1 + 0.132\left(\frac{\mathcal{K}}{\mathcal{K}_{\rm max}}\right)^2 + 0.071\left(\frac{\mathcal{K}}{\mathcal{K}_{\rm max}}\right)^4\right], \quad (63)$$

where $\mathcal{K}_{\rm max} = 0.68$ is the Kerr parameter at the mass-shedding limit, for two limiting cases: (a) $M_{\rm TOV} = 2.08\ M_\odot$ and (b) $M_{\rm TOV} = 2.3\ M_\odot$ [139]. The limiting cases correspond to the minimum and maximum possible mass for a neutron star [139], along with the maximum value of the Kerr parameter (considering the minimum possible mass) [139]. The area marked by the intersection of the gravitational mass, $M = 2.59^{+0.08}_{-0.09}\ M_\odot$, with the Kerr parameter, $\mathcal{K} = [0.49, 0.68]$ [139], notes the area where the compact object can exist. Figures 1 and 7 show that the pure MDI-APR EoS, which is well-defined in the above limits, is a suitable hadronic EoS to describe the $\sim$2.6 $M_\odot$ compact object.

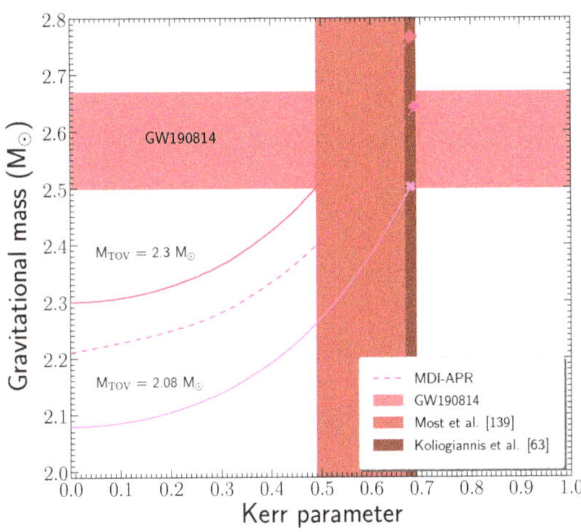

Figure 7. Dependence of the gravitational mass on the Kerr parameter. The lower solid line represents Equation (63) with $M_{\rm TOV} = 2.08\ M_\odot$, while the upper solid line represents Equation (63) with $M_{\rm TOV} = 2.3\ M_\odot$. The dashed line corresponds to the MDI-APR EoS. In addition, the horizontal shaded region notes the mass of the second component of GW190814 event, and the vertical wide shaded region marks the Kerr parameter, $\mathcal{K} = [0.49, 0.68]$, according to Reference [139]. Furthermore, the narrow vertical shaded region indicates the Kerr parameter, $\mathcal{K}_{\rm max} = [0.67, 0.69]$, extracted from Reference [63] by assuming that the low mass component was rotating at its mass-shedding limit. The cross, the plus sign, and the diamond show the maximum mass configuration at the mass-shedding limit.

Furthermore, by assuming that the second merger component is rotating at its mass-shedding limit, possible constraints are available through the Kerr parameter, the equatorial radius, and the central energy/baryon density. Specifically, by employing the relation found in Reference [63]

$$\mathcal{K}_{\rm max} = 0.488 + 0.074\left(\frac{M_{\rm max}}{M_\odot}\right), \quad (64)$$

for the observable gravitational mass, the maximum Kerr parameter is evaluated in the range $\mathcal{K}_{\rm max} = [0.67, 0.69]$, a feature that is also noted in Figure 7. Moreover, taking into consideration the relation from Reference [64] that connects the Kerr parameter with the compactness parameter at the mass-shedding limit, namely

$$\mathcal{K}_{\max} = 1.34\sqrt{\beta_{\max}} \quad \text{and} \quad \beta_{\max} = \frac{G}{c^2}\frac{M_{\max}}{R_{\max}}, \tag{65}$$

the equatorial radius is calculated in the range $R_{\max} = [14.77, 14.87]$ km.

Finally, we focused on the central energy/baryon density, a property that is connected to the time evolution of pulsars and the appearance of a possible phase transition. The above dependence is presented in Figure 8 as the dependence of the maximum gravitational mass on both the central energy density and the central baryon density. Specifically, Figure 8 contains a wide range of hadronic EoSs (23 EoSs) [63] both at the nonrotating and maximally rotating configurations, the analytical solution of Tolman VII, Equation (66), denoted as

$$\frac{M}{M_\odot} = 4.25\sqrt{\frac{10^{15}\ \text{g cm}^{-3}}{\mathcal{E}_c/c^2}}, \tag{66}$$

according to Reference [63], the calculation data from Cook et al. [140] and Salgado et al. [141], and the recent data for the nonrotating and maximally rotating configurations both in the cases of $(v_s/c)^2 = 1/3$ and $(v_s/c)^2 = 1$ with the corresponding transition densities.

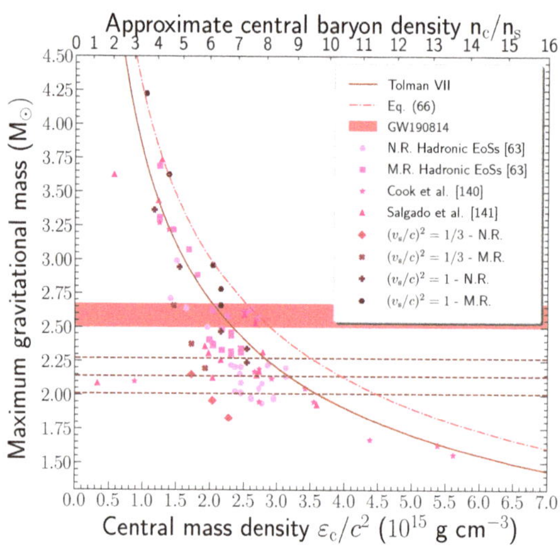

Figure 8. Dependence of the maximum gravitational mass on the central energy/baryon density both at nonrotating and rapidly rotating with the Kepler frequency configurations. Circles and squares correspond to 23 hadronic EoSs [63] at the nonrotating (N.R.) and maximally-rotating (M.R.) cases, respectively, and stars and triangles correspond to data of Cook et al. [140] and Salgado et al. [141], respectively. Furthermore, diamonds and plus signs note the nonrotating configuration, while crosses and polygons show the maximally-rotating one, in the cases of the two limiting values of the sound speed. The horizontal dashed lines mark the current observed neutron star mass limits (2.01 $M_\odot$ [16], 2.14 $M_\odot$ [17], and 2.27 $M_\odot$ [18]). Equation (66) is noted with the dashed-dotted line, while for comparison, the Tolman VII analytical solution [63] is added with the solid line. The horizontal shaded region notes the mass range of the second component of GW190814 event.

In addition, via Equation (66), which is used for the description of the upper bound for the density of cold baryonic matter [63], the central energy density can be constrained in the narrow range $\mathcal{E}_c/c^2 = [2.53, 2.89]\ 10^{15}$ g cm^{-3}. The latter indicates that neutron stars with higher values of central energy density cannot exist. Furthermore, Figure 8 provides us the tools to extract the corresponding region for the central baryon density, which is $n_c = [7.27, 8.09]\ n_s$. It is worth mentioning that the cases in this review meet the limit for the central energy/baryon density as they are included in the region described under Equation (66).

4.3. *The Case of a Very Massive Neutron Star*

4.3.1. Isolated Non-Rotating Neutron Star

In this study, we extended our previous work in Reference [65] for an isolated non-rotating neutron star by using two transition densities $n_{\rm tr} = [1.5, 2]n_s$ and eight values of speed of sound bounds $(v_s/c)^2 = [1/3, 0.4, 0.5, 0.6, 0.7, 0.8, 0.9, 1]$ [66]. The values of transition density were taken to be close to the constraints of Reference [65]. By solving numerically the system of TOV equations, combined with the previous bounds for the speed of sound, we obtained the M-R diagram, displayed in Figure 9a.

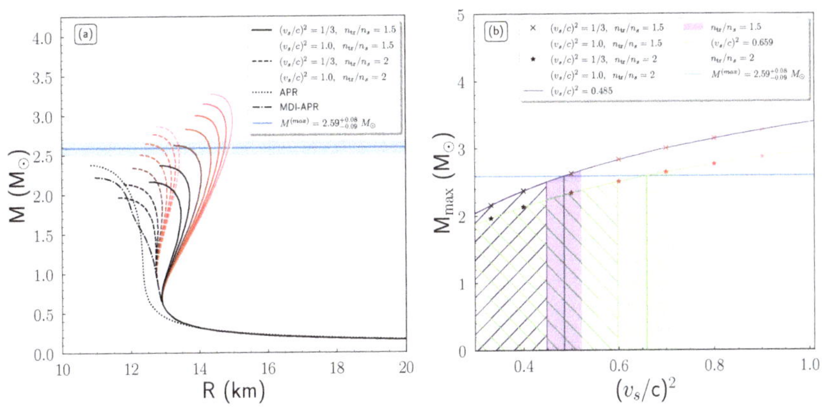

Figure 9. (**a**) Mass vs. radius for an isolated nonrotating neutron star, for each transition density $n_{\rm tr}$ and all speed of sound cases. The darker curves' color corresponds to the lower values of speed of sound. The blue horizontal line and region indicate the mass estimation of the massive compact object of Reference [19]. The dashed-dotted and dotted curves correspond to the MDI-APR and APR EoS, respectively. (**b**) The maximum mass M_{max} of a nonrotating neutron star as a relation to the bounds of the speed of sound $(v_s/c)^2$ for each transition density $n_{\rm tr}$ (in units of saturation density n_s). The purple vertical shaded region corresponds to the $n_{\rm tr} = 1.5n_s$ case, while the green one corresponds to the $n_{\rm tr} = 2n_s$ case. The purple (green) vertical line indicates the corresponding value of the speed of sound for a massive object with $M = 2.59\ M_\odot$.

At first sight, there are two main branches in Figure 9a related to the transition density. The solid (dashed) curves correspond to the $n_{\rm tr} = 1.5n_s$ ($n_{\rm tr} = 2n_s$) case. Depending on each value for the bound of the speed of sound, there are bifurcations in the families of EoSs. As we move to higher values for the speed of sound, the color of the curves lightens. The blue solid horizontal line, with the accompanying shaded region, represents the estimation of the recently observed massive compact object of Reference [19]. In general, the branch of EoSs with $n_{\rm tr} = 1.5n_s$ provides stiffer EoSs compared to the $n_{\rm tr} = 2n_s$ branch. Hence, the EoSs of the $n_{\rm tr} = 1.5n_s$ case are more likely to provide such a massive nonrotating neutron star than the $n_{\rm tr} = 2n_s$ case in which three EoSs lie outside of the shaded region. Specifically, between the same kind of transition density $n_{\rm tr}$ the EoSs with higher bounds of the speed of sound lead to higher values of neutron star mass and radius. Therefore, a generally high bound of the speed of sound (as the transition density has higher values, the speed of sound would be closer to the causal scenario) is needed for the description of such a massive compact object.

As one can observe in Figure 9a, there is a trend across the maximum masses. In order to study this behavior, we constructed the diagram of Figure 9b. The cross and star marks represent the maximum masses of $n_{\rm tr} = 1.5n_s$ and $n_{\rm tr} = 2n_s$ cases, respectively. The color of the marks is lighter for higher values of the speed of sound. The blue solid horizontal line, with the accompanying shaded region, indicates the estimation of the recently observed massive compact object of Reference [19]. The purple and green curves represent the following expression for the $n_{\rm tr} = 1.5n_s$ and $n_{\rm tr} = 2n_s$ cases, respectively, given below

$$M_{max} = c_1 d^{c_2} + c_3 d + c_4, \tag{67}$$

where $d = (v_s/c)^2$. The coefficients are given in Table 3.

Table 3. Parameters of Equation (67) and bounds of speed of sound value of Figure 9b. The parameters c_1, c_3, and c_4 are in solar mass units $M_\odot$.

$n_{\rm tr}$	c_1	c_2	c_3	c_4	$(v_s/c)^2_{min}$	$(v_s/c)^2$	$(v_s/c)^2_{max}$
$1.5n_s$	-1.6033×10^3	-7.56×10^{-4}	-1.64×10^{-1}	1.6068×10^3	0.448	0.485	0.52
$2n_s$	5.5754	0.2742	−0.6912	−1.9280	0.597	0.659	0.72

By applying the formula mentioned above, we were able to obtain estimations of the speed of sound values for each transition density $n_{\rm tr}$. In particular, for a nonrotating massive neutron star with $M = 2.59\ M_\odot$ the speed of sound must be (a) $(v_s/c)^2 = 0.485$ ($n_{\rm tr} = 1.5n_s$), and (b) $(v_s/c)^2 = 0.659$ ($n_{\rm tr} = 2n_s$). The exact values' interval is given in Table 3. We note that in the case of higher values of transition density $n_{\rm tr}$, the fitted expression and marks are shifted downwards; i.e., the higher the point of the transition in density, the smaller the provided maximum mass. In addition, as one can see in Figure 9b, the higher values of the speed of sound are better able to predict such massive neutron stars, until a specific boundary value of transition density $n_{\rm tr}$ (higher than those we adopted in our study), in which even the causality could not lead to such a massive neutron star.

Hence, a very massive nonrotating neutron star favors higher values of the speed of sound than the $v_s = c/\sqrt{3}$ limit. Based on our previous work, the current observation of neutron star mergers leads to a lower bound on the transition density $n_{\rm tr}$ [65]. At this point, a contradiction arises; the transition density $n_{\rm tr}$ must be above a specific lower value but not big enough to predict very massive masses.

In Figure 10, we display the tidal parameters for the single neutron star case that we study as a function of the mass. The vertical blue shaded region and line indicate the mass estimation for the second compact object of Reference [19].

One can observe that in both diagrams there are two main families of EoSs, distinguished by the transition density $n_{\rm tr}$. The EoSs with higher speed of sound bounds lead to bigger values on both tidal parameters. This means that a neutron star with a higher speed of sound value is more deformable than a more compact one (with a lower speed of sound value). Moreover, the EoSs with smaller transition density $n_{\rm tr}$ and higher $(v_s/c)^2$ values are more likely to predict a very massive neutron star of $M = 2.59\ M_\odot$.

We notice that taking into consideration cases with higher transition density $n_{\rm tr}$ could lead to smaller values of tidal parameters, i.e., more compact and less deformable stars. Hence, a very high value of the speed of sound, close to the causality, would be necessary to provide such a massive nonrotating neutron star.

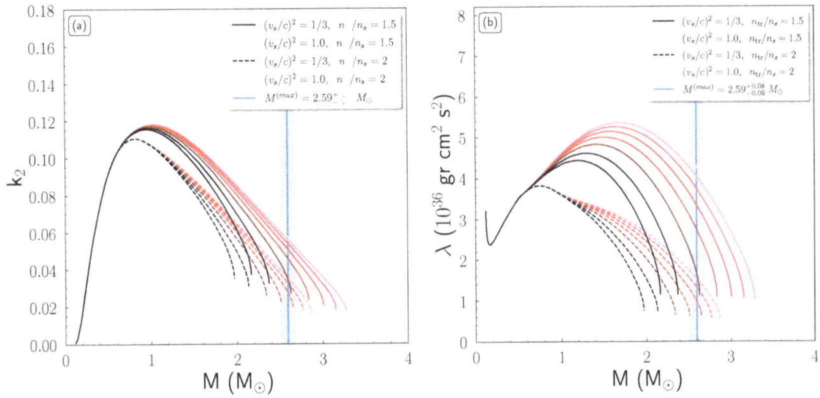

Figure 10. Tidal parameters (**a**) k_2 and (**b**) λ as a function of an neutron star's mass. The blue vertical line and shaded region indicate the estimation of the recently observed massive compact object of Reference [19]. The solid (dashed) curves correspond to the $n_{tr} = 1.5n_s$ ($n_{tr} = 2n_s$) case. The lower values of the speed of sound correspond to the darker-colored curves.

4.3.2. A Very Massive Neutron Star in a Binary Neutron Stars System

Beyond the hypothetical scenario of a single nonrotating neutron star, it is of particular interest to consider the binary case of two neutron stars, with the heavier having a mass of $m_1 = 2.59\ M_{\odot}$ and letting the secondary lighter neutron star fluctuate within the range $m_2 \in (1, 2.59)\ M_{\odot}$. By subtracting the component masses m_1, m_2 in Equation (18), we obtain the corresponding values of $\mathcal{M}_c$. Then, since the masses are defined, from the Equations (19) and (20), the effective tidal deformability $\tilde{\Lambda}$ can be determined.

In Figure 11a, we display the effective tidal deformability $\tilde{\Lambda}$ as a function of the chirp mass $\mathcal{M}_c$ of the system, for all the possible binary neutron star systems with such a massive neutron star component. We have to underline that from all the EoSs that we studied in the single-neutron star case previously, in the binary case, we used only those that can provide a neutron star with 2.59 $M_{\odot}$ mass. As one can see in Figure 11, there are two families of EoSs, distinguished by the transition density n_{tr}. For each family of EoSs, the EoSs with higher values of the speed of sound predict higher values of $\tilde{\Lambda}$. We notice that for a binary system with $m_1 = 2.59\ M_{\odot}$ and $m_2 = 1.4\ M_{\odot}$, the chirp mass is $\mathcal{M}_c = 1.642\ M_{\odot}$. Another remark is that binary neutron star systems with both heavy components, meaning higher $\mathcal{M}_c$, lead to smaller values of $\tilde{\Lambda}$. In this case, a possible lower limit on $\tilde{\Lambda}$ might provide useful constraints on the EoS.

In Figure 11b, we display the dependence of $\tilde{\Lambda}$ on the corresponding binary mass ratio q.

We have to underline that this kind of $\tilde{\Lambda} - q$ diagram is different from the usual ones (see in comparison Figure 3) because the chirp mass $\mathcal{M}_c$ has no specific value. In particular, in this work, the $\mathcal{M}_c$ is treated as a variable, and each point of Figure 11b corresponds to a different binary neutron star system with the heavier component in all cases being a very massive neutron star with mass 2.59 $M_{\odot}$. Similarly to Figure 11a, there are two main families for the curves, and the EoSs with higher speed of sound provide higher values of $\tilde{\Lambda}$. The more symmetric binary neutron star systems ($q \to 1$) lead to smaller values of $\tilde{\Lambda}$. On the contrary, the highest values of $\tilde{\Lambda}$ correspond to the most asymmetric binary neutron star systems. We notice that for a binary system with $m_1 = 2.59\ M_{\odot}$ and $m_2 = 1.4\ M_{\odot}$, the asymmetry ratio is $q = 0.541$.

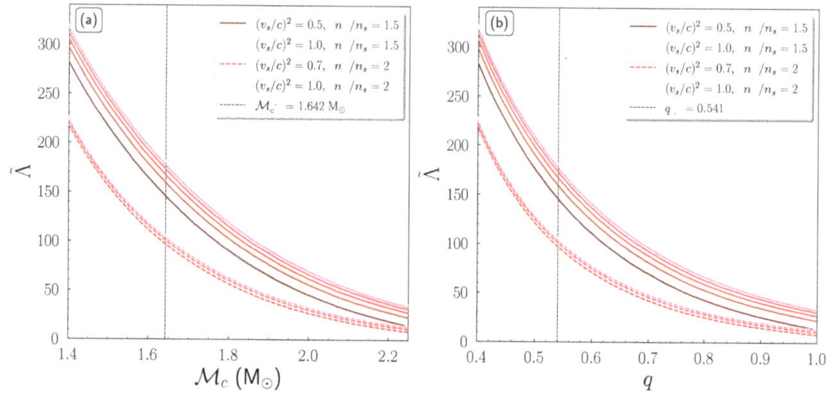

Figure 11. The effective tidal deformability $\tilde{\Lambda}$ as a function of (**a**) the chirp mass $\mathcal{M}_c$ and (**b**) binary mass ratio q, in the case of a very massive neutron star component, identical to Reference [19]. The darker colored curves correspond to lower values of speed of sound. The black dashed vertical line shows (**a**) the corresponding chirp mass $\mathcal{M}_c$ and (**b**) mass ratio q, of a binary neutron star system with $m_1 = 2.59\ M_\odot$ and $m_2 = 1.4\ M_\odot$, respectively.

Moreover, we studied the effective tidal deformability $\tilde{\Lambda}$ and the $R_{1.4}$ case of an $m_2 = 1.4\ M_\odot$ secondary component neutron star, with the heavier component neutron star taken to be $m_1 = 2.59\ M_\odot$. In Figure 12, we display this dependence. To be more specific, the EoSs are in five main groups, characterized by the transition density $n_{\rm tr}$. Our study has been expanded to transition densities $n_{\rm tr} = [1.25, 1.75, 2.25]\ n_s$ so that the calculations could be more accurate The higher speed of sound values correspond to lighter color. In analog to the observations of the previous Figure 11, the high speed of sound bounds lead to higher $\tilde{\Lambda}$ and $R_{1.4}$.

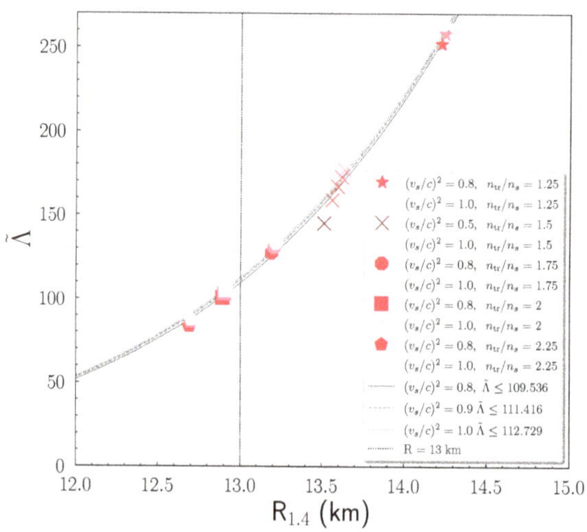

Figure 12. The effective tidal deformability $\tilde{\Lambda}$ as a function of the radius $R_{1.4}$ of an $m_2 = 1.4\ M_\odot$ neutron star. The heavier component of the system was taken to be $m_1 = 2.59\ M_\odot$. The darker colors correspond to lower values of speed of sound bounds. The grey lines show the expression of Equation (68). The black dotted vertical line indicates the proposed upper limit of Reference [132].

In addition, we applied a fitting formula to the $(v_s/c)^2 = [0.8, 0.9, 1]$ cases. The formula was taken to be in the kind of the proposed relations of References [133,134],

$$\tilde{\Lambda} = c_5 R_{1.4}^{c_6} \tag{68}$$

where the coefficients for each case are given in Table 4. According to a recent study, a similar power-law relation connects the tidal deformability of a single neutron star to the $R_{1.4}$ [142]. The significance of the tidal deformability $\Lambda_{1.4}$ and $R_{1.4}$ in order to obtain information about microscopic quantities was studied in Reference [143]. By imposing an upper limit on $R_{1.4}$, one can obtain an upper limit on $\tilde{\Lambda}$. Hence, by adopting the general limit of Reference [132], we obtained the constraints of Table 4.

Table 4. Parameters of Equation (68) and bounds of $\tilde{\Lambda}$ of Figure 12.

$(v_s/c)^2$	c_5 (km^{-1})	c_6	$\tilde{\Lambda}$
0.8	4.1897×10^{-9}	9.3518	109.536
0.9	5.3213×10^{-9}	9.2652	111.416
1	6.1109×10^{-9}	9.2159	112.729

4.4. *Finite Temperature Effects on Rapidly Rotating Neutron Stars*

4.4.1. Sequences of Constant Baryon Mass

A way to study the effects of finite temperature on the rapidly rotating remnant after a binary neutron star merger is the sequences of constant baryon mass. The sequences provide us with information about the evolution and instabilities of hot neutron stars. In particular, in the case of the isothermal EoSs, we considered the same baryon mass configuration for the EoSs and constructed a sequence related to the cooling of a hot neutron star [64].

Figure 13a displays the Kepler frequency as a function of the temperature for four baryon masses in the range $[1.6, 2.2]$ $M_\odot$. Specifically, in the range $[0, 15]$ MeV, the Kepler frequency decreases sharply with the temperature, while for higher temperatures, a smoother behavior is presented. This dependence can be described as

$$f(T) = \alpha_0 + \alpha_1 T^3 + \alpha_2 exp[\alpha_3 T] \quad \text{(Hz)}, \tag{69}$$

where f and T are in units of Hz and MeV, respectively, and the coefficients α_i with i = 0–3 are given in Table 5.

Table 5. Coefficients α_i with i = 0–3 for the empirical relation (69) and baryon masses in the range [1.6–2.2] $M_\odot$.

Coefficients	**Baryon Mass**			
	1.6 $M_\odot$	**1.8** $M_\odot$	**2.0** $M_\odot$	**2.2** $M_\odot$
a_0 $(\times 10^2)$	4.259	5.284	6.414	7.863
a_1 $(\times 10^{-3})$	−4.787	−3.202	−2.099	−1.443
a_2 $(\times 10^2)$	5.401	4.929	4.363	3.530
a_3 $(\times 10^{-1})$	−1.468	−1.443	−1.424	−1.636

This behavior suggests that the effects of temperature are more pronounced in the range $[0, 15]$ MeV, leading to significant lower Kepler frequencies, where for higher temperatures than $T > 15$ MeV, the effects are moderated.

Afterwards, in Figure 13b, the dependence of Kepler frequency on the central baryon density is presented, for four baryon masses in the range $[1.6, 2.2]$ $M_\odot$. While for low values of temperature the central baryon density increases with increasing temperature, for high values of temperature, a reduction in the values of the central baryon density is noted. In addition, for low values of temperature ($T < 30$ MeV) there is the appearance of a linear relation between the Kepler frequency and the central baryon density, assuming always a constant temperature. The significance of Figure 13b is focused on temperatures higher than $T = 30$ MeV, where a linear relation described as

$$f(n_b^c) = -473.144 + 2057.271 n_b^c \quad (\text{Hz}), \tag{70}$$

with f and n_b^c given in units of Hz and fm^{-3}, respectively, interprets the dependence of the Kepler frequency on the central baryon density independent from the specific baryon mass. Henceforth, Equation (70) defines the allowed region that a neutron star can exist with rotation at its mass-shedding limit for a specific central baryon density and vice versa.

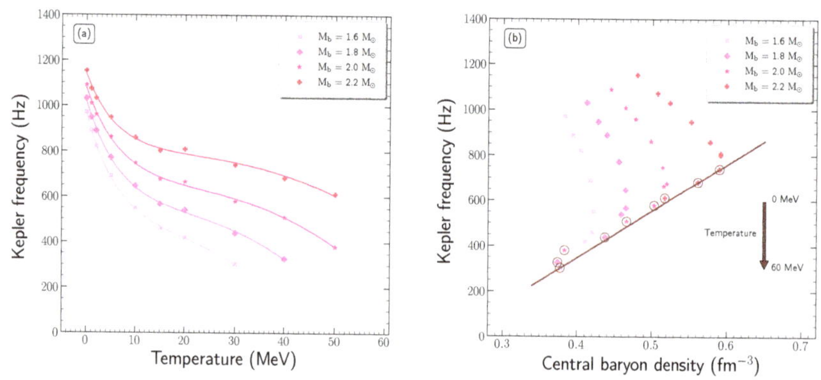

Figure 13. Dependence of the Kepler frequency on (**a**) the temperature and (**b**) the central baryon density for baryon masses in the range $[1.6, 2.2]\ M_\odot$. (**a**) Solid lines represent the fits originated from Equation (69). (**b**) The solid line represents Equation (70), while open circles note the high-temperature region ($T \geq 30$ MeV).

4.4.2. Moment of Inertia, Kerr Parameter, and Ratio T/W

Figure 14 displays the dimensionless moment of inertia as a function of the compactness parameter for neutron stars at the mass-shedding limit in (a) isothermal and (b) isentropic profile. In both cases, the increase of the temperature or the entropy per baryon (assuming a constant lepton fraction) leads to lower values of moment of inertia and lesser compact stars than the cold star. However, for low values of temperature ($T < 2$ MeV) or low values of entropy per baryon ($S = 1$ with $Y_l = 0.2$ and 0.3), the dimensionless moment of inertia and the compactness parameter exceed the values of the cold neutron star.

Afterwards, in Figure 15 we present the Kerr parameter as a function of the gravitational mass for neutron stars at the mass-shedding limit in (a) isothermal and (b) isentropic profile. In addition, we display the constraints for the Kerr parameter of neutron stars with the shaded region [64] and the Kerr bound for astrophysical Kerr black holes [97]. As the temperature or the entropy per baryon increases in the neutron star, the Kerr parameter decreases, except for high values of temperature ($T = 60$ MeV) where a slightly increase is observed. Furthermore, Figure 15c displays the Kerr parameter as a function of the temperature for constant gravitational mass in the isothermal profile. After $T = 30$ MeV, it is observed that the Kerr parameter creates a plate, meaning that the increase in temperature does not affect the Kerr parameter.

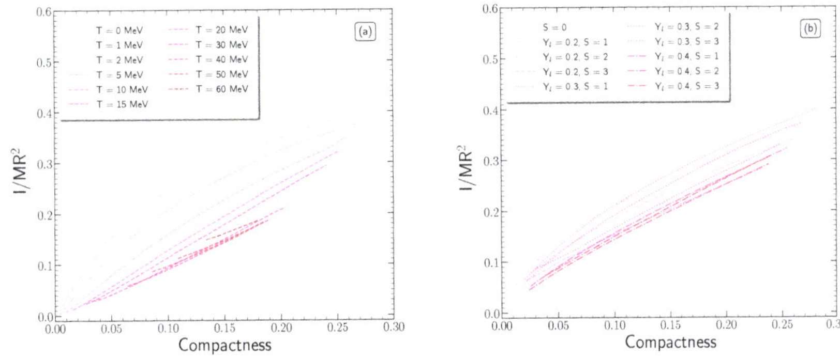

Figure 14. Dependence of the dimensionless moment of inertia on the compactness parameter at the mass-shedding limit in the case of (**a**) isothermal and (**b**) isentropic profiles. Dashed lines note the hot configurations, while the solid line notes the cold configuration.

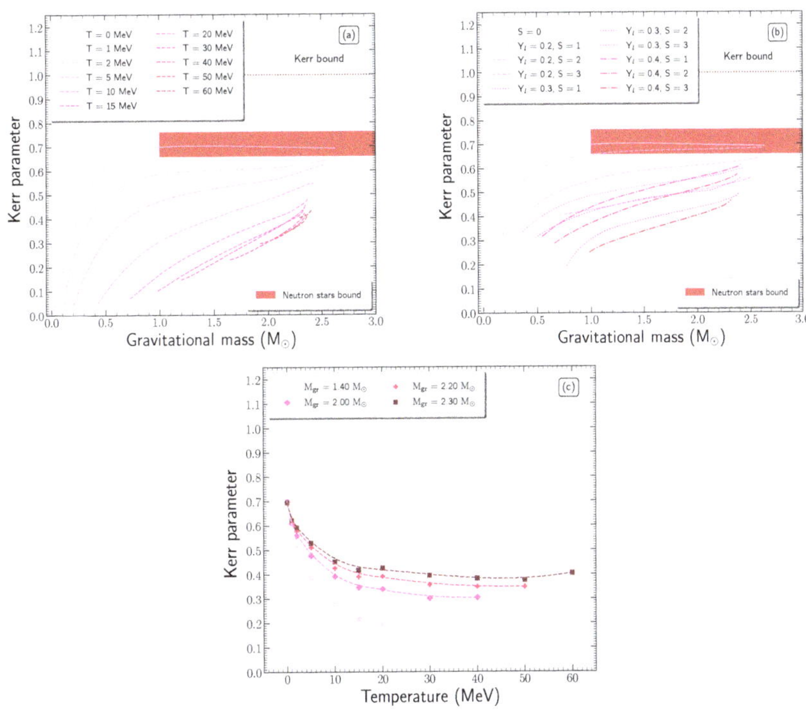

Figure 15. Dependence of the Kerr parameter on the gravitational mass at the mass-shedding limit in the case of (**a**) isothermal and (**b**) isentropic profiles. Dashed lines note the hot configurations, while the solid line notes the cold configuration. The horizontal dotted line corresponds to the Kerr bound for astrophysical Kerr black holes, $\mathcal{K}_{\text{B.H.}} = 0.998$ [144], while the shaded region corresponds to the neutron star limits from Reference [64]. (**c**) Dependence of the Kerr parameter on the temperature for gravitational masses in the range $[1.4, 2.3]$ $M_{\odot}$ and in the case of isothermal profile.

The introduction of temperature in neutron stars cannot violate the proposed limit for Kerr black holes [97] and the one from cold neutron stars [64]. Therefore, the gravitational collapse of a hot and uniformly rotating neutron star, constrained to mass–energy and angular momentum conservation, cannot lead to a maximally rotating Kerr black hole.

In addition, it is worth mentioning that, while in the cold neutron star, for $M_{gr} > 1\ M_\odot$, the Kerr parameter is almost independent of the gravitational mass, in hot configurations, Kerr parameter is an increasing function of the gravitational mass. The latter leads to the conclusion that while the interplay between the angular momentum and the gravitational mass in cold neutron stars is imperceptible, that is not the case in hot neutron stars, where a significant dependence is suggested.

Finally, Figure 16 displays the angular velocity as a function of the ratio of kinetic to gravitational binding energy T/W for neutron stars at the mass-shedding limit in (a) isothermal and (b) isentropic profile. Gravitational waves can be produced from neutron stars through the nonaxisymmetric perturbations. The vertical dotted line notes the limit for nonaxisymmetric instabilities from gravitational radiation, located at $T/W \sim 0.08$ for models with $M_{gr} = 1.4\ M_\odot$ [145]. The introduction of temperature leads to the conclusion that nonaxisymmetric instabilities cannot exist in hot neutron stars. However, for cases with low values of temperature ($T \leq 1$ MeV and $S = 1$ with $Y_l = 0.2$), the nonaxisymmetric instability would set in before the mass-shedding limit is reached. The latter indicates that both the maximum gravitational mass and the angular velocity will be lowered.

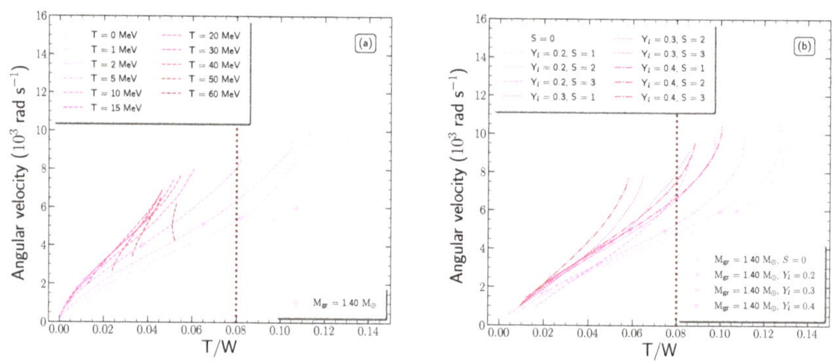

Figure 16. Dependence of the angular velocity on the ratio of rotational kinetic to gravitational binding energy at the mass-shedding limit in the case of (**a**) isothermal and (**b**) isentropic profiles. Dashed lines note the hot configurations, while the solid line notes the cold configuration. Markers correspond to the $M_{gr} = 1.4\ M_\odot$ configuration. The vertical dotted line notes the critical value, $T/W = 0.08$, for gravitational radiation instabilities.

The aftermath from the analysis on the compactness parameter, Kerr parameter, and ratio T/W is the insight for the hot and rapidly rotating remnant after a neutron star merger. Actually, the remaining object is a compact object consisting of neutron star matter. By assuming a remnant with at least $T \geq 30$ MeV for isothermal neutron stars and $S = 1$ with $Y_l = 0.2$ for isentropic ones, rotating at its mass-shedding limit, possible constraints are available through the mentioned quantities. In particular,

- compactness parameter: $\beta^{iso}_{rem} \leq 0.19$ and $\beta^{ise}_{rem} \leq 0.27$,
- Kerr parameter: $\mathcal{K}^{iso}_{rem} \leq 0.42$ and $\mathcal{K}^{ise}_{rem} \leq 0.68$,
- ratio T/W: $(T/W)^{iso}_{rem} \leq 0.05$ and $(T/W)^{ise}_{rem} \leq 0.127$,

where the superscripts "iso" and "ise" correspond to isothermal and isentropic profiles [for more details see Reference [64]]. Specifically, in the case that the remnant follows the isothermal profile, the remaining object is a lesser compact object than the cold neutron star, with lower values of maximum gravitational mass and frequency, and stable toward the dynamical instabilities. In the case that the remnant follows the isentropic profile, the remaining object is comparable to the cold neutron star.

5. Concluding Remarks

In this review we have presented a suitable EoS parameterized to reproduce specific values of the speed of sound and gravitational mass of neutron stars. In addition, we introduced the effects of finite temperature both in isolated neutron stars and in matters of merging. In particular, we have constructed equilibrium sequences of both nonrotating and rapidly rotating with the Kepler frequency neutron stars and paid special attention to the gravitational and baryon mass, the radius, the transition baryon density, the Kerr parameter, the moment of inertia, the ratio T/W, and the tidal deformability. This study is applied in several gravitational wave events, GW170817, GW190425, and GW190814, and possible constraints for the EoS are extracted.

Firstly, we studied the EoS and especially imposed constraints on the speed of sound (which affects the stiffness of the EoS) and the transition density by using recent observations of two binary neutron stars mergers (GW170817 and GW190425 events). The implemented method that we developed was based on the upper limits of the effective tidal deformability (derived from the mentioned events), combined with measurements of the maximum neutron star mass. As a base in our study, we used the MDI-APR EoS, for two cases of speed of sound bounds [63,74]; the conformal case $v_s = c/\sqrt{3}$ and the causal one of $v_s = c$.

The treatment of the effective tidal deformability as a function of the transition density allowed us to extract constraints on the bounds of the speed of sound. Specifically, for the GW170817 event, we found that the speed of sound must be lower than the value $v_s = c/\sqrt{3}$, at least up to densities $n_{\rm tr} \approx 1.6n_s$, and lower than $v_s = c$ up to densities $n_{\rm tr} \approx 1.8n_s$. For the GW190425 event, the respective values are $n_{\rm tr} \approx n_s$ for the lower speed of sound bound and $n_{\rm tr} \approx 1.2n_s$ for the upper one. These constraints are less rigorous than those derived from the GW170817 event.

Moreover, we studied the effective tidal deformability as a function of the maximum mass for both cases of speed of sound bounds. For the GW170817, we obtained that the maximum mass should be $M_{\rm max} \leq 2.106\ M_\odot$ for the $v_s = c/\sqrt{3}$ bound and $M_{\rm max} \leq 3.104\ M_\odot$ for the upper bound $v_s = c$. We notice that the limit of $M_{\rm max} \approx 2.11\ M_\odot$ corresponds to a transition density equal to $n_{\rm tr} \approx 1.5n_s$. Hence, according to this finding, the conformal limit $v_s = c/\sqrt{3}$ is in contradiction with the observational estimations of the $M_{\rm max}$ of neutron stars. Therefore, it must be violated in order to be able to simultaneously describe small values of the effective tidal deformability and high values for neutron star mass. The reason for this contradiction is based on two different points of view that antagonize one another: the upper limit on $\tilde{\Lambda}$ favors softer EoSs (higher values of $n_{\rm tr}$), while the maximum mass observations require stiffer EoSs (smaller values of $n_{\rm tr}$). For higher values of the speed of sound, this contradiction becomes less severe (i.e., $M_{\rm max} \approx 3.1\ M_\odot$ for the case $v_s = c$). We notice that the GW190425 was not able to offer further information.

Furthermore, from the study of the effective tidal deformability and the radius $R_{1.4}$ of a 1.4 $M_\odot$ neutron star, we observed that all the EoSs follow a common trend. This trend is affected mainly by the chirp mass of the binary system. To be more specific, as the chirp mass reaches higher values, the trend moves downwards. From the event GW170817 we obtained an upper limit $R_{1.4} \approx 13$ km for both cases, which is consistent with other estimations. The event GW190425 provided an upper limit $R_{1.4} \approx 14.712$ km for the $v_s = c/\sqrt{3}$ bound and $R_{1.4} \approx 14.53$ km for the $v_s = c$ bound.

We postulate that the discovery of future events of binary neutron stars mergers will provide rich information and further constraints on the bound of the speed of sound. In particular, the detection of future events could lead to more stringent constraints on the upper limit of $\tilde{\Lambda}$ and therefore more rigorous constraints on $n_{\rm tr}$ and bounds of the sound speed. Based on our approach, the more useful events for the lower limit of $n_{\rm tr}$ would be those with lighter component masses. Moreover, it would be of great interest to probe the lower limit of $\tilde{\Lambda}$. Such a lower limit might lead to an upper value of the transition density $n_{\rm tr}$. We make the assumption that heavier neutron star mergers would be suitable in the direction of a possible upper limit on n_{tr}. In any case, further detection of neutron stars mergers will assist in these open problems.

The baryon mass of the postmerger remnant is considered approximately conserved, a feature that gives rise to the significance of the temperature. In particular, in the case of hot neutron stars, the baryon mass is lower than the cold ones. As remnants are considered rapidly rotating, we study them at the mass-shedding limit. Specifically, in the cold case, the baryon mass is 3.085 $M_{\odot}$, while for a hot one at $T = 30$ MeV is 2.427 $M_{\odot}$ and for one at $S = 1$ is 3.05 $M_{\odot}$. By considering that the merger components have approximately equal masses, the above limits correspond to merger components with $\sim$1.5425, $\sim$1.2135, and $\sim$1.525 $M_{\odot}$ baryon masses, respectively. Furthermore, the immediate aftermaths of GW170817 [13] and GW190425 [14] events created hot and rapidly rotating remnants probably at the mass-shedding limit. In the case of GW170817 event, the remnant with $\sim$2.7 $M_{\odot}$ can be supported under the uniform rotation of cold and isentropic neutron stars with respect to the baryon mass of MDI+APR1 EoS. In contrast, isothermal neutron stars cannot support these values of mass. As far as the GW190425 event, the assumption of uniform rotation cannot be used to interpret the remnant of $\sim$3.7 $M_{\odot}$. It has to be noted that the postmeger remnant is assumed to rotate differentially. However, uniform rotation is a valid candidate to provide us with useful information about neutron stars.

In the GW190814 event [19], a compact object with a mass of $\sim$2.6 $M_{\odot}$ was observed as a merger component. It is believed to be either the lightest black hole or the most massive neutron star [139]. Nonetheless, Most et al. [139] suggest that the compact object could be a neutron star rapidly spinning with $\mathcal{K}$ in the range $[0.49, 0.68]$. In this case, the relevant postulation is in accordance with this study. More specifically, the values of the gravitational mass and Kerr parameter coincide with the ones from the MDI+APR1 EoS in both cold catalyzed matter and isentropic matter with $S = 1$ and $Y_l = 0.2$. Following the latter conclusion, there is a possibility that the observed star was rotating close to or at its mass-shedding limit and provide us additional constraints on the high density region of the nuclear EoS. In addition, possible constraints can be extracted for the corresponding equatorial radius. The Kerr parameter at the mass-shedding limit of the MDI+APR1 EoS lies in the region of $\mathcal{K}_{max} = [0.67, 0.69]$. This region also includes the upper limit of the relevant region from Reference [139] in a narrow range. Furthermore, by exploiting the relation between the Kerr parameter and the compactness parameter, a possible tight region for the equatorial radius of the star is implied as $R_{max} = [14.77, 14.87]$ km.

The Kerr parameter also has the role of an indicator of the collapse to a black hole. In hot neutron stars, the Kerr parameter decreases as the temperature inside the neutron star increases and never exceeds that of the cold neutron star. In conclusion, thermal support cannot lead a rapidly rotating star to collapse into a maximally rotating Kerr black hole. In addition, after $\sim$1 $M_{\odot}$, while the Kerr parameter is almost constant for the cold neutron stars, for hot neutron stars, the Kerr parameter is increasing with respect to the gravitational mass. The latter leads to a specific maximum value.

The ratio T/W is explicitly linked to the gravitational collapse to a black hole and the existence of stable supramassive neutron stars. We consider in the present study only the first scenario. Instabilities originating from the gravitational radiation, in which the critical value is at $T/W \sim 0.08$ for the $M_{gr} = 1.4$ $M_{\odot}$ configuration [145], do not exist for hot neutron stars. Nonetheless, for low values of temperatures, as the ratio T/W exceeds the critical value, this limit sets the upper value for the maximum gravitational mass and angular velocity. It has to be noted that studies concerning the Kerr parameter as well as the ratio T/W and the corresponding effect of the temperature are very rare. The existence of the latter studies may open a new window in neutron star studies.

A way to manifest the significance of the thermal support in neutron stars is the evolutionary sequences of constant baryon mass. In this configuration, the dependence of the Kepler frequency on the central baryon density presents a linear relation for temperatures higher than $T = 30$ MeV. The existence of such a relation, independent of the baryon mass, can define the allowed region of the pair of the central baryon density and corresponding Kepler frequency for a rotating hot neutron star at its mass-shedding limit.

Central baryon/energy density can also be of great interest in cold neutron stars, as it is connected with the evolution of the neutron star and the possible appearance of a phase transition. The end point from our study is that the central energy density must be lower than the values in the range $\varepsilon_c/c^2 = [2.53, 2.89]\ 10^{15}$ g cm^{-3}, while for the central baryon density, the corresponding range is $n_c = [7.27, 8.09]\ n_s$. The latter can inform us about the stability of the neutron star, as a neutron star with higher values of central energy/baryon density cannot exist, as well as can the appearance of the back-bending process.

Moreover, we examined the hypothetical scenario of a very massive neutron star with mass equal to $\sim 2.59^{+0.08}_{-0.09}\ M_\odot$, such as the secondary component of the GW190814 system [19]. The study of the maximum mass of each EoS as a function of the speed of sound bounds (for each value of transition density $n_{\rm tr}$), provide us constraints in the speed of sound. To be more specific, the $n_{\rm tr} = 1.5n_s$ case leads to $(v_s/c)^2 \in [0.448, 0.52]$, while the $n_{\rm tr} = 2n_s$ case leads to $(v_s/c)^2 \in [0.597, 0.72]$. We postulate that as the transition density $n_{\rm tr}$ is getting higher values, it is more difficult to achieve such a massive nonrotating neutron star. In particular, above a specific high-transition-density $n_{\rm tr}$ value, the speed of sound should be close to the causality to provide such a massive neutron star.

By studying the tidal parameters for the single case, we observed that the lower transition densities $n_{\rm tr}$ lead to higher tidal parameters. Hence, the transition density $n_{\rm tr} = 2n_s$ corresponds to a more compact neutron star (less deformation). Furthermore, between the same kind of transition density $n_{\rm tr}$, the EoSs with higher bounds on the speed of sound predict higher tidal parameters. Therefore, for the same transition density, the higher speed of sound bound means that the neutron star is less compact (more deformation).

Moving on to the binary neutron star system case, we adopted the hypothesis of a very massive component with $m_1 = 2.59\ M_\odot$, allowing us to investigate a variety of hypothetical binary neutron star systems with such a heavy component neutron star. In the case of heavy components of binary neutron star systems, meaning high value for the system's chirp mass $\mathcal{M}_c$, the effective tidal deformability $\tilde{\Lambda}$ has smaller values (smaller deformation). This behavior was noticed also in the $\tilde{\Lambda} - q$ diagram, in which the increasing binary mass symmetric ratio q leads to smaller values of $\tilde{\Lambda}$. For a binary neutron system with the heavier component equal to $m_1 = 2.59\ M_\odot$ and the lighter one equal to $m_2 = 1.4\ M_\odot$, the chirp mass $\mathcal{M}_c$ and the ratio q were estimated to be $\mathcal{M}_c = 1.642\ M_\odot$ and $q = 0.541$, respectively.

Lastly, we studied the case of a binary neutron star system with $m_1 = 2.59\ M_\odot$ with a secondary component $m_2 = 1.4\ M_\odot$. We especially concentrated on the radius $R_{1.4}$. A general remark is that the transition density $n_{\rm tr} = 1.5n_s$ provides higher values of $R_{1.4}$ and $\tilde{\Lambda}$ than the $n_{\rm tr} = 2n_s$ case. We extended our study to further transition densities $n_{\rm tr}$, which confirmed this general behavior. Additionally, the higher bounds of the speed of sound provide higher values on both parameters. By imposing an upper limit on the radius, we extracted some upper limits on the $\tilde{\Lambda}$ for each value of the speed of sound. In particular, this upper limit on $\tilde{\Lambda}$ shifts to higher values as the bound of the speed of sound increases.

This hypothetical scenario of a very massive neutron star demonstrated the key role of a microscopic quantity of the EoS, the speed of sound, which dramatically affects the EoS in combination with the changes in the transition density. We notice that the existence of such a massive nonrotating neutron star would mean a significant difference from all the cases known so far, which is a challenge for physics.

The underlying physics of neutron star mergers and the hot, rapidly rotating remnant should be investigated by considering differential rotation and cooling mechanisms, as these are the main features in the early postmerger phase. In addition, special emphasis should be given in the phase transition region, the existence of exotic degrees of freedom in the interior of neutron stars, as well as the accurate measurement of the tidal deformability. Finally, the observation of binary neutron star mergers and black-hole–neutron-star mergers, combined with the above studies, may provide significant constraints for the construction of the EoS.

Author Contributions: Conceptualization, P.S.K., A.K.-P. and C.C.M.; Methodology, P.S.K. and A.K.-P.; Software, P.S.K. and A.K.-P.; Validation, P.S.K., A.K.-P. and C.C.M.; Formal Analysis, P.S.K. and A.K.-P.; Investigation, P.S.K., A.K.-P. and C.C.M.; Data Curation, P.S.K. and A.K.-P.; Writing—Original Draft Preparation, P.S.K., A.K.-P. and C.C.M.; Writing—Review and Editing, P.S.K., A.K.-P. and C.C.M.; Visualization, P.S.K., A.K.-P. and C.C.M.; Supervision, C.C.M. All authors contributed equally to this work. All authors have read and agreed to the published version of the manuscript.

Funding: This research received no external funding.

Institutional Review Board Statement: Not applicable.

Informed Consent Statement: Not applicable.

Data Availability Statement: Not applicable.

Acknowledgments: The authors would like to thank K. Kokkotas for their constructive comments and insight during the preparation of the manuscript and also L. Rezzolla for useful correspondence and clarifications. The authors also thank B. Farr and Chatziioannou for their assistance providing computational tools regarding the kernel density estimation of the sample data and D. Radice, N. Stergioulas, and N. Minkov for the useful correspondence. We also acknowledge P. Meszaros and A. Sedrakian for their useful comments regarding the hot equations of state. In addition, the authors wish to thank the Bulgarian Academy of Science and the organizers of the International Workshop "Shapes and Dynamics of Atomic Nuclei: Contemporary Aspects" (SDANCA-21) for the their hospitality while this was completed. We would like to thank the financial support of the Bulgarian National Science Fund under contracts No. KP-06-N48/1 and No. KP-06-N28/6.

Conflicts of Interest: The authors declare no conflict of interest.

Abbreviations

The following abbreviations are used in this manuscript:

EoS	Equation of state
MDI	Momentum dependent interaction
APR	Akmal, Pandharipande, and Ravenhall
SNM	Symmetric nuclear matter
ANM	Asymmetric nuclear matter
PNM	Pure neutron matter
NM	Nuclear model
TOV	Tolman-Oppenheimer-Volkoff
N.R.	Non-rotating configuration
M.R.	Maximally-rotating configuration

References

1. Shapiro, S.; Teukolsky, S. *Black Holes, White Dwarfs, and Neutron Stars*; John Wiley and Sons: New York, NY, USA, 1983.
2. Glendenning, N. *Compact Stars: Nuclear Physics, Particle Physics, and General Relativity*; Springer: Berlin, Germany, 2000.
3. Haensel, P.; Potekhin, A.; Yakovlev, D. *Neutron Stars 1: Equation of State and Structure*; Springer: New York, NY, USA, 2007.
4. Weinberg, S. *Gravitational and Cosmology: Principle and Applications of the General Theory of Relativity*; Wiley: New York, NY, USA, 1972.
5. Zel'dovich, Y.B. The equation of state at ultrahigh densities and its relativistic limitations. *Sov. Phys. JETP* **1962**, *14*, 1143–1147.
6. Zel'dovich, Y.; Novikov, I. *Stars and Relativity*; Dover Publications, Inc.: Mineapolis, MN, USA; New York, NY, USA, 1971.
7. Hartle, J.B. Bounds on the mass and moment of inertia of non-rotating neutron stars. *Phys. Rep.* **1978**, *46*, 201–247. [CrossRef]
8. Bedaque, P.; Steiner, A.W. Sound Velocity Bound and Neutron Stars. *Phys. Rev. Lett.* **2015**, *114*, 031103. [CrossRef]
9. Moustakidis, C.C.; Gaitanos, T.; Margaritis, C.; Lalazissis, G.A. Bounds on the speed of sound in dense matter, and neutron star structure. *Phys. Rev. C* **2017**, *95*, 045801. [CrossRef]
10. Reed, B.; Horowitz, C.J. Large sound speed in dense matter and the deformability of neutron stars. *Phys. Rev. C* **2020**, *101*, 045803. [CrossRef]
11. Van Oeveren, E.D.; Friedman, J.L. Upper limit set by causality on the tidal deformability of a neutron star. *Phys. Rev. D* **2017**, *95*, 083014. [CrossRef]
12. Ma, Y.L.; Rho, M. Sound velocity and tidal deformability in compact stars. *Phys. Rev. D* **2019**, *100*, 114003. [CrossRef]
13. Abbott, B.P.; Abbott, R.; Abbott, T.D.; Acernese, F.; Ackley, K.; Adams, C.; Adams, T.; Addesso, P.; Adhikari, R.X.; Adya, V.B.; et al. GW170817: Observation of Gravitational Waves from a Binary Neutron Star Inspiral. *Phys. Rev. Lett.* **2017**, *119*, 161101. [CrossRef]

14. Abbott, B.P.; Abbott, R.; Abbott, T.D.; Abraham, S.; Acernese, F.; Ackley, K.; Adams, C.; Adhikari, R.X.; Adya, V.B.; Affeldt, C.; et al. GW190425: Observation of a Compact Binary Coalescence with Total Mass $\sim 3.4 M_\odot$. *Astrophys. J.* **2020**, *892*, L3. [CrossRef]
15. Arzoumanian, Z.; Brazier, A.; Burke-Spolaor, S.; Chamberlin, S.; Chatterjee, S.; Christy, B.; Cordes, J.M.; Cornish, N.J.; Crawford, F.; Cromartie, H.T.; et al. The NANOGrav 11-year Data Set: High-precision Timing of 45 Millisecond Pulsars. *Astrophys. J. Suppl. Ser.* **2018**, *235*, 37. [CrossRef]
16. Antoniadis, J.; Freire, P.; Wex, N.; Tauris, T.; Lynch, R.; van Kerkwijk, M.H.; Kramer, M.; Bassa, C.; Dhillon, V.S.; Driebe, T.; et al. A massive pulsar in a compact relativistic binary. *Science* **2013**, *340*. [CrossRef]
17. Cromartie, H.; Fonseca, E.; Ransom, S.; Demorest, P.B.; Arzoumanian, Z.; Blumer, H.; Brook, P.R.; DeCesar, M.E.; Dolch, T.; Ellis, J.A.; et al. Relativistic Shapiro delay measurements of an extremely massive millisecond pulsar. *Nat. Astron.* **2020**, *4*, 72–76. [CrossRef]
18. Linares, M.; Shahbaz, T.; Casares, J. Peering into the Dark Side: Magnesium Lines Establish a Massive Neutron Star in PSR J2215+5135. *Astrophys. J.* **2018**, *859*, 54. [CrossRef]
19. Abbott, R.; Abbott, T.D.; Abraham, S.; Acernese, F.; Ackley, K.; Adams, C.; Adhikari, R.X.; Adya, V.B.; Affeldt, C.; Agathos, M.; et al. GW190814: Gravitational Waves from the Coalescence of a 23 Solar Mass Black Hole with a 2.6 Solar Mass Compact Object. *Astrophys. J.* **2020**, *896*, L44. [CrossRef]
20. Tan, H.; Noronha-Hostler, J.; Yunes, N. Neutron Star Equation of State in Light of GW190814. *Phys. Rev. Lett.* **2020**, *125*, 261104. [CrossRef]
21. Datta, S.; Phukon, K.S.; Bose, S. Recognizing black holes in gravitational-wave observations: Challenges in telling apart impostors in mass-gap binaries. *arXiv* **2021**, arXiv:gr-qc/2004.05974.
22. Alsing, J.; Silva, O.H.; Berti, E. Evidence for a maximum mass cut-off in the neutron star mass distribution and constraints on the equation of state. *Mon. Not. R. Astron. Soc.* **2018**, *478*, 1377. [CrossRef]
23. Farr, W.M.; Chatziioannou, K. A Population-Informed Mass Estimate for Pulsar J0740+6620. *Res. Not. Am. Astron. Soc.* **2020**, *4*, 65. [CrossRef]
24. Abbott, B.P.; Abbott, R.; Abbott, T.D.; Acernese, F.; Ackley, K.; Adams, C.; Adams, T.; Addesso, P.; Adhikari, R.X.; Adya, V.B.; et al. GW170817: Measurements of Neutron Star Radii and Equation of State. *Phys. Rev. Lett.* **2018**, *121*, 161101. [CrossRef]
25. Bethe, H.; Brown, G.; Applegate, J.; Lattimer, J. Equation of state in the gravitational collapse of stars. *Nucl. Phys. A* **1979**, *324*, 487–533. [CrossRef]
26. Brown, G.; Bethe, H.; Baym, G. Supernova theory. *Nucl. Phys. A* **1982**, *375*, 481–532. [CrossRef]
27. Lamb, D.Q.; Lattimer, J.M.; Pethick, C.J.; Ravenhall, D.G. Hot Dense Matter and Stellar Collapse. *Phys. Rev. Lett.* **1978**, *41*, 1623–1626. [CrossRef]
28. Lattimer, J.M.; Ravenhall, D.G. Neutron star matter at high temperatures and densities. I. Bulk properties of nuclear matter. *Astrophys. J.* **1978**, *223*, 314–323. [CrossRef]
29. Lattimer, J.M. The Equation of State of Hot Dense Matter and Supernovae. *Ann. Rev. Nucl. Part. Sci.* **1981**, *31*, 337–374.
30. Lattimer, J.M.; Swesty, F.D. A generalized equation of state for hot, dense matter. *Nucl. Phys. A* **1991**, *535*, 331–376. [CrossRef]
31. Shen, H.; Toki, H.; Oyamatsu, K.; Sumiyoshi, K. Relativistic equation of state of nuclear matter for supernova and neutron star. *Nucl. Phys. A* **1998**, *637*, 435–450. [CrossRef]
32. Shen, G.; Horowitz, C.J.; O'Connor, E. Second relativistic mean field and virial equation of state for astrophysical simulations. *Phys. Rev. C* **2011**, *83*, 065808. [CrossRef]
33. Wellenhofer, C.; Holt, J.W.; Kaiser, N. Thermodynamics of isospin-asymmetric nuclear matter from chiral effective field theory. *Phys. Rev. C* **2015**, *92*, 015801. [CrossRef]
34. Constantinou, C.; Muccioli, B.; Prakash, M.; Lattimer, J.M. Thermal properties of supernova matter: The bulk homogeneous phase. *Phys. Rev. C* **2014**, *89*, 065802. [CrossRef]
35. Constantinou, C.; Muccioli, B.; Prakash, M.; Lattimer, J.M. Thermal properties of hot and dense matter with finite range interactions. *Phys. Rev. C* **2015**, *92*, 025801. [CrossRef]
36. Sammarruca, F.; Machleidt, R.; Millerson, R. Temperature effects on the neutron matter equation of state obtained from chiral effective field theory. *Mod. Phys. Lett. A* **2020**, *35*, 2050156,
37. Nicotra, O.E.; Baldo, M.; Burgio, G.F.; Schulze, H.-J. Protoneutron stars within the Brueckner-Bethe-Goldstone theory. *Astrophys. J.* **2006**, *451*, 213–222.:20053670. [CrossRef]
38. Burgio, G.F.; Schulze, H.-J. The maximum and minimum mass of protoneutron stars in the Brueckner theory. *Astron. Astrophys.* **2010**, *518*, A17. [CrossRef]
39. Baldo, M.; Burgio, G. The nuclear symmetry energy. *Prog. Part. Nucl. Phys.* **2016**, *91*, 203–258. [CrossRef]
40. Fortin, M.; Taranto, G.; Burgio, G.F.; Haensel, P.; Schulze, H.J.; Zdunik, J.L. Thermal states of neutron stars with a consistent model of interior. *Mon. Not. R. Astron. Soc.* **2018**, *475*, 5010–5022. Available online: https://academic.oup.com/mnras/article-pdf/475/4/5010/24056722/sty147.pdf (accessed on 24 September 2021).
41. Lu, J.J.; Li, Z.H.; Burgio, G.F.; Figura, A.; Schulze, H.J. Hot neutron stars with microscopic equations of state. *Phys. Rev. C* **2019**, *100*, 054335. [CrossRef]
42. Li, F.; Lu, J.J.; Li, Z.H.; Chen, C.Y.; Burgio, G.F.; Schulze, H.J. Accurate nuclear symmetry energy at finite temperature within a Brueckner-Hartree-Fock approach. *Phys. Rev. C* **2021**, *103*, 024307. [CrossRef]

43. Figura, A.; Lu, J.J.; Burgio, G.F.; Li, Z.H.; Schulze, H.J. Hybrid equation of state approach in binary neutron-star merger simulations. *Phys. Rev. D* **2020**, *102*, 043006. [CrossRef]
44. Shang, X.L.; Li, A.; Miao, Z.Q.; Burgio, G.F.; Schulze, H.J. Nucleon effective mass in hot dense matter. *Phys. Rev. C* **2020**, *101*, 065801. [CrossRef]
45. Wei, J.B.; Burgio, G.F.; Schulze, H.J.; Zappalà, D. Cooling of hybrid neutron stars with microscopic equations of state. *Mon. Not. R. Astron. Soc.* **2020**, *498*, 344–354. Available online: https://academic.oup.com/mnras/article-pdf/498/1/344/33710758/staa1879.pdf (accessed on 24 September 2021).
46. Raithel, C.A.; Özel, F.; Psaltis, D. Finite-temperature Extension for Cold Neutron Star Equations of State. *Astrophys. J.* **2019**, *875*, 12. [CrossRef]
47. Pons, J.A.; Reddy, S.; Prakash, M.; Lattimer, J.M.; Miralles, J.A. Evolution of Proto–Neutron Stars. *Astrophys. J.* **1999**, *513*, 780–804. [CrossRef]
48. Prakash, M.; Lattimer, J.; Pons, J.; Steiner, A.; Reddy, S. Evolution of a Neutron Star From its Birth to Old Age. *Lect. Not. Phys.* **2001**, *578*, 364–423. [CrossRef]
49. Lattimer, J.M.; Prakash, M. The equation of state of hot, dense matter and neutron stars. *Phys. Rep.* **2016**, *621*, 127–164. [CrossRef]
50. Perego, A.; Bernouzzi, S.; Radie, D. Thermodynamics conditions of matter in neutron star mergers. *Eur. Phys. J. A* **2019**, *55*, 124. [CrossRef]
51. Bauswein, A.; Janka, H.T.; Oechslin, R. Testing approximations of thermal effects in neutron star merger simulations. *Phys. Rev. D* **2010**, *82*, 084043. [CrossRef]
52. Kaplan, J.D.; Ott, C.D.; O'Connor, E.P.; Kiuchi, K.; Roberts, L.; Duez, M. The Influence of Thermal Pressure on Equilibrium Models of Hypermassive Neutron Star Merger Remnants. *Astrophys. J.* **2014**, *790*, 19. [CrossRef]
53. Tsokaros, A.; Ruiz, M.; Shapiro, S.L.; Sun, L.; Uryū, K.b.o. Great Impostors: Extremely Compact, Merging Binary Neutron Stars in the Mass Gap Posing as Binary Black Holes. *Phys. Rev. Lett.* **2020**, *124*, 071101. [CrossRef]
54. Yasin, H.; Schäfer, S.; Arcones, A.; Schwenk, A. Equation of State Effects in Core-Collapse Supernovae. *Phys. Rev. Lett.* **2020**, *124*, 092701. [CrossRef]
55. Radice, D.; Bernuzzi, S.; Perego, A. The Dynamics of Binary Neutron Star Mergers and GW170817. *Ann. Rev. Nucl. Part. Sci.* **2020**, *70*, 95–119.
56. Sarin, N.; Lasky, P.D.; Ashton, G. Gravitational waves or deconfined quarks: What causes the premature collapse of neutron stars born in short gamma-ray bursts? *Phys. Rev. D* **2020**, *101*, 063021. [CrossRef]
57. Soma, S.; Bandyopadhyay, D. Properties of Binary Components and Remnant in GW170817 Using Equations of State in Finite Temperature Field Theory Models. *Astrophys. J.* **2020**, *890*, 139. [CrossRef]
58. Sen, D. Nuclear matter at finite temperature and static properties of proto-neutron star. *J. Phys. G* **2020**, *48*, 025201. [CrossRef]
59. Gale, C.; Bertsch, G.; Das Gupta, S. Heavy-ion collision theory with momentum-dependent interactions. *Phys. Rev. C* **1987**, *35*, 1666–1671. [CrossRef]
60. Prakash, M.; Bombaci, I.; Prakash, M.; Ellis, P.J.; Lattimer, J.M.; Knorren, R. Composition and structure of protoneutron stars. *Phys. Rep.* **1997**, *280*, 1–77. [CrossRef]
61. Li, B.A.; Schröder, W.U. *Isospin Physics in Heavy-Ion Collisions at Intermediate Energies*; Nova Science: New York, NY, USA, 2001.
62. Li, B.A.; Chen, L.W.; Ko, C.M. Recent progress and new challenges in isospin physics with heavy-ion reactions. *Phys. Rep.* **2008**, *464*, 113–281. [CrossRef]
63. Koliogiannis, P.S.; Moustakidis, C.C. Effects of the equation of state on the bulk properties of maximally rotating neutron stars. *Phys. Rev. C* **2020**, *101*, 015805. [CrossRef]
64. Koliogiannis, P.S.; Moustakidis, C.C. Thermodynamical Description of Hot, Rapidly Rotating Neutron Stars, Protoneutron Stars, and Neutron Star Merger Remnants. *Astrophys. J.* **2021**, *912*, 69. [CrossRef]
65. Kanakis-Pegios, A.; Koliogiannis, P.S.; Moustakidis, C.C. Speed of sound constraints from tidal deformability of neutron stars. *Phys. Rev. C* **2020**, *102*, 055801. [CrossRef]
66. Kanakis-Pegios, A.; Koliogiannis, P.S.; Moustakidis, C.C. Probing the Nuclear Equation of State from the Existence of a $\sim 2.6\ M_{\odot}$ Neutron Star: The GW190814 Puzzle. *Symmetry* **2021**, *13*, 183,
67. Sieniawska, M.; Bejger, M. Continuous Gravitational Waves from Neutron Stars: Current Status and Prospects. *Universe* **2019**, *5*, 217. [CrossRef]
68. Glampedakis, K.; Gualtieri, L. Gravitational Waves from Single Neutron Stars: An Advanced Detector Era Survey. In *The Physics and Astrophysics of Neutron Stars*; Rezzolla, L., Pizzochero, P., Jones, D.I., Rea, N., Vidaña, I., Eds.; Springer International Publishing: Cham, Switzerland, 2018; pp. 673–736._12. [CrossRef]
69. Andersson, N.; Kokkotas, K.D. The r-mode instability in rotating neutron stars. *Int. J. Mod. Phys. D* **2001**, *10*, 381–441.
70. Özel, F.; Freire, P. Masses, Radii, and the Equation of State of Neutron Stars. *Annu. Rev. Astron. Astrophys.* **2016**, *54*, 401–440.
71. Chatziioannou, K. Neutron-star tidal deformability and equation-of-state constraints. *Gen. Relativ. Gravit.* **2020**, *52*, 109. [CrossRef]
72. Moustakidis, C.C. Effects of the nuclear equation of state on the *r*-mode instability and evolution of neutron stars. *Phys. Rev. C* **2015**, *91*, 035804. [CrossRef]
73. Akmal, A.; Pandharipande, V.R.; Ravenhall, D.G. Equation of state of nucleon matter and neutron star structure. *Phys. Rev. C* **1998**, *58*, 1804–1828. [CrossRef]

74. Margaritis, C.; Koliogiannis, P.S.; Moustakidis, C.C. Speed of sound constraints on maximally rotating neutron stars. *Phys. Rev. D* **2020**, *101*, 043023. [CrossRef]
75. Rhoades, C.E.; Ruffini, R. Maximum Mass of a Neutron Star. *Phys. Rev. Lett.* **1974**, *32*, 324–327. [CrossRef]
76. Kalogera, V.; Baym, G. The maximum mass of a neutron star. *Astrophys. J.* **1996**, *470*, L61. [CrossRef]
77. Koranda, S.; Stergioulas, N.; Friedman, J.L. Upper Limits Set by Causality on the Rotation and Mass of Uniformly Rotating Relativistic Stars. *Astrophys. J.* **1997**, *488*, 799. [CrossRef]
78. Chamel, N.; Haensel, P.; Zdunik, J.; Fantina, A. On the Maximum Mass of Neutron Stars. *Int. J. Mod. Phys. E* **2013**, *22*, 1330018. [CrossRef]
79. Podkowka, D.M.; Mendes, R.F.P.; Poisson, E. Trace of the energy-momentum tensor and macroscopic properties of neutron stars. *Phys. Rev. D* **2018**, *98*, 064057. [CrossRef]
80. Xia, C.; Zhu, Z.; Zhou, X.; Li, A. Sound velocity in dense stellar matter with strangeness and compact stars. *Chin. Phys. C* **2021**, *45*, 055104. [CrossRef]
81. Tews, I.; Carlson, J.; Gandolfi, S.; Reddy, S. Constraining the Speed of Sound inside Neutron Stars with Chiral Effective Field Theory Interactions and Observations. *Astrophys. J.* **2018**, *860*, 149. [CrossRef]
82. Feynman, R.; Metropolis, N.; Teller, E. Equations of State of Elements Based on the Generalized Fermi-Thomas Theory. *Phys. Rev.* **1949**, *75*, 1561. [CrossRef]
83. Baym, G.; Pethik, C.; Sutherland, P. The Ground State of Matter at High Densities: Equation of State and Stellar Models. *Astrophys. J.* **1971**, *170*, 299. [CrossRef]
84. Moustakidis, C.C. The stability of relativistic stars and the role of the adiabatic index. *Gen. Rel. Grav.* **2017**, *49*, 68. [CrossRef]
85. Lattimer, J.M.; Prakash, M. Neutron Star Structure and the Equation of State. *Astrophys. J.* **2001**, *550*, 426–442. [CrossRef]
86. Postnikov, S.; Prakash, M.; Lattimer, J.M. Tidal Love numbers of neutron and self-bound quark stars. *Phys. Rev. D* **2010**, *82*, 024016. [CrossRef]
87. Lattimer, J.M.; Prakash, M. Ultimate Energy Density of Observable Cold Baryonic Matter. *Phys. Rev. Lett.* **2005**, *94*, 111101. [CrossRef] [PubMed]
88. Lattimer, J.M.; Prakash, M. Nuclear matter and its role in supernovae, neutron stars and compact object binary mergers. *Phys. Rep.* **2000**, *333–334*, 121–146. [CrossRef]
89. Baiotti, L. Gravitational waves from neutron star mergers and their relation to the nuclear equation of state. *Prog. Part. Nucl. Phys.* **2019**, *109*, 103714. [CrossRef]
90. Flanagan, E.E.; Hinderer, T. Constraining neutron-star tidal Love numbers with gravitational-wave detectors. *Phys. Rev. D* **2008**, *77*, 021502. [CrossRef]
91. Hinderer, T. Tidal Love Numbers of Neutron Stars. *Astrophys. J.* **2008**, *677*, 1216–1220. [CrossRef]
92. Damour, T.; Nagar, A. Relativistic tidal properties of neutron stars. *Phys. Rev. D* **2009**, *80*, 084035. [CrossRef]
93. Hinderer, T.; Lackey, B.D.; Lang, R.N.; Read, J.S. Tidal deformability of neutron stars with realistic equations of state and their gravitational wave signatures in binary inspiral. *Phys. Rev. D* **2010**, *81*, 123016. [CrossRef]
94. Fattoyev, F.J.; Carvajal, J.; Newton, W.G.; Li, B.A. Constraining the high-density behavior of the nuclear symmetry energy with the tidal polarizability of neutron stars. *Phys. Rev. C* **2013**, *87*, 015806. [CrossRef]
95. Lackey, B.D.; Wade, L. Reconstructing the neutron-star equation of state with gravitational-wave detectors from a realistic population of inspiralling binary neutron stars. *Phys. Rev. D* **2015**, *91*, 043002. [CrossRef]
96. Takátsy, J.; Kovács, P. Comment on "Tidal Love numbers of neutron and self-bound quark stars". *Phys. Rev. D* **2020**, *102*, 028501. [CrossRef]
97. Thorne, K.S. Tidal stabilization of rigidly rotating, fully relativistic neutron stars. *Phys. Rev. D* **1998**, *58*, 124031. [CrossRef]
98. Abbott, B.P.; Abbott, R.; Abbott, T.D.; Acernese, F.; Ackley, K.; Adams, C.; Adams, T.; Addesso, P.; Adhikari, R.X.; Adya, V.B.; et al. Properties of the Binary Neutron Star Merger GW170817. *Phys. Rev. X* **2019**, *9*, 011001. [CrossRef]
99. Friedman, J.L.; Stergioulas, N. *Rotating Relativistic Stars*; Cambridge Monographs on Mathematical Physics, Cambridge University Press: Cambridge, UK, 2013; doi:10.1017/CBO9780511977596. [CrossRef]
100. Takami, K.; Rezzolla, L.; Yoshida, S. A quasi-radial stability criterion for rotating relativistic stars. *Mon. Not. R. Astron. Soc. Lett.* **2011**, *416*, L1–L5. Available online: https://academic.oup.com/mnrasl/article-pdf/416/1/L1/6140636/416-1-L1.pdf (accessed on 24 September 2021).
101. Weih, L.R.; Most, E.R.; Rezzolla, L. On the stability and maximum mass of differentially rotating relativistic stars. *Mon. Not. R. Astron. Soc. Lett.* **2017**, *473*, L126–L130. Available online: https://academic.oup.com/mnrasl/article-pdf/473/1/L126/24841378/slx178.pdf (accessed on 24 September 2021).
102. Stergioulas, N. 1996. Available online: http://www.gravity.phys.uwm.edu/rns/ (accessed on 24 September 2021).
103. Stergioulas, N.; Friedman, J.L. Comparing Models of Rapidly Rotating Relativistic Stars Constructed by Two Numerical Methods. *Astrophys. J.* **1995**, *444*, 306. [CrossRef]
104. Komatsu, H.; Eriguchi, Y.; Hachisu, I. Rapidly rotating general relativistic stars. I-Numerical method and its application to uniformly rotating polytropes. *Mon. Not. R. Astron. Soc.* **1989**, *237*, 355. [CrossRef]
105. Cook, G.; Shapiro, S.; Teukolsky, S. Rapidly rotating polytropes in general relativity. *Astrophys. J.* **1994**, *422*, 227. [CrossRef]
106. Goodstein, D.L. *States of Matter*; Dove: New York, NY, USA, 1985.
107. Fetter, A.L.; Walecka, J.D. *Quantum Theory of Many-Particle Systems*; Dover: Mineola, NY, USA, 2003.

108. Prakash, M. *The Equation of State and Neutron Star*; Unpublished, 1994.
109. Burgio, G.F.; Baldo, M.; Nicotra, O.E.; Schulze, H.J. A microscopic equation of state for protoneutron stars. *Astrophys. J. Suppl.* **2007**, *308*, 387–394. [CrossRef]
110. Xu, J.; Chen, L.W.; Li, B.A.; Ma, H.R. Effects of isospin and momentum dependent interactions on liquid–gas phase transition in hot asymmetric nuclear matter. *Phys. Lett. B* **2007**, *650*, 348–353. [CrossRef]
111. Moustakidis, C.C. Temperature and momentum dependence of single-particle properties in hot asymmetric nuclear matter. *Phys. Rev. C* **2008**, *78*, 054323. [CrossRef]
112. Moustakidis, C.C.; Panos, C.P. Equation of state for β-stable hot nuclear matter. *Phys. Rev. C* **2009**, *79*, 045806. [CrossRef]
113. Tan, N.H.; Loan, D.T.; Khoa, D.T.; Margueron, J. Mean-field study of hot β-stable protoneutron star matter: Impact of the symmetry energy and nucleon effective mass. *Phys. Rev. C* **2016**, *93*, 035806. [CrossRef]
114. Takatsuka, T.; Nishizaki, S.; Hiura, J. Particular Properties of Dense Supernova Matter. *Prog. Theor. Phys.* **1994**, *92*, 779–802. Available online: https://academic.oup.com/ptp/article-pdf/92/4/779/5358408/92-4-779.pdf (accessed on 24 September 2021).
115. Takatsuka, T. Hot Neutron Stars at Birth and Related Problems*). *Prog. Theor. Phys.* **1996**, *95*, 901–912. Available online: https://academic.oup.com/ptp/article-pdf/95/5/901/5280763/95-5-901.pdf (accessed on 24 September 2021).
116. Friedman, J.L.; Ipser, J.R.; Sorkin, R.D. Turning Point Method for Axisymmetric Stability of Rotating Relativistic Stars. *Astrophys. J.* **1988**, *325*, 722. [CrossRef]
117. Sorkin, R.D. A Stability Criterion for Many Parameter Equilibrium Families. *Astrophys. J.* **1982**, *257*, 847. [CrossRef]
118. Goussard, J.O.; Haensel, P.; Zdunik, J.L. Rapid uniform rotation of protoneutron stars. *Astron. Astrophys.* **1997**, *321*, 822–834. Available online: http://aa.springer.de/papers/7321003/2300822.pdf (accessed on 24 September 2021).
119. Marques, M.; Oertel, M.; Hempel, M.; Novak, J. New temperature dependent hyperonic equation of state: Application to rotating neutron star models and $I-Q$ relations. *Phys. Rev. C* **2017**, *96*, 045806. [CrossRef]
120. LORENE. LORENE: Langage Objet pour la RElativité NumériquE. 1998. Available online: https://lorene.obspm.fr/ (accessed on 24 September 2021).
121. Miller, M.C.; Lamb, F.K.; Dittmann, A.J.; Bogdanov, S.; Arzoumanian, Z.; Gendreau, K.C.; Guillot, S.; Harding, A.K.; Ho, W.C.G.; Lattimer, J.M.; et al. PSR J0030+0451 Mass and Radius from NICER Data and Implications for the Properties of Neutron Star Matter. *Astrophys. J.* **2019**, *887*, L24. [CrossRef]
122. Greif, S.K.; Raaijmakers, G.; Hebeler, K.; Schwenk, A.; Watts, A.L. Equation of state sensitivities when inferring neutron star and dense matter properties. *Mon. Not. R. Astron. Soc.* **2019**, *485*, 5363–5376. Available online: https://academic.oup.com/mnras/article-pdf/485/4/5363/28262089/stz654.pdf (accessed on 24 September 2021).
123. Radice, D.; Perego, A.; Zappa, F.; Bernuzzi, S. GW170817: Joint Constraint on the Neutron Star Equation of State from Multimessenger Observations. *Astrophys. J.* **2018**, *852*, L29. [CrossRef]
124. Tews, I.; Margueron, J.; Reddy, S. Critical examination of constraints on the equation of state of dense matter obtained from GW170817. *Phys. Rev. C* **2018**, *98*, 045804. [CrossRef]
125. Radice, D.; Dai, L. Multimessenger parameter estimation of GW170817. *Eur. Phys. J. A* **2019**, *55*, 50,
126. Kiuchi, K.; Kyutoku, K.; Shibata, M.; Taniguchi, K. Revisiting the Lower Bound on Tidal Deformability Derived by AT 2017gfo. *Astrophys. J.* **2019**, *876*, L31. [CrossRef]
127. Coughlin, M.W.; Dietrich, T.; Doctor, Z.; Kasen, D.; Coughlin, S.; Jerkstrand, A.; Leloudas, G.; McBrien, O.; Metzger, B.D.; O'Shaughnessy, R.; et al. Constraints on the neutron star equation of state from AT2017gfo using radiative transfer simulations. *Mon. Not. R. Astron. Soc.* **2018**, *480*, 3871–3878. Available online: https://academic.oup.com/mnras/article-pdf/480/3/3871/33043575/sty2174.pdf (accessed on 24 September 2021).
128. Coughlin, M.W.; Dietrich, T.; Margalit, B.; Metzger, B.D. Multimessenger Bayesian parameter inference of a binary neutron star merger. *Mon. Not. R. Astron. Soc. Lett.* **2019**, *489*, L91–L96. Available online: https://academic.oup.com/mnrasl/article-pdf/489/1/L91/30032497/slz133.pdf (accessed on 24 September 2021).
129. Most, E.R.; Weih, L.R.; Rezzolla, L.; Schaffner-Bielich, J. New Constraints on Radii and Tidal Deformabilities of Neutron Stars from GW170817. *Phys. Rev. Lett.* **2018**, *120*, 261103. [CrossRef]
130. Foley, R.J.; Coulter, D.A.; Kilpatrick, C.D.; Piro, A.L.; Ramirez-Ruiz, E.; Schwab, J. Updated parameter estimates for GW190425 using astrophysical arguments and implications for the electromagnetic counterpart. *Mon. Not. R. Astron. Soc.* **2020**, *494*, 190–198. Available online: https://academic.oup.com/mnras/article-pdf/494/1/190/32982848/staa725.pdf (accessed on 24 September 2021).
131. Rezzolla, L.; Most, E.R.; Weih, L.R. Using Gravitational-wave Observations and Quasi-universal Relations to Constrain the Maximum Mass of Neutron Stars. *Astrophys. J.* **2018**, *852*, L25. [CrossRef]
132. Raithel, C.A.; Özel, F.; Psaltis, D. Tidal Deformability from GW170817 as a Direct Probe of the Neutron Star Radius. *Astrophys. J.* **2018**, *857*, L23. [CrossRef]
133. Zhao, T.; Lattimer, J.M. Tidal deformabilities and neutron star mergers. *Phys. Rev. D* **2018**, *98*, 063020. [CrossRef]
134. De, S.; Finstad, D.; Lattimer, J.M.; Brown, D.A.; Berger, E.; Biwer, C.M. Tidal Deformabilities and Radii of Neutron Stars from the Observation of GW170817. *Phys. Rev. Lett.* **2018**, *121*, 091102. [CrossRef]
135. Annala, E.; Gorda, T.; Kurkela, A.; Vuorinen, A. Gravitational-Wave Constraints on the Neutron-Star-Matter Equation of State. *Phys. Rev. Lett.* **2018**, *120*, 172703. [CrossRef]

136. Fasano, M.; Abdelsalhin, T.; Maselli, A.; Ferrari, V. Constraining the Neutron Star Equation of State Using Multiband Independent Measurements of Radii and Tidal Deformabilities. *Phys. Rev. Lett.* **2019**, *123*, 141101. [CrossRef]
137. Fattoyev, F.J.; Piekarewicz, J.; Horowitz, C.J. Neutron Skins and Neutron Stars in the Multimessenger Era. *Phys. Rev. Lett.* **2018**, *120*, 172702. [CrossRef]
138. Landry, P.; Essick, R.; Chatziioannou, K. Nonparametric constraints on neutron star matter with existing and upcoming gravitational wave and pulsar observations. *Phys. Rev. D* **2020**, *101*, 123007. [CrossRef]
139. Most, E.R.; Papenfort, L.J.; Weih, L.R.; Rezzolla, L. A lower bound on the maximum mass if the secondary in GW190814 was once a rapidly spinning neutron star. *Mon. Not. R. Astron. Soc. Lett.* **2020**, *499*, L82–L86. [CrossRef]
140. Cook, G.; Shapiro, S.; Teukolsky, S. Rapidly Rotating Neutron Stars in General Relativity: Realistic Equations of State. *Astrophys. J.* **1994**, *424*, 823. [CrossRef]
141. Salgado, M.; Bonazzola, S.; Gourgoulhon, E.; Haensel, P. High precision rotating neutron star models 1: Analysis of neutron star properties. *Astron. Astrophys.* **1994**, *291*, 155.
142. Tsang, C.; Tsang, M.; Danielewicz, P.; Fattoyev, F.; Lynch, W. Insights on Skyrme parameters from GW170817. *Phys. Lett. B* **2019**, *796*, 1–5. [CrossRef]
143. Weih, J.B.; Lu, J.J.; Burgio, G.F.; Li, Z.H.; Schulze, H.J. Are nuclear matter properties correlated to neutron star observables? *Eur. Phys. J. A* **2020**, *56*, 63. [CrossRef]
144. Thorne, K.S. Disk-Accretion onto a Black Hole. II. Evolution of the Hole. *Astrophy. J.* **1974**, *191*, 507–520. [CrossRef]
145. Morsink, S.M.; Stergioulas, N.; Blattnig, S.R. Quasi-normal Modes of Rotating Relativistic Stars: Neutral Modes for Realistic Equations of State. *Astrophys. J.* **1999**, *510*, 854–861. [CrossRef]

foundations

Review

Fundamentals of Diatomic Molecular Spectroscopy

Christian G. Parigger

Physics and Astronomy Department, University of Tennessee, University of Tennessee Space Institute, Center for Laser Applications, 411 B.H. Goethert Parkway, Tullahoma, TN 37388-9700, USA; cparigge@tennessee.edu; Tel.: +1-(931)-841-5690

Abstract: The interpretation of optical spectra requires thorough comprehension of quantum mechanics, especially understanding the concept of angular momentum operators. Suppose now that a transformation from laboratory-fixed to molecule-attached coordinates, by invoking the correspondence principle, induces reversed angular momentum operator identities. However, the foundations of quantum mechanics and the mathematical implementation of specific symmetries assert that reversal of motion or time reversal includes complex conjugation as part of anti-unitary operation. Quantum theory contraindicates sign changes of the fundamental angular momentum algebra. Reversed angular momentum sign changes are of a heuristic nature and are actually not needed in analysis of diatomic spectra. This review addresses sustenance of usual angular momentum theory, including presentation of straightforward proofs leading to falsification of the occurrence of reversed angular momentum identities. This review also summarizes aspects of a consistent implementation of quantum mechanics for spectroscopy with selected diatomic molecules of interest in astrophysics and in engineering applications.

Keywords: foundations of quantum mechanics; molecular spectroscopy; diatomic molecules; symmetry transformations; optical emission spectroscopy; astrophysics

Citation: Parigger, C.G. Fundamentals of Diatomic Molecular Spectroscopy. *Foundations* **2021**, *1*, 208–216. https://doi.org/10.3390/foundations1020016

Academic Editor: Eugene Oks

Received: 19 September 2021
Accepted: 29 October 2021
Published: 2 November 2021

1. Introduction

Identification of diatomic molecular spectra necessitates a clear description of angular momentum (AM) in order to demarcate the various features that comprise optical fingerprints. Quantum mechanics theory (QMT) asserts that not all three components of AM can be measured simultaneously, usually the total AM and one projection of the total AM describe upper and lower states of molecular transitions. The components of AM are formally described by a set of operator equations.

Classical mechanics (CM) description and associated quantization of the asymmetric top [1] suggests occurrence of commutator relations with different signs when computing momenta with respect to the principal axes of inertia. In other words, a laboratory-fixed system shows standard AM commutators, but with respect to the molecule-attached coordinate system, there is a sign change that carries the name "reversed" internal AM [2]. The derivation by Klein in 1929 [1] is based on the correspondence principle that in essence emphasizes that QMT reproduces classical physics in the limit of large quantum numbers. From a CM point of view, reversal of motion occurs when transforming from a lab-fixed to a molecule-attached coordinate system, akin to experience of motion reversal when jumping onto a moving merry-go-around. However, reversal of motion in quantum mechanics (QM) is described by an anti-unitary transformation, requiring sign change and complex conjugation. The reversed internal AM concept [2] and applications actually are communicated and applied in analysis of molecular spectra by Van Vleck in 1951 in his review article on coupling angular momenta, i.e., AM, referring to axes mounted on the molecule, adheres to opposite-sign commutator algebra. This evolved into so-called reversed angular momentum (RAM) concepts for prediction of molecular spectra.

However, orthodox or classic QM abides by strict mathematical rules associated with the theory. Use of RAM techniques is contraindicated, especially since Nöther-type symmetry transformation [3] sustains the standard commutator relations, viz., reversal of motion is an anti-unitary transformation, just like in the Schrödinger wave equation that is invariant with respect to motion-reversal or time-reversal due to anti-unitary operation, as expected. It is important to recognize that a transformation from laboratory-fixed to molecular-attached coordinates within standard QM does not condone anomalous AM operator identities.

This review communicates proofs that the quantum-mechanic AM equations remain the same in a transition from laboratory-fixed to molecular-attached coordinates. Methods that invoke RAM for the prediction of molecular spectra are misleading. Application of standard QM establishes within the concept of line strengths [4] consistent computation of diatomic spectra [5]; examples include hydroxyl, cyanide, and diatomic carbon spectra [6]. First, Oscar Klein's paper [1] is discussed showing his original argumentation. This is followed by presenting proofs consistent with QMT opposing RAM concepts and occurrence of a minus sign in unitary and anti-unitary transformations. The "new" aspect of this review is the emphasis of invoking mathematics consistent with QMT. Subsequently, this review summarizes the approach for prediction of selected diatomic spectra including presentation of computed diatomic spectra of OH and C_2 molecules.

2. Materials and Methods

The premise of this article is Oscar Klein's work [1] "Zur Frage der Quantelung des asymmetrischen Kreisels" or "On the question of the quantization of the asymmetric top." This particular work is in German without an available translation; the essential contents are in the Einleitung, viz., the introduction, and on the page following the introduction. Klein's paper reflects the initial argumentation of the RAM method, and essential aspects of this paper are discussed below, up to Equation (6).

The purpose of the 1929 work is, as O. Klein writes, to reduce quantization of the asymmetric top to simple algebra for the components of the angular momentum "... that were developed by Dirac [7] and as well by Born, Heisenberg and Jordan [8]." For a solid body, the main moments of inertia are labeled as A, B, and C, the angular momenta are labeled P, Q, R, and one finds the CM energy of rotation, E,

$$E = \frac{1}{2}\left(\frac{P^2}{A} + \frac{Q^2}{B} + \frac{R^2}{C}\right), \tag{1}$$

or perhaps with convenient notation, using for operators $\tilde{J}_1 = P$, $\tilde{J}_2 = Q$, $\tilde{J}_3 = R$, where the tilde-symbol indicates that angular momenta (that would be AM operators in QM) are referred to the main axis of the ellipse of inertia (or in molecules, referred to molecular-fixed coordinates), and for moments of inertia $I_1 = A$, $I_2 = B$, $I_3 = C$,

$$E = \frac{1}{2}\sum_{k=1}^{k=3} \frac{1}{I_k}\tilde{J}_k. \tag{2}$$

Subsequently, O. Klein writes that P, Q, R can be understood to describe matrices satisfying QM equations of motion, with $i = \sqrt{-1}$ and using the standard $\hbar$ for Planck's constant divided by 2π,

$$\frac{dP}{dt} = \frac{i}{\hbar}(EP - PE), \quad \frac{dQ}{dt} = \frac{i}{\hbar}(EQ - QE), \quad \frac{dR}{dt} = \frac{i}{\hbar}(ER - RE). \tag{3}$$

In terms of operators, using the Hamilton operator $\mathcal{H}$ instead of E and writing the equation in the Heisenberg-picture for an abstract observable (operator), $\mathcal{O}$, without explicit time-dependence of the observable, i.e., $\frac{\partial \mathcal{O}}{\partial t} = 0$, and using the commutator $[\mathcal{H}, \mathcal{O}] = \mathcal{H}\mathcal{O} - \mathcal{O}\mathcal{H}$, gives

$$\frac{d\mathcal{O}}{dt} = \frac{i}{\hbar}[\mathcal{H}, \mathcal{O}] + \frac{\partial \mathcal{O}}{\partial t}. \tag{4}$$

The hypothesis of O. Klein comprises the requirement of utilizing Equation (3) in Equation (1). Consequently, O. Klein assumes commutator relations for P, Q, R,

$$i\hbar P = RQ - QR, \quad i\hbar Q = PR - RP, \quad i\hbar R = QP - PQ, \tag{5}$$

or using abbreviated nomenclature and the Levi-Civita symbol, with $\varepsilon_{klm} = 1$ for even permutations, and $\varepsilon_{klm} = -1$ for odd ones, otherwise $\varepsilon_{klm} = 0$ for identical indices, $k, l, m = 1, 2, 3$,

$$\left[\tilde{J}_k, \tilde{J}_l\right] = -i\hbar\varepsilon_{klm}\tilde{J}_m. \tag{6}$$

With the commutator relations in Equation (5), the correspondence principle leads to the equations of motion, and as O. Klein writes, "... as we overlook occurrence of the action-quant ...," viz. overlook $\hbar$. Further, O. Klein remarks that Equation (5) differs only by the sign of i from the well-known quantum-mechanical commutators for a laboratory-fixed system. In summary, O. Klein's work concludes that a minus sign is required for consistency with classical mechanics and a result of the application of the correspondence principle.

Clearly, writing Equation (5) in the compact form of Equation (6) highlights the minus sign that differs from the standard equations of AM operators J_k, $k = 1, 2, 3$,

$$[J_k, J_l] = \ i\hbar\varepsilon_{klm}J_m. \tag{7}$$

The minus sign in Equation (6) is labeled "anomalous" by some authors, e.g., J. Van Vleck [2], but there is no justification for the anomalous minus sign to occur within QMT. Usually, one considers right-hand systems, so Equation (7) is termed as the standard quantum-mechanic AM operator identity. Sustenance of RAM concepts may appear convenient, even calling the negative sign an "anomaly" but without QMT support. In scientific approach and in spite of the initial success in explaining spectra within various approximations, one usually avoids starting with an "anomaly" and/or inaccurate presuppositions that are readily falsified [9]. However, several textbooks and works continue support of RAM in the theory of molecular spectra [10–22], in spite of obvious falsification by QMT. This work emphasizes that there is no need to resort to RAM "cook book" [22] methods.

The methods in this work utilize standard QMT [23,24] and standard mathematical methods [25], showing that there is no sign change of the standard commutator relations when transforming from a laboratory-fixed to a molecule-attached coordinate system. Consistent application of standard AM algebra in the establishment of computed spectra yield nice agreement with laboratory experimental results [5] and agreement in analysis of astrophysical C_2 Swan data from the white dwarf Procyon B [5], including agreement in comparisons with computed spectra that are obtained with other molecular fitting programs such as PGOPHER [26].

Methods for measurement of optical emission signals from diatomic molecules are comprised of standard molecular spectroscopy experimental arrangements such as in laser-induced plasma or breakdown spectroscopy [27–34], encountered as well in stellar plasma physics or astrophysics to name other areas of interest. Particular interests in astrophysics include "cool" stars, brown dwarfs, and extra-solar planets, and the associated need for accurate theoretical models for ab initio calculations of diatomic molecular spectra, nicely reviewed recently [35].

3. Results

3.1. Angular Momentum Commutators

The invariance of standard QMT commutator relations (see Equation (7)) is communicated in this section.

3.1.1. Invariance for Unitary Transformations

Application of unitary transformation, viz., transforming from one coordinate system to another, leaves the AM commutator relations invariant [36]. A unitary transformation operator, $\mathcal{U}$, acting on an operator $\mathcal{O} \longrightarrow \mathcal{O}'$, with $\mathcal{U}^\dagger = \mathcal{U}^{-1}$, is defined by

$$\mathcal{O}' = \mathcal{U}\mathcal{O}\mathcal{U}^\dagger \quad \text{or} \quad \mathcal{O} = \mathcal{U}^\dagger\mathcal{O}'\mathcal{U}. \tag{8}$$

The invariance of the AM commutators with respect to a unitary transformation, Equation (8),

$$[J_k, J_l] = i\varepsilon_{klm}J_m \quad \longrightarrow \quad [J'_k, J'_l] = i\varepsilon_{klm}J'_m, \tag{9}$$

can be derived by inserting $J_k = \mathcal{U}^\dagger J'_k\mathcal{U}$ and $J_l = \mathcal{U}^\dagger J'_l\mathcal{U}$ in Equation (9) to obtain the intermediate step,

$$\mathcal{U}^\dagger J'_k\mathcal{U}\mathcal{U}^\dagger J'_l\mathcal{U} - \mathcal{U}^\dagger J'_l\mathcal{U}\mathcal{U}^\dagger J'_k\mathcal{U} = \mathcal{U}^\dagger J'_kJ'_l\mathcal{U} - \mathcal{U}^\dagger J'_lJ'_k\mathcal{U} = i\varepsilon_{klm}\mathcal{U}^\dagger J'_m\mathcal{U}. \tag{10}$$

Multiplying from left with $\mathcal{U}$ and from right with $\mathcal{U}^{-1}$ yields the transformed identity in Equation (9). In other words, a unitary transformation preserves the quantum-mechanic AM commutators. For example, the Euler rotation matrix is easily demonstrated to be unitary [6]. In other words, there is no anomaly when going from a laboratory-fixed to a molecule-attached coordinate system.

3.1.2. Invariance for Time Reversal or Reversal of Motion

Time reversal or reversal of motion in QMT requires sign changes of the operators and complex conjugation, leaving the QMT commutators invariant,

$$[J_k, J_l] = i\varepsilon_{klm}J_m \quad \longleftrightarrow \quad [(-J_k), (-J_l)] = (-i)\varepsilon_{klm}(-J_m). \tag{11}$$

CM would indicate a reversal of motion when going from a laboratory-fixed to a molecular-fixed coordinate system; however, reversal of motion requires complex conjugation due to the anti-unitary requirement. In other words, the sign is preserved. QMT so-to-speak opposes the hypothesis by O. Klein.

The invariance regarding time reversal or reversal of motion of course also would apply to the abstract form of the time-dependent Schrödinger equation,

$$i\hbar\frac{\partial}{\partial t}\psi = \mathcal{H}\psi \quad \longleftrightarrow \quad (-i)\hbar\frac{\partial}{\partial(-t)}\psi = \mathcal{H}\psi, \tag{12}$$

where ψ describes an abstract vector in Hilbert space, and $\mathcal{H}$ is a Hamiltonian. Changing time $t \longrightarrow -t$ and applying conjugate complex of i preserves the left-hand side of the equation. For example, for a free particle of mass m and momentum $\mathcal{P}$, the Hamiltonian is $\mathcal{H} = \mathcal{P}^2/2m$, and the form of Schrödinger's equation is preserved.

Equally, the operator equation in the Heisenberg picture, see Equation (4), preserves form under time reversal or reversal of motion,

$$\frac{d\mathcal{O}}{dt} = \frac{i}{\hbar}[\mathcal{H}, \mathcal{O}] + \frac{\partial\mathcal{O}}{\partial t} \quad \longleftrightarrow \quad \frac{d(-\mathcal{O})}{d(-t)} = \frac{(-i)}{\hbar}[\mathcal{H}, (-\mathcal{O})] + \frac{\partial(-\mathcal{O})}{\partial(-t)}. \tag{13}$$

A change of sign for the operators and complex conjugation leaves the equation invariant. The mentioned symmetry can also be associated with usual Nöther symmetries [3].

3.2. *Diatomic Wave Function*

For diatomic molecules, symmetry properties allow one to invoke simplifications when evaluating the laboratory wave-function in terms of rotated coordinates [5]. For internuclear geometry, the spherical polar coordinates are r, ϕ, and θ, and one (arbitrary) electron is described by cylindrical coordinates ρ, χ, ζ. For coordinate rotation, one uses

Euler angles α, β, γ, and without loss of generality, one can choose $\alpha = \phi, \beta = \theta, \chi = \gamma$ [5]. The result is the Wigner–Witmer eigenfunction (WWE) for diatomic molecules [37,38],

$$\langle \rho, \zeta, \chi, \mathbf{r}_2, \ldots, \mathbf{r}_N, r, \theta, \phi \,| nvJM\rangle = \sum_{\Omega=-J}^{J} \langle \rho, \zeta, \mathbf{r}'_2, \ldots, \mathbf{r}'_N, r \,| nv\rangle \, D^{J^*}_{M\Omega}(\phi, \theta, \chi). \tag{14}$$

The usual total AM quantum numbers are J and M, and the electronic–vibrational eigenfunction is explicitly written by extracting v from the collection of quantum numbers, n. The WWE exactly separates ϕ, θ, χ. The quantum numbers J, M, Ω refer to the total AM. The sum over Ω in Equation (14) originates from the usual abstract transformation,

$$|JM\rangle = \sum_{\Omega=-J}^{J} |J\Omega\rangle \, \langle J\Omega \,| JM\rangle, \tag{15}$$

where Ω is the magnetic quantum number along the rotated, or new, z'-axis. The sum in Equation (15) ensures that the quantum numbers for total AM are J and M. In Hund's case *a* [39], Ω describes the projection of the total AM, within L-S coupling. Hund's case *a* eigenfunctions form a basis; therefore, from a computational point of view, these eigenfunctions form a complete (sufficient) set. In various approximate descriptions and for specific diatomic molecules, it may be desirable to use other Hund cases.

From the rotation operator $\mathcal{R}(\alpha, \beta, \gamma)$, with the Euler angles α, β, γ, one finds for D-matrix elements,

$$D^{J^*}_{M\Omega}(\alpha, \beta, \gamma) = \langle JM| \, \mathcal{R}(\alpha, \beta, \gamma) \, |J\Omega\rangle^*. \tag{16}$$

D-matrices are the usual mathematical tool for transformation from one basis to another, but the D-matrix cannot represent an eigenfunction due to presence of two magnetic quantum numbers M and Ω, so the sum over Ω is needed in the transformed coordinates.

Diatomic spectra composed of line positions and line strengths are based on WWE [5] instead of eigenfunctions used for the Born-Oppenheimer approximation. Extensive experimental studies confirm agreement of computed spectra with measured emission spectra from laser-induced optical plasma [5].

3.3. Selected Diatomic Spectra

Typical spectra of some diatomic molecules of general interest are presented. Figure 1 illustrates OH molecular spectra for different spectral resolutions. Figures 2 and 3 show computed C_2 Swan spectra for the vibrational sequences $\Delta v = -1, +1$. The OH spectra, Figure 1, are a superposition of 0-0 (band head near 306 nm), 1-1 (band head near 312 nm), and 2-2 (band head near 318 nm) vibrational transition along with rotational contributions. Four C_2 vibrational peaks, Figures 2 and 3, are clearly discernible. Rotational contributions for the selected spectral resolution, $\Delta\lambda$, appear to have beats (especially Figure 2) that, however, are purely coincidental.

The details for the computation, line strength data for C_2 Swan bands, and programs are published [40]. Computation of diatomic spectra utilizes high-resolution data for determination of molecular constants of selected molecular transitions from an upper to a lower energy level. Numerical solution of the Schrödinger equation for potentials yield r-centroids and transition-factors associated with vibrational transitions, viz. Franck–Condon factors. Calculated rotational factors are interpreted as selection rules because these factors are zero for forbidden transitions, viz. Hönl–London factors. Hönl–London factors in traditional molecular spectroscopy involve selection rules that may require use of anomalous commutators and use of two magnetic quantum numbers M and Ω for a given total angular momentum J. Anomalous selection rules and two quantum numbers for angular momentum J appear to be associated with approximations. The published line strength data [40,41] are derived consistent with standard quantum mechanics, in other words, without anomalous commutators and without states that have two magnetic quantum numbers associated with angular momentum.

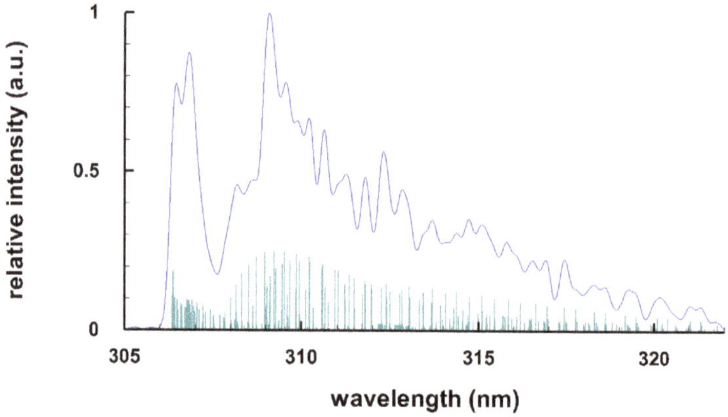

Figure 1. Computed spectrum of the $A^2\Sigma \rightarrow X^2\Pi$ uv band of OH, T = 4 k K, (top) spectral resolutions of $\Delta\lambda = 0.32$ nm ($\Delta\tilde{\nu} = 32$ cm^{-1}) and (bottom) idealized resolution for the stick spectrum $\Delta\lambda = 0.002$ nm ($\Delta\tilde{\nu} = 0.2$ cm^{-1}) of the $\Delta v = 0$ sequence (adapted from [6]).

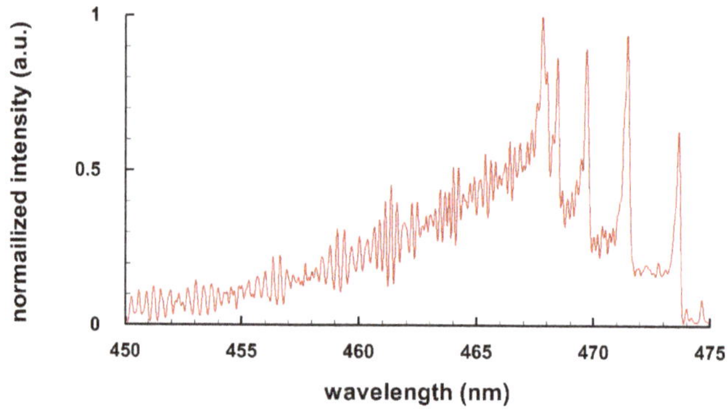

Figure 2. C_2 Swan $d^3\Pi_g \rightarrow a^3\Pi_u$ band $\Delta v = -1$ sequence, T = 8 kK, $\Delta\lambda = 0.13$ nm ($\Delta\tilde{\nu} = 6$ cm^{-1}) (adapted from [6]).

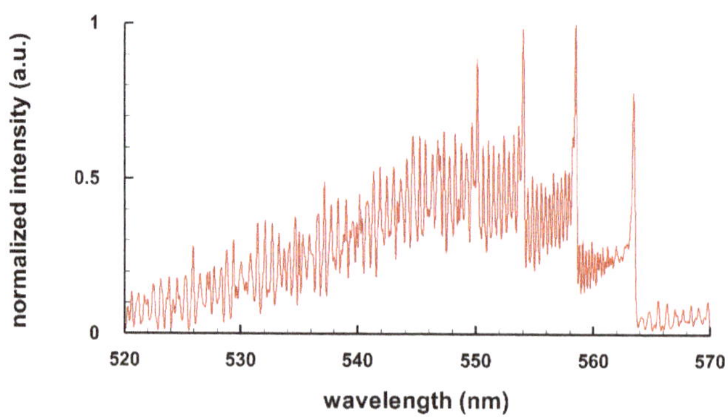

Figure 3. C_2 Swan $d^3\Pi_g \rightarrow a^3\Pi_u$ band $\Delta v = +1$ sequence, T = 8 kK, $\Delta\lambda = 0.18$ nm ($\Delta\tilde{\nu} = 6$ cm^{-1}) (adapted from [6]).

The published program package [40] also includes a worked high-temperature cyanide example, the Boltzmann equilibrium spectrum program (BESP) for computation of equilibrium spectra, and the Nelder–Mead temperature (NMT) routine that utilizes a non-linear fitting algorithm. The OH line strength data have been made available recently [41].

Various reported studies of plasma spectra, including astrophysics plasma, and of molecular laser-induced breakdown spectroscopy (LIBS) [40–43] illustrate nice comparisons of recorded and of computed diatomic spectra. In LIBS, plasma generated by focusing coherent radiation is analyzed primarily in visible/optical or in near-uv to near-ir regions. After initiation of optical breakdown with typically 10 nanoseconds, 100 mJ laser pulses focused in standard ambient temperature and pressure (SATP) air or in gas mixtures [42], molecule formation including, for example, OH in air, C_2 in carbon monoxide, and CN in 1:1 molar N_2:CO_2 mixture, leads to recombination radiation that is typically measured using time-resolved optical emission laser spectroscopy. When using a metallic target, other diatomic molecules can be investigated, e.g., TiO or AlO, and molecular spectra can be computed from line strength data [40].

4. Discussion and Conclusions

Angular momentum operators are well defined in quantum mechanics theory, including the fact that there is an inherent limit in measurement of its components. Another way of formulating this could be: There are only two quantum numbers needed for description of angular momentum, usually the total angular momentum and its projection onto a quantization axis. The use of the correspondence principle to ensure compatibility with classical mechanics equations of motion brings about an ad hoc hypothesis of a negative sign for the commutators, as originally communicated by Oscar Klein in 1929. Subsequent application of reversed angular momentum coupling continues to find support in analytic description of molecules that also includes modeling of quantum mechanic vector-operators as vectors.

However, quantum mechanics theory already ensures how to mathematically describe angular momentum, not supporting heuristic conclusions involving reversed angular momentum concepts, nor occurrence of more than two quantum numbers for the total angular momentum of diatomic molecules. This review emphasizes that there is no mathematical justification of reversed angular momentum algebra, and it also discusses applications in diatomic molecular spectroscopy. Consistent application of standard quantum mechanics theory is preferred, including avoidance of a priori use of separating electronic, vibrational, rotational wave functions. Subsequent to implementation of diatomic molecular symmetries, line strengths for selected diatomic molecules that contain effects of spin splitting and lambda-doubling as function of wavelength are in agreement with results from optical emission spectroscopy. The computed and fitted diatomic spectra nicely match within reasonable error bars, but without invoking heuristic selection rules that may be affected by initial approximations or by spurious use of reversal of angular momentum.

Funding: This research received no specific grant-number external funding.

Institutional Review Board Statement: Not applicable.

Informed Consent Statement: Not applicable.

Data Availability Statement: The data for Figures 1–3 can be computed using provided programs and published line strengths for C_2 and OH in Refs. [39,40], respectively.

Acknowledgments: The author acknowledges support in part from the State of Tennessee funded Center for Laser Applications at the University of Tennessee Space Institute.

Conflicts of Interest: The author declares no conflict of interest. The funders had no role in the design of the study; in the collection, analyses, or interpretation of data; in the writing of the manuscript; or in the decision to publish the results.

Abbreviations

The following abbreviations are used in this manuscript:

AM	Angular Momentum
BESP	Boltzmann Equilibrium Spectrum Program
CM	Classical Mechanics
NMT	Nelder–Mead Temperature
PGOPHER	Program for simulating rotational, vibrational, and electronic spectra, or "Program Gopher"
QM	Quantum Mechanics
QMT	Quantum Mechanics Theory
RAM	Reversed Angular Momentum
WWE	Wigner–Witmer Eigenfunction

References

1. Klein, O. Zur Frage der Quantelung des asymmetrischen Kreisels. *Z. Phys.* **1929**, *58*, 730–734. [CrossRef]
2. Van Vleck, J.H. The Coupling of Angular Momentum Vectors in Molecules. *Rev. Mod. Phys.* **1951**, *23*, 213–227. [CrossRef]
3. Nöther, E. Invariante Variationsprobleme. *Nachr. D. König. Gesellsch. D. Wiss. Göttingen* **1918**, *918*, 235–257; Invariant variation problems. *Transp. Theory Statist. Phys.* **1971**, *1*, 183–207.
4. Condon, E.U.; Shortley, G. *The Theory of Atomic Spectra*; Cambridge University Press: Cambridge, UK, 1953.
5. Parigger, C.G.; Hornkohl, J.O. *Quantum Mechanics of the Diatomic Molecule with Applications*; IOP Publishing: Bristol, UK, 2020.
6. Parigger, C.G.; Hornkohl, J.O. Diatomic Molecular Spectroscopy with Standard and Anomalous Commutators. *Int. Rev. At. Mol. Phys.* **2010**, *1*, 25–43.
7. Dirac, P.A.M. The Elimination of Nodes in Quantum Mechanics. *Proc. Roy. Soc. Lond. A* **1926**, *111*, 281–305.
8. Heisenberg, W.; Born, M.; Jordan, P. Zur Quantenmechanik II. *Z. Phys.* **1926**, *35*, 557–615.
9. Popper, K. The Logic and Evolution of Scientific Theory. In *All Life is Problem Solving*; Routledge: London, UK, 1999; Chapter 1, pp. 3–22.
10. Brown, J.; Carrington, A. *Rotational Spectroscopy of Diatomic Molecules*; Cambridge University Press: Cambridge, UK, 2003.
11. Lefebvre-Brion, H.; Field, R.W. *The Spectra and Dynamics of Diatomic Molecules*; Elsevier: Amsterdam, NL, USA, 2004.
12. Bunker, P.R.; Jensen, P. *Fundamentals of Molecular Spectroscopy*; IOP Publishing: Bristol, UK, 2005.
13. Gottfried, K. *Quantum Mechanics*; Addison-Wesley: Reading, UK, 1989.
14. Baym, G. *Lectures on Quantum Mechanics*; Benjamin/Cummings: Reading, UK, 1969.
15. Shore, B.W.; Menzel, D.H. *Principles of Atomic Spectra*; Addison-Wesley: Reading, UK, 1968.
16. Judd, B.R. *Angular Momentum Theory for Diatomic Molecules*; Academic Press: New York, NY, USA, 1975.
17. Mizushima, M. *The Theory of Rotating Diatomic Molecules*; John Wiley & Sons: New York, NY, USA, 1975.
18. Kovacs, I. *Rotational Structure in the Spectra of Diatomic Molecules*; Elsevier: New York, NY, USA, 1969.
19. Hougen, J.T. *The Calculation of Rotational Energy Levels and Rotational Line Intensities in Diatomic Molecules*; NBS Monograph 115; U.S. Government Printing Office: Washington, DC, USA, 1970.
20. Kroto, H.W. *Molecular Rotation Spectra*; Dover: New York, NY, USA, 1992.
21. Carrington, A.; Levy, D.H.; Miller, T.A. Electron Resonance of Gaseous Diatomic Molecules. *Adv. Chem. Phys.* **1970**, *18*, 149–248.
22. Freed, K.F. Theory of the Hyperfine Structure of Molecules: Application to $^3\Pi$ States of Diatomic Molecules Intermediate between Hund's Cases (a) and (b). *J. Chem. Phys.* **1966**, *45*, 4214–4241. [CrossRef]
23. Cohen-Tannoudji, C.; Diu, B.; Laloe, F. *Quantum Mechanics, Volume 1: Basic Concepts, Tools, and Application*, 2nd ed.; Wiley-VCH: Weinheim, Germany, 2019.
24. Cohen-Tannoudji, C.; Diu, B.; Laloe, F. *Quantum Mechanics, Volume 2: Angular Momentum, Spin, and Approximation Methds*, 2nd ed.; Wiley-VCH: Weinheim, Germany, 2019.
25. Arfken, G.B.; Weber, H.J.; Harris, F.E. *Mathematical Methods for Physicists, A Comprehensive Guide*, 7th ed.; Academic Press: New York, NY, USA, 2012.
26. Western, C.M. PGOPHER: A program for simulating rotational, vibrational and electronic spectra. *J. Quant. Spectrosc. Radiat. Transf.* **2017**, *186*, 221–242. [CrossRef]
27. Kunze, H.-J. *Introduction to Plasma Spectroscopy*; Springer: Heidelberg, Germany, 2009.
28. Demtröder, W. *Laser Spectroscopy 1: Basic Principles*, 5th ed.; Springer: Heidelberg, Germany, 2014.
29. Demtröder, W. *Laser Spectroscopy 2: Experimental Techniques*, 5th ed.; Springer: Heidelberg, Germany, 2015.
30. Hertel, I.V.; Schulz, C.-P. *Atoms, Molecules and Optical Physics 1, Atoms and Spectroscopy*; Springer: Heidelberg, Germany, 2015.
31. Hertel, I.V.; Schulz, C.-P. *Atoms, Molecules and Optical Physics 2, Molecules and Photons—Spectroscopy and Collisions*; Springer: Heidelberg, Germany, 2015.
32. Cremers, D.E.; Radziemski, L.J. *Handbook of Laser-Induced Breakdown Spectroscopy*; Wiley: New York, NY, USA, 2006.
33. Miziolek, A.W.; Palleschi, V.; Schechter, I. (Eds.) *Laser Induced Breakdown Spectroscopy*; Cambridge University Press: New York, NY, USA, 2006.

34. Singh, J.P.; Thakur, S.N. (Eds.) *Laser-Induced Breakdown Spectroscopy*, 2nd ed.; Elsevier: New York, NY, USA, 2020.
35. Tennyson, J.; Lodi, L.; McKemmish, L.K.; Yurchenko, S.N. The *ab initio* calculation of spectra of open shell diatomic molecules. *arXiv* **2016**, arXiv:1605.02301v1.
36. Davydov, A.S. *Quantum Mechanics*; Pergamon Press: Oxford, UK, 1965.
37. Wigner, E.; Witmer, E.E. Über die Struktur der zweiatomigen Molekelspectren nach der Quantenmechanik. *Z. Physik* **1928**, *51*, 859–886 [CrossRef]
38. Hettema, H., Ed. On the Structure of the Spectra of Two-Atomic Molecules According to Quantum Mechanics. In *Quantum Chemistry: Classic Scientific Papers*; World Scientific: Singapore, 2000; pp. 287–311.
39. Bransden, B.H.; Joachain, C.J. *Physics of Atoms and Molecules*, 2nd ed.; Prentice Hall: Essex, UK, 2003.
40. Parigger, C.G.; Woods, A.C.; Surmick, D.M.; Gautam, G.; Witte, M.J.; Hornkohl, J.O. Computation of diatomic molecular spectra for selected transitions of aluminum monoxide, cyanide, diatomic carbon, and titanium monoxide. *Spectrochim. Acta Part B At. Spectrosc.* **2015**, *107*, 132–138. [CrossRef]
41. Parigger, C.G.; Helstern, C.M.; Jordan, B.S.; Surmick, D.M.; Splinter, R. Laser-Plasma Spectroscopy of Hydroxyl with Applications. *Molecules* **2020**, *25*, 988. [CrossRef] [PubMed]
42. Parigger, C.G.; Surmick, D.M.; Helstern, C.M.; Gautam G.; Bol'shakov, A.A.; Russo, R.E. Molecular laser-induced breakdown spectroscopy. In *Laser-Induced Breakdown Spectroscopy*, 2nd ed.; Singh, J.P., Thakur, S.N., Eds.; Elsevier: Amsterdam, The Netherlands, 2020; Chapter 7, pp. 167–209.
43. Parigger, C.G.; Helstern, C.M.; Jordan, B.S.; Surmick, D.M.; Splinter, R. Laser-Plasma Spatiotemporal Cyanide Spectroscopy and Applications. *Molecules* **2020**, *25*, 615. [CrossRef] [PubMed]

foundations

Review

Information Geometric Measures of Complexity with Applications to Classical and Quantum Physical Settings

Carlo Cafaro [1,*] and Sean A. Ali [2]

[1] SUNY Polytechnic Institute, Albany, NY 12203, USA
[2] Albany College of Pharmacy and Health Sciences, Albany, NY 12208, USA; sean.ali@acphs.edu
* Correspondence: CCafaro@sunypoly.edu

Abstract: We discuss the fundamental theoretical framework together with numerous results obtained by the authors and colleagues over an extended period of investigation on the Information Geometric Approach to Chaos (IGAC).

Keywords: information geometry; complexity; classical and quantum physics

PACS: chaos (05.45.-a); complexity (89.70.Eg); entropy (89.70.Cf); inference methods (02.50.Tt); information theory (89.70.+c); probability theory (02.50.Cw); riemannian geometry (02.40.Ky)

Citation: Cafaro, C.; Ali, S.A. Information Geometric Measures of Complexity with Applications to Classical and Quantum Physical Settings . *Foundations* **2021**, *1*, 45–62. https://doi.org/10.3390/foundations1010006

Academic Editor: Eugene Oks

Received: 20 July 2021
Accepted: 25 August 2021
Published: 31 August 2021

1. Theoretical Background

Statistical models are employed to formulate probabilistic descriptions of systems of arbitrary nature when only partial knowledge about the system is available. Indeed, in recent years, methods of entropic inference [1] have been utilized in conjunction with information geometry (IG) [2] for the purpose of developing complexity indicators of statistical models. From the perspective of this hybrid framework, such complexity indicators can be understood as being quantitative measures that describe the complication of inferring macroscopic predictions about statistical models. In this context, the difficulty of making macroscopic predictions is attributed to the fact that statistical models intrinsically reflect only partial information about the microscopic degrees of freedom of the system being modeled. Initial theoretical investigation in this direction, quoted as the *Information Geometric Approach to Chaos* (IGAC), was originally proposed by Cafaro in his physics Ph.D. doctoral dissertation in [3].

A general summary of the IGAC framework is described as follows [4,5]: upon identifying the microscopic degrees of freedom of a complex system, one must obtain data and choose important information constraints on the system. Entropic methods are then utilized to obtain an initial, static statistical model of the system. In this way, the system is described by a statistical model specified in terms of probability distributions that are characterized by statistical macrovariables. The statistical macrovariables are determined by the data and the specific functional expression of the information constraints used to implement statistical inferences. The next step in the theoretical scheme is concerned with the temporal evolution of the system. If it is assumed that the system changes, then the corresponding statistical model evolves from its initial to final configurations in a manner specified by Entropic Dynamics (ED, [6]). The ED framework can be viewed as a form of constrained information dynamics that is formulated on statistical manifolds, the elements of which are probability distributions. These distributions, in turn, are in one-to-one relation with an appropriate set of statistical macrovariables that determine a parameter space, where the latter serves to provide a suitable parameterization of points on the original statistical manifold.

Within the context of ED, the change of probability distributions is described in terms of a principle of entropic inference. Specifically, beginning with a known initial configuration, change toward the final configuration happens by the maximization of the logarithmic relative entropy (known as the Maximum relative Entropy method—or MrE method in brief, [1]) between any two successive intermediate configurations. We emphasize that ED specifies the *expected* rather than the *actual* dynamical paths of the system. Inferences within the ED framework depends on the data and functional form of the selected information constraints used in the MrE algorithm. Indeed, modeling strategies of this kind can only be corroborated *a posteriori*. This fact implies that in the event inferred predictions fail to match experimental measurements, a new set of information constraints should be chosen. This feature of the MrE algorithm is of critical significance and was recently re-examined by Cafaro and Ali in [7].

The change of probability distributions characterized by the maximization algorithm outlined above prescribes a geodesic evolution for the statistical macrovariables [1]. The Fisher–Rao information metric [2] yields a measure of distance between any two dissimilar probability distributions on a statistical manifold. The notion of distance between elements of a statistical manifold can be regarded as the degree of distinguishability between any two different probability distribution functions. Once the information metric has been obtained, differential geometric techniques can be readily applied to study the geometry of the curved statistical manifold. Broadly speaking, standard Riemannian geometric quantities, such as Christoffel connection coefficients of the second kind, Riemannian curvature tensor, Ricci tensor, Ricci scalar curvature, Weyl anisotropy tensor, sectional curvatures, Killing fields and Jacobi fields (including the IG analogue of Lyapunov exponents) can be calculated in the usual fashion. In particular, the *chaoticity* (i.e., *temporal complexity*) of such statistical models can be analyzed via appropriately selected indicators, such as the signs of the Ricci scalar and sectional curvatures of the statistical manifold, non-vanishing Weyl anisotropy tensor, the asymptotic temporal behavior of Jacobi fields and the existence of Killing vectors. Along with the various indicators mentioned above, the notion of complexity within the IGAC framework can also be characterized by the *Information Geometric Entropy* (IGE), originally proposed in [3]. We make reference to the Ali–Cafaro effort in [8] for a more extensive summary of the IGAC framework that incorporates a set of remarks on entropic evolution and the MrE algorithm. For a presentation of alternative information geometric descriptions of complexity, we suggest the investigation by Felice, Cafaro and Mancini in [5]. While we certainly appreciate the power of the synthetic, non-component approach to tensor analysis commonly used in theoretical physics (for instance, see [9]), we have nevertheless chosen to employ the component approach in the present paper. In our opinion, the applied nature of our works can be formulated and analyzed more efficiently (and transparently) within the component approach to tensor calculus.

In the next section, we introduce suitable indicators of complexity within the IGAC.

2. Indicators of Complexity

In this section, we introduce three indicators of complexity within the context of the IGAC framework. We present the IGE, the curvature of the statistical manifold, and finally, the notion of Jacobi fields arising from the equation of geodesic deviation.

2.1. Information Geometric Entropy

We begin this subsection by discussing the IGE. Assuming the elements $\{p(x;\theta)\}$ of an n-dimensional statistical manifold $\mathcal{M}_s$ are parameterized by n, *real*-valued variables $(\theta^1,\ldots,\theta^n)$, the statistical manifold is defined by the set

$$\mathcal{M}_s \stackrel{\text{def}}{=} \left\{p(x;\theta) : \theta = \left(\theta^1,\ldots,\theta^n\right) \in \mathcal{D}_\theta^{(\text{tot})}\right\}. \quad (1)$$

We point out that the quantity x in Equation (1) denotes microvariables belonging to the microspace $\mathcal{X}$, whereas the macrovariables θ appearing in Equation (1) are elements of the parameter space $\mathcal{D}_{\theta}^{(\text{tot})}$ defined by

$$\mathcal{D}_{\boldsymbol{\theta}}^{(\text{tot})} \stackrel{\text{def}}{=} \bigotimes_{j=1}^{n} \mathcal{I}_{\theta^j} = (\mathcal{I}_{\theta^1} \otimes \mathcal{I}_{\theta^2} \ldots \otimes \mathcal{I}_{\theta^n}) \subseteq \mathbb{R}^n. \tag{2}$$

The quantity $\mathcal{I}_{\theta^j}$ with $1 \leq j \leq n$ in Equation (2) is a subset of $\mathbb{R}^n$ that denotes the full range of admissible values of the statistical macrovariables θ^j. The IGE serves as an indicator of temporal complexity associated with geodesic paths in the IGAC framework. The IGE is given by

$$\mathcal{S}_{\mathcal{M}_s}(\tau) \stackrel{\text{def}}{=} \log \widetilde{vol}[\mathcal{D}_{\boldsymbol{\theta}}(\tau)], \tag{3}$$

where the average dynamical statistical volume $\widetilde{vol}[\mathcal{D}_{\boldsymbol{\theta}}(\tau)]$ is defined as

$$\widetilde{vol}[\mathcal{D}_{\boldsymbol{\theta}}(\tau)] \stackrel{\text{def}}{=} \frac{1}{\tau} \int_0^{\tau} vol[\mathcal{D}_{\boldsymbol{\theta}}(\tau')] d\tau'. \tag{4}$$

We remark that the temporal averaging operation is denoted by the tilde symbol in Equation (4). Furthermore, the volume $vol[\mathcal{D}_{\boldsymbol{\theta}}(\tau')]$ appearing on the RHS of Equation (4) is given by

$$vol[\mathcal{D}_{\boldsymbol{\theta}}(\tau')] \stackrel{\text{def}}{=} \int_{\mathcal{D}_{\boldsymbol{\theta}}(\tau')} \rho\left(\theta^1, \ldots, \theta^n\right) d^n\theta, \tag{5}$$

where $\rho(\theta^1, \ldots, \theta^n)$ is known as the Fisher density and is equal to the square root of the determinant $g(\theta)$ of the Fisher–Rao information metric tensor $g_{\mu\nu}(\theta)$,

$$\rho\left(\theta^1, \ldots, \theta^n\right) \stackrel{\text{def}}{=} \sqrt{g(\theta)}. \tag{6}$$

The Fisher–Rao information metric tensor $g_{\mu\nu}(\theta)$ is defined as

$$g_{\mu\nu}(\theta) \stackrel{\text{def}}{=} \int p(x|\theta) \partial_{\mu} \log p(x|\theta) \partial_{\nu} \log p(x|\theta) dx, \tag{7}$$

where $\mu, \nu = 1, \ldots, n$ for an n-dimensional manifold and $\partial_{\mu} \stackrel{\text{def}}{=} \frac{\partial}{\partial \theta^{\mu}}$. The volume $vol[\mathcal{D}_{\boldsymbol{\theta}}(\tau')]$ in Equation (5) can be recast in a more crystalline manner for cases involving statistical manifolds whose information metric tensor has a determinant that can be expressed in a factorized form as follows,

$$g(\theta) = g\left(\theta^1, \ldots, \theta^n\right) = \prod_{j=1}^{n} g_j\left(\theta^j\right). \tag{8}$$

By using the factorized form of the determinant, the IGE appearing in Equation (3) can be expressed as

$$\mathcal{S}_{\mathcal{M}_s}(\tau) = \log\left\{\frac{1}{\tau} \int_0^{\tau} \left[\prod_{j=1}^{n} \left(\int_{\tau_0}^{\tau_0+\tau'} \sqrt{g_j[\theta^j(\xi)]} \frac{d\theta^j}{d\xi} d\xi\right)\right] d\tau'\right\}. \tag{9}$$

Within the IGAC framework, the leading asymptotic behavior of $\mathcal{S}_{\mathcal{M}_s}(\tau)$ in Equation (9) is employed to specify the complexity of the statistical model under investigation. Therefore, it is quite instructive to take into consideration the quantity

$$\mathcal{S}_{\mathcal{M}_s}^{(\text{asymptotic})}(\tau) \approx \lim_{\tau \to \infty} [\mathcal{S}_{\mathcal{M}_s}(\tau)], \tag{10}$$

that is, the leading asymptotic term in the expression of the IGE. The integration space $\mathcal{D}_{\theta}(\tau')$ in Equation (5) is defined by

$$\mathcal{D}_{\theta}(\tau') \stackrel{\text{def}}{=} \left\{\theta : \theta^{j}(\tau_0) \leq \theta^{j} \leq \theta^{j}(\tau_0 + \tau')\right\}, \tag{11}$$

where $\theta^{j} = \theta^{j}(\xi)$ with $\tau_0 \leq \xi \leq \tau_0 + \tau'$ and τ_0 denotes the initial value of the affine parameter ξ such that

$$\frac{d^2\theta^{j}(\xi)}{d\xi^2} + \Gamma^{j}_{ik}\frac{d\theta^{i}}{d\xi}\frac{d\theta^{k}}{d\xi} = 0. \tag{12}$$

The domain of integration $\mathcal{D}_{\theta}(\tau')$ in Equation (11) is an n-dimensional subspace of $\mathcal{D}_{\theta}^{(\text{tot})}$. The elements of $\mathcal{D}_{\theta}^{(\text{tot})}$ are n-dimensional macrovariables $\{\theta\}$ whose components θ^{j} are bounded by the limits of integration $\theta^{j}(\tau_0)$ and $\theta^{j}(\tau_0 + \tau')$. The temporal functional form of such limits is determined by the integration of the geodesic equations in Equation (12).

The IGE evaluated at a particular instant is specified by the logarithm of the volume of the effective parameter space probed by the system at that instant. In order to coarse-grain the possibly very complex details of the entropic dynamical characterization of the system, however, the process of temporal averaging has been employed. Moreover, in order to remove the effects of potential transient features that may enter the calculation of the expected value of the volume of the effective parameter space, only its asymptotic temporal behavior is taken into consideration. For these reasons, it is evident that the IGE serves as an asymptotic, coarse-grained complexity indicator of dynamical systems in the presence of partial information. For additional specifics concerning the IGE, we refer the interested reader to [8,10].

As a conclusive side remark, we emphasize that it would be interesting to characterize the tendency to increase of the entropy of a physical system that approaches equilibrium as specified by the Boltzmann **H** theorem and the second law of thermodynamics [11] from an information geometric perspective. For a recent information geometric interpretation of the entropy production, we refer to [12]. In particular, to understand the possible link between the IGE and the Boltzmann–Shannon entropy, it would be important to study the Kaniadakis $\mathcal{S}_{\kappa}$ entropy (with κ being the so-called deformation parameter) and comprehend how the statistical mechanics based on $\mathcal{S}_{\kappa}$ can be regarded as a natural generalization of the equilibrium Boltzmann–Gibbs statistical mechanics [13]. We leave these intriguing lines of investigations to future efforts.

2.2. Curvature

We present here the notion of curvature of statistical manifolds. We begin by recalling that an n-dimensional, $\mathbb{C}^{\infty}$ differentiable manifold is defined by a set of points $\mathcal{M}$ endowed with coordinate systems $\mathcal{C}_{\mathcal{M}}$ fulfilling the following two requirements: (1) each element $c \in \mathcal{C}_{\mathcal{M}}$ is a one-to-one mapping from $\mathcal{M}$ to an open subset of $\mathbb{R}^{n}$; (2) given any one-to-one mapping $\eta : \mathcal{M} \to \mathbb{R}^{n}$, we have that $\forall c \in \mathcal{C}_{\mathcal{M}}, \eta \in \mathcal{C}_{\mathcal{M}} \Leftrightarrow \eta \circ c^{-1}$ is a $\mathbb{C}^{\infty}$ diffeomorphism.

In this paper, we focus on Riemannian manifolds $(\mathcal{M}, g)$ where the points of $\mathcal{M}$ are probability distribution functions. It is worth noting that the manifold structure of $\mathcal{M}$ is insufficient to specify in a unique manner the Riemannian metric g. On a formal level, an infinite number of Riemannian metrics can be defined on the manifold $\mathcal{M}$. In the context of information geometry however, the selection of the Fisher–Rao information metric (see Equation (7)) as the metric underlying the Riemannian geometry of probability distributions [2,14,15] serves as a primary working assumption. The characterization theorem attributed to Cencov [16] gives significant support for this particular choice of metric. In this characterization theorem, Cencov demonstrates that, up to any arbitrary constant scale factor, the information metric is the only Riemannian metric that is invariant under congruent embeddings (that is, under a family of probabilistically meaningful mappings) of the Markov morphism [16,17].

Upon introducing the Fisher–Rao information metric $g_{\mu\nu}(\theta)$ in Equation (7), standard differential geometric techniques can be used on the space of probability distributions to describe the geometry of the statistical manifold $\mathcal{M}_s$. The Ricci scalar curvature $\mathcal{R}_{\mathcal{M}_s}$ is one example of such a geometric property, where $\mathcal{R}_{\mathcal{M}_s}$ is defined as [18]

$$\mathcal{R}_{\mathcal{M}_s} \stackrel{\text{def}}{=} g^{\mu\nu}\mathcal{R}_{\mu\nu}, \tag{13}$$

where $g^{\mu\nu}g_{\nu\rho} = \delta^{\mu}_{\rho}$ and $g^{\mu\nu} = (g_{\mu\nu})^{-1}$. The Ricci tensor $\mathcal{R}_{\mu\nu}$ appearing in Equation (13) is given as [18]

$$\mathcal{R}_{\mu\nu} \stackrel{\text{def}}{=} \partial_{\gamma}\Gamma^{\gamma}_{\mu\nu} - \partial_{\nu}\Gamma^{\lambda}_{\mu\lambda} + \Gamma^{\gamma}_{\mu\nu}\Gamma^{\eta}_{\gamma\eta} - \Gamma^{\eta}_{\mu\gamma}\Gamma^{\gamma}_{\nu\eta}. \tag{14}$$

The Christoffel connection coefficients $\Gamma^{\rho}_{\mu\nu}$ of the second kind that specify the Ricci tensor in Equation (14) are [18]

$$\Gamma^{\rho}_{\mu\nu} \stackrel{\text{def}}{=} \frac{1}{2}g^{\rho\sigma}(\partial_{\mu}g_{\sigma\nu} + \partial_{\nu}g_{\mu\sigma} - \partial_{\sigma}g_{\mu\nu}). \tag{15}$$

Next we consider geodesic curves on statistical manifolds. A geodesic on an n-dimensional statistical manifold $\mathcal{M}_s$ can be interpreted as the maximum probability trajectory explored by a complex system during its change from an initial θ_{initial} to final macrostates θ_{final}, respectively. Each point along a geodesic path corresponds to a macrostate specified by the macroscopic variables $\theta = (\theta^1, \ldots, \theta^n)$. In the context of ED, each component θ^j with $j = 1, \ldots, n$ is a solution of the geodesic equation [6],

$$\frac{d^2\theta^k}{d\xi^2} + \Gamma^k_{lm}\frac{d\theta^l}{d\xi}\frac{d\theta^m}{d\xi} = 0. \tag{16}$$

At this juncture, we reiterate the fact that each macrostate θ is in one-to-one correspondence with the probability distribution $p(x|\theta)$, with the latter characterizing a distribution of the microstates x. It is useful to recognize that the scalar curvature $\mathcal{R}_{\mathcal{M}_s}$ can be readily recast as the sum of sectional curvatures $\mathcal{K}(e_\rho, e_\sigma)$ of all tangent space planes $T_p\mathcal{M}_s$ with $p \in \mathcal{M}_s$ spanned by pairs of orthonormal basis vectors $\left\{e_\rho = \partial_{\theta_\rho(p)}\right\}$,

$$\mathcal{R}_{\mathcal{M}_s} \stackrel{\text{def}}{=} \mathcal{R}^{\alpha}_{\alpha} \stackrel{\text{def}}{=} \sum_{\rho\neq\sigma}\mathcal{K}(e_\rho, e_\sigma), \tag{17}$$

where $\mathcal{K}(a, b)$ is given by [18]

$$\mathcal{K}(a, b) \stackrel{\text{def}}{=} \frac{\mathcal{R}_{\mu\nu\rho\sigma}a^{\mu}b^{\nu}a^{\rho}b^{\sigma}}{(g_{\mu\sigma}g_{\nu\rho} - g_{\mu\rho}g_{\nu\sigma})a^{\mu}b^{\nu}a^{\rho}b^{\sigma}}, \tag{18}$$

with

$$a \stackrel{\text{def}}{=} \sum_{\rho}\langle a, e^{\rho}\rangle e_\rho,\ b \stackrel{\text{def}}{=} \sum_{\rho}\langle b, e^{\rho}\rangle e_\rho, \text{ and } \langle e_\rho, e^{\sigma}\rangle \stackrel{\text{def}}{=} \delta^{\sigma}_{\rho}. \tag{19}$$

We observe that the Riemann curvature tensor $\mathcal{R}_{\alpha\beta\rho\sigma}$ [18] is fully determined by the sectional curvatures $\mathcal{K}(e_\rho, e_\sigma)$ where

$$\mathcal{R}^{\alpha}{}_{\beta\rho\sigma} \stackrel{\text{def}}{=} g^{\alpha\gamma}\mathcal{R}_{\gamma\beta\rho\sigma} \stackrel{\text{def}}{=} \partial_{\sigma}\Gamma^{\alpha}{}_{\beta\rho} - \partial_{\rho}\Gamma^{\alpha}{}_{\beta\sigma} + \Gamma^{\alpha}{}_{\lambda\sigma}\Gamma^{\lambda}{}_{\beta\rho} - \Gamma^{\alpha}{}_{\lambda\rho}\Gamma^{\lambda}{}_{\beta\sigma}. \tag{20}$$

The negativity of the Ricci scalar curvature $\mathcal{R}_{\mathcal{M}_s}$ is a strong (i.e., a sufficient but not necessary) criterion of local dynamical instability. Moreover, the compactness of the manifold $\mathcal{M}_s$ is required to specify genuine chaotic (that is, temporally complex) dynamical systems. In particular, it is evident from Equation (17) that the negativity of $\mathcal{R}_{\mathcal{M}_s}$ implies that negative principal curvatures (i.e., extrema of sectional curvatures) are more dominant than positive ones. For this reason, the negativity of $\mathcal{R}_{\mathcal{M}_s}$ is a sufficient but not necessary requirement for local instability of geodesic flows on statistical manifolds. It is worth mentioning the possible circumstance of scenarios in which negative sectional curvatures are present, but the positive curvatures dominate in the sum of Equation (17) such that $\mathcal{R}_{\mathcal{M}_s}$ is a non-negative quantity despite flow instability in those directions. For additional mathematical considerations related to the concept of curvature in differential geometry, we refer to [19].

2.3. Jacobi Fields

We introduce here the concept of the Jacobi vector field. It is worth noting that the analysis of stability/instability arising in natural (geodesic) evolutions is readily accomplished by means of the Jacobi–Levi–Civita (JLC) equation for geodesic deviation. This equation is familiar in both theoretical physics (for example, in the case of General Relativity) as well as in Riemannian geometry. The JLC equation describes in a covariant manner, the degree to which neighboring geodesics locally scatter. In particular, the JLC equation effectively connects the curvature properties of an underlying manifold to the stability/instability of the geodesic flow induced thereupon. Indeed, the JLC equation provides a window into a diverse and mostly unexplored field of study concerning the connections among topology, geometry and geodesic instability, and thus to complexity and chaoticity. The use the JLC equation in the setting of information geometry originally appeared in [20].

In what follows, we take into consideration two neighboring geodesic paths $\theta^{\alpha}(\xi)$ and $\theta^{\alpha}(\xi)+\delta\theta^{\alpha}(\xi)$, where the quantity ξ denotes an affine parameter satisfying the geodesic equations,

$$\frac{d^2\theta^{\alpha}}{d\xi^2}+\Gamma^{\alpha}_{\beta\gamma}(\theta)\frac{d\theta^{\beta}}{d\xi}\frac{d\theta^{\gamma}}{d\xi}=0, \tag{21}$$

and

$$\frac{d^2[\theta^{\alpha}+\delta\theta^{\alpha}]}{d\xi^2}+\Gamma^{\alpha}_{\beta\gamma}(\theta+\delta\theta)\frac{d\left[\theta^{\beta}+\delta\theta^{\beta}\right]}{d\xi}\frac{d[\theta^{\gamma}+\delta\theta^{\gamma}]}{d\xi}=0, \tag{22}$$

respectively. Noting that to first order in $\delta\theta^{\alpha}$,

$$\Gamma^{\alpha}_{\beta\gamma}(\theta+\delta\theta)\approx\Gamma^{\alpha}_{\beta\gamma}(\theta)+\partial_{\eta}\Gamma^{\alpha}_{\beta\gamma}\delta\theta^{\eta}, \tag{23}$$

after some algebraic calculations, to first order in $\delta\theta^{\alpha}$, Equation (22) becomes

$$\frac{d^2\theta^{\alpha}}{d\xi^2}+\frac{d^2(\delta\theta^{\alpha})}{d\xi^2}+\Gamma^{\alpha}_{\beta\gamma}(\theta)\frac{d\theta^{\beta}}{d\xi}\frac{d\theta^{\gamma}}{d\xi}+2\Gamma^{\alpha}_{\beta\gamma}(\theta)\frac{d\theta^{\beta}}{d\xi}\frac{d(\delta\theta^{\gamma})}{d\xi}+\partial_{\eta}\Gamma^{\alpha}_{\beta\gamma}(\theta)\delta\theta^{\eta}\frac{d\theta^{\beta}}{d\xi}\frac{d\theta^{\gamma}}{d\xi}=0. \tag{24}$$

The equation of geodesic deviation can be found by subtracting Equation (21) from Equation (24),

$$\frac{d^2(\delta\theta^{\alpha})}{d\xi^2}+2\Gamma^{\alpha}_{\beta\gamma}(\theta)\frac{d\theta^{\beta}}{d\xi}\frac{d(\delta\theta^{\gamma})}{d\xi}+\partial_{\eta}\Gamma^{\alpha}_{\beta\gamma}(\theta)\delta\theta^{\eta}\frac{d\theta^{\beta}}{d\xi}\frac{d\theta^{\gamma}}{d\xi}=0. \tag{25}$$

Equation (25) can be conveniently recast via the covariant derivatives (see [21], for instance) along the curve $\theta^{\alpha}(\xi)$,

$$\frac{D^2(\delta\theta^{\alpha})}{D\xi^2}=\frac{d^2(\delta\theta^{\alpha})}{d\xi^2}+\partial_{\beta}\Gamma^{\alpha}_{\rho\sigma}\frac{d\theta^{\beta}}{d\xi}\delta\theta^{\rho}\frac{d\theta^{\sigma}}{d\xi}+2\Gamma^{\alpha}_{\rho\sigma}\frac{d(\delta\theta^{\rho})}{d\xi}\frac{d\theta^{\sigma}}{d\xi}+$$

$$-\Gamma^{\alpha}_{\rho\sigma}\Gamma^{\sigma}_{\kappa\lambda}\delta\theta^{\rho}\frac{d\theta^{\kappa}}{d\xi}\frac{d\theta^{\lambda}}{d\xi}+\Gamma^{\alpha}_{\rho\sigma}\Gamma^{\rho}_{\kappa\lambda}\delta\theta^{\kappa}\frac{d\theta^{\lambda}}{d\xi}\frac{d\theta^{\sigma}}{d\xi}. \tag{26}$$

The covariant derivative is defined as $D_{\xi}\delta\theta^{\alpha} \overset{\text{def}}{=} \partial_{\xi}\delta\theta^{\alpha} + \Gamma^{\alpha}_{\xi\kappa}\delta\theta^{\kappa}$ with $D_{\xi} \overset{\text{def}}{=} D/D\xi$ and $\partial_{\xi} \overset{\text{def}}{=} \partial/\partial\xi$, respectively. By combining Equations (25) and (26), and performing some tensor algebra calculations, we obtain

$$\frac{D^2(\delta\theta^{\alpha})}{D\xi^2} = \left(\partial_{\rho}\Gamma^{\alpha}_{\eta\sigma} - \partial_{\eta}\Gamma^{\alpha}_{\rho\sigma} + \Gamma^{\alpha}_{\lambda\sigma}\Gamma^{\lambda}_{\eta\rho} - \Gamma^{\alpha}_{\eta\lambda}\Gamma^{\lambda}_{\rho\sigma}\right)\delta\theta^{\eta}\frac{d\theta^{\rho}}{d\xi}\frac{d\theta^{\sigma}}{d\xi}. \tag{27}$$

Finally, the geodesic deviation equation expressed in component form becomes

$$\frac{D^2 J^{\alpha}}{D\xi^2} + \mathcal{R}^{\alpha}_{\rho\eta\sigma}\frac{d\theta^{\rho}}{d\xi}J^{\eta}\frac{d\theta^{\sigma}}{d\xi} = 0, \tag{28}$$

where $J^{\alpha} \overset{\text{def}}{=} \delta\theta^{\alpha}$ is the α-component of the Jacobi vector field [18]. Equation (28) is known formally as the JLC equation. We observe from the JLC equation in Equation (28) that neighboring geodesics accelerate relative to each other at a rate measured in a direct manner by the Riemannian curvature tensor $R_{\alpha\beta\gamma\delta}$. The quantity J^{α} is defined as,

$$J^{\alpha} = \delta\theta^{\alpha} \overset{\text{def}}{=} \delta_{\phi}\theta^{\alpha} = \left(\frac{\partial\theta^{\alpha}(\xi;\phi)}{\partial\phi}\right)_{\tau=\text{constant}}\delta\phi, \tag{29}$$

where $\{\theta^{\mu}(\xi;\phi)\}$ denotes the one-parameter ϕ family of geodesics whose evolution is specified by means of the affine parameter ξ. The Jacobi vector field intensity $J_{\mathcal{M}_s}$ on the manifold $\mathcal{M}_s$ is given by

$$J_{\mathcal{M}_s} \overset{\text{def}}{=} \left(J^{\alpha}g_{\alpha\beta}J^{\beta}\right)^{1/2}. \tag{30}$$

In general, the JLC equation is intractable even at low dimensions. However, in the case of isotropic manifolds, it reduces to

$$\frac{D^2 J^{\mu}}{D\xi^2} + \mathcal{K}J^{\mu} = 0. \tag{31}$$

The sectional curvature $\mathcal{K}$ in Equation (31) assumes a constant value throughout the manifold. In particular, when $\mathcal{K} < 0$, unstable solutions of Equation (31) become

$$J^{\mu}(\xi) = \frac{\omega_0^{\mu}}{\sqrt{-\mathcal{K}}}\sinh\left(\sqrt{-\mathcal{K}}\xi\right), \tag{32}$$

with initial conditions $J^{\mu}(0) = 0$ and $\frac{dJ^{\mu}(0)}{d\xi} = \omega^{\mu}(0) = \omega_0^{\mu} \neq 0$, respectively, for any $1 \leq \mu \leq n$ with n being the dimensionality of the underlying manifold. For additional remarks concerning the JLC equation, we refer to [18,21,22].

We point out that it would be intriguing to understand the behavior of the Jacobi vector fields within the geometry of the Kaniadakis statistical mechanics emerging from a one deformation parameter κ [13]. We leave this fascinating line of study to future scientific inquiry. For a schematic description of the behavior of the IGE and the Jacobi field for two-dimensional surfaces with distinct (Gaussian) curvatures, we refer to Table 1 and Figure 1.

In the next section, making use of the complexity quantifiers introduced in Equations (3), (17), and (30), we present numerous illustrative examples within the IGAC framework.

Table 1. Schematic description of the behavior of the IGE and the Jacobi field for different types of two-dimensional surfaces characterized by distinct constant values of their Gaussian curvature. For such surfaces, the sectional and the scalar curvatures coincide, while the Gaussian curvature is simply one-half of the scalar curvature. In particular, positive curvature causes geodesics to converge while negative curvature causes geodesics to spread out. More specifically, in flat, positively, and negatively curved manifolds, the geodesic deviation equation yields deviations of nearby geodesics that exhibit linear, oscillatory, and exponential behaviors, respectively. Moreover, the volumes of the manifolds regions explored during the entropic motion tend to increase while transitioning from positively to negatively curved manifolds. Correspondingly, the IGE exhibits its maximum growth (that is, linear growth) in the presence of exponential instability on negatively curved manifolds.

Surface	Curvature	Jacobi Field Behavior	IGE Behavior
sphere	positive	oscillatory	sublogarithmic
cylinder	zero	linear	logarithmic
hyperboloid	negative	exponential	linear

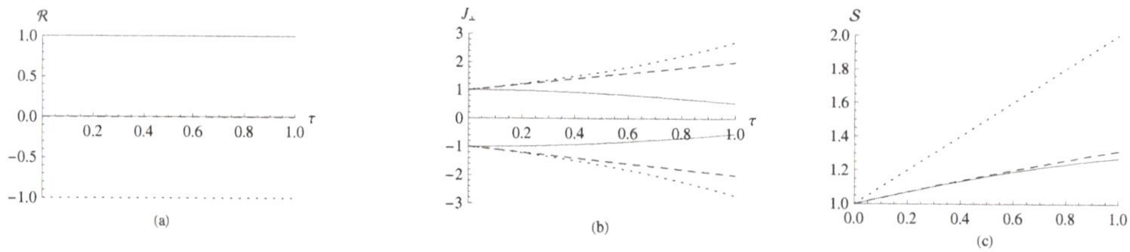

Figure 1. Graphical depictions of the links among curvature, Jacobi fields, and IGE. In (**a**), we depict the constant scalar curvature of a positively curved manifold (solid line), a flat manifold (dashed line), and a negatively curved manifold (dotted line). In (**b**), we illustrate the behavior of the normal components of the Jacobi fields quantifying how nearby geodesics are changing in the normal direction (that is, the direction that is orthogonal to the unit tangent vector of the geodesic) as we move along the geodesics. In the positive, flat, and negative curvature cases, we observe oscillatory behavior (solid line), linear behavior (dashed line), and exponential behavior (dotted line), respectively. Finally, in (**c**), we plot the temporal behavior of the IGE in the positive (sublogarithmic behavior, solid line), flat (logarithmic behavior, dashed line), and negative (linear behavior, dotted line) curvature cases.

3. Applications

In this section, with the help of the three complexity quantifiers introduced above, we report the results of several applications of the IGAC in which the complexity of geodesic trajectories on statistical manifolds are quantified. We present these illustrative examples in a chronological order, from the first one to the last one. For brevity, we omit technical details and confine the presentation to our own information geometric approach to complexity. Early notions and applications of the IGAC originally appeared in [23–25]. For a recent review of the IGAC framework, we refer to [4,8,26,27] and [10], respectively.

3.1. Uncorrelated Gaussian Statistical Models

In [20,28], the IGAC framework was employed to study the information geometric features of a system of arbitrary nature, characterized by l degrees of freedom. Each of these degrees of freedom is described by two relevant pieces of information, namely its mean and variance. The infinitesimal line element for this model is given by [28],

$$ds^2 \overset{\text{def}}{=} \sum_{k=1}^{l} \frac{1}{\sigma_k^2} d\mu_k^2 + \frac{2}{\sigma_k^2} d\sigma_k^2, \tag{33}$$

with μ_k and σ_k denoting the expectation value and the square root of the variance of the microvariable x_k, respectively. It was found that the family of statistical models associated to such a system is Gaussian in form. Specifically, it was determined that this set of Gaussian distributions yields a non-maximally symmetric $2l$-dimensional statistical manifold $\mathcal{M}_s$ whose scalar curvature $\mathcal{R}_{\mathcal{M}_s}$ assumes a constant negative value that is proportional to the number of degrees of freedom of the system,

$$\mathcal{R}_{\mathcal{M}_s} = -l. \tag{34}$$

It was determined that the system explores volume elements on $\mathcal{M}_s$ at an exponential rate. In particular, the IGE $\mathcal{S}_{\mathcal{M}_s}$ was found to increase in a linear fashion in the asymptotic temporal limit (more precisely, in asymptotic limit of the statistical affine parameter τ) and is proportional to the number of degrees of freedom l,

$$\mathcal{S}_{\mathcal{M}_s}(\tau) \overset{\tau\to\infty}{\sim} l\lambda\tau. \tag{35}$$

The quantity λ in Equation (35) denotes the maximal positive Lyapunov exponent that specifies the statistical model. Geodesic trajectories on $\mathcal{M}_s$ were found to be hyperbolic curves. Finally, it was determined that in the asymptotic limit, the Jacobi vector field intensity $J_{\mathcal{M}_s}$ is exponentially divergent and is proportional to the number of degrees of freedom l,

$$J_{\mathcal{M}_s}(\tau) \overset{\tau\to\infty}{\sim} l\exp(\lambda\tau). \tag{36}$$

Given that the exponential divergence of the Jacobi vector field intensity $J_{\mathcal{M}_s}$ is an established classical feature of chaos, based on the results displayed in Equations (34)–(36), the authors suggest that $\mathcal{R}_{\mathcal{M}_s}$, $\mathcal{S}_{\mathcal{M}_s}$ and $J_{\mathcal{M}_s}$ each behave as legitimate measures of chaoticity, with each indicator being proportional to the number of Gaussian-distributed microstates of the system. Although this result was verified in the context of this special scenario, the proportionality among $\mathcal{R}_{\mathcal{M}_s}$, $\mathcal{S}_{\mathcal{M}_s}$ and $J_{\mathcal{M}_s}$ constitutes the first known example appearing in the literature of a possible connection among information geometric indicators of chaoticity obtained from probabilistic modeling of dynamical systems.In this first example, we have compared all three measures $\mathcal{R}_{\mathcal{M}_s}$, $\mathcal{S}_{\mathcal{M}_s}$ and $J_{\mathcal{M}_s}$. Although we have not performed such a comparative analysis in all applications, we shall attempt to mention curvature and/or Jacobi vector field intensity behaviors whenever possible. Our emphasis here is especially on our entropic measure of complexity. For more details on the other types of complexity indicators, we refer to our original works cited in this manuscript.

3.2. Correlated Gaussian Statistical Models

In [29], the IGAC framework was used to analyze the information constrained dynamics of a system comprised of two correlated, Gaussian-distributed microscopic degrees of freedom each having the same variance. The infinitesimal line element for this model is given by [29]

$$ds^2 \overset{\text{def}}{=} \frac{1}{\sigma^2}\left[\frac{1}{1-r^2}d\mu_x^2 + \frac{1}{1-r^2}d\mu_y^2 - \frac{2r}{1-r^2}d\mu_x d\mu_y + 4d\sigma^2\right], \tag{37}$$

with μ_x and μ_y denoting the expectation values of the microvariables x and y. The quantity σ^2, instead, is the variance while r is the usual correlation coefficient between x and y. The scalar curvature $\mathcal{R}_{\mathcal{M}_s}$of the manifold with line element in Equation (37) is $\mathcal{R}_{\mathcal{M}_s} = -3/2$. The inclusion of microscopic correlations give rise to asymptotic compression of the statistical macrostates explored by the system at a faster rate than that observed in the absence of microscopic correlations. Specifically, it was determined that in the asymptotic limit

$$[\exp(\mathcal{S}_{\mathcal{M}_s}(\tau))]_{\text{correlated}} \overset{\tau\to\infty}{\sim} \mathcal{F}(r)\cdot[\exp(\mathcal{S}_{\mathcal{M}_s}(\tau))]_{\text{uncorrelated}}, \tag{38}$$

where the function $\mathcal{F}(r)$ in Equation (38) with $0 \le \mathcal{F}(r) \le 1$ is defined as [29]

$$\mathcal{F}(r) \stackrel{\text{def}}{=} \frac{1}{2^{\frac{5}{2}}}\left[\sqrt{\frac{4(4-r^2)}{(2-2r^2)^2}}\left(\frac{2+r}{4(1-r^2)}\right)^{-\frac{3}{2}}\right]. \tag{39}$$

The function $\mathcal{F}(r)$ is a monotone decreasing compression envelope $\forall r \in (0,1)$. This result provides an explicit link between correlations at the *microscopic* level and complexity at the *macroscopic* level. It also furnishes a transparent and concise description of the functional change of the macroscopic complexity of the underlying statistical manifold caused by the occurrence of microscopic correlations.

3.3. Inverted Harmonic Oscillators

Generally speaking, the fundamental issues addressed by the General Theory of Relativity are twofold: firstly, one wishes to understand how the geometry of spacetime evolves in response to the presence of mass–energy distributions; secondly, one seeks to investigate how configurations of mass–energy move in dynamical spacetime geometry. By contrast, within the IGAC framework, one is concerned only with the manner in which systems move within a given statistical geometry, while the evolution of the statistical manifold itself is neglected. The recognition that there exist two separate and distinct characteristics to consider regarding the interplay between mass–energy and spacetime geometry served as a catalyst in the development of the IGAC framework, ultimately leading to a rather interesting finding. The first result obtained in this novel research direction was proposed by Caticha and Cafaro in [30]. In that article, the possibility of utilizing well established principles of inference to obtain Newtonian dynamics from relevant prior information encoded in a suitable statistical manifold was investigated. The primary working assumption in that derivation was the assumed existence of an irreducible uncertainty in the location of particles. This uncertainty requires the state of a particle to be described by a probability distribution. The resulting configuration space is therefore a statistical manifold whose Riemannian geometry is specified by the Fisher–Rao information metric. The expected trajectory is a consequence of the MrE method, with the latter being regarded as a principle of inference. An unexpected consequence of this approach is that no additional physical postulates such as an equation of motion, principle of least action, nor the concept of momentum, mass, phase space or external time are required. Newton's mechanics involving any number of self-interacting particles as well as particles interacting with external fields is entirely recovered by the resulting entropic dynamics. Indeed, a powerful result of this approach is the fact that interactions among particles as well as particle masses are all justified in terms of the underlying statistical manifold.

Our next example will be of a more applied nature. In [31,32], Zurek and Paz explored the effects of decoherence in quantum chaos by analyzing a single unstable harmonic oscillator with frequency Ω and potential $V(x)$,

$$V(x) \stackrel{\text{def}}{=} -\frac{\Omega^2 x^2}{2}, \tag{40}$$

coupled to an external environment. They determined that in the reversible classical limit, the von Neumann entropy of such a system increases linearly at a rate determined by the Lyapunov exponent Ω according to

$$\mathcal{S}_{\text{quantum}}^{(\text{chaotic})}(\tau) \overset{\tau \to \infty}{\sim} \Omega\tau. \tag{41}$$

Building upon the results obtained in [30], an information geometric analogue of the Zurek–Paz quantum chaos criterion in the classical reversible limit was proposed in [33,34]. In these works, the IGAC framework was employed to study a set of l, three-dimensional, anisotropic, uncoupled, inverted harmonic oscillators (IHO) with an Ohmic distributed frequency spectrum.In this example, the infinitesimal line element is given by

$$ds^2 \stackrel{\text{def}}{=} [1 - \Phi(\theta)]\delta_{\mu\nu}(\theta)d\theta^\mu d\theta^\nu, \tag{42}$$

where $\Phi(\theta)$ is defined as

$$\Phi(\theta) = \sum_{k=1}^{l} u_k(\theta), \tag{43}$$

with $u_k(\theta) \stackrel{\text{def}}{=} -(1/2)\omega_k^2\theta_k^2$and $\omega_k$being the frequency of the k-th inverted harmonic oscillator. Neglecting mathematical details, it was demonstrated in [33,34] that the asymptotic temporal behavior of the IGE for such a system becomes

$$\mathcal{S}_{\mathcal{M}_{\text{IHO}}^{(l)}}(\tau; \omega_1, \ldots, \omega_l) \overset{\tau\to\infty}{\sim} \Omega\tau, \tag{44}$$

where,

$$\Omega \stackrel{\text{def}}{=} \sum_{i=1}^{l} \omega_i, \tag{45}$$

and ω_i with $1 \leq i \leq l$ is the frequency of the ith IHO. Equation (44) indicates an asymptotic, linear IGE growth for the set of IHOs and can be regarded as an extension of the result of Zurek and Paz appearing in [31,32] to an ensemble of anisotropic, uncoupled, inverted harmonic oscillators in the context of the IGAC. We remark that Equation (44) was proposed as the classical IG analogue of Equation (41) in [33,34].

3.4. Quantum Spin Chains

In [35,36], the IGAC was used to study the ED on statistical manifolds whose elements are classical probability distribution functions routinely employed in the study of regular and chaotic quantum energy level statistics. Specifically, an IG description of the chaotic (integrable) energy level statistics of a quantum antiferromagnetic Ising spin chain immersed in a tilted (transverse) external magnetic field was presented. The IGAC of a Poisson distribution coupled to an Exponential bath (that specifies a spin chain in a *transverse* magnetic field and corresponds to the integrable case) along with that of a Wigner–Dyson distribution coupled to a Gaussian bath (that specifies a spin chain in a *tilted* magnetic field and corresponds to the chaotic case) were investigated. The line elements in the integrable and chaotic cases are given by

$$ds^2_{\text{integrable}} \stackrel{\text{def}}{=} ds^2_{\text{Poisson}} + ds^2_{\text{Exponential}} = \frac{1}{\mu_A^2}d\mu_A^2 + \frac{1}{\mu_B^2}d\mu_B^2, \tag{46}$$

and,

$$ds^2_{\text{chaotic}} \stackrel{\text{def}}{=} ds^2_{\text{Wigner-Dyson}} + ds^2_{\text{Gaussian}} = \frac{4}{\mu_A'^2}d\mu_A'^2 + \frac{1}{\sigma_B'^2}d\mu_B'^2 + \frac{2}{\sigma_B'^2}d\sigma_B'^2, \tag{47}$$

respectively. In Equation (46), μ_A and $\mu_B$are the average spacing of the energy levels and the average intensity of the magnetic field, respectively. A similar notation is employed for the second scenario described in Equation (47) where, clearly, $\sigma_B'^2$denotes the variance of the intensity of the magnetic field. Remarkably, it was determined that in the former case, the IGE shows asymptotic logarithmic growth,

$$\mathcal{S}_{\mathcal{M}_s}^{(\text{integrable})}(\tau) \overset{\tau\to\infty}{\sim} c\log(\tau) + \tilde{c}, \tag{48}$$

whereas in the latter case, the IGE shows asymptotic linear growth,

$$\mathcal{S}_{\mathcal{M}_s}^{(\text{chaotic})}(\tau) \overset{\tau\to\infty}{\sim} \mathcal{K}\tau. \tag{49}$$

We emphasize that the quantities c and $\tilde{c}$ in Equation (48) are integration constants that depend upon the dimensionality of the statistical manifold and the boundary constraint conditions on the statistical variables, respectively. The quantity $\mathcal{K}$ appearing in Equation (49) denotes a model parameter describing the asymptotic temporal rate of change of the IGE. The findings described above suggest that the IGAC framework may prove useful in the analysis of applications involving quantum energy level statistics. It is worth noting that in such cases, the IGE effectively serves the role of the standard entanglement entropy used in quantum information science [37,38].

3.5. Statistical Embedding and Complexity Reduction

Expanding upon the analysis presented in [39], Cafaro and Mancini utilized the IGAC framework in [40] to study the $2l$-dimensional Gaussian statistical model $\mathcal{M}_s$ induced by an appropriate embedding within a larger $4l$-dimensional Gaussian manifold. The geometry of the $4l$-dimensional Gaussian manifold is defined by a Fisher–Rao information metric $g_{\mu\nu}$ with non-vanishing off-diagonal elements. It should be noted that these non-vanishing off-diagonal terms arise due to the occurrence of macroscopic correlation coefficients ρ_k with $1 \leq k \leq l$ that specify the embedding constraints among the statistical variables on the larger manifold. The infinitesimal line element is given by [40]

$$ds^2 \overset{\text{def}}{=} \sum_{k=1}^{l} \frac{1}{\sigma_{2k-1}^2}\left[d\mu_{2k-1}^2 + 2\rho_{2k-1} d\mu_{2k-1} d\sigma_{2k-1} + 2 d\sigma_{2k-1}^2\right], \tag{50}$$

with ρ_{2k-1}defined as

$$\rho_{2k-1} \overset{\text{def}}{=} \frac{\frac{\partial \mu_{2k}}{\partial \mu_{2k-1}} \frac{\partial \mu_{2k}}{\partial \sigma_{2k-1}}}{\left[1 + \left(\frac{\partial \mu_{2k}}{\partial \mu_{2k-1}}\right)^2\right]^{1/2} \left[2 + \frac{1}{2}\left(\frac{\partial \mu_{2k}}{\partial \sigma_{2k-1}}\right)^2\right]^{1/2}}, \tag{51}$$

where $\sigma_{2k} = \sigma_{2k-1}$ and $\mu_{2k} = \mu_{2k}(\mu_{2k-1}, \sigma_{2k-1})$for any $1 \leq k \leq l$. Two significant results were obtained. First, a power law decay of the IGE at a rate determined by the correlation coefficients ρ_k was observed

$$\mathcal{S}_{\mathcal{M}_s}(\tau; l, \lambda_k, \rho_k) \overset{\tau \to \infty}{\sim} \log\left[\Lambda(\rho_k) + \frac{\tilde{\Lambda}(\rho_k, \lambda_k)}{\tau}\right]^l, \tag{52}$$

with $\rho_k = \rho_s$ $\forall k$ and $s = 1, \ldots, l$, where

$$\Lambda(\rho_k) \overset{\text{def}}{=} \frac{2\rho_k\sqrt{2-\rho_k^2}}{1+\sqrt{\Delta(\rho_k)}}, \tilde{\Lambda}(\rho_k, \lambda_k) \overset{\text{def}}{=} \frac{\sqrt{\Delta(\rho_k)(2-\rho_k^2)}\log[\Sigma(\rho_k, \lambda_k, \alpha_\pm)]}{\rho_k\lambda_k},$$
$$\text{and } \alpha_\pm(\rho_k) \overset{\text{def}}{=} \frac{1}{2}\left(3 \pm \sqrt{\Delta(\rho_k)}\right). \tag{53}$$

The quantity $\Sigma(\rho_k, \lambda_k, \alpha_\pm)$ is a strictly positive function of its arguments for $0 \leq \rho_k < 1$ and is given by [40]

$$\Sigma(\rho_k, \lambda_k, \alpha_\pm) \overset{\text{def}}{=} -\frac{\Xi_k}{4\lambda_k} \frac{1 + \sqrt{\Delta(\rho_k)}}{1 - \sqrt{\Delta(\rho_k)}} \sqrt{\frac{2\alpha_-(\rho_k)}{\alpha_+(\rho_k)}}, \tag{54}$$

where Ξ_k and λ_k are real, positive constants of integration, and

$$\Delta(\rho_k) \overset{\text{def}}{=} 1 + 4\rho_k^2. \tag{55}$$

Equation (52) represents the first main finding reported in [40] and can be interpreted as a quantitative indication that the IGC of a system decreases in response to the emergence of correlational structures. Second, it was demonstrated that the presence of embedding constraints among the Gaussian macrovariables of the larger $4l$-dimensional manifold results in an attenuation of the asymptotic exponential divergence of the Jacobi field intensity on the embedded $2l$-dimensional manifold. Neglecting mathematical details, it was determined in [40] that in the asymptotic limit $\tau \gg 1$,

$$0 \leq \frac{J_{\mathcal{M}_s}^{2l\text{-embedded}}(\tau)}{J_{\mathcal{M}_s}^{4l\text{-larger}}(\tau)} < 1. \tag{56}$$

Equation (56) constitutes the second main finding reported in [40]. The observed attenuation of the asymptotic exponential divergence of the Jacobi vector field associated with the larger $4l$-manifold, suggests that the occurrence of such embedding constraint relations results in an asymptotic compression of the macrostates explored on the statistical manifold $\mathcal{M}_s$. These two findings serve to advance, in a non-trivial manner, the goal of developing a description of complexity of either macroscopically or microscopically correlated, multi-dimensional Gaussian statistical models relevant in the modeling of complex systems.

3.6. Entanglement Induced via Scattering

Guided by the original study appearing in [41], the IGAC framework was employed to furnish an IG viewpoint on the phenomena of quantum entanglement emerging via s-wave scattering between interacting Gaussian wave packets in [42,43]. Within the IGAC framework, the pre and post quantum scattering scenarios associated with elastic, head-on collision are hypothesized to be macroscopic manifestations arising from underlying microscopic statistical structures. By exploiting this working hypothesis, the pre and post quantum scattering scenarios were modeled by uncorrelated and correlated Gaussian statistical models, respectively.Using the standard notation used so far in this article, the infinitesimal line elements in the absence and presence of correlations are given by

$$ds^2_{\text{no-correlations}} = \frac{1}{\sigma^2}\left[d\mu_x^2 + d\mu_y^2 + 4d\sigma^2\right], \tag{57}$$

and,

$$ds^2_{\text{correlations}} \stackrel{\text{def}}{=} \frac{1}{\sigma^2}\left[\frac{1}{1-r^2}d\mu_x^2 + \frac{1}{1-r^2}d\mu_y^2 - \frac{2r}{1-r^2}d\mu_x d\mu_y + 4d\sigma^2\right], \tag{58}$$

respectively. The scalar curvature $\mathcal{R}_{\mathcal{M}_s}$of the manifolds with line elements in Equations (57) and (58) is $\mathcal{R}_{\mathcal{M}_s} = -3/2$. Using such a hybrid modeling approach enabled the authors to express the entanglement strength in terms of the scattering potential and incident particle energy. Moreover, the manner in which the entanglement duration is related to the scattering potential and incident particle energy was furnished with a possible explanation. Finally, the link between complexity of informational geodesic paths and entanglement was discussed. In particular, it was demonstrated that in the asymptotic limit,

$$[\exp(\mathcal{S}_{\mathcal{M}_s}(\tau))]_{\text{correlated}} \stackrel{\tau\to\infty}{\sim} \mathcal{F}(r) \cdot [\exp(\mathcal{S}_{\mathcal{M}_s}(\tau))]_{\text{uncorrelated}}, \tag{59}$$

where the function $\mathcal{F}(r)$ in Equation (59) with $0 \leq \mathcal{F}(r) \leq 1$ is defined as

$$\mathcal{F}(r) \stackrel{\text{def}}{=} \sqrt{\frac{1-r}{1+r}}. \tag{60}$$

The function $\mathcal{F}(r)$ is a monotone decreasing compression factor with $0 < r < 1$. The analysis proposed in [42,43] is a significant progress toward the understanding among the concepts of entanglement and statistical micro-correlations, as well as the impact of micro-correlations on the complexity of informational geodesic paths. The finding appearing in

Equation (59) suggests that the IGAC construct may prove useful in developing a sound IG perspective of the phenomenon of quantum entanglement.

3.7. Softening of Classical Chaos by Quantization

Expanding upon the original analysis presented in [44–46], the IGAC framework was utilized to investigate the entropic dynamics and information geometry of a three-dimensional Gaussian statistical model as well as the two-dimensional Gaussian statistical model derived from the former model by introducing the following macroscopic information constraint,

$$\sigma_x \sigma_y = \Sigma^2, \tag{61}$$

where $\Sigma^2 \in \mathbb{R}_0^+$. The quantities x and y label the microscopic degrees of freedom of the system. The constraint given by Equation (61) resembles the standard minimum uncertainty relation encountered in quantum mechanics [47]. The infinitesimal line elementsin the $3D$- and $2D$-Gaussian statistical models are given by

$$ds_{3D}^2 \stackrel{\text{def}}{=} \frac{1}{\sigma_x^2} d\mu_x^2 + \frac{2}{\sigma_x^2} d\sigma_x^2 + \frac{2}{\sigma_y^2} d\sigma_y^2, \tag{62}$$

and,

$$ds_{2D}^2 \stackrel{\text{def}}{=} \frac{1}{\sigma^2} d\mu_x^2 + \frac{4}{\sigma^2} d\sigma^2, \tag{63}$$

respectively. Note that the expectation value μ_y of the microvariable y is set equal to zero in Equation (62), while $\sigma_x = \sigma$ with $\sigma_x \sigma_y = \Sigma^2$ in Equation (63). Furthermore, the scalar curvatures corresponding to the $3D$ and $2D$ cases are equal to $\mathcal{R}_{3D} = -1$ and $\mathcal{R}_{2D} = -1/2$, respectively. It was determined that the complexity of the $2D$-Gaussian statistical model specified by the IGE is relaxed when compared with the complexity of the $3D$-Gaussian statistical model,

$$\mathcal{S}_{\mathcal{M}_s}^{(2D)}(\tau) \overset{\tau \to \infty}{\sim} \left(\frac{\lambda_{2D}}{\lambda_{3D}} \right) \cdot \mathcal{S}_{\mathcal{M}_s}^{(3D)}(\tau), \tag{64}$$

with λ_{2D} and λ_{3D} being both positive model parameters (satisfying the condition $\lambda_{2D} \leq \lambda_{3D}$) that express the asymptotic temporal rates of change of the IGE in the $2D$ and $3D$ cases, respectively. Motivated by the connection between the macroscopic information constraint (61) on the variances and the phase-space coarse-graining due to the Heisenberg uncertainty relations, the authors suggest their work may shed light on the phenomenon of classical chaos suppression arising from the process of quantization when expressed in an IG setting. It is worth noting that a similar analysis was implemented in [48] where the work in [47] was generalized to a scenario where—in conjunction with the macroscopic constraint in Equation (61)—the microscopic degrees of freedom x and y of the system are also correlated.

3.8. Topologically Distinct Correlational Structures

In [49], the asymptotic behavior of the IGE associated with either bivariate or trivariate Gaussian statistical models, with or without micro-correlations, was analyzed by Felice and coworkers. For correlated cases, several correlational configurations among the microscopic degrees of freedom of the system were taken into consideration. It was found that the complexity of macroscopic inferences is dependent on the quantity of accessible microscopic information, as well as on how such microscopic information is correlated. Specifically, in the mildly connected case defined by a trivariate statistical model with two correlations among the three degrees of freedom of the system, the infinitesimal line element is

$$\left(\left[ds^2 \right]_{\text{trivariate}}^{\text{(mildly connected)}} \right)_{\text{correlated}} \stackrel{\text{def}}{=} \frac{1}{\sigma^2} \frac{3 - 4r}{1 - 2r^2} d\mu^2 + \frac{6}{\sigma^2} d\sigma^2. \tag{65}$$

Moreover, the infinitesimal line element in the uncorrelated trivariate case is given by

$$\left(\left[ds^2\right]_{\text{trivariate}}\right)_{\text{uncorrelated}} \stackrel{\text{def}}{=} \frac{3}{\sigma^2}d\mu^2 + \frac{6}{\sigma^2}d\sigma^2. \tag{66}$$

In Equations (65) and (66), μ, σ, and r denote the expectation value, the standard deviation, and the correlation coefficient, respectively. It was determined that in the asymptotic limit,

$$\left(\exp\left[\mathcal{S}_{\text{trivariate}}^{(\text{mildly connected})}(\tau)\right]\right)_{\text{correlated}} \stackrel{\tau\to\infty}{\sim} \tilde{\mathcal{R}}_{\text{trivariate}}^{(\text{mildly connected})}(r)\left(\exp\left[\mathcal{S}_{\text{trivariate}}^{(\text{mildly connected})}(\tau)\right]\right)_{\text{uncorrelated}}, \tag{67}$$

where

$$\tilde{\mathcal{R}}_{\text{trivariate}}^{(\text{mildly connected})}(r) \stackrel{\text{def}}{=} \sqrt{\frac{3(1-2r^2)}{3-4r}}. \tag{68}$$

In Equation (67), the quantity r is the micro-correlation coefficient. The function $\tilde{\mathcal{R}}_{\text{trivariate}}^{(\text{mildly connected})}(r)$ shows non-monotone behavior in the correlation parameter r and assumes a value of zero at the extrema of the permitted range $r \in \left(-\sqrt{2}/2, \sqrt{2}/2\right)$. By contrast, for closed bivariate configurations where all microscopic variables are correlated with each other, the complexity ratio between correlated and uncorrelated cases presents monotone behavior in the correlation parameter r. For example, in the fully connected bivariate Gaussian case with $\mu_x = \mu_y = \mu$ and $\sigma_x = \sigma_y = \sigma$, the infinitesimal line element is

$$\left(\left[ds^2\right]_{\text{bivariate}}^{(\text{fully connected})}\right)_{\text{correlated}} \stackrel{\text{def}}{=} \frac{2}{\sigma^2}\frac{1}{1+r}d\mu^2 + \frac{4}{\sigma^2}d\sigma^2. \tag{69}$$

It was found that

$$\left(\exp\left[\mathcal{S}_{\text{bivariate}}^{(\text{fully connected})}(\tau)\right]\right)_{\text{correlated}} \stackrel{\tau\to\infty}{\sim} \tilde{\mathcal{R}}_{\text{bivariate}}^{(\text{fully connected})}(r)\left(\exp\left[\mathcal{S}_{\text{bivariate}}^{(\text{fully connected})}(\tau)\right]\right)_{\text{uncorrelated}}, \tag{70}$$

where

$$\tilde{\mathcal{R}}_{\text{bivariate}}^{(\text{fully connected})}(r) \stackrel{\text{def}}{=} \sqrt{1+r}. \tag{71}$$

Finally, in the fully connected trivariate Gaussian case with trivariate models having all microscopic variables correlated with each other, the infinitesimal line element is

$$\left(\left[ds^2\right]_{\text{trivariate}}^{(\text{fully connected})}\right)_{\text{correlated}} \stackrel{\text{def}}{=} \frac{3}{\sigma^2}\frac{1}{1+2r}d\mu^2 + \frac{6}{\sigma^2}d\sigma^2. \tag{72}$$

It was determined in this case that

$$\left(\exp\left[\mathcal{S}_{\text{trivariate}}^{(\text{fully connected})}(\tau)\right]\right)_{\text{correlated}} \stackrel{\tau\to\infty}{\sim} \tilde{\mathcal{R}}_{\text{trivariate}}^{(\text{fully connected})}(r)\left(\exp\left[\mathcal{S}_{\text{trivariate}}^{(\text{fully connected})}(\tau)\right]\right)_{\text{uncorrelated}}, \tag{73}$$

where

$$\tilde{\mathcal{R}}_{\text{trivariate}}^{(\text{fully connected})}(r) \stackrel{\text{def}}{=} \sqrt{1+2r}. \tag{74}$$

These results imply that in the fully connected bivariate and trivariate configurations, the ratios $\tilde{\mathcal{R}}_{\text{bivariate}}^{(\text{fully connected})}(r)$ and $\tilde{\mathcal{R}}_{\text{trivariate}}^{(\text{fully connected})}(r)$ both present monotone behavior in r over the open intervals $(-1, 1)$ and $(-1/2, 1)$, respectively. On the other hand, in the mildly connected trivariate scenario appearing in Equation (67), an extremum in the function $\tilde{\mathcal{R}}_{\text{trivariate}}^{(\text{mildly connected})}(r)$ occurs at $r_{\text{peak}} = 1/2 \geq 0$. Such a distinctly different behavior between mildly and fully connected trivariate configurations can be attributed to the fact that when making statistical inferences subject to the hypothesis of three positively correlated Gaussian random variables, the system becomes frustrated because the maximum entropy favorable state—characterized by minimum complexity—is incompatible with the initial working hypothesis. Guided by these results, it was suggested in [49] that the impossibility of realizing the maximally favorable state for specific correlational configurations among microscopic degrees of freedom, viewed from an entropic inference perspective, yields an information geometric analogue of the statistical physics frustration effect that arises when loops are present [50].

4. Final Remarks

In this paper, we discussed the primary results obtained by the authors and colleagues over an extended period of work on the IGAC framework. A summary of the IGAC applications can be found in Table 2. For ease of readability, we have chosen to omit technicalities in our discussion. We are aware of several unresolved issues within the IGAC framework, including a deep understanding of the foundational aspects of the IGE measure of complexity. Further developments of the framework are necessary, especially within a fully quantum mechanical setting. For a more detailed list on limitations and future directions of the IGAC approach, we refer the interested reader to [8]. In particular, we mentioned there that one of our main objectives in the near future is to extend our comprehension of the relationship between the IGE and the Kolmogorov–Sinai dynamical entropy [51], the coarse-grained Boltzmann entropy [51] and the von Neumann entropy [52], depending upon the peculiarity of the system being investigated. Despite its limitations, we are pleased that our theoretical modeling approach is steadily gaining interest in the community of researchers. Indeed, there appears to be an increasing number of scientists who either actively use, or who's work is linked to the theoretical framework described in the present brief feature review article [53–78].

Table 2. Schematic description of existing mathematical, classical, and quantum investigations within the IGAC.

Math & IGAC	Classical & IGAC	Quantum & IGAC
Micro and macro correlations	Geometrization of Newtonian mechanics	Spin chains and energy levels statistics
Statistical embeddings	Inverted harmonic oscillators	Scattering induced entanglement
Topology and correlational structures	Macro effects from micro information	Softening chaoticity by quantization

Author Contributions: The authors contributed equally to this work. Both authors have read and agreed to the published version of the manuscript.

Funding: This research received no external funding.

Institutional Review Board Statement: Not applicable.

Informed Consent Statement: Not applicable.

Data Availability Statement: No new data were created or analyzed in this study. Data sharing is not applicable to this article.

Acknowledgments: C. C. acknowledges the hospitality of the *United States Air Force Research Laboratory* in Rome where part of his initial contribution to this work was completed.

Conflicts of Interest: The authors declare no conflict of interest.

References

1. Caticha, A. *Entropic Inference and the Foundations of Physics*; USP Press: São Paulo, Brazil, 2012. Available online: http://www.albany.edu/physics/ACaticha-EIFP-book.pdf (accessed on 20 July 2021).
2. Amari, S.; Nagaoka, H. *Methods of Information Geometry*; Oxford University Press: Oxford, UK, 2000.
3. Cafaro, C. The Information Geometry of Chaos. Ph.D. Thesis, State University of New York, Albany, NY, USA, 2008.
4. Ali, S.A.; Cafaro, C.; Gassner, S.; Giffin, A. An information geometric perspective on the complexity of macroscopic predictions arising from incomplete information. *Adv. Math. Phys.* **2018**, *2018*, 2048521. [CrossRef]
5. Felice, D.; Cafaro, C.; Mancini, S. Information geometric methods for complexity. *Chaos* **2018**, *28*, 032101. [CrossRef]
6. Caticha, A. Entropic Dynamics. *AIP Conf. Proc.* **2002**, *617*, 302.
7. Cafaro, C.; Ali, S.A. Maximum caliber inference and the stochastic Ising model. *Phys. Rev.* **2016**, *E94*, 052145. [CrossRef]
8. Ali, S.A.; Cafaro, C. Theoretical investigations of an information geometric approach to complexity. *Rev. Math. Phys.* **2017**, *29*, 1730002. [CrossRef]
9. Cafaro, C.; Ali, S.A. The spacetime algebra approach to massive classical electrodynamics with magnetic monopoles. *Adv. Appl. Clifford Algebr.* **2007**, *17*, 23. [CrossRef]

10. Cafaro, C.; Giffin, A.; Ali, S.A.; Kim, D.-H. Reexamination of an information geometric construction of entropic indicators of complexity. *Appl. Math. Comput.* **2010**, *217*, 2944. [CrossRef]
11. Kittel, C. *Elementary Statistical Physics*; John Wiley & Sons, Inc.: Hoboken, NJ, USA, 1958.
12. Ito, S.; Oizumi, M.; Amari, S. Unified framework for the entropy production and the stochastic interaction based on information geometry. *Phys. Rev. Res.* **2020**, *2*, 033048. [CrossRef]
13. Kaniadakis, G. Statistical mechanics in the context of special relativity. *Phys. Rev.* **2002**, *E66*, 056125. [CrossRef]
14. Fisher, R.A. Theory of statistical estimation. *Proc. Cambridge Philos. Soc.* **1925**, *122*, 700. [CrossRef]
15. Rao, C.R. Information and accuracy attainable in the estimation of statistical parameters. *Bull. Calcutta Math. Soc.* **1945**, *37*, 81.
16. Cencov, N.N. Statistical decision rules and optimal inference. *Transl. Math. Monogr. Amer. Math. Soc.* **1981**, *53*.
17. Campbell, L.L. An extended Cencov characterization of the information metric. *Proc. Am. Math. Soc.* **1986**, *98*, 135.
18. Weinberg, S. *Gravitation and Cosmology*; John Wiley & Sons, Inc.: Hoboken, NJ, USA, 1972.
19. Lee, J.M. *Riemannian Manifolds: An Introduction to Curvature*; Springer: Berlin, Germany, 1997.
20. Cafaro, C.; Ali, S.A. Jacobi fields on statistical manifolds of negative curvature. *Physica* **2007**, *D70*, 234. [CrossRef]
21. Ohanian, H.C.; Ruffini, R. *Gravitation and Spacetime*; W. W. Norton & Company: New York, NY, USA, 1994.
22. Carmo, M.P.D. *Riemannian Geometry*; Birkhauser: Basel, Switzerland, 1992.
23. Cafaro, C.; Ali, S.A.; Giffin, A. An application of reversible entropic dynamics on curved statistical manifolds. *AIP Conf. Proc.* **2006**, *872*, 243.
24. Cafaro, C. Information geometry and chaos on negatively curved statistical manifolds. *AIP Conf. Proc.* **2007**, *954*, 175.
25. Cafaro, C. Recent theoretical progress on an information geometrodynamical approach to chaos. *AIP Conf. Proc.* **2008**, *1073*, 16.
26. Ali, S.A.; Cafaro, C.; Giffin, A.; Kim, D.-H. Complexity characterization in a probabilistic approach to dynamical systems through information geometry and inductive inference. *Phys. Scr.* **2012**, *85*, 025009. [CrossRef]
27. Cafaro, C. Information geometric complexity of entropic motion on curved statistical manifolds. In Proceedings of the 12th Joint European Thermodynamics Conference, Brescia, Italy, 1–5 July 2013; Pilotelli, M., Beretta, G.P., Eds.; Cartolibreria Snoopy: Brescia, Italy, 2013; pp. 110–118.
28. Cafaro, C. Information-geometric indicators of chaos in Gaussian models on statistical manifolds of negative Ricci curvature. *Int. J. Theor. Phys.* **2008**, *47*, 2924. [CrossRef]
29. Ali, S.A.; Cafaro, C.; Kim, D.-H.; Mancini, S. The effect of microscopic correlations on the information geometric complexity of Gaussian statistical models. *Physica* **2010**, *A389*, 3117. [CrossRef]
30. Caticha, A.; Cafaro, C. From information geometry to Newtonian dynamics. *AIP Conf. Proc.* **2007**, *954*, 165.
31. Zurek, W.H.; Paz, J.P. Decoherence, chaos, and the second law. *Phys. Rev. Lett.* **1994**, *72*, 2508. [CrossRef]
32. Zurek, W.H.; Paz, J.P. Quantum chaos: A decoherent definition. *Physica* **1995**, *D83*, 300. [CrossRef]
33. Cafaro, C.; Ali, S.A. Geometrodynamics of information on curved statistical manifolds and its applications to chaos. *EJTP* **2008**, *5*, 139.
34. Cafaro, C. Works on an information geometrodynamical approach to chaos. *Chaos Solitons Fractals* **2009**, *41*, 886. [CrossRef]
35. Cafaro, C. Information geometry, inference methods and chaotic energy levels statistics. *Mod. Phys. Lett.* **2008**, *B22*, 1879. [CrossRef]
36. Cafaro, C.; Ali, S.A. Can chaotic quantum energy levels statistics be characterized using information geometry and inference methods? *Physica* **2008**, *A387*, 6876. [CrossRef]
37. Prosen, T.; Znidaric, M. Is the efficiency of classical simulations of quantum dynamics related to integrability? *Phys. Rev.* **2007**, *E75*, 015202. [CrossRef]
38. Prosen, T.; Pizorn, I. Operator space entanglement entropy in transverse Ising chain. *Phys. Rev.* **2007**, *A76*, 032316. [CrossRef]
39. Cafaro, C.; Mancini, S. On the complexity of statistical models admitting correlations. *Phys. Scr.* **2010**, *82*, 035007. [CrossRef]
40. Cafaro, C.; Mancini, S. Quantifying the complexity of geodesic paths on curved statistical manifolds through information geometric entropies and Jacobi fields. *Phys. D Nonlinear Phenom.* **2011**, *240*, 607. [CrossRef]
41. Kim, D.-H.; Ali, S.A.; Cafaro, C.; Mancini, S. An information geometric analysis of entangled continuous variable quantum systems. *J. Phys. Conf. Ser.* **2011**, *306*, 012063. [CrossRef]
42. Kim, D.-H.; Ali, S.A.; Cafaro, C.; Mancini, S. Information geometric modeling of scattering induced quantum entanglement. *Phys. Lett.* **2011**, *A375*, 2868. [CrossRef]
43. Kim, D.-H.; Ali, S.A.; Cafaro, C.; Mancini, S. Information geometry of quantum entangled wave-packets. *Physica* **2012**, *A391*, 4517. [CrossRef]
44. Cafaro, C.; Giffin, A.; Lupo, C.; Mancini, S. Insights into the softening of chaotic statistical models by quantum considerations. *AIP Conf. Proc.* **2012**, *1443*, 366.
45. Ali, S.A.; Cafaro, C.; Giffin, A.; Lupo, C.; Mancini, S. On a differential geometric viewpoint of Jaynes' MaxEnt method and its quantum extension. *AIP Conf. Proc.* **2012**, *1443*, 120.
46. Giffin, A.; Ali, S.A.; Cafaro, C. Local softening of chaotic statistical models with quantum consideration. *AIP Conf. Proc.* **2013**, *1553*, 238.
47. Cafaro, C.; Giffin, A.; Lupo, C.; Mancini, S. Softening the complexity of entropic motion on curved statistical manifolds. *Open Syst. Inf. Dyn.* **2012**, *19*, 1250001. [CrossRef]

48. Giffin, A.; Ali, S.A.; Cafaro, C. Local softening of information geometric indicators of chaos in statistical modeling in the presence of quantum-like considerations. *Entropy* **2013**, *15*, 4622. [CrossRef]
49. Felice, D.; Cafaro, C.; Mancini, S. Information geometric complexity of a trivariate Gaussian statistical model. *Entropy* **2014**, *16*, 2944. [CrossRef]
50. Sadoc, J.F.; Mosseri, R. *Geometrical Frustration*; Cambridge University Press: Cambridge, UK, 2006.
51. Greven, A.; Keller, G.; Warnecke, G. *Entropy*; Princeton University Press: Princeton, NJ, USA, 2003.
52. Peres, A. *Quantum Theory: Concepts and Methods*; Kluwer Academic Publishers: London, UK, 1995
53. Peng, L.; Sun, H.; Xu, G. Information geometric characterization of the complexity of fractional Brownian motion. *J. Math. Phys.* **2012**, *53*, 123305. [CrossRef]
54. Peng, L.; Sun, H.; Sun, D.; Yi, J. The geometric structures and instability of entropic dynamical models. *Adv. Math.* **2011**, *227*, 459. [CrossRef]
55. Semarak, O.; Sukova, P. Free motion around black holes with discs or rings: between integrability and chaos-I. *Mon. Not. R. Astron. Soc.* **2010**, *404*, 545. [CrossRef]
56. Li, C.; Sun, H.; Zhang, S. Characterization of the complexity of an ED model via information geometry. *Eur. Phys. J. Plus* **2013**, *128*, 70. [CrossRef]
57. Cao, L.; Li, D.; Zhang, E.; Sun, H. A statistical cohomogeneity one metric on the upper plane with constant negative curvature. *Adv. Math. Phys.* **2014**, *2014*, 832683. [CrossRef]
58. Felice, D.; Mancini, S.; Pettini, M. Quantifying networks complexity from information geometry viewpoint. *J. Math. Phys.* **2014**, *55*, 043505. [CrossRef]
59. Abtahi, S.M.; Sadati, S.H.; Salarieh, H. Ricci-based chaos analysis for roto-translatory motion of a Kelvin-type gyrostat satellite. *J.-Multi-Body Dyn.* **2014**, *228*, 34. [CrossRef]
60. Mikes, J.; Stepanova, E. A five-dimensional Riemannian manifold with an irreducible SO(3)-structure as a model of abstract statistical manifold. *Ann. Glob. Anal. Geom.* **2014**, *45*, 111. [CrossRef]
61. Weis, S. Continuity of the maximum-entropy inference. *Commun. Math. Phys.* **2014**, *330*, 1263.
62. Li, C.; Peng, L.; Sun, H. Entropic dynamical models with unstable Jacobi fields. *Rom. Journ. Phys.* **2015**, *60*, 1249.
63. Itoh, M.; Satoh, H. Geometry of Fisher information metric and the barycenter map. *Entropy* **2015**, **17**, 1814. [CrossRef]
64. Franzosi, R.; Felice, D.; Mancini, S.; Pettini, M. A geometric entropy detecting the Erdös-Rényi phase transition. *Eur. Phys. Lett.* **2015**, *111*, 20001. [CrossRef]
65. Martins, A.C.R. Opinion particles: Classical physics and opinion dynamics. *Phys. Lett.* **2015**, *A379*, 89. [CrossRef]
66. Muhammad, S.A.; Zhang, E.; Sun, H. Jacobi fields on the manifold of Freund. *Ital. J. Pure Appl. Math.* **2015**, *34*, 181.
67. Felice, D.; Mancini, S. Gaussian network's dynamics reflected into geometric entropy. *Entropy* **2015**, *17*, 5660. [CrossRef]
68. Wen-Haw, C. A review of geometric mean of positive definite matrices. *Br. J. Math. Comput.* **2015**, *5*, 1.
69. Weis, S.; Knauf, A.; Ay, N.; Zhao, M.-J. Maximizing the divergence from a hierarchical model of quantum states. *Open Syst. Inf. Dyn.* **2015**, *22*, 1550006. [CrossRef]
70. Weis, S. Maximum-entropy inference and inverse continuity of the numerical range. *Rep. Math. Phys.* **2015**, *77*, 251. [CrossRef]
71. Shalymov, D.S.; Fradkov, A.L. Dynamics of non-stationary processes that follow the maximum of the Rényi entropy principle. *Proc. R. Soc.* **2016**, *A472*, 20150324. [CrossRef] [PubMed]
72. Henry, G.; Rodriguez, D. On the instability of two entropic dynamical models. *Chaos Solitons Fractals* **2016**, *91*, 604. [CrossRef]
73. Gomez, I.S.; Portesi, M. Ergodic statistical models: Entropic dynamics and chaos. *AIP Conf. Proc.* **2017**, *1853*, 100001.
74. Gomez, I.S. Notions of the ergodic hierarchy for curved statistical manifolds. *Physica* **2017**, *A484*, 117. [CrossRef]
75. Gassner, S.; Cafaro, C. Information geometric complexity of entropic motion on curved statistical manifolds under different metrizations of probability spaces. *Int. J. Geom. Methods Mod.* **2019**, *16*, 1950082. [CrossRef]
76. Gomez, I.S.; Portesi, M.; Borges, E.P. Universality classes for the Fisher metric derived from relative group entropy. *Physica* **2020**, *A547*, 123827. [CrossRef]
77. Summers, R.L. *Experiences in the Biocontinuum: A New Foundation for Living Systems*; Cambridge Scholar Publishing: Cambridge, UK, 2020.
78. Deshmukh, S.; Ishan, A.; Al-Shaik, S.B.; Özgür, C. A note on Killing calculus on Riemannian manifolds. *Mathematics* **2021**, *9*, 307. [CrossRef]

MDPI
St. Alban-Anlage 66
4052 Basel
Switzerland
Tel. +41 61 683 77 34
Fax +41 61 302 89 18
www.mdpi.com

Foundations Editorial Office
E-mail: foundations@mdpi.com
www.mdpi.com/journal/foundations

www.ingramcontent.com/pod-product-compliance
Lightning Source LLC
Chambersburg PA
CBHW061556220926
51850CB00005B/355
* 9 7 8 3 0 3 6 5 5 7 4 5 8 *